# APPLICATION OF APOPTOSIS TO CANCER TREATMENT

# Application of Apoptosis to Cancer Treatment

*by*

MELS SLUYSER

*The Netherlands Cancer Institute,*
*Amsterdam, The Netherlands*

Springer

A C.I.P. Catalogue record for this book is available from the Library of Congress.

ISBN-10 1-4020-3303-6 (HB) Springer Dordrecht, Berlin, Heidelberg, New York
ISBN-10 1-4020-3302-8 (e-book) Springer Dordrecht, Berlin, Heidelberg, New York
ISBN-13 978-1-4020-3303-2 (HB) Springer Dordrecht, Berlin, Heidelberg, New York
ISBN-13 978-1-4020-3302-5 (e-book) Springer Dordrecht, Berlin, Heidelberg, New York

---

Published by Springer,
P.O. Box 17, 3300 AA Dordrecht, The Netherlands.

Cover image:
Please refer to chapter 4, figure 16.

*Printed on acid-free paper*



Printed in the Netherlands.

# TABLE OF CONTENTS

# FOREWORD

Apoptosis, or programmed cell death, plays an essential natural physiological role in removing cells that are superfluous, diseased, or otherwise have served their useful purpose. During apoptosis (a term derived from the Greek word for the falling of leaves from trees in autumn) living cells go through a predictable, well-choreographed series of events in which they are broken down by enzymes internally and then fall apart. The remaining pieces are devoured by still vital neighboring cells. Apoptosis can also be induced in cancer cells artificially, by treating tumors with drugs or radiation. This break-through finding has led to the idea that a completely novel way of cancer therapy might be developed using drugs that directly switch on the cell death machinery in tumors. Certain drugs with pro-apoptotic potential are now being tested for possible use in the clinic.

One of the major problems in oncology is the fact that tumors, which initially respond to a certain therapy modality, develop resistance after a time. Selective resistance to cell death is recognized as a major health-related problem now that it has become clear that the effects of therapy on genetically unstable, rapidly dividing groups of tumor cells usually leads to only temporary relief of the tumor burden, because this is followed by the outgrowth of a subpopulation that carries advantageous mutations that make them non-responsive to the therapy. Loss of ability to undergo apoptosis may be one of the key factors in the clonal selection leading to treatment resistance.

This book is a state-of-the art presentation of the achievements of research made in this area. We hope that it will be a useful guide to workers in the field and to clinicians.

Mels Sluyser

Chapter 1

# DISFUNCTION OF THE APOPTOTIC PATHWAY IN CANCER CELLS

Lily Yang

*Deparment of Surgery and Winship Cancer Institute, Emory University School of Medicine, Atlanta, GA 30322*

Abstract: Apoptosis is an important physiological process for maintaining homeostasis, remodeling and eliminating abnormal cells in normal tissues. Development of human cancer is a multistage process involving various genetic alternations and cellular abnormalities. Cellular changes should lead to activation of the apoptotic pathway and induction of cell death, which prevents tumor growth and progression. However, during tumorigenesis, some tumor cells develop apoptosis-resistant mechanisms that allow the cancer cells to avoid apoptotic cell death, resulting in the initiation and progression of human cancers. Defects in the apoptotic signaling pathway have been detected in many cancer cells and cancer tissues. A deregulated apoptotic signal pathway confers a high survivability and resistance of the tumor cells to therapeutic reagents. Understanding the alterations in apoptotic signaling in human cancer cells should provide important information for the development of novel cancer therapies directly targeting the apoptotic signal pathway in cancer cells.

Key words: Apoptosis resistance, apoptotic signal pathway, human cancer cell, apoptosis signal defects

## 1. INTRODUCTION

Programmed cell death (apoptosis) is an important cellular process that allows proper development and remodeling of normal tissues, generating immune responses and destroying abnormal cells. A regulated apoptotic pathway ensures homeostasis and integrity of the normal tissues [1,2]. It is well known that malignant transformation of human cancer cells is a multi-stage



*M. Sluyser (ed.), Application of Apoptosis to Cancer Treatment,* 1-28.

process involving mutations or deletions of various tumor suppressor genes, activation of oncogenes and alterations in the levels of expression of key regulatory genes, providing growth advantages and metastatic potential for tumor cells [3]. Those genetic alterations result in abnormalities in cellular and nuclear morphology and signal transduction pathways which would normally activate a suicidal pathway and induce apoptosis in the cells [2,4]. Increasing evidence shows that impairments in apoptotic signaling enable tumor cells to avoid apoptotic cell death and grow into tumor masses that are resistant to apoptosis [5-8]. Defects in regulation of apoptosis have been detected in both upstream and downstream of the apoptotic signal pathway in many types of human tumor cells [6-9]. Recent studies have also revealed molecular targets in the apoptotic pathway that play important roles in the apoptosis resistance of cancer cells. The feasibility of targeting the apoptotic pathway as a novel cancer therapy has also been examined.

Apoptosis is the most common type of cell death, characterized as chromatin condensation, nuclear fragmentation, cell shrinkage and membrane blebbing. Apoptotic cells then break into small membrane-surrounded apoptotic bodies that are removed by phagocytosis [1]. In normal cells, apoptosis is induced under some physiological conditions such as tissue and organ development in fetus, menstrual cycle, and involution of breast ducts after lactation [10-13]. Apoptotic cell death is also induced in the cells with viral infection, DNA damage or other genomic alterations and regulation of cell-mediated immune responses [2,14,15]. Regulated apoptosis therefore maintains tissue integrity as well as a balance between cell proliferation and death in normal tissues.

During the last decade, the identification and characterization of cellular factors in the apoptotic signal pathway have been an intensive research area. Many cellular factors involved in apoptotic signaling were discovered and their roles in the regulation of the apoptotic pathway have been elucidated. Apoptosis is initiated when the cells receive negative signaling, such as growth factor withdrawal, DNA damage by oxidants, ultraviolet light and x-rays, and chemotherapy drug treatment [2,16-18]. Activation of apoptotic signaling is achieved by either an extrinsic or an intrinsic pathway [19]. The extrinsic pathway is triggered by ligation of cell surface death receptors with their specific ligands, such as Fas Ligand, tumor necrosis factor $\alpha$ (TNF-$\alpha$) and tumor necrosis factor-related apoptosis inducing ligand (TRAIL). Binding of apoptosis inducing ligands to their corresponding receptors activates an intracellular domain (the death domain) of the receptor to attract an adaptor protein, Fas-associated death domain protein (FADD). FADD then recruits inactive caspase 8 to form a death-inducing signaling complex (DISC), resulting in the activation of caspase-8. Active caspase 8 then cleaves and activates caspase-3 and -7 [19]. The intrinsic pathway is activated when the cells are under severe stresses such as growth factor deprivation, oxidants and DNA-damaging agents by leakage of cytochrome c from mitochondria. This results in the activation of caspase-9 and then caspase-3, -6, and -7 [19-21]. The crosstalk between cell death receptors and mitochondrial pathways is also present in some conditions. Death receptor activated caspase-8 cleaves Bid, which then translocates to the mitochondria to amplify the apoptosis signal by activating the mitochondrial pathway [22].

Caspases can be divided into two groups based on the length of their prodomain and substrate specificity. Caspase-2, -8, -9 and -10 are initiator caspases using their long N-terminal prodomains to interact with adapter molecules and form a death inducing signal complex (DISC). Downstream caspases, including caspase-3, -6, and -7, are executioner caspases that remain dormant until the initiator caspases activate them by proteolysis [22]. Activated executioner caspase-3, -6 and -7 recognize specific substrate sequences in targeting cellular proteins and cleave a number of structural and regulatory proteins such as Poly (ADP-ribose) polymerase (PARP), lamins, DNA fragmentation factor-45 (DFF45/ICA) and cytokeratins, leading to apoptotic cell death [23].

Examination of the levels and activity of apoptotic effectors, inhibitors and regulators in human cancer cells and tissues has demonstrated that deregulation of apoptotic signal pathway is present in most human cancer cells. Human tumor cells escape apoptotic cell death by avoiding the activation of upstream apoptotic signals and/or by upregulation of inhibitory factors in the apoptotic signal pathway [6-9].

## 2. CELL DEATH RECEPTOR-MEDIATED APOPTOSIS

### 2.1 Fas and Fas ligand (Fas L)

Fas (APO-1 or CD 95) is a widely expressed transmembrane protein in the tumor necrosis receptor family. Interaction of Fas with its legend, FasL, initiates the death receptor-mediated cell death pathway [22,24]. However, dysfunction of the Fas-mediated apoptotic signal has been found in several tumor types. It has been shown that many tumor cells are resistant to FasL or Fas antibody induced apoptosis [25,26]. Further studies indicate that human cancer cells have developed resistant mechanisms to avoid Fas-mediated apoptosis. Somatic deletions and mutations of Fas receptor were first discovered in human lymphoid-lineage malignancies [26,27]. Later, Fas mutations were detected in small percentage of solid tumors, such as in gastric (11.6%), non-small cell lung (7.7-20%), and malignant melanomas (6.8%)[28,29]. Although Fas mutation is not a common phenomenon in solid tumors, a reduced level of expression of cell surface Fas receptor is found in many tumor types either by downregulating Fas gene expression or by decreasing cell surface transportation [30-33]. In addition, some tumor cells also produce a high level of soluble Fas to block interactions between cell surface Fas receptor and FasL [34,35].

Although downregulation of Fas levels or function could explain the insensitivity to Fas-mediated apoptosis in some tumor cells, many tumor cells do not have Fas mutations and an adequate level of Fas expression is detected in tumor cells that are resistant to Fas-mediated apoptosis [36-40]. Interestedly, those tumor cells also co-express a high level of FasL, an activating ligand for Fas receptor. In normal tissues, FasL is only expressed at a low level in cytotoxic T lymphocyte, natural killer cells, sertoli cells of testis, ocular cells and normal breast ductal epithelial cells [41-43]. However, upregulation of FasL has been found in many tumor cells as well as tumor tissues [44-47]. Co-expression of Fas and FasL in tumor cells resistant to Fas-mediated apoptosis suggests the presence of intrinsic anti-apoptotic factors downstream of the death receptor that block the apoptotic signal pathway and prevent apoptosis [31]. Consistent with this notion, it has been shown that tumor cells resistant to Fas-induced cell death also showed a low sensitivity to chemotherapy drugs or to TRAIL induced apoptosis [48-50]. Moreover, upregulation of cell surface FasL provides a growth advantage to the cells by counteracting tumor-infiltration immune cells and/or facilitating the destruction of surrounding tissues to increase the invasiveness of the tumor cells [25,44].

## 2.2 TRAIL and TRAIL receptors

TRAIL is a member of the tumor necrosis factor (TNF) family of cytokines that binds to its death receptors, DR4 and DR5, and activates the apoptotic pathway [51]. Although TRAIL is constitutively expressed in many tissue types, apoptotic cell death is selectively induced in cancer cells but not in normal cells [52-54]. This selectivity may be due to a higher level of TRAIL receptors in cancer cells than in normal cells. In addition, TRAIL also interacts with "decoy" receptors DcR1 and DcR2, which lack functional death domains and do not induce apoptosis [55]. The role of the decoy receptors in protecting normal cells from TRAIL-induced apoptosis has yet to be determined.

Although activation of TRAIL-mediated apoptotic pathway has great potential for developing tumor-specific therapy, further studies of the ant-tumor effects of TRAIL in different tumor cell lines indicate that human tumor cells have a wide range of sensitivity to TRAIL-induced apoptosis [56]. A large fraction of tumor cells display a low level of TRAIL expression or activity. Some tumor cells have completely lost the expression of TRAIL receptor [57,58]. Additionally, several studies demonstrate that high levels of both TRAIL receptor and ligand are found in some TRAIL-resistant tumor cells, suggesting other downstream anti-apoptotic factors may contribute to lack of TRAIL-induced apoptosis in those cells. However, it has been shown that treatment of TRAIL resistant tumor cell lines with subtoxic concentrations of chemotherapy drugs sensitizes TRAIL-induced apoptosis [57,59,60].

## 2.3 TNF-α and receptors

TNF-α, a cytokine produced by macrophages/monocytes during acute inflammation, regulates inflammation, survival, proliferation and apoptosis of cells. TNF-α binds to cell surface receptor TNFR-1 or TNFR-2 and trimerizes the receptors [61]. The activated receptors further recruit adaptor proteins TRADD and TRAF2, and death effect domain protein FADD to form DISC and then cleaves procaspase 8 to active caspase 8 [62,63]. Unlike other TNF-α family receptors, recruiting TRAF2 to TNF-R1 triggers the activation of cell survival factor NF-κB resulting in the activation of anti-apoptosis factors such as c-FLIP or cIAPs, which are inhibitors for caspase 8 [62,64]. Since the level of TRAF2 is elevated in various human tumors, this may cause the formation of the TNF-R, TRADD and TRAF2 complex and activate the cell survival pathway, resulting in resistance of the tumor cells to TNF-α mediated apoptosis [65,66].

# 3. CASPASE ACTIVATION

## 3.1 Downregulation of caspases in tumor cells

Caspases are synthesized as inactive zymogenes with a prodomain followed by a large (p20) and a small (p17) subunit. Activation of the procaspases by a series of cleavage events is a critical process for execution of apoptosis. Deficiency in the levels of expression of procaspase genes is detected in some tumor cell lines and tissues. For example, deletion or silencing of the caspase 8 gene was discovered in neuroblastoma and non-small lung carcinomas [67-69]. Deficiency in caspase 3 was also found in some human tumor cell lines and tissues such as human breast cancer cells, drug resistant human cervical cancer cells, human neuroblastoma, hepatocellular and renal cell carcinomas tissues [68,70,71]. Results from examination of levels of caspase expression using immunohistochemistry staining further showed that 46% to 85% of human colon cancer tissues have low levels of caspase-7 and -9 [72].

## 3.2 Apoptotic protease activating factor 1

Apoptotic protease activating factor 1 (APAF-1) is a cytoplasm protein that binds to cytochrome C after its release from mitochondria and forms an apoptosome with cytochrome C and procaspase 9. At the apoptosome, procaspase 9 is activated, resulting in the cleavage and activation of caspase 3. Functional of APAF-1 is required for activation of caspase 9 in the intrinsic pathway [73]. However, tumor cells, such as metastatic melanomas have developed a way to avoid the mitochondrial-mediated apoptosis by downregualting expression of the APAF-1 gene through allelic loss or gene methylation [74]. Those APAF 1-negative cells are highly resistant to chemotherapy [4].

## 3.3 Upregulation of levels of gene expression and caspase activity in human tumor cells

Although impairments in the levels of caspases greatly affect the apoptotic response in human tumor cells, recent studies have demonstrated the presence of higher levels of expression of procaspase genes and/or active caspases in some tumor cells and tissues as compared to normal cells [75,76,77 #143,78,79]. Examination of levels of procaspases and active caspases in breast carcinoma tissues from 440 breast cancer patients at different stages of the disease in five independent studies yielded surprising results demonstrating a high level of procaspases and/or active forms of caspases in most human breast cancer tissues [75-79]. A high level of procaspase-3 expression is found in 58% of ductal carcinoma *in situ* (DCIS) and ~90% of invasive breast cancer tissues but is not found in normal breast ductal cells. A strong expression of procaspase-3, -6 and -8 is significantly associated with the extent of apoptosis and high grade of DCIS lesions [76]. It has also been shown that over 80% of breast cancer tissues display high levels of active caspase-3 and -6 detected by immunohistochemical staining using antibodies specific for active forms of caspase -3 and -6. In these patient samples, apoptosis is highly correlated with the level of proliferation but not with the level of active caspases [77].

Overexpression of caspase-3 gene is also detected in pancreatic cancer but not in normal pancreas tissues[80]. High levels of caspase-8, -3 and -6 activity are found in pancreatic and colon cancer cells that are not undergoing apoptotic cell death [79]. Analysis of the expression of caspase-3, -8, -9 and -10 in 60 advanced gastric adenocarcinomas by immunohistochemistry using a tissue microarray approach showed that over 90% of the gastric cancer tissues express high levels of caspase-3, -8, -9 and -10. However, normal gastric mucosal cells show no or weak expression of caspases. A high level of active caspase-3 in gastric cancer tissues is significantly correlated with lymph node metastasis and a worse prognosis of the patient but not with the extent of apoptosis [81]. At present, the significance of caspase activation in human tumor tissues is still under investigation. Activation of the caspase cascade, especially caspase 3, has been considered as an irreversible process that leads to "point of no-return" apoptotic death in the cells. An important question to be answered is that how those tumor cells with active caspases are still alive and maintaining proliferative ability. Recent studies showed that in addition to their function in apoptosis, limited activation of caspases is required for some normal cell functions such as proliferation of T and B lymphocytes and differentiation of several cell types [82-84]. In tumor cells, activation of caspases may result from the activation of the apoptotic pathway due to the presence of abnormalities in cancer cells. In addition to a high level of caspase and/or active caspase, upregulation of FasL, an important activator for death receptor-mediated apoptosis, is seen in many tumor cells and tissues. Expression of FasL has been associated with counteracting the cytotoxic T cell immune response and the invasiveness of the tumor cells [85]. It seems that those FasL-mediated effects are the results of activation of apoptotic signal in tumor cells. Development of anti-apoptosis mechanisms, especially factors inhibiting caspase activity, allows survival and progression of human tumor cells.

## 4. ANTI-APOPTOTIC FACTORS IN TUMOR CELLS

### 4.1 Upstream inhibitors for cell death receptor-mediated apoptotic pathway

Death receptor activated apoptosis is negatively regulated by FADD-like interleukin-1 β-converting enzyme-like protease (cFLIP) [86]. cFLIP protein has homology with procaspase 8 but lacks the catalytic domain of the enzyme. Binding of cFLIPs to the DISC interferes with the processing and activation of caspase 8, which inhibits initiation of death receptor-mediated apoptosis. It has been shown that cFLIP, potently inhibits death signaling mediated by all known death receptors, including Fas, TNF-R, and TRAIL-Rs. cFLIP is constitutively expressed at a high level in many human tumor types including heptocellular carcinomas, malignant melanomas, gastric, ovary and prostate cancers [87-89]. The anti-apoptosis function of cFLIP is further demonstrated by the attenuating cisplatin-induced cleavage of caspase-8 and -3 and apoptosis in chemosensitive ovary cells after overexpression of cFLIP, and by increased apoptosis after downregulating cFLIP in chemoresistant cells [90].

## 4.2 Bcl-2 family proteins

The mitochondria-dependent apoptosis pathway is regulated by anti- and pro-apoptotic proteins of the Bcl-2 family. About twenty proteins have been identified as members of the Bcl-2 family [91]. The anti-apoptotic Bcl-2 family includes proteins such as Bcl-2, Bcl-$X_L$, Bcl-w, Mcl-1, A1/BFL1, which contain Bcl-2 homology (BH) domains 1, 2 and 4. The Bcl-2 family proteins with a proapoptotic function can be further divided into Bax subfamily (Bax, Bak and Bok), and BH3 subfamily with such members as Bik, Bim, Bad, HRK/DP5, NOX, Puma, NIP3, Bid and BMF [92].

Overexpression of anti-apoptotic Bcl-2 proteins inhibits apoptosis induced by various apoptosis stimuli including chemotherapy drugs, $\gamma$-radiation, FasL and TNF-$\alpha$ [93]. In normal tissues, maintaining homeostasis requires a balance between the anti-apoptotic and proapoptotic Bcl-2 family proteins. When cells are under stress, Bax and Bak translocate from the cytoplasm to the outer mitochondria membrane and undergo oligomeriztion. Oligomerized Bax or Bak then inserts into the membrane to induce cytochrome c release. Bcl-2 selectively binds to Bax and prevents insertion of Bax into the mitochondrial membrane. Therefore, the interaction of pro- and anti-apoptotic Bcl-2 family proteins determines mitochondrial membrane permeability suppression or promotion, which controls the release of cytochrome C and other apoptosis activating proteins from the mitochondria [19,92]. Anti-apoptotic proteins Bcl-2 and Bcl-$X_L$ are overexpressed in many tumor types [6,94]. Upregulation of Bcl-2 or Bcl-$X_L$ has been demonstrated to block the apoptotic response and to be a key factor in tumorigenesis and apoptosis resistance in several tumor types [95,96]. Downrgulation of Bcl-2 function or expression by anti-sense or synthetic BH3 peptides has been shown to induce apoptosis and sensitize tumor cells to chemotherapy [97-99].

Although the role of Bcl-2 in apoptosis resistance has been demonstrated in several tumor types, especially in lymphomas, whether Bcl-2 protein plays an important role in breast cancer has yet to be determined. It is clear that over 80% of breast cancer tissues express a high level of Bcl-2 [94,100]. Overexpression of the BCL-2 protein enhanced resistance to apoptosis in human breast cancer cell lines [96,101]. However, expression of Bcl-2 in human breast cancer tissues correlates with a favorable prognosis and an overall better survival rate [100,102-104]. This intriguing observation may be interpreted in part by the effect of Bcl-2 prolonging the transition from $G_0$ or $G_1$ to S phase of the cycle [92]. However, further studies are needed to determine the significance of Bcl-2 expression in the apoptosis or survival of breast cancers.

Increasing evidence demonstrates that apoptosis resistance in cancer cells is as a result of impairment of the mitochondria-mediated apoptotic pathway by downregulating the function or levels of proapoptotic Bcl-2 family proteins in cancer cells [105]. It has been shown that transgenic mice deficient in Bax have accelerated onset of tumor growth [106]. Bax frameshift mutations are found in over 50% of colon and gastric cancers of the microsatellite mutator phenotype [107]. Bax deficiency has been shown to promote drug resistant and oncogenic transformation of cells. Results from analysis of the level of Bax expression in breast cancer tissues show that most breast cancer tissues weakly express Bax gene and about one-third of the cancer tissues have lost this gene expression. Moreover, a reduced Bax level is associated with a poor response to therapy, faster tumor progression, and an overall poorer prognosis for the patient [108-110]. On the other hand, overexpression of the BAX gene induces apoptotic cell death and enhances the effect of chemotherapy drugs on cancer cell lines [108].

## 4.3 Inhibitor of apoptosis protein family

In addition to upstream apoptotic inhibitory factors that control the activation of cell death receptor or mitochondria pathway, the apoptotic signal is also regulated by the inhibitor of apoptosis protein (IAP) family. IAPs are a family of proteins containing one or more conserved, cysteine and histidine-rich baculoviral IAP repeat (BIR) N-terminal domains and a C-terminal RING domain. About seven IAP proteins, including NAIP, XIAP, c-IAP1, c-IAP2, survivin, Livin and Ts-IAP, have been identified and their roles in inhibiting caspase activity have been elucidated [7,9,111-113]. The BIR domains of the IAPs form the zinc-figure-like structures that bind to the surface of caspases to block caspase activity. The RING domain acts as an ubiquitin ligase to facilitate the proteasomal degradation of caspases [114]. Specific interactions of BIR domains with different caspases have been determined by studying the structures of caspases and IAPs. The results from crystallography and mutagenesis studies of XIAP show that the proximal link region of BIR2 binds and blocks the active site of caspase-3 and -7. The interaction of the BIR2 domain with the amino-terminal of the small subunit of caspase 7 further stabilizes the binding. The BIR3 domains of XIAP, c-IAP1 and C-IAP2 are able to bind and inhibit active caspase-9.

Single BIR domain IAP proteins such as livin and Ts-IAP have been demonstrated to bind and inhibit caspase-9. However, the role of another single BIR domain protein, survivin, in the inhibition of caspase-3 and -7 is still controversial. Although physical interactions between survivin and caspases, and inhibition of caspase-3 and -7 activities have been reported, a structural basis for a direct interaction between survivin and caspase-3 has not been defined [115]. Increasing evidence suggests that survivin is closely associated with mitochondria-dependent apoptosis. Downregulation of survivin expression or function results in the activation of caspase-9. A recent study shows that survivin is able to associate with XIAP through the BIR domain and form a survivin-XIAP complex that promotes increased XIAP stability and synergistic inhibition of apoptosis [116].

It has been shown that Smac/DIABLO (second mitochondria activator of caspases), a proapoptotic protein released together with cytochrome C from mitochondria into the cytosol, interacts with all mammalian IAP proteins on both BIR 2 and BIR 3 domains. Binding of Smac to IAPs inactivates the function of IAPs and enhances the apoptotic response by releasing caspases from the IAP-inhibition [117,118].

Upregulation of IAPs is found in many tumor cell lines as well as in primary tumor tissues. Although XIAP is expressed at a low level in normal cells and tissues, a high level of XIAP is detected in many human tumor cells. Increases in XIAP expression have been associated with apoptosis resistance and low sensitivity to chemotherapy drugs in several tumor types. Downregulation of XIAP releases its inhibition on caspase-3 and induces apoptotic cell death in tumor cell lines as well as *in vivo* in a mouse tumor model. In addition to increasing the XIAP level, tumor cells also downregulate cellular factors that inhibit XIAP function. In normal cells, expression of XIAP associated factor 1 (XAF1) counteracts the anti-apoptotic function of XIAP by competing with active caspases for XIAP binding sites and releasing caspases from XIAP inhibition [119]. However, the level of XAF 1 is decreased or lost in many tumor cell types [7,119-121].

Unlike other IAPs, survivin is expressed broadly in embryonic and fetal tissues but is undetectable in most differentiated normal adult tissues, except thymocytes, CD 34+ stem cells and basal colonic epithelial cells [122,123]. However, survivin is expressed in most common tumor types including brain, lung, breast, liver, pancreas, gastric, colon, uterus, ovary, lymphoma, leukemia, melanoma and soft tissue sarcomas [123-125]. For example, survivin is found in over 70% of human breast or pancreatic cancer tissues and in 64% of human colon tissues [80,126,127]. Expression of the survivin gene in human tumor cells is regulated at a transcriptional level through increasing survivin promoter activity, amplification of the survivin locus on 17q25, demethylation of survivin exon 1, and releasing transcriptional repression by p53 mutation [123,128-131]. Recent studies also demonstrate that survivin is a reliable marker for aggressive disease, resistant to chemo- or radio-therapy and indicative of a poor prognosis for human cancers [132-134]. Overexpression of survivin in human tumor cells reduces the apoptotic response induced by various apoptosis stimuli [115]. Transgenic expression of survivin in the skin inhibits UVB-induced apoptosis in skin epidermal cells in the mice whereas it does not affect Fas-induced cell death [135]. On the other hand, downregulation of survivin function with anti-sense, siRNA, dominant negative mutant or the ribozyme for survivin induces apoptotic cell death and sensitizes cancer cells to chemotherapy drugs [79,135-137].

In addition to its anti-apoptotic function, survivin is also linked to mitotic progression and cell division. Expression of survivin is increased in cells undergoing mitosis. Disrupting survivin function results in cells with centrosome deregulation, multipolar mitotic spindles and multinucleated nuclei [138]. Therefore, survivin has a dual function in regulating the cell cycle progression and blocking apoptotic signaling.

## 5. REGULATION OF APOPTOSIS SIGNAL BY OTHER CELL SIGNAL TRANSDUCTION PATHWAYS

### 5.1 Tumor suppressor gene p53

Mutation of tumor suppressor gene p53 is one of the most common types of genetic alterations in human tumors. p53 suppresses tumor growth through multiple pathways that involve gene transcription, DNA synthesis and repair, cell cycle arrest, senescence and apoptosis. Mutations of p53 gene or loss of p53 function results in tumor progression, genetic instability and apoptosis resistance [139-141]. It has been shown that p53 regulates both extrinsic and intrinsic apoptotic pathways through the transactivating transcription of proapoptotic factors and suppressing expression of anti-apoptotic genes. For example, upregulation of cell death receptors such as Fas and TRAIL-Rs is detected in tumor cells following DNA-damaging or chemotherapy drug-induced p53 expression [142]. Induction of transcription of proapoptotic Bcl-2 family genes including Bid, Bax, and Puma, and APAF-1 by p53 further enhances the mitochondria-mediated apoptosis [4,143,144]. Importantly, p53 also acts as a transrepressor for anti-apoptosis factors. It binds to survivin promoter and inhibits survivin gene transcription [131]. In addition, p53 itself can activate apoptosis without utilization of its transcription function. For example, p53 protein directly localizes to mitochondria following DNA damage and interacts with anti-apoptotic protein Bcl-2 and Bcl-$X_L$ to promote apoptosis [145,146]. Mutations in p53 have been found in more than half of human tumors [139]. p53 mutations in human cancer cells confer apoptosis resistance and promote survival and progression of the tumors.

## 5.2 PI3 kinase/AKT pathway

The Phosphatidylinositol 3-kinase (PI3K) pathway is a major cell survival pathway activated by growth factors, cytokins, and hormones [147]. PI3K is a heterodimer composed of a p85 regulatory and a p110 catalytic subunit. Active PI3K phosphorylates 3-phosphorylated lipid phosphatidylinositol-3,4,5-trisphophosphate (PtdIns(3,4,5,)P3), which then recruits the phosphoinositide-dependent protein kinases (PDK 1 and PDK 2) and protein kinase B (AKT) to the cellular membrane [148]. In the complex, PDKs activate AKT by phosphorylation [149]. Activation of AKT mediates a series of downstream effects to promote cell survival, such as phosphorylation and inhibition of proapoptotic Bad and caspase 9 and decreasing p53-mediated transcription of proapoptotic genes [150,151]. Phosphorylation of XIAP by Akt protects XIAP from ubiquitination and degradation in response to apoptosis stimuli [152]. Moreover, activation of the PI3K/AKT pathway after VEGF treatment increases the level of IAP protein survivin in endothelial cells [153].

The role of the PI3K/AKT pathway in the survival, growth and metastasis of tumor cells has been extensively studied. It has been shown that the PI3K/AKT pathway is highly activated in many tumor types [154,155]. The presence of an activated PI3K/AKT signal confers tumor cell resistance to apoptosis induction by growth factor withdrawal or chemotherapy drugs. Further, inhibition of PI3K/AKT activity greatly increases apoptotic cell death and drug sensitivity [156].

PI3K activity is negatively regulated by a tumor suppressor gene PTEN. PTEN antagonizes PI3K function by removing the 3-phosphate from (PtdIns(3,4,5,)P3). PTEN gene is frequently mutated or lost in several human tumor types [157]. Loss of PTEN function releases the inhibition on the PI3K/AKT pathway and increases the cell survival.

## 5.3 NFκB pathway

Nuclear factor κB (NFκB) is a transcriptional factor regulating apoptosis and cell survival. NFκB is present in cytoplasm in an inactive state by binding with its inhibitor protein, IκB. Upon receiving external stimuli, such as stress, cytokines, DNA damaging reagents or pathogens, IκB is phosphorylated and then degraded by ubiquitinylation, resulting in migration of DNA-bound subunit NFκB into the nucleus and activation of transcription of target genes [158]. It has been shown that NFκB functions as either an anti-apoptotic or a proapoptotic factor [159-162]. Recent findings have provided important insights into the role of the NFκB in regulating life and death decision. In the TNF-α activated cell death pathway, recruiting TRAF2 into TNFR, TRADD and RIP1 complex activates NFκB resulting in transcriptional activation of the caspase 8 inhibitor cFLIP. However, binding of the same complex to FADD activates caspase-8 and -10 and induces apoptosis [163]. NFκB-dependent transcription of anti-apoptotic Bcl-2 family proteins such as Bcl-2 and Bcl-$X_L$ confers protection against hypoxia and nitric oxide-induced apoptosis [114,164,165]. Upregualtion of the expression of the IAP genes such as c-IAP1, c-IAP2 and XIAP further enhances the anti-apoptotic effect of the NFκB [114,166,167]. Although it is clear that NFκB is a critical cell survival factor, there are a number of reports showing that under certain circumstances, activation of NFκB promotes apoptosis. NFκB induces expression of proapoptotic factor genes such as p53, FasL, TRAIL, cell death receptors and proapoptotic Bcl-Xs protein [168].

The anti-apoptotic activity of NFκB has been shown to be an important factor for tumorigenesis [161]. A high level of constitutive nuclear NFκB activity has been found in many human leukemias, lymphomas and solid tumors [169-171]. Suppression of the NFκB function results in apoptosis and/or sensitization of tumor cells to TNF-α or chemotherapy drug-induced apoptosis [166,172].

# 6. TUMOR ENVIRONMENT AND APOPTOSIS RESISTANCE

## 6.1 Hypoxia

It is well known that human tumors contain regions that are deficient in oxygen due to a rapid growth rate of the tumor cells and the presence of an abnormal vasculature [173]. Studies have shown that there are significant associations between intratumoral hypoxia and tumor metastasis, response to chemotherapy or radiotherapy, and prognosis of cancer patients [174-177]. Hypoxia induces upregulation of a key transcription factor, HIF-1 α, which mediates transcription of hypoxia-inducible genes in the cells [173]. It has been shown that hypoxia upregulates either anti-apoptotic or proapoptotic factors in cancer cells depending on the cell types and experimental conditions [178,179].

Evidence indicates that hypoxia suppresses the apoptosis induced by chemotherapy drugs or γ-irradiation. A recent study demonstrates that hypoxia-induced HIF-1 α expression protects HepG2 cells from apoptosis induction [180]. Resistance to staurosporine-induced apoptosis in hypoxic cells is mediated by an HIF-1 α independent upregulation of c-IAP2 [181]. Treatment of cancer cells with chronic hypoxia results in selective growth of apoptosis resistant cells that express a high level of anti-apoptotic Bcl-2 family protein BCL-$X_L$ [182]. It has also been shown that hypoxia activates the PI3K/Akt/NFκB and the MAPK(Erk) signaling pathways, resulting in the resistance of pancreatic cancer cells to gemcitabine treatment [183]. Hypoxia also increases the level of survivin expression in human tumor cells through HIF-1 α– dependent transcription. However, survivin is not expressed in normal cells either under normoxic or hypoxic conditions [184]; unpublished results, Lily Yang).

Despite its anti-apoptotic effects, hypoxia also activates proapoptotic factors and induces apoptosis in cancer cells. It has been demonstrated that hypoxia-induced apoptosis mainly relies on mitochondrial pathways. In human tumors, hypoxia may lead to the selection of hypoxia-resistant cells with defects in mitochondrial apoptosis signaling pathways [179]. Expression of a proapoptotic Bcl-2 family protein, BNIP3, is increased in hypoxia through transcriptional activation of BNIP 3 by HIF-1 α. The presence of a hypoxia-responsive element in the BNip3 promoter that activates the level of BNIP 3 gene transcription by HIF-1 α has been demonstrated in many types of human cancer cell lines [185]. In pancreatic cancer tissues, methylation of BNIP 3 promoter inhibits the expression of the BNIP 3 gene despite the upregulation of other hypoxia-inducible genes, resulting in resistance to hypoxia-induced apoptosis. Restoration of BNIP 3 expression increases the sensitivity of the pancreatic cancer cells to hypoxia-induced cell death [186].

Moreover, hypoxia also induces the stabilization of p53 protein, which is a key transcription factor for promoting apoptosis [187]. It is possible that the dual effects of hypoxia on apoptosis are influenced by the severity of hypoxia in the cells. The proapoptotic function of HIF-1 α is activated in the cells under extreme hypoxia when the cellular protective function is not sufficient to protect cells from hypoxia damage [178].

## 6.2 Extracellular matrix

It is well established that extracellar matrix (ECM) is a critical regulator for signal transduction pathways. Interactions between cancer cells and ECM also contribute to the survival and apoptosis resistance in the cells [188,189]. Loss of contact between ECM and cells has been associated with apoptosis induction and lumen formation during normal tissue development [12]. A pecial form of apoptotic cell death, anoikis, is induced in the cells that have ost contact with ECM and surrounding cells [190]. A recent study further lemonstrates that ECM increases expression of antiapoptotic proteins Bcl-2 and $Bcl\text{-}X_L$ and reduced drug-induced apoptosis in small lung cancer cells, nyelomas and gliomas [188,189,191]. Additionally, upregulation of matrix netalloproteinase (MMP) is a common phenomenon in human tumors and nas been associated with tumor progression, metastasis and angiogenesis [91,192]. It has been shown that MMP-7, which is produced by tumor cells, specifically cleaves Fas and FasL, resulting in inhibition of Fas-mediated apoptosis [188,193,194]. Overexpression of the MMP-7 gene in the mouse nammary gland promotes mammary hyperplasia and accelerates the onset of oncogene-induced mammary tumors [195]. Cell-ECM interactions are mediated by adhesion receptors such as integrins on the cell surface. Studies have shown that integrins are expressed in some human tumor cells as well as in angiogenic tumor endothelial cells. Interaction of $\beta 1$ integrin with ECM in breast cancer cells significantly inhibits apoptosis induced by chemotherapy drugs paclitaxel and vincristine [196]. A recent study reports that loss of cell attachment to ECM induces caspase independent apoptosis through releasing a mitochondria protein Bit-1 into the cytosol and inducing apoptosis in the cell. Tumor cells expressing the $\alpha v\beta 5$ integrin initiates signals capable of blocking Bit-induced apoptosis [197].

## 7. MOLECULAR TARGETS IN APOPTOSIS SIGNAL PATHWAY FOR CANCER THERAPY

Understanding molecular alterations in apoptosis signal pathway helps to identify novel therapeutic targets. Results from the examination of apoptotic effectors and regulators in the apoptotic signal pathway in various tumor types demonstrate the presence of a deregulated apoptosis signal pathway in human cancer cells. Those defects confer apoptosis resistance and provide growth advantage for the tumor cells.

Strategies for targeting upstream defects in apoptosis pathways are developed and the feasibility of those approaches has been evaluated in human tumor cell lines and animal tumor models. For example, it has been shown that overexpression of death receptor-ligands, such as Fas L and TRAIL, with adenoviral vectors or delivery of recombinant FasL and TRAIL induces apoptotic cell death and sensitizes the response to chemotherapy drugs in some tumor cells [198]. However, extensive investigations of this approach on various human tumor cells reveals that many human tumor cells are resistant to FasL or TRAIL-induced apoptosis despite the expression of cell-death receptors on the cells [25,26,56].

Since activation of caspases is a hallmark for apoptosis induction, a logical approach for activating apoptosis is to express procaspase or active caspase genes in tumor cell. The feasibility of apoptosis induction by overexpression of procaspase-3, -7, -8 and -9, and an engineered autocatalytic caspase-3 have been examined in several laboratories using various tumor cells [199-201]. The results of those studies have shown that expression of procaspase or active caspase gene is able to induce apoptotic cell death in many human tumor cell lines. However, as compared to apoptosis induction in normal cells, tumor cell lines are less sensitive to caspase-induced apoptosis than normal cells[79].

Deregulation of Bcl-2 family proteins is found in many cancer types. Strategies downregulating anti-apoptotic or increasing the levels of proapoptotic Bcl-2 family proteins have been developed and some of them are already in clinical trails to determine the toxicity and efficiency. It has been shown that a decrease in Bcl-2 expression using Bcl-2 antisense induces apoptosis and sensitizing the cells to chemotherapy drugs [98]. Expression of Bax or Bak genes from adenoviral vectors shows anti-tumor effects both *in vitro* and in animal tumor models [202,203]. Small peptides targeting Bcl-2 and Bcl-$X_L$ are capable of inhibiting activity of Bcl-2 and Bcl-$X_L$ and have shown therapeutic potential as anticancer drugs for treating cancers overexpressing Bcl-2 and/or Bcl-X(L) proteins [97].

Results from dissecting deregulated apoptotic signals in human tumor cells further show that although different upstream deficiencies, such as Fas mutation and defects in caspase expression, are found in tumor cell lines and tissues, they are limited to small percentage of tumor cells in several cancer types [29,67,68,204]. On the other hand, upregulation of IAPs is a common feature for the majority tumor types [7,123]. Novel approaches targeting the IAP proteins should provide new ways to treat most human cancers.

A recent study has shown that inhibition of XIAP with small molecular antagonists stimulates an increase in the level of caspase activity and induces apoptotic cell death both in tumor cell lines and in established animal tumor models. Interestedly, apoptosis induction through inhibition of XIAP is tumor specific and there is very litter toxicity in normal cell lines as well as in normal tissues [205]. Inhibition of XIAP function could also be achieved through expression of a XIAP-counteracting protein gene, XAF1. Overexpression of XAF1 using an adenoviral vector selectively increases caspase 3 activity and induces apoptotic cell death in human breast and pancreatic cancer cells but not in normal cells [79].

Survivin is not expressed in normal cells but it is highly expressed in most tumor cells. Direct inhibition of survivin expression or function may have greater impact on the survival of tumor cells than for normal cells. Several reports have shown that inhibition of survivin function with xpression of a dominant negative mutant survivin (T34A), survivin antiense or siRNA increases caspase 9 activity and results in apoptotic cell leath in human tumor cells and xenografted tumor models [123,136,206]. mportantly, downregulation of survivin specifically induces apoptotic cell leath in tumor cells without obvious toxic effects on various normal cell ines [79,136]. Downregulation of survivin function also enhances the effects of chemotherapy drugs on the tumor cells [136,137]. Therefore targeting IAP proteins is a promising approach for the development of cancer-cell specific herapy.

At present, the mechanisms for tumor-specific induction of apoptosis by nhibiting IAP function are still under investigation. As discussed above, apoptosis is the physiological cell death process for the removal of abnormal cells. Human tumor cells are generated from multiple genetic alterations and have enormous abnormalities. These should cause activation of the apoptotic signal and induction of apoptosis. Evaluation of the process of tumor development suggests that this is the case for most transformed preneoplastic cells. Induction of apoptotic cell death in pre-neoplastic lesions has been found in early stage of human cancers such as breast ductal carcinoma *in situ* (DCIS) [207]. It has been shown that many DCIS lesions keep a balance between cell proliferation and apoptosis for many years without developing into invasive breast cancers. It is well established that treating rats with chemical carcinogens initiates many pre-neoplastic nodules in the liver but only a few of these develop into hepatocellular carcinomas [208]. It is possible that most pre-neoplastic cells are destroyed by apoptosis and only a small fraction of transformed cells that have upregulated their anti-apoptotic mechanisms, such as IAPs and Bcl-2, are able to survive and develop into a tumor mass. The selective growth of tumor cells with a high level of antiapoptotic factors confers apoptosis resistance and a poor response to therapeutic reagents (Figure). Co-existence of high levels of active caspase 3 and IAP proteins, survivin and XIAP, has been demonstrated in human pancreas, colon and breast cancer cell lines that are not apoptotic [79]. The results from analysis of human breast cancer tissues further supported the presence of activated apoptotic signals and upregulated anti-apoptotic factors in cancer tissues. A positive correlation between the levels of active caspase-3 and -6, and the IAP proteins survivin and XIAP has been established using either immunostaining or Western blots with antibodies specific for active caspase-3, caspase 6, XIAP and survivin [77,79].

It has been shown that overexpression of procaspase 3 gene in ovary cancer cells increases survivin gene expression. It is possible that one of the cellular responses to a high level of caspase or caspase activity is to upregulate cell survival factors such as survivin and XIAP (Figure). The presence of a high balance between pro- and anti-apoptotic factors in human tumor cells but not in normal cells suggests that targeting IAP protein provides a selective advantage, inducing apoptotic cell death in human tumo cells while minimizing the effects on normal cells (Figure 1).

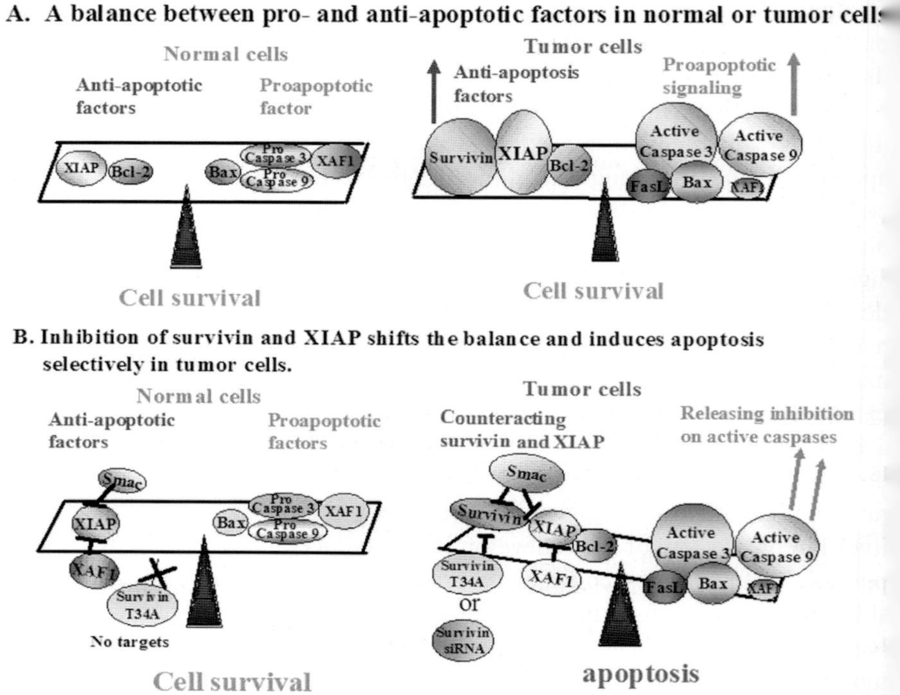


*Figure 1.* Targeting IAP proteins for cancer specific therapy. In normal cells, the absence of an apoptotic signal keeps a low balance between pro- and anti-apoptotic factors. However, molecular changes associated with malignant transformation of human tumor cells lead to activation of the apoptotic signals such as expression of FasL and activation of caspases. The tumor cells are able to block apoptosis by upregulating IAPs that inhibit active caspases. Therefore, cancer cells have high levels of both pro- and anti-apoptosis factors. The apoptotic process could be restored selectively in tumor cells by inhibiting IAP functions such as the expression of dominant negative survivinT34A gene and survivin siRNA, XIAP counteracting protein XAF1 and active Smac protein or peptides.

## REFERENCES


1. Wyllie AH. Apoptosis: cell death under homeostatic control. *Arch Toxicol Suppl* 1987; **11:** 3-10.
2. Rich T, Allen RL, Wyllie AH. Defying death after DNA damage. *Nature* 2000; **407:** 777-783.
3. Hanahan D, Weinberg RA. The hallmarks of cancer. *Cell* 2000; **100:** 57-70.
4. Hickman ES, Helin K. The regulation of APAF1 expression during development and tumourigenesis. *Apoptosis* 2002; **7:** 167-171.
5. Wyllie AH *et al.* Apoptosis and carcinogenesis. *Br J Cancer* 1999; **80 Suppl 1:** 34-37.
6. Reed JC. Mechanisms of apoptosis avoidance in cancer. *Curr Opin Oncol* 1999; **11:** 68-75.
7. Holcik M, Gibson H, Korneluk RG. XIAP: apoptotic brake and promising therapeutic target. *Apoptosis* 2001; **6:** 253-261.
8. Igney FH, Krammer PH. Death and anti-death: tumour resistance to apoptosis. *Nat Rev Cancer* 2002; **2:** 277-288.
9. Altieri DC. Survivin and apoptosis control. *Adv Cancer Res* 2003; **88:** 31-52.
10. Zakeri Z, Lockshin RA. Cell death during development. *J Immunol Methods* 2002; **265:** 3-20.
11. Spencer SJ, Cataldo NA, Jaffe RB. Apoptosis in the human female reproductive tract. *Obstet Gynecol Surv* 1996; **51:** 314-323.
12. Debnath J *et al.* The role of apoptosis in creating and maintaining luminal space within normal and oncogene-expressing mammary acini. *Cell* 2002; **111:** 29-40.
13. Hahm HaD, NE. Apoptosis in the mammary gland and breast cancer. *Endocrine-Related Cancer* 1998; **5:** 199-211.
14. Collins M. Potential roles of apoptosis in viral pathogenesis. *Am J Respir Crit Care Med* 1995; **152:** S20-24.
15. Winoto A. Cell death in the regulation of immune responses. *Curr Opin Immunol* 1997; **9:** 365-370.
16. Kamarajan P, Chao CC. UV-induced apoptosis in resistant HeLa cells. *Biosci Rep* 2000; **20:** 99-108.
17. Park IC *et al.* Ionizing radiation and nitric oxide donor sensitize Fas-induced apoptosis via up-regulation of Fas in human cervical cancer cells. *Oncol Rep* 2003; **10:** 629-633.
18. Kaufmann SH, Earnshaw WC. Induction of apoptosis by cancer chemotherapy. *Exp Cell Res* 2000; **256:** 42-49.
19. Reed JC. Mechanisms of apoptosis. *Am J Pathol* 2000; **157:** 1415-1430.
20. Suzuki A *et al.* Mitochondrial regulation of cell death: mitochondria are essential for procaspase 3-p21 complex formation to resist Fas-mediated cell death. *Mol Cell Biol* 1999; **19:** 3842-3847.
21. Henry-Mowatt J, Dive C, Martinou JC, James D. Role of mitochondrial membrane permeabilization in apoptosis and cancer. *Oncogene* 2004; **23:** 2850-2860.
22. Budihardjo I *et al.* Biochemical pathways of caspase activation during apoptosis. *Annu Rev Cell Dev Biol* 1999; **15:** 269-290.
23. Stroh C, Schulze-Osthoff K. Death by a thousand cuts: an ever increasing list of caspase substrates. *Cell Death Differ* 1998; **5:** 997-1000.
24. Peter ME, Krammer PH. The CD95(APO-1/Fas) DISC and beyond. *Cell Death Differ* 2003; **10:** 26-35.
25. Barnhart BC *et al.* CD95 ligand induces motility and invasiveness of apoptosis-resistant tumor cells. *Embo J* 2004; **23:** 3175-3185.

26. Landowski TH *et al.* CD95 antigen mutations in hematopoietic malignancies. *Leu Lymphoma* 2001; **42:** 835-846.
27. Beltinger C *et al.* CD95 (APO-1/Fas) mutations in childhood T-lineage acut lymphoblastic leukemia. *Blood* 1998; **91:** 3943-3951.
28. Lee SH *et al.* Alterations of Fas (Apo-1/CD95) gene in non-small cell lung cancer *Oncogene* 1999; **18:** 3754-3760.
29. Boldrini L *et al.* Identification of Fas (APO-1/CD95) and p53 gene mutations in non small cell lung cancer. *Int J Oncol* 2002; **20:** 155-159.
30. Bullani RR *et al.* Frequent downregulation of Fas (CD95) expression and function in melanoma. *Melanoma Res* 2002; **12:** 263-270.
31. Lee SH *et al.* Expression of Fas and Fas-related molecules in human hepatocellular carcinoma. *Hum Pathol* 2001; **32:** 250-256.
32. Nambu Y *et al.* Lack of cell surface Fas/APO-1 expression in pulmonary adenocarcinomas. *J Clin Invest* 1998; **101:** 1102-1110.
33. Viard-Leveugle I *et al.* Frequent loss of Fas expression and function in human lung tumours with overexpression of FasL in small cell lung carcinoma. *J Pathol* 2003; **201:** 268-277.
34. Lee SH *et al.* In vivo expression of soluble Fas and FAP-1: possible mechanisms of Fas resistance in human hepatoblastomas. *J Pathol* 1999; **188:** 207-212.
35. Liu JH *et al.* Blockade of Fas-dependent apoptosis by soluble Fas in LGL leukemia. *Blood* 2002; **100:** 1449-1453.
36. Elnemr A *et al.* Human pancreatic cancer cells disable function of Fas receptors at several levels in Fas signal transduction pathway. *Int J Oncol* 2001; **18:** 311-316.
37. Gerharz CD *et al.* Resistance to CD95 (APO-1/Fas)-mediated apoptosis in human renal cell carcinomas: an important factor for evasion from negative growth control. *Lab Invest* 1999; **79:** 1521-1534.
38. Muschen M *et al.* Resistance to CD95-mediated apoptosis in breast cancer is not due to somatic mutation of the CD95 gene. *Int J Cancer* 2001; **92:** 309-310.
39. Keane MM *et al.* Fas expression and function in normal and malignant breast cell lines. *Cancer Res* 1996; **56:** 4791-4798.
40. Ungefroren H *et al.* Human pancreatic adenocarcinomas express Fas and Fas ligand yet are resistant to Fas-mediated apoptosis. *Cancer Res* 1998; **58:** 1741-1749.
41. Suda T, Takahashi T, Golstein P, Nagata S. Molecular cloning and expression of the Fas ligand, a novel member of the tumor necrosis factor family. *Cell* 1993; **75:** 1169-1178.
42. Bellgrau D *et al.* A role for CD95 ligand in preventing graft rejection. *Nature* 1995; **377:** 630-632.
43. Mullauer L *et al.* Fas ligand is expressed in normal breast epithelial cells and is frequently up-regulated in breast cancer. *J Pathol* 2000; **190:** 20-30.
44. Abrahams VM, Kamsteeg M, Mor G. The Fas/Fas ligand system and cancer: immune privilege and apoptosis. *Mol Biotechnol* 2003; **25:** 19-30.
45. Bennett MW *et al.* Fas ligand upregulation is an early event in colonic carcinogenesis. *J Clin Pathol* 2001; **54:** 598-604.
46. Herrnring C *et al.* Expression of the apoptosis-inducing ligands FasL and TRAIL in malignant and benign human breast tumors. *Histochem Cell Biol* 2000; **113:** 189-194.
47. Sheehan KM *et al.* Prognostic relevance of Fas (APO-1/CD95) ligand in human colorectal cancer. *Eur J Gastroenterol Hepatol* 2003; **15:** 375-380.
48. Boesen-de Cock JG *et al.* Common regulation of apoptosis signaling induced by CD95 and the DNA-damaging stimuli etoposide and gamma-radiation downstream from caspase-8 activation. *J Biol Chem* 1999; **274:** 14255-14261.

49. Fulda S *et al.* The CD95 (APO-1/Fas) system mediates drug-induced apoptosis in neuroblastoma cells. *Cancer Res* 1997; **57:** 3823-3829.
50. Landowski TH, Gleason-Guzman MC, Dalton WS. Selection for drug resistance results in resistance to Fas-mediated apoptosis. *Blood* 1997; **89:** 1854-1861.
51. Wang S, El-Deiry WS. TRAIL and apoptosis induction by TNF-family death receptors. *Oncogene* 2003; **22:** 8628-8633.
52. Kelley SK, Ashkenazi A. Targeting death receptors in cancer with Apo2L/TRAIL. *Curr Opin Pharmacol* 2004; **4:** 333-339.
53. LeBlanc HN, Ashkenazi A. Apo2L/TRAIL and its death and decoy receptors. *Cell Death Differ* 2003; **10:** 66-75.
54. Marsters SA, Pitti RA, Sheridan JP, Ashkenazi A. Control of apoptosis signaling by Apo2 ligand. *Recent Prog Horm Res* 1999; **54:** 225-234.
55. Ashkenazi A, Dixit VM. Apoptosis control by death and decoy receptors. *Curr Opin Cell Biol* 1999; **11:** 255-260.
56. Ibrahim SM *et al.* Pancreatic adenocarcinoma cell lines show variable susceptibility to TRAIL-mediated cell death. *Pancreas* 2001; **23:** 72-79.
57. Singh TR *et al.* Synergistic interactions of chemotherapeutic drugs and tumor necrosis factor-related apoptosis-inducing ligand/Apo-2 ligand on apoptosis and on regression of breast carcinoma in vivo. *Cancer Res* 2003; **63:** 5390-5400.
58. Jin Z, McDonald ER, 3rd, Dicker DT, El-Deiry WS. Deficient TRAIL death receptor transport to the cell surface in human colon cancer cells selected for resistance to TRAIL-induced apoptosis. *J Biol Chem* 2004.
59. Odoux C *et al.* TRAIL, FasL and a blocking anti-DR5 antibody augment paclitaxel-induced apoptosis in human non-small-cell lung cancer. *Int J Cancer* 2002; **97:** 458-465.
60. Shankar S, Srivastava RK. Enhancement of therapeutic potential of TRAIL by cancer chemotherapy and irradiation: mechanisms and clinical implications. *Drug Resist Updat* 2004; **7:** 139-156.
61. Ashkenazi A, Dixit VM. Death receptors: signaling and modulation. *Science* 1998; **281:** 1305-1308.
62. Morgan M, Thorburn J, Pandolfi PP, Thorburn A. Nuclear and cytoplasmic shuttling of TRADD induces apoptosis via different mechanisms. *J Cell Biol* 2002; **157:** 975-984.
63. Varfolomeev EE, Ashkenazi A. Tumor necrosis factor: an apoptosis JuNKie? *Cell* 2004; **116:** 491-497.
64. Micheau O *et al.* NF-kappaB signals induce the expression of c-FLIP. *Mol Cell Biol* 2001; **21:** 5299-5305.
65. Murray PG *et al.* Expression of the tumour necrosis factor receptor-associated factors 1 and 2 in Hodgkin's disease. *J Pathol* 2001; **194:** 158-164.
66. Zapata JM *et al.* TNFR-associated factor family protein expression in normal tissues and lymphoid malignancies. *J Immunol* 2000; **165:** 5084-5096.
67. Hopkins-Donaldson S *et al.* Silencing of death receptor and caspase-8 expression in small cell lung carcinoma cell lines and tumors by DNA methylation. *Cell Death Differ* 2003; **10:** 356-364.
68. Iolascon A *et al.* Caspase 3 and 8 deficiency in human neuroblastoma. *Cancer Genet Cytogenet* 2003; **146:** 41-47.
69. Teitz T *et al.* Caspase 8 is deleted or silenced preferentially in childhood neuroblastomas with amplification of MYCN. *Nat Med* 2000; **6:** 529-535.
70. Fujikawa K *et al.* Reduced expression of ICE/caspase1 and CPP32/caspase3 in human hepatocellular carcinoma. *Anticancer Res* 2000; **20:** 1927-1932.

71. Kolenko V *et al.* Dead or dying: necrosis versus apoptosis in caspase-deficient huma renal cell carcinoma. *Cancer Res* 1999; **59:** 2838-2842.
72. Palmerini F *et al.* Caspase 7 downregulation as an immunohistochemical marker ( colonic carcinoma. *Hum Pathol* 2001; **32:** 461-467.
73. Chu ZL *et al.* A novel enhancer of the Apafl apoptosome involved in cytochrome dependent caspase activation and apoptosis. *J Biol Chem* 2001; **276:** 9239-9245.
74. Soengas MS *et al.* Inactivation of the apoptosis effector Apaf-1 in malignant melanom *Nature* 2001; **409:** 207-211.
75. Nakopoulou L *et al.* Immunohistochemical expression of caspase-3 as an advers indicator of the clinical outcome in human breast cancer. *Pathobiology* 2001; **69:** 266 273.
76. Vakkala M, Paakko P, Soini Y. Expression of caspases 3, 6 and 8 is increased i parallel with apoptosis and histological aggressiveness of the breast lesion. *Br J Cance* 1999; **81:** 592-599.
77. Parton M *et al.* Coordinate Expression of Apoptosis-associated Proteins in Huma Breast Cancer before and during Chemotherapy. *Clin Cancer Res* 2002; **8:** 2100-2108.
78. O'Donovan N *et al.* Caspase 3 in breast cancer. *Clin Cancer Res* 2003; **9:** 738-742.
79. Yang L, Cao Z, Yan H, Wood WC. Coexistence of high levels of apoptotic signalin and inhibitor of apoptosis proteins in human tumor cells: implication for cancer specifi therapy. *Cancer Res* 2003; **63:** 6815-6824.
80. Satoh K *et al.* Expression of survivin is correlated with cancer cell apoptosis and i involved in the development of human pancreatic duct cell tumors. *Cancer* 2001; **92** 271-278.
81. Yoo NJ *et al.* Stomach cancer highly expresses both initiator and effector caspases; a immunohistochemical study. *Apmis* 2002; **110:** 825-832.
82. Kennedy NJ, Kataoka T, Tschopp J, Budd RC. Caspase activation is required for T cel proliferation. *J Exp Med* 1999; **190:** 1891-1896.
83. Mouhamad S *et al.* Differential modulation of interleukin-2-and interleukin-4-mediatec early activation of normal human B lymphocytes by the caspase inhibitor zVAD-fmk *Eur Cytokine Netw* 2002; **13:** 439-445.
84. Rendl M *et al.* Caspase-14 expression by epidermal keratinocytes is regulated by retinoids in a differentiation-associated manner. *J Invest Dermatol* 2002; **119:** 1150- 1155.
85. Reichmann E. The biological role of the Fas/FasL system during tumor formation and progression. *Semin Cancer Biol* 2002; **12:** 309-315.
86. Scaffidi C, Schmitz I, Krammer PH, Peter ME. The role of c-FLIP in modulation of CD95-induced apoptosis. *J Biol Chem* 1999; **274:** 1541-1548.
87. Abedini MR, Qiu Q, Yan X, Tsang BK. Possible role of FLICE-like inhibitory protein (FLIP) in chemoresistant ovarian cancer cells in vitro. *Oncogene* 2004.
88. Lee SH *et al.* Increased expression of FLIP, an inhibitor of Fas-mediated apoptosis, in stomach cancer. *Apmis* 2003; **111:** 309-314.
89. Griffith TS *et al.* Intracellular regulation of TRAIL-induced apoptosis in human melanoma cells. *J Immunol* 1998; **161:** 2833-2840.
90. Kamarajan P, Sun NK, Chao CC. Up-regulation of FLIP in cisplatin-selected HeLa cells causes cross-resistance to CD95/Fas death signalling. *Biochem J* 2003; **376:** 253-260.
91. Reed JC. Bcl-2 family proteins. *Oncogene* 1998; **17:** 3225-3236.
92. Cory S, Huang DC, Adams JM. The Bcl-2 family: roles in cell survival and oncogenesis. *Oncogene* 2003; **22:** 8590-8607.

93. Reed JC. Bcl-2: prevention of apoptosis as a mechanism of drug resistance. *Hematol Oncol Clin North Am* 1995; **9:** 451-473.
94. Gee JM *et al.* Immunocytochemical localization of BCL-2 protein in human breast cancers and its relationship to a series of prognostic markers and response to endocrine therapy. *Int J Cancer* 1994; **59:** 619-628.
95. Tu Y *et al.* Upregulated expression of BCL-2 in multiple myeloma cells induced by exposure to doxorubicin, etoposide, and hydrogen peroxide. *Blood* 1996; **88:** 1805-1812.
96. Fulda S, Meyer E, Debatin KM. Inhibition of TRAIL-induced apoptosis by Bcl-2 overexpression. *Oncogene* 2002; **21:** 2283-2294.
97. Wang S, Yang D, Lippman ME. Targeting Bcl-2 and Bcl-XL with nonpeptidic small-molecule antagonists. *Semin Oncol* 2003; **30:** 133-142.
98. Reed JC. Promise and problems of Bcl-2 antisense therapy. *J Natl Cancer Inst* 1997; **89:** 988-990.
99. Reed JC. Apoptosis-based therapies. *Nat Rev Drug Discov* 2002; **1:** 111-121.
100. Krajewski S *et al.* Prognostic significance of apoptosis regulators in breast cancer. *Endocr Relat Cancer* 1999; **6:** 29-40.
101. Feuerhake F *et al.* Immunohistochemical analysis of Bcl-2 and Bax expression in relation to cell turnover and epithelial differentiation markers in the non-lactating human mammary gland epithelium. *Cell Tissue Res* 2000; **299:** 47-58.
102. Joensuu H, Pylkkanen L, Toikkanen S. Bcl-2 protein expression and long-term survival in breast cancer. *Am J Pathol* 1994; **145:** 1191-1198.
103. Zhang GJ *et al.* Apoptotic index correlates to bcl-2 and p53 protein expression, histological grade and prognosis in invasive breast cancers. *Anticancer Res* 1998; **18:** 1989-1998.
104. Villar E *et al.* bcl-2 Expression and apoptosis in primary and metastatic breast carcinomas. *Tumour Biol* 2001; **22:** 137-145.
105. Gutierrez MI *et al.* Bax is frequently compromised in Burkitt's lymphomas with irreversible resistance to Fas-induced apoptosis. *Cancer Res* 1999; **59:** 696-703.
106. Knudson CM, Johnson GM, Lin Y, Korsmeyer SJ. Bax accelerates tumorigenesis in p53-deficient mice. *Cancer Res* 2001; **61:** 659-665.
107. Rampino N *et al.* Somatic frameshift mutations in the BAX gene in colon cancers of the microsatellite mutator phenotype. *Science* 1997; **275:** 967-969.
108. Bargou RC *et al.* Expression of the bcl-2 gene family in normal and malignant breast tissue: low bax-alpha expression in tumor cells correlates with resistance towards apoptosis. *Int J Cancer* 1995; **60:** 854-859.
109. Krajewski S *et al.* Reduced expression of proapoptotic gene BAX is associated with poor response rates to combination chemotherapy and shorter survival in women with metastatic breast adenocarcinoma. *Cancer Res* 1995; **55:** 4471-4478.
110. Redondo M *et al.* Expression of bax and p53 proteins in the tumorigenesis and progression of breast carcinomas. *Tumour Biol* 2003; **24:** 23-31.
111. Roy N *et al.* The c-IAP-1 and c-IAP-2 proteins are direct inhibitors of specific caspases. *Embo J* 1997; **16:** 6914-6925.
112. Suzuki Y *et al.* X-linked inhibitor of apoptosis protein (XIAP) inhibits caspase-3 and -7 in distinct modes. *J Biol Chem* 2001; **276:** 27058-27063.
113. Kasof GM, Gomes BC. Livin, a novel inhibitor of apoptosis protein family member. *J Biol Chem* 2001; **276:** 3238-3246.
114. Zhang HG *et al.* Regulation of apoptosis proteins in cancer cells by ubiquitin. *Oncogene* 2004; **23:** 2009-2015.

115. Tamm I *et al.* IAP-family protein survivin inhibits caspase activity and apoptosi induced by Fas (CD95), Bax, caspases, and anticancer drugs. *Cancer Res* 1998; **58** 5315-5320.
116. Dohi T *et al.* An IAP-IAP complex inhibits apoptosis. *J Biol Chem* 2004; **279:** 34087 34090.
117. Du C *et al.* Smac, a mitochondrial protein that promotes cytochrome c-dependen caspase activation by eliminating IAP inhibition. *Cell* 2000; **102:** 33-42.
118. Liu Z *et al.* Structural basis for binding of Smac/DIABLO to the XIAP BIR3 domain *Nature* 2000; **408:** 1004-1008.
119. Liston P *et al.* Identification of XAF1 as an antagonist of XIAP anti-Caspase activity *Nat Cell Biol* 2001; **3:** 128-133.
120. Fong WG *et al.* Expression and genetic analysis of XIAP-associated factor 1 (XAF1) ir cancer cell lines. *Genomics* 2000; **70:** 113-122.
121. Byun DS *et al.* Hypermethylation of XIAP-associated factor 1, a putative tumoı suppressor gene from the 17p13.2 locus, in human gastric adenocarcinomas. *Cancer Res* 2003; **63:** 7068-7075.
122. Adida C *et al.* Developmentally regulated expression of the novel cancer anti-apoptosis gene survivin in human and mouse differentiation. *Am J Pathol* 1998; **152:** 43-49.
123. Altieri DC. Survivin, versatile modulation of cell division and apoptosis in cancer. *Oncogene* 2003; **22:** 8581-8589.
124. Ambrosini G, Adida C, Altieri DC. A novel anti-apoptosis gene, survivin, expressed in cancer and lymphoma. *Nat Med* 1997; **3:** 917-921.
125. Li F. Survivin study: what is the next wave? *J Cell Physiol* 2003; **197:** 8-29.
126. Tanaka K *et al.* Expression of survivin and its relationship to loss of apoptosis in breast carcinomas. *Clin Cancer Res* 2000; **6:** 127-134.
127. Sarela AI *et al.* Expression of survivin, a novel inhibitor of apoptosis and cell cycle regulatory protein, in pancreatic adenocarcinoma. *Br J Cancer* 2002; **86:** 886-892.
128. Islam A *et al.* Role of survivin, whose gene is mapped to 17q25, in human neuroblastoma and identification of a novel dominant-negative isoform, survivin-beta/2B. *Med Pediatr Oncol* 2000; **35:** 550-553.
129. Hattori M, Sakamoto H, Satoh K, Yamamoto T. DNA demethylase is expressed in ovarian cancers and the expression correlates with demethylation of CpG sites in the promoter region of c-erbB-2 and survivin genes. *Cancer Lett* 2001; **169:** 155-164.
130. Bao R *et al.* Activation of cancer-specific gene expression by the survivin promoter. *J Natl Cancer Inst* 2002; **94:** 522-528.
131. Hoffman WH *et al.* Transcriptional repression of the anti-apoptotic survivin gene by wild type p53. *J Biol Chem* 2002; **277:** 3247-3257.
132. Asanuma K *et al.* A role for survivin in radioresistance of pancreatic cancer cells. *Jpn J Cancer Res* 2002; **93:** 1057-1062.
133. Kawasaki H *et al.* Inhibition of apoptosis by survivin predicts shorter survival rates in colorectal cancer. *Cancer Res* 1998; **58:** 5071-5074.
134. Kami K *et al.* Survivin expression is a prognostic marker in pancreatic cancer patients. *Surgery* 2004; **136:** 443-448.
135. Grossman D *et al.* Transgenic expression of survivin in keratinocytes counteracts UVB-induced apoptosis and cooperates with loss of p53. *J Clin Invest* 2001; **108:** 991-999.
136. Mesri M *et al.* Cancer gene therapy using a survivin mutant adenovirus. *J Clin Invest* 2001; **108:** 981-990.
137. Zangemeister-Wittke U. Antisense to apoptosis inhibitors facilitates chemotherapy and TRAIL-induced death signaling. *Ann N Y Acad Sci* 2003; **1002:** 90-94.

138. Li F *et al.* Pleiotropic cell-division defects and apoptosis induced by interference with survivin function. *Nat Cell Biol* 1999; **1:** 461-466.
139. Vogelstein B, Kinzler KW. p53 function and dysfunction. *Cell* 1992; **70:** 523-526.
140. Vogelstein B, Lane D, Levine AJ. Surfing the p53 network. *Nature* 2000; **408:** 307-310.
141. Polyak K *et al.* A model for p53-induced apoptosis. *Nature* 1997; **389:** 300-305.
142. Muller M *et al.* p53 activates the CD95 (APO-1/Fas) gene in response to DNA damage by anticancer drugs. *J Exp Med* 1998; **188:** 2033-2045.
143. Yin C, Knudson CM, Korsmeyer SJ, Van Dyke T. Bax suppresses tumorigenesis and stimulates apoptosis in vivo. *Nature* 1997; **385:** 637-640.
144. Yu J *et al.* PUMA mediates the apoptotic response to p53 in colorectal cancer cells. *Proc Natl Acad Sci U S A* 2003; **100:** 1931-1936.
145. Marchenko ND, Zaika A, Moll UM. Death signal-induced localization of p53 protein to mitochondria. A potential role in apoptotic signaling. *J Biol Chem* 2000; **275:** 16202-16212.
146. Mihara M *et al.* p53 has a direct apoptogenic role at the mitochondria. *Mol Cell* 2003; **11:** 577-590.
147. Danielsen AJ, Maihle NJ. The EGF/ErbB receptor family and apoptosis. *Growth Factors* 2002; **20:** 1-15.
148. Ram TG, Ethier SP. Phosphatidylinositol 3-kinase recruitment by p185erbB-2 and erbB-3 is potently induced by neu differentiation factor/heregulin during mitogenesis and is constitutively elevated in growth factor-independent breast carcinoma cells with c-erbB-2 gene amplification. *Cell Growth Differ* 1996; **7:** 551-561.
149. Toker A, Newton AC. Akt/protein kinase B is regulated by autophosphorylation at the hypothetical PDK-2 site. *J Biol Chem* 2000; **275:** 8271-8274.
150. Datta SR *et al.* Akt phosphorylation of BAD couples survival signals to the cell-intrinsic death machinery. *Cell* 1997; **91:** 231-241.
151. Sabbatini P, McCormick F. Phosphoinositide 3-OH kinase (PI3K) and PKB/Akt delay the onset of p53-mediated, transcriptionally dependent apoptosis. *J Biol Chem* 1999; **274:** 24263-24269.
152. Dan HC *et al.* Akt phosphorylation and stabilization of X-linked inhibitor of apoptosis protein (XIAP). *J Biol Chem* 2004; **279:** 5405-5412.
153. Tran J *et al.* Marked induction of the IAP family antiapoptotic proteins survivin and XIAP by VEGF in vascular endothelial cells. *Biochem Biophys Res Commun* 1999; **264:** 781-788.
154. Fry MJ. Phosphoinositide 3-kinase signalling in breast cancer: how big a role might it play? *Breast Cancer Res* 2001; **3:** 304-312.
155. Hutchinson J *et al.* Activation of Akt (protein kinase B) in mammary epithelium provides a critical cell survival signal required for tumor progression. *Mol Cell Biol* 2001; **21:** 2203-2212.
156. O'Gorman DM, McKenna SL, McGahon AJ, Cotter TG. Inhibition of PI3-kinase sensitises HL60 human leukaemia cells to both chemotherapeutic drug- and Fas-induced apoptosis by a JNK independent pathway. *Leuk Res* 2001; **25:** 801-811.
157. Haas-Kogan D *et al.* Protein kinase B (PKB/Akt) activity is elevated in glioblastoma cells due to mutation of the tumor suppressor PTEN/MMAC. *Curr Biol* 1998; **8:** 1195-1198.
158. Karin M, Ben-Neriah Y. Phosphorylation meets ubiquitination: the control of NF-[kappa]B activity. *Annu Rev Immunol* 2000; **18:** 621-663.
159. Kimura K, Gelmann EP. Propapoptotic effects of NF-kappaB in LNCaP prostate cancer cells lead to serine protease activation. *Cell Death Differ* 2002; **9:** 972-980.

160. Dudley E *et al.* NF-kappaB regulates Fas/APO-1/CD95- and TCR- mediated apoptosis of T lymphocytes. *Eur J Immunol* 1999; **29:** 878-886.
161. Karin M, Cao Y, Greten FR, Li ZW. NF-kappaB in cancer: from innocent bystander to major culprit. *Nat Rev Cancer* 2002; **2:** 301-310.
162. Kucharczak J, Simmons MJ, Fan Y, Gelinas C. To be, or not to be: NF-kappaB is the answer--role of Rel/NF-kappaB in the regulation of apoptosis. *Oncogene* 2003; **22:** 8961-8982.
163. Micheau O, Tschopp J. Induction of TNF receptor I-mediated apoptosis via two sequential signaling complexes. *Cell* 2003; **114:** 181-190.
164. Grumont RJ, Rourke IJ, Gerondakis S. Rel-dependent induction of A1 transcription is required to protect B cells from antigen receptor ligation-induced apoptosis. *Genes Dev* 1999; **13:** 400-411.
165. Viatour P *et al.* NF- kappa B2/p100 induces Bcl-2 expression. *Leukemia* 2003; **17:** 1349-1356.
166. Notarbartolo M *et al.* Expression of the IAPs in multidrug resistant tumor cells. *Oncol Rep* 2004; **11:** 133-136.
167. Wang Q, Wang X, Evers BM. Induction of cIAP-2 in human colon cancer cells through PKC delta/NF-kappa B. *J Biol Chem* 2003; **278:** 51091-51099.
168. Harwood FG *et al.* Regulation of FasL by NF-kappaB and AP-1 in Fas-dependent thymineless death of human colon carcinoma cells. *J Biol Chem* 2000; **275:** 10023-10029.
169. Lin A, Karin M. NF-kappaB in cancer: a marked target. *Semin Cancer Biol* 2003; **13:** 107-114.
170. Cao Y, Karin M. NF-kappaB in mammary gland development and breast cancer. *J Mammary Gland Biol Neoplasia* 2003; **8:** 215-223.
171. Eid MA, Lewis RW, Abdel-Mageed AB, Kumar MV. Reduced response of prostate cancer cells to TRAIL is modulated by NFkappaB-mediated inhibition of caspases and Bid activation. *Int J Oncol* 2002; **21:** 111-117.
172. Meli M *et al.* NF-kappaB inhibition restores sensitivity to Fas-mediated apoptosis in lymphoma cell lines. *Ann N Y Acad Sci* 2003; **1010:** 232-236.
173. Semenza GL. Involvement of hypoxia-inducible factor 1 in human cancer. *Intern Med* 2002; **41:** 79-83.
174. Shibaji T *et al.* Prognostic significance of HIF-1 alpha overexpression in human pancreatic cancer. *Anticancer Res* 2003; **23:** 4721-4727.
175. Akakura N *et al.* Constitutive expression of hypoxia-inducible factor-1alpha renders pancreatic cancer cells resistant to apoptosis induced by hypoxia and nutrient deprivation. *Cancer Res* 2001; **61:** 6548-6554.
176. Bardos JI, Ashcroft M. Hypoxia-inducible factor-1 and oncogenic signalling. *Bioessays* 2004; **26:** 262-269.
177. Weinmann M, Belka C, Plasswilm L. Tumour hypoxia: impact on biology, prognosis and treatment of solid malignant tumours. *Onkologie* 2004; **27:** 83-90.
178. Piret JP, Mottet D, Raes M, Michiels C. Is HIF-1alpha a pro- or an anti-apoptotic protein? *Biochem Pharmacol* 2002; **64:** 889-892.
179. Weinmann M *et al.* Molecular ordering of hypoxia-induced apoptosis: critical involvement of the mitochondrial death pathway in a FADD/caspase-8 independent manner. *Oncogene* 2004; **23:** 3757-3769.
180. Piret JP *et al.* Hypoxia and CoCl2 protect HepG2 cells against serum deprivation- and t-BHP-induced apoptosis: a possible anti-apoptotic role for HIF-1. *Exp Cell Res* 2004; **295:** 340-349.

181. Dong Z, Wang JZ, Yu F, Venkatachalam MA. Apoptosis-resistance of hypoxic cells: multiple factors involved and a role for IAP-2. *Am J Pathol* 2003; **163:** 663-671.
182. Dong Z, Wang J. Hypoxia selection of death-resistant cells. A role for Bcl-X(L). *J Biol Chem* 2004; **279:** 9215-9221.
183. Yokoi K, Fidler IJ. Hypoxia increases resistance of human pancreatic cancer cells to apoptosis induced by gemcitabine. *Clin Cancer Res* 2004; **10:** 2299-2306.
184. Yang L *et al.* Tumor-specific gene expression using the survivin promoter is further increased by hypoxia. *Gene Ther* 2004; **11:** 1215-1223.
185. Sowter HM *et al.* HIF-1-dependent regulation of hypoxic induction of the cell death factors BNIP3 and NIX in human tumors. *Cancer Res* 2001; **61:** 6669-6673.
186. Okami J, Simeone DM, Logsdon CD. Silencing of the hypoxia-inducible cell death protein BNIP3 in pancreatic cancer. *Cancer Res* 2004; **64:** 5338-5346.
187. An WG *et al.* Stabilization of wild-type p53 by hypoxia-inducible factor 1alpha. *Nature* 1998; **392:** 405-408.
188. Mitsiades N, Poulaki V, Mitsiades CS, Anderson KC. Induction of tumour cell apoptosis by matrix metalloproteinase inhibitors: new tricks from a (not so) old drug. *Expert Opin Investig Drugs* 2001; **10:** 1075-1084.
189. Sethi T *et al.* Extracellular matrix proteins protect small cell lung cancer cells against apoptosis: a mechanism for small cell lung cancer growth and drug resistance in vivo. *Nat Med* 1999; **5:** 662-668.
190. Valentijn AJ, Zouq N, Gilmore AP. Anoikis. *Biochem Soc Trans* 2004; **32:** 421-425.
191. Hojilla CV, Mohammed FF, Khokha R. Matrix metalloproteinases and their tissue inhibitors direct cell fate during cancer development. *Br J Cancer* 2003; **89:** 1817-1821.
192. Duffy MJ *et al.* Metalloproteinases: role in breast carcinogenesis, invasion and metastasis. *Breast Cancer Res* 2000; **2:** 252-257.
193. Strand S *et al.* Cleavage of CD95 by matrix metalloproteinase-7 induces apoptosis resistance in tumour cells. *Oncogene* 2004; **23:** 3732-3736.
194. Mitsiades N *et al.* Matrix metalloproteinase-7-mediated cleavage of Fas ligand protects tumor cells from chemotherapeutic drug cytotoxicity. *Cancer Res* 2001; **61:** 577-581.
195. Vargo-Gogola T, Fingleton B, Crawford HC, Matrisian LM. Matrilysin (matrix metalloproteinase-7) selects for apoptosis-resistant mammary cells in vivo. *Cancer Res* 2002; **62:** 5559-5563.
196. Aoudjit F, Vuori K. Integrin signaling inhibits paclitaxel-induced apoptosis in breast cancer cells. *Oncogene* 2001; **20:** 4995-5004.
197. Jan Y *et al.* A mitochondrial protein, Bit1, mediates apoptosis regulated by integrins and Groucho/TLE corepressors. *Cell* 2004; **116:** 751-762.
198. Ambar BB *et al.* Treatment of experimental glioma by administration of adenoviral vectors expressing Fas ligand. *Hum Gene Ther* 1999; **10:** 1641-1648.
199. Komata T *et al.* Caspase-8 gene therapy using the human telomerase reverse transcriptase promoter for malignant glioma cells. *Hum Gene Ther* 2002; **13:** 1015-1025.
200. Shinoura N *et al.* Transduction of Apaf-1 or caspase-9 induces apoptosis in A-172 cells that are resistant to p53-mediated apoptosis. *Biochem Biophys Res Commun* 2000; **272:** 667-673.
201. Shinoura N *et al.* Adenovirus-mediated transfer of caspase-3 with Fas ligand induces drastic apoptosis in U-373MG glioma cells. *Exp Cell Res* 2000; **256:** 423-433.
202. Li X *et al.* Adenovirus-mediated Bax overexpression for the induction of therapeutic apoptosis in prostate cancer. *Cancer Res* 2001; **61:** 186-191.
203. Pataer A *et al.* Adenovirus-mediated Bak gene transfer induces apoptosis in mesothelioma cell lines. *J Thorac Cardiovasc Surg* 2001; **121:** 61-67.

204. Park WS *et al.* Somatic mutations in the death domain of the Fas (Apo-1/CD95) gene in gastric cancer. *J Pathol* 2001; **193:** 162-168.
205. Schimmer AD *et al.* Small-molecule antagonists of apoptosis suppressor XIAP exhibit broad antitumor activity. *Cancer Cell* 2004; **5:** 25-35.
206. Grossman D, Kim PJ, Schechner JS, Altieri DC. Inhibition of melanoma tumor growth in vivo by survivin targeting. *Proc Natl Acad Sci U S A* 2001; **98:** 635-640.
207. Bodis S *et al.* Extensive apoptosis in ductal carcinoma in situ of the breast. *Cancer* 1996; **77:** 1831-1835.
208. Pitot HC *et al.* Quantitation of multistage carcinogenesis in rat liver. *Toxicol Pathol* 1996; **24:** 119-128.

# Chapter 2

# CALCIUM SIGNALING AND APOPTOSIS RESISTANCE OF CANCER CELLS

Natalia Prevarskaya[1], Roman Skryma[1] and Yaroslav Shuba[2]
[1]*Laboratoire de Physiologie Cellulaire, INSERM EPI-9938, USTL, Bat. SN3, 59655 Villeneuve d'Ascq Cedex, FRANCE;* [2]*Bogomoletz Institute of Physiology, NASU, Bogomoletz Str., 4, 01024 Kiev-24, UKRAINE*

Abstract: $Ca^{2+}$ is a universal messenger regulating not only specific cellular functions during a cell's lifetime, but also the progression through the life cycle per se, which is concluded by the execution of an orderly self-elimination process upon completion of its physiological mission. The ability of the cell to commit self-destruction in a controlled fashion through a series of precisely regulated, biochemically and morphologically defined steps, called apoptosis, is an integral part of normal tissue homeostasis. When this ability is impaired, aberrant cells might ultimately dominate the tissue, thereby transforming it into a cancer. $Ca^{2+}$ signaling is crucial for apoptosis. Therefore, understanding how $Ca^{2+}$ signals govern apoptotic events, what molecular determinants are involved and how they change during acquisition of apoptotic resistance, provides a good opportunity for controlling tumor growth. In the present chapter, we by concentrating mainly on the example of prostate cancer examine the major molecular determinants of $Ca^{2+}$ homeostasis in cancer cells and establish their role in the transformation to apoptotic resistant cell phenotypes, which are typical of advanced androgen-independent prostate cancer. We show that the hallmark of such transformation is the loss of apoptotic pathway associated with endoplasmic reticulum $Ca^{2+}$ stores depletion.

Key words: cancer; prostate cancer; apoptosis; calcium signaling; androgen-independency; endoplasmic reticulum; store-operated channels; Bcl-2



*M. Sluyser (ed.), Application of Apoptosis to Cancer Treatment,* 29-49.

# 1. INTRODUCTION

The balance between the rate of cells proliferation and apoptosis is crucial for normal tissue homeostasis. Enhanced proliferation together with an impaired ability to die is a prime reason for abnormal tissue growth, which can eventually turn into uncontrolled expansion and invasion, characteristic of cancer. Although molecular machineries of proliferation and apoptosis are different, with the first relying on cyclin-dependent protein kinases (CDKs) – regulators of cell division cycle [1] – and the second mainly on caspases – cysteine proteases executing cell death program [2] – they both require intense $Ca^{2+}$ signaling, and $Ca^{2+}$ homeostasis provides the critical environment for their operation [3].

Despite considerable progress in cancer screening, prevention and treatment achieved over the recent decades, with ten million new cancer patients diagnosed worldwide each year [4] cancer remains one of the major health problems for humanity. There are many types of cancers, each with its own genetic, environmental, metabolic and biochemical reasons and specific genesis. However, it is becoming more and more evident that misregulation of apoptosis may be a common feature that contributes to the development of all cancer types. As apoptosis is a $Ca^{2+}$-dependent process [5], it seems that cancer cells acquire the ability to disobey calcium signals to die. Being unable to cover all types of cancer cells, in the present chapter we will examine how and why this happens in the example of prostate cancer. Prostate cancer (PCa) is one of the most commonly diagnosed malignancies and after lung cancer is the second cause of cancer-related death in men. The distinguishing feature of PCa among other malignancies, is that it progresses through two clearly separated stages – the early one, which depends on androgens for growth and survival, and the late, androgen-independent one (e.g., [6]). The dependence of early PCa on androgens makes androgen ablation therapy effective at this time in causing tumors to regress, largely by triggering the programmed cell death pathway [7] – a property which is completely lost in the late androgen-independent stage of PCa, for which there is currently no successful therapy. Needless to say, understanding what drives the progression to androgen independence is imperative for successful PCa treatment. Transition to androgen-independence is characterized by the appearance of new cell phenotypes, for which apoptosis inhibition rather than enhanced proliferation [8, 9] becomes the primary distinctive feature. In this chapter, we describe the major molecular and structural determinants of $Ca^{2+}$ homeostasis in PCa epithelial cells, establish their role in apoptosis-related $Ca^{2+}$ signaling and track down how they evolve during transformation to apoptotic resistant cell phenotypes typical of advanced androgen-independent PCa. We also demonstrate some common features of

$Ca^{2+}$ signaling for the apoptosis of other types of cancer cells. Although the role of $Ca^{2+}$ signaling at molecular level in PCa cells apoptosis is only beginning to be explored, there is no doubt that our understanding of fundamental processes will eventually make its way to practical implications.

## 2. CALCIUM AND APOPTOSIS: HOW DO THEY RELATE TO EACH OTHER?

Apoptosis is a precisely regulated form of physiological cell death destined for harmless elimination of the cells that have completed, or for any reason are incapable of performing their physiological function and are therefore no longer needed. It is an integral part of normal tissue homeostasis and abnormalities in apoptosis are believed to be involved in the pathogenesis of many diseases. The molecular machinery of apoptosis from its initiation point till the final phagocytosis of the cellular remnants is complex, involving many molecular players and signaling pathways. For an in-depth analysis of its different aspects, we redirect the reader to a number of excellent specialized recent reviews (e.g., [5, 10-16]), whereas here we only briefly outline the crucial points for $Ca^{2+}$ involvement in apoptotic events.

It has long been established that the initial stimulus for apoptosis is the lethal influx of $Ca^{2+}$ into the cells. However, how this influx consequently translates into an orderly sequence of events ultimately leading to cell demise, has only recently become evident. It is clear now that cellular $Ca^{2+}$ overload may recruit three major apoptotic mechanisms. Although interrelated and interdependent, these mechanisms can be tentatively subdivided into mitochondrial, cytoplasmic and endoplasmic reticulum(ER)-mediated. As organelles possessing the necessary apparatus for initiating suicidal process, mitochondria represent a central integration point for the signals regulating a cell's destiny. $Ca^{2+}$ overload, which may be triggered by various initial stimuli, promotes mitochondrial $Ca^{2+}$ uptake. Excessive $Ca^{2+}$ accumulation within the mitochondrial matrix, is one of the primary reasons for mitochondrial permeability transition, which is at least partly mediated by the opening of a permeability transition pore (PTP) – a multi-protein complex located at the contact sites between the inner and the outer mitochondrial membranes. PTP opening permits the release of mitochondrial apoptogenic factors, such as cytochrome c (Cyt-c) and an apoptosis-inducing factor (AIF) into the cytoplasm, where they in turn activate a death-executing caspase cascade. Mitochondrial permeability in general and the PTP complex in particular, are regulated by the members of Bcl-2 family of proteins, of which those preventing apoptogenic factors release, Bcl-2 per se, Bcl-$x_L$, and Mcl-1, protect against apoptosis, whereas those promoting it, Bax and Bak, act as apoptosis enhancers (e.g., [17]).

Initial cytosolic $Ca^{2+}$ overload may also take a different cell death row, without directly involving $Ca^{2+}$-mediated mitochondrial permeability transition. It mainly relies on the activation of calcium/calmodulin(CaM)-dependent phosphatase, calcineurin. Calcineurin-catalyzed dephosphorylation promotes apoptosis by regulating the activity of a number of downstream targets, including the pro-apoptotic Bcl-2 family member, Bad [18], and transcription factors from the NFAT (nuclear factor of activated T cells) family [19]. There are also some other $Ca^{2+}$-dependent enzymes taking part in the apoptotic events, among which, of special note, are several DNA-degrading endonucleases [20] and $Ca^{2+}$-activated cystein proteases of the calpain family essential for the enzymatic activation of the crucial pro-apoptotic effectors (e.g., [21]).

ER is an important, dynamic organelle providing a storage place for the majority of intracellular $Ca^{2+}$ and the machinery for the maturation of the nascent proteins. So-called ER stress, which may result from any disturbance in its $Ca^{2+}$ homeostasis and/or newly synthesized protein processing, is also a critical determinant in apoptosis (e.g., [13, 16]). The role of the calcium component in the ER stress-mediated apoptosis became especially evident with the introduction into research practice of thapsigargin (TG) – a plant sesquiterpene lactone that irreversibly blocks the sarco(endo)plasmic reticulum $Ca^{2+}$-ATPase (SERCA) pump providing for ER $Ca^{2+}$ uptake. TG-induced SERCA blockade allows the ER $Ca^{2+}$ store to be depleted and cytosolic $Ca^{2+}$ concentration to be elevated due to $Ca^{2+}$ release from the store, as well as due to the initiation of store-dependent $Ca^{2+}$ influx. TG-induced apoptosis represents popular experimental model for the study of $Ca^{2+}$-dependent apoptosis mechanisms.

It has been shown that ER stress activates specific ER-localized caspase-12 [22], which in turn triggers a cascade of reactions leading to the activation of other caspases in a mitochondria- and Cyt-c-independent manner [23]. Recent evidence also suggests the crucial role in ER-mediated apoptosis of extramitochondrial Bcl-2 family members particularly localized in the ER membrane (reviewed in Schinzel et al., 2004, Annis et al., 2004, Distelhorst and Shore, 2004). ER-localized pro- and anti-apoptotic Bcl-2 members have been implicated in the regulation of the luminal $Ca^{2+}$ levels and the expression of ER resident proteins, thereby strongly affecting the apoptotic status of the cells [24-26].

# 3. CALCIUM IN PROSTATE CANCER CELLS APOPTOSIS

Understanding the mechanisms of PCa cells apoptosis, including their $Ca^{2+}$-dependency, is important for developing new approaches to PCa treatment. Prostate cancer is a very heterogeneous entity, generally composed of a mixture of cells expressing an androgen receptor (AR), which confers the property of androgen dependency, and androgen-independent cells. At the early stage most of the neoplastic mass is represented by androgen-dependent cells whose pro-apoptotic potential is regulated by AR. It is well established that this type of cells can be easily induced to undergo apoptotic cell death after androgen ablation [27, 28]. However, associated tumor enrichment with a propagating androgen-independent cell population, as well as the emergence of the new cell phenotypes with enhanced resistance to apoptosis due to genetic transformations, will eventually lead to the complete loss of hormone-sensitivity and the curability characteristic of the advanced stage of cancer. Fortunately, androgen-independent PCa cells retain the appropriate molecular machinery to undergo apoptosis. Thus, the problem is to explore all its specific features and then find the means for its activation.

It has long been considered that the primary cause of apoptosis induction in response to androgen ablation is the sustained elevation of intracellular $Ca^{2+}$ ($[Ca^{2+}]_{in}$) [29], with a consequent activation of $Ca^{2+}$-$Mg^{2+}$-dependent endonuclease involved in genomic DNA fragmentation [30]. The molecular mechanisms underlying $[Ca^{2+}]_{in}$ elevation in response to androgens withdrawal is not very clear, but the reduction in the percentage of epithelial cells from rat ventral prostate undergoing apoptosis in response to the administration of voltage-gated $Ca^{2+}$ channels (VGCCs) blockers, such as nifedipine and verapamil, suggested $Ca^{2+}$ entry from extracellular space [29, 31]. However, in view of the fact that VGCCs are not the major $Ca^{2+}$ entry pathway in most types of PCa epithelial cells, as is evident from direct membrane currents measurements, the effectiveness of these agents in counteracting pro-apoptotic effects of androgen ablation is unlikely to be directly related to $Ca^{2+}$ entry inhibition, prompting its further exploration from other perspectives.

Interestingly, androgen ablation not only promotes androgen-dependent PCa cells apoptosis via an elevation of cytosolic $Ca^{2+}$, but increased $[Ca^{2+}]_{in}$ per se leads to the down-regulation of AR expression, as demonstrated in the androgen-dependent LNCaP cell line subjected to the treatments with $Ca^{2+}$ ionophores or the SERCA pump inhibitor TG [32]. This suggests that androgen ablation therapy, although effective in providing temporary apoptosis enhancement, may simultaneously serve as a facilitating factor in cells transformation into androgen-independent phenotypes.

The strong androgen-dependence of prostate cells apoptosis necessarily involves up- or down-regulation of androgen-response genes. One of the androgen-inducible genes identified in the prostate encodes calreticulin [33, 34] – a high capacity ER luminal $Ca^{2+}$ binding protein. This identification provided the first strong indication of the potential involvement of not only cytosolic $Ca^{2+}$, but also of ER-dependent mechanisms in PCa cells apoptosis. The importance of ER mechanisms was also evident from studies in other cell models (e.g., [35]). Nevertheless, the type, extent and manner of ER involvement in PCa cells apoptosis still remains a subject of ongoing research.

The first studies of TG-induced apoptosis in an androgen-independent rat, AT-3, and human, TSU-Pr1, DU-145 and PC-3, prostatic cell lines, while confirming the importance for sustained $[Ca^{2+}]_{in}$ rise in apoptosis induction, ruled out any substantial role for ER store depletion [36]. The concept of a primary role of cytosolic $Ca^{2+}$ received further development in the experiments on prolonged $[Ca^{2+}]_{in}$ monitoring during the whole apoptosis cycle. The long-term $[Ca^{2+}]_{in}$ changes induced in prostatic AT-3 and TSU cell lines by chronic TG exposure were dissected into an early (1-12 h) phase of moderate increase (<0.5 μM) and a delayed phase (>12 h) of strong increase (~10 μM) [37]. The early phase was manifested by the induction of a GADD153 transcription factor, whereas the delayed phase was characterized by the execution of calcineurin-mediated apoptotic pathway, resulting in Bad translocation to mitochondria, Cyt-c release, activation of the initiator caspase-9, and DNA fragmentation [37].

In contrast, characterization of TG-induced apoptosis in androgen-dependent LNCaP cells, with the use of experimental conditions that permitted ER store release without elevating $[Ca^{2+}]_{in}$, showed that store depletion is by itself sufficient to induce LNCaP cells apoptosis [38]. Furthermore, ER store depletion alone resulted in even earlier caspase-9 activation than the dual insult of store depletion and $[Ca^{2+}]_{in}$ elevation, although further downstream apoptotic pathways appeared to be the same, irrespective of the initiating insult [38]. In our own study, conducted on LNCaP cells, we for the first time directly measured the store-operated $Ca^{2+}$ current ($I_{SOC}$) underlying TG-induced $Ca^{2+}$ entry in the cells of prostatic origin and showed that in order for TG to trigger apoptosis, activation of this current and associated sustained $Ca^{2+}$ entry is not required [39]. Moreover, TG-induced apoptosis of LNCaP cells was even enhanced under inhibited SOCE, especially during the initial 24 h [39], suggesting that ER-stress-initiated and $[Ca^{2+}]_{in}$-initiated pathways may compete for some common molecular determinants, which agrees well with the earlier caspase-9 activation in response to ER store depletion, as observed by Wertz and Dixit [38]. Interestingly, LNCaP cells apoptosis could be slightly stimulated by caffeine, whose major action is to release ER store via ryanodine receptors (RyRs) [40]. However, caffeine action in LNCaP cells was also accompanied by $Ca^{2+}$ entry of as yet a not very clear nature, not permitting clear dissection of the ER-dependent component in its pro-apoptotic effects [40].

The ER $Ca^{2+}$ content of androgen-dependent LNCaP cells has been proven to be important in determining the mode of action of such growth regulators as the insulin growth factor (IGF) and the tumor necrosis factor-alpha (TNF-α). The action of the growth-promoting IGF correlated with high ER $Ca^{2+}$ content, whereas the action of the pro-apoptotic agent TNF-α was associated with reduced ER $Ca^{2+}$ content [41].

Interesting interplay between ER $Ca^{2+}$ mobilization and mitochondrial uptake has recently been demonstrated for the staurosporine-induced apoptosis of androgen-independent PC-3 cells [42]. In this cell model, the action of some, but not all pro-apototic stimuli involved the reduction of ER $Ca^{2+}$ content, which preceded caspase activation [42]. In particular, the staurosporine-induced apoptosis was associated with initial ER $Ca^{2+}$ pool emptying followed by an increase in mitochondrial uptake and Cyt-c release – the process, which seemed to require Bax as signal mediator between ER and mitochondria [42].

Thus, it seems that both ER-stress-mediated and $[Ca^{2+}]_{in}$-mediated apoptotic pathways are operational in PCa epithelial cells. However, more studies are needed for the elucidation and dissection of underlying molecular mechanisms. In particular, the role of ER-resident caspase-12 still remains undetermined. There is also a distinct possibility that the primary $Ca^{2+}$-dependent apoptotic mechanisms may differ in androgen-dependent and androgen-independent cell phenotypes.

TG, $Ca^{2+}$ ionophores and agonists acting via receptors coupled to phospholipase C (PLC)-catalyzed inositol phospholipids breakdown signaling pathway can mobilize intracellular $Ca^{2+}$ and induce sustained $Ca^{2+}$ entry. Yet, there are number of pro-apoptotic agents and insults that do not produce an obvious immediate $[Ca^{2+}]_{in}$ rise. In a recent study by Tombal et al. [43] it was shown that their representatives, such as the transforming growth factor-β (TGF-β), chemotherapeutic anticancer agents, 5-fluorouridine (5-FdUR) and doxorubicin, as well as radiation still induced a delayed $[Ca^{2+}]_{in}$ rise in prostatic AT-3, TSU, and PC-3 cell lines, which accompanied morphological changes characteristic of apoptosis progression. In response to any insult, $[Ca^{2+}]_{in}$ started to increase after an initial lag of about 28-48 h and progressed in two consecutive steps of 0.5-2 h and 4-8 h in duration bringing total $[Ca^{2+}]_{in}$ to supramicromolar levels [43]. This

$[Ca^{2+}]_{in}$ elevation occurred due to $Ca^{2+}$ influx, as was evidenced by the delay of apoptotic events in a low $Ca^{2+}$ medium, and was required for Cyt-c release, caspase-3 activation and DNA fragmentation [43]. However, the exact PM $Ca^{2+}$ entry pathways involved in these actions remain unclear.

TGF-β was also reported to induce a transient (within minutes) submicromolar $[Ca^{2+}]_{in}$ increase in prostatic PC-3U cells due to TGF-β-mediated mitochondria inhibition and an acute release of $Ca^{2+}$ from the mitochondria as a result of mitochondrial uncoupling [44]. It was suggested that these effects underlie TGF-β pro-apoptotic actions [44].

Cardiac glycosides are mostly known for their positive inotropic effects on heart function. However, they have also been found to exert a potent pro-apoptotic action on PCa cells, involving a perturbation of $Ca^{2+}$ homeostasis [45, 46]. Using the PC-3 cell line as a model system, it was demonstrated that cardiac glycosides induce $Ca^{2+}$ influx [45, 46], which in analogy to cardiac myocytes was attributed to reversed $Na^{+}$-$Ca^{2+}$ exchange activity due to $Na^{+}$ overload resulting from cardiac glycoside-induced $Na^{+}$-$K^{+}$ ATPase blockade [46]. Following the $Ca^{2+}$ influx, the cell death row took a clear mithochondrial direction, manifested by the loss of the mithochondrial membrane potential [46] and by Cyt-c release followed by caspase-3 and -8 processing [45]. These results are very intriguing, although the $Ca^{2+}$ entry pathway involved needs further elucidation, as functional $Na^{+}$-$Ca^{2+}$ exchange mechanism so far has been not directly demonstrated in PCa epithelial cells.

Taken together, the results obtained so far suggest that $Ca^{2+}$ signaling involving all three: cytoplasmic, mitochondrial, and ER-dependent mechanisms play a role in PCa epithelial cells apoptosis. However, the relative contribution of each mechanism depends on the nature of the pro-apoptotic stimulus and initial apoptotic status of the cell.

## 4. MOLECULAR PLAYERS INVOLVED

Prostate cancer epithelial cells are representatives of non-excitable secretory-type cells and in that respect, they are not unique in terms of what sub-cellular and molecular determinants participate in the regulation of their $Ca^{2+}$ homeostasis. In general, there are three major $Ca^{2+}$ pools that living cells can utilize for producing various types of cytosolic $Ca^{2+}$ signals: extracellular space, ER and mitochondria. Each of these pools is separated by its own membrane incorporating diverse $Ca^{2+}$ handling proteins. The specific representation of these proteins largely determines the relative contribution of each pool in the creation of a $Ca^{2+}$ signal as well as its spatial-temporal characteristics.

In PCa epithelial cells, as in other non-excitable cell types, $Ca^{2+}$ entry from extracellular space is mainly supported by the mechanism called "capacitative calcium entry" (CCE) or "store-operated calcium entry" (SOCE) [47]. This mechanism is capable of monitoring ER $Ca^{2+}$ filling permitting the influx only when ER content essentially decreases. It is mediated via specialized plasma membrane (PM) store-operated $Ca^{2+}$-permeable channels (SOC). The common physiological trigger for the activation of these channels is $IP_3$-induced $Ca^{2+}$ release from the ER in response to stimulation of surface receptors coupled to PLC-catalyzed inositol phospholipids breakdown signaling pathway. This is why upon discovery, they were termed "$Ca^{2+}$ release-activated channels" (CRAC) [48].

Although the molecular nature of SOC and CRAC, as well as the mechanisms linking them to the ER $Ca^{2+}$ store depletion, still remain a mystery, the widely investigated family of mammalian homologues of the *Drosophila* TRP (Transient Receptor Potential) channels is currently viewed as the most likely source of candidate proteins for the role of SOCs (for recent reviews see [47, 49-51]). Our own studies conducted on the androgen-dependent LNCaP human prostate cancer epithelial cell line have implicated the members of the "canonical" TRP subfamily, TRPC1 [52, 53], TRPC4 [53], and "vanilloid" TRP subfamily, TRPV6 [53, 54], in prostate-specific endogenous SOCs. In this respect, TRPV6, also known as $Ca^{2+}$ transporter type 1 (CaT1) or epithelial calcium channel 2 (ECaC2) [55], attracts special attention, as its expression was shown to correlate with PCa grade [56-58]. Moreover, TRPV6 expression seems to be under negative control from AR, as pharmacological AR suppression by anti-androgens not only up-regulated TRPV6 mRNA levels in LNCaP cells [54, 56], but also augmented store-operated membrane current ($I_{SOC}$) [54]. A recent study conducted on tissue samples from 140 patients with PCa, further demonstrated the association of TRPV6 with PCa progression and suggested it as a prognostic molecular marker in cancer classification [58]. It seems, however, that the functional role of endogenous TRPV6 in prostatic $I_{SOC}$ is tightly linked to other potential SOC constituents and/or regulators, as artificial TRPV6 overexpression in LNCaP cells resulted in the appearance of additional membrane current with distinct properties to endogenous $I_{SOC}$ [59]. At any rate, molecular bases for SOCE in PCa epithelial cells is still far from being clear, and the fact that aside from TRPV6 it may also include TRPC1 and TRPC4, suggests its complexity and probable multi-channel nature. Hopefully, future studies will bring more clarity to this issue.

ER is an important, dynamic organelle responsible for the storage of the majority of intracellular $Ca^{2+}$. The primary molecular determinants involved in $Ca^{2+}$ handling by the ER include: $IP_3$ ($IP_3R$) and ryanodine (RyR) receptors permitting active $Ca^{2+}$ release, the ER leak channels underlying passive loss of $Ca^{2+}$, the SERCA $Ca^{2+}$ pump providing for $Ca^{2+}$ reuptake back to the ER, and $Ca^{2+}$-binding proteins (chaperones) participating in the intraluminal $Ca^{2+}$ storage. By functioning in concert, ER leak channels, SERCA pump and intraluminal $Ca^{2+}$-binding chaperones maintain optimal basal $Ca^{2+}$ concentration within the ER required for protein synthesis and processing.

The data on representation of specific isoforms of the key ER $Ca^{2+}$ handling proteins in the PCa epithelial cells are extremely limited, and again are mostly restricted to various prostatic cell lines. Out of three known $IP_3R$ isoforms ($IP_3R1$, $IP_3R2$, $IP_3R3$, e.g., [60]) LNCaP cells seem to preferentially express $IP_3R1$ with almost 3-fold lower $IP_3R3$ expression and none of $IP_3R2$ [53]. Nearly the same $IP_3Rs$ ratio was found for the androgen-independent DU-145 prostate cancer cell line [61]. Considering that, on the one hand, direct $IP_3R$-SOC conformational coupling is implicated in SOCE activation (e.g., [53, 62]), and, on the other hand, $IP_3R$ subtypes represent an important determinant in apoptosis [63-65], this information may be useful both for establishing precise interactions governing SOCE in PCa epithelial cells and for better understanding ER-dependence of PCa cells apoptosis. LNCaP cells have also been shown to express functional RyRs with dominant representation of RyR1 and RyR2 isoforms capable of mobilizing $Ca^{2+}$ from intracellular stores [40], although the physiological stimulus leading to RyRs activation as yet remains unclear. Unfortunately, the molecular nature of passive $Ca^{2+}$ exit pathway from the ER, namely the ER leak channels is not known. It has been suggested that $Ca^{2+}$ leak may occur via the translocon [66] – an aqueous pore that regulates the movement of polypeptides through the ER membrane. Our most recent data are consistent with potential role of the translocon as an ER leak channel in LNCaP cells [67].

$Ca^{2+}$ uptake back into ER is controlled by the family of SERCA pumps [68] located in the ER membrane, whereas the luminal environment of the ER consists of $Ca^{2+}$-binding chaperones, which are involved in protein folding, post-translational modification and $Ca^{2+}$ storage. Out of several tissue-specific SERCA-pump isoforms, only the widely distributed phospholamban-dependent SERCA2b one was detected in LNCaP cells [69, 70]. Prostate cells, including LNCaP, also express high levels of calreticulin – one of the most ubiquitous ER $Ca^{2+}$-binding (storage) chaperones [71]. The significance of calreticulin in $Ca^{2+}$ homeostasis is heightened by the fact that it represents one of androgen-response genes in the prostate [33, 34], thus making it particularly important in cancer-related growth and death regulation.

Mitochondria are endowed with multiple $Ca^{2+}$ transport mechanisms by which they take up and release $Ca^{2+}$ across their inner membrane [72]. These transport processes function to control local and global cytosolic $Ca^{2+}$ concentration, thereby regulating a number of $Ca^{2+}$-sensitive cellular mechanisms. The PTP together with $H^+$-$Ca^{2+}$ and $Na^+$-$Ca^{2+}$ exchange mechanisms form $Ca^{2+}$ efflux pathways from mitochondria, whereas mitochondrial $Ca^{2+}$ uptake mainly occurs via a channel-like uniporter [73].

Thus, although the major determinants of $Ca^{2+}$ homeostasis in PCa epithelial cells have been outlined, a lot of work still has to be done to identify their specific molecular nature, which is imperative for the practical perspectives of a targeted influence on their function in PCa. Nevertheless, the data already available provide sufficient baseline knowledge for further study into how major determinants of $Ca^{2+}$ homeostasis may evolve during acquisition of apoptosis resistance in androgen-independent PCa.

## 5. HOW IT ALL CHANGES IN ANDROGEN-INDEPENDENT PROSTATE CANCER

Progression of prostate cancer to the stage of androgen-independence is accompanied by the appearance of new apoptotic-resistance cell phenotypes. One of them is closely associated with overexpression of the common anti-apoptotic oncoprotein, Bcl-2 ([8], reviewed in [74]). Non-pathologic prostate epithelial cells normally do not express Bcl-2, but its expression has been found to increase in hormone-untreated prostate adenocarcinoma and even more so in hormone-refractory adenocarcinoma following a hormone ablation regimen [75-77]. Aside from Bcl-2, the development of hormone-insensitive prostate cancer has also been shown to correlate with the increased levels of other survival proteins from Bcl-2 family such as Bcl-$x_L$, and Mcl-1 [78].

Of special note is the enrichment of androgen-independent tumors with malignant neuroendocrine (NE) cells. Fully differentiated, non-proliferating, neuron-like NE cells are a normal component of the prostate epithelium, which by releasing a variety of neurosecretory products, regulate the development and secretory activity of the prostate in the endocrine/paracrine manner [79, 80]. However, expanding their population beyond normal proportions due to malignant transformation of epithelial/basal cells is a common characteristic of prostate cancer progression [80]. NE cells lack nuclear AR [81], thereby representing an androgen-insensitive cell phenotype in the prostate. They also exhibit a high apoptotic resistance [9], which according to the existing evidence is unrelated to Bcl-2 [82], but is rather conferred by new survival proteins, survivin [83] and clusterin [84].

Androgen-dependent prostatic LNCaP cells serve as a good experimental model for studying both the Bcl-2- and the NE differentiation-conferred phenotypes of androgen-independence and apoptotic resistance due to the creation of LNCaP clones with stable Bcl-2 overexpression [8] and possibility to induce NE differentiation of LNCaP cells by such simple experimental maneuvers as cAMP elevation or androgen deprivation (e.g., [85, 86]). Our own studies of the alterations in $Ca^{2+}$ homeostasis of Bcl-2-overexpressing (LNCaP/Bcl-2) and NE-differentiated (NE-LNCaP) LNCaP cells have established not only the independence of these alterations from specific anti-apoptotic mechanisms, but also the commonality of the molecular events that underlie them [70, 87]. The major features of $Ca^{2+}$ homeostasis in LNCaP/Bcl-2 and NE-LNCaP cells compared to the wild-type androgen-dependent LNCaP cells were: 1) reduced basal $Ca^{2+}$ filling of the ER pool, and 2) reduced store-operated $Ca^{2+}$ entry. These changes were accompanied by an increased resistance to TG- and TNF-α-induced apoptosis, with clear shift to a higher importance of $Ca^{2+}$ influx vs. ER store depletion in apoptosis induction [70, 87] compared to the wild-type androgen-dependent LNCaP cells [38, 39]. The observed shift in apoptosis induction from ER stress-dependence to $Ca^{2+}$ influx-dependence is consistent with the early data on androgen ablation-induced apoptosis [29, 31], as well as with the bulk of data obtained in androgen-independent prostatic cell lines (e.g., [36, 37]), suggesting that it may signify the distinguishing feature of androgen-independence in general, irrespective of what specific anti-apoptotic mechanisms it may involve.

Chronic underfilling of the ER $Ca^{2+}$ pool in LNCaP/Bcl-2 and NE-LNCaP cells was found to result from enhanced ER leak accompanied by the decreased expression of ER luminal $Ca^{2+}$ binding protein, calreticulin, and SRCA2b $Ca^{2+}$ pump isoform, whereas reduced SOCE most likely resulted from the diminished number of functional SOCs [70, 87]. So far, out of $Ca^{2+}$-handling proteins found to be affected, only calreticulin was demonstrated to represent an androgen-response gene in the prostate [33, 34]. Moreover, calreticulin seems to be important determinant in androgen-dependent control of PCa cells apoptosis [88]. Therefore, our view is that calreticulin underexpression and the related decrease in the ER $Ca^{2+}$ storage capacity are the primary reasons for the diminished role of the ER stress-mediated apoptosis pathway and consequent general enhancement of androgen-independent PCa cells capability to survive. Indeed, androgen ablation would allow only those cells to survive, for which low ER filling due to decreased storage capacity accompanied by adaptive enhancement of leak, decreased uptake (SERCA pump underexpression) and impaired refilling (PM SOCs underexpression), would represent a new level of equilibrium. Such cells would emerge as resistant to the ER stress-related apoptosis, which, thus, could be only triggered by cytosolic or mitochondrial $Ca^{2+}$-dependent mechanisms.

The molecular bases for the enhanced ER leak in androgen-independent PCa cells is not clear. It may be due to an enhanced activity or density of the putative leak channels, for which translocon is suggested [66, 67], or may be related to the channel-forming capability of ER-localized pro-apoptotic proteins per se, which at least in the case of Bcl-2 members is a distinct possibility (reviewed in [89]). It should be noted, however, that the role of Bcl-2 in controlling ER $Ca^{2+}$ filling is quite controversial, with a nearly equal number of reports showing that Bcl-2 decreases or increases luminal $Ca^{2+}$ (for critical review see [26]). In addition to directly affecting ER leak thereby contributing to elimination of the ER stress-related apoptotic pathway, Bcl-2 overexpression in androgen-independent PCa cells counteracts apoptosis via its usual mitochondrial site of action, which altogether produces strong resultant anti-apoptotic effect.

Although the studies in androgen-dependent LNCaP cells have already implicated three TRP members – TRPC1, TRPC4 and TRPV6 in PCa cell-specific endogenous SOCs [52-54], this data is still insufficient to judge on precise molecular organization of SOCs and even more so on its transformation during the transition to androgen-independence, which underlies reduced SOCE. The fact that TRPV6 expression is enhanced in LNCaP cells in response to pharmacological AR inhibition [54, 56], makes it potentially an important determinant in apoptosis, although such change seems to be in conflict with the observed reduced SOCE in androgen-independent LNCaP/Bcl-2 and NE-LNCaP cells [70, 87]. In this respect, it is also important to assess the role of $IP_3Rs$ in the downregulation of SOCE and $I_{SOC}$, as it may be related to the alterations in $IP_3R$-SOC coupling, which is suggested in SOC activation. Indeed, overexpression of anti-apoptotic Bcl-$x_L$ protein and Cyt-c release have been shown to directly affect $IP_3Rs$ [64, 65], which is consistent with a potential contribution of indirect mechanisms in $I_{SOC}$ reduction. At any rate, decreased SOCE and $I_{SOC}$, in androgen-independent PCa cells will also essentially impair both mitochondrial and cytosolic $Ca^{2+}$-dependent apoptotic pathways.

TRPV6 expression is also known to be strongly vitamin D dependent (e.g., [90]), suggesting that this TRP member may play an important role in the pro-apoptotic action of vitamin D receptor ligands on PCa cells [91]. By acting via a vitamin D receptor, these ligands were shown to suppress the expression of Bcl-2-related anti-apoptotic proteins and to stimulate Cyt-c release in LNCaP and ALVA-31 cell lines [91]. It is quite plausible that TRPV6 may be involved in providing the $Ca^{2+}$ entry required for the mitochondrial permeability transition leading to Cyt-c release.

Prostatic NE cells generally express PM ion channels that are typical for excitable cells [92]. In particular, one of characteristic features of NE differentiation of LNCaP cells is the strong overexpression of voltage-gated

T-type $Ca^{2+}$ channels represented by their α1H isoform [86]. By providing constant $Ca^{2+}$ influx around the resting potential these channels seem to be involved in the formation of neuronal-like morphological features (i.e., neurites outgrowth). However, whether or not these channels contribute to the enhanced anti-apoptotic potential of NE cells is not yet clear.

# 6. OTHER TYPES OF CANCER

Chronic exposure of MCF-7 and MDA MB 468 breast carcinoma cell lines to TG resulted in biphasic elevation of intracellular $Ca^{2+}$ [93], just as was reported for prostate cancer cells [37]. The initial 4-fold transient submicromolar rise lasting for several hours was followed by 12-36 hours delayed rise above 5 μM. Morphological changes and DNA fragmentation, typical of apoptotic cell death, were detected in MCF-7 cells only subsequent to the secondary $[Ca^{2+}]_{in}$ rise, after about 48 hours of TG exposure [93]. These observations suggested that the secondary rise in $[Ca^{2+}]_{in}$ to micromolar levels is not prostate-specific, but rather reflects a general feature of $Ca^{2+}$ signaling required for endonucleases activation during TG-induced cancer cells apoptosis. In contrast, an activation of $Ca^{2+}/Mg^{2+}$-dependent endonuclease was detected as an early event in MCF-7 cells apoptosis induced by a widely used anticancer drug, etoposide (VP-16) [94]. $[Ca^{2+}]_{in}$ elevation associated with vitamin D-induced apoptosis of MCF-7 cells was shown to result not from entry, but from $Ca^{2+}$ release from the ER [95]. Interestingly, increased cytosolic $Ca^{2+}$ in this case stimulated the caspase-independent apoptotic pathway, which relied solely on the activation of $Ca^{2+}$-dependent cysteine proteases of the calpain family [95].

Sustained increase in intracellular $Ca^{2+}$ most likely via the entry mechanism, which correlated with enhanced activities of caspase-3, -8 and -9, was detected in the LK-2 and LU65A human lung cancer cell lines subjected to hyperthermia-induced apoptosis [96]. The disruption of $Ca^{2+}$ homeostasis, mitochondrial swelling and redistribution of Bax to the mitochondrial membrane, characterized ciprofloxacin-induced apoptosis in HTB9 bladder cancer cells [97]. The pro-apoptotic efficacy of arsenic trioxide in various human malignant cell lines correlated with the ability of this agent to increase the intracellular $Ca^{2+}$ concentration, with MGC-803 gastric cancer cell line being the most susceptible to both apoptosis induction and $[Ca^{2+}]_{in}$ elevation [98].

Thus, the results available so far suggest that intracellular $Ca^{2+}$ elevation is a common feature of apoptosis in response to most pro-apoptotic stimuli in the majority of cancer cell types. However, the exact mechanism(s) of $Ca^{2+}$ elevation, as well as apoptotic pathway(s) it may recruit may be different, requiring individual elucidation in each particular case. There is evidence, however, that enhanced resistance to apoptosis of cancer cells with increased malignancy may be related to the profound deficiency of major $Ca^{2+}$-dependent apoptosis execution mechanisms. Indeed, using a panel of nine murine and human cancer cell lines with different metastatic potential it has been shown that higher metastatic potential is associated with diminished level of nuclear $Ca^{2+}$-dependent endonucleases, and significantly reduced activity of caspase-3 (CPP32/Yama) death protease [99].

## 7. CONCLUSIONS

Although much has been learned about the role of $Ca^{2+}$ in PCa cells apoptosis, a lot of questions still await their answer. In particular, it is still not clear how the specificity of the $Ca^{2+}$ signal is achieved. It is well established that various cellular $Ca^{2+}$-dependent processes rely on the specific spatial and temporal characteristics of $Ca^{2+}$ signaling. However, the type and manner of its organization during various stages of PCa cells apoptosis is not known. In this respect, it is important to place more emphasis on the study of primary PCa epithelial cells, which preserve their morphological polarization. These problems are equally relevant to other types of cancer.

## REFERENCES

1. Nigg, E.A., 1995, Cyclin-dependent protein kinases: key regulators of the eukaryotic cell cycle, *Bioessays* **17:**471-480.
2. Nicholson, D.W., 1999, Caspase structure, proteolytic substrates, and function during apoptotic cell death, *Cell Death Differ.* **6:**1028-1042.
3. Berridge, M.J., Bootman, M.D., and Lipp, P., 1998, Calcium - a life and death signal, *Nature* **395:**645-648.
4. Parkin, D.M., 2001, Global cancer statistics in the year 2000, *Lancet Oncol.* **2:**533-543.
5. Orrenius, S., Zhivotovsky, B., and Nicotera, P., 2003, Regulation of cell death: the calcium-apoptosis link, *Nat. Rev. Mol. Cell Biol.* **4:**552-565.
6. Feldman, B.J., and Feldman, D., 2001, The development of androgen-independent prostate cancer, *Nat. Rev. Cancer.* **1:**34-45.
7. Denmeade, S.R., Lin, X.S., and Isaacs, J.T., 1996, Role of programmed (apoptotic) cell death during the progression and therapy for prostate cancer, *Prostate* **28:**251-265.
8. Raffo, A.J., Perlman, H., Chen, M.W., Day, M.L., Streitman, J.S., and Buttyan, R., 1995, Overexpression of bcl-2 protects prostate cancer cells from apoptosis in vitro and confers resistance to androgen depletion in vivo, *Cancer Res.* **55:**4438-4445.
9. Fixemer, T., Remberger, K., and Bonkhoff, H., 2002, Apoptosis resistance of neuroendocrine phenotypes in prostatic adenocarcinoma, *Prostate* **53:**118-123.
10. Hengartner, M.O., 2000, The biochemistry of apoptosis, *Nature* **407:**770-776.

11. Kroemer, G., and Reed, J.C., 2000, Mitochondrial control of cell death, *Nat. Med.* **6:**513-519.
12. Zimmermann, K.C., Bonzon, C., and Green, D.R., 2001, The machinery of programmed cell death, *Pharmacol. Ther.* **92:**57-70.
13. Breckenridge, D.G., Germain, M., Mathai, J.P., Nguyen, M., and Shore, G.C., 2003, Regulation of apoptosis by endoplasmic reticulum pathways, *Oncogene* **22:**8608-8618.
14. Rizzuto, R., Pinton, P., Ferrari, D., Chami, M., Szabadkai, G., Magalhaes, P.J., Di Virgilio, F., and Pozzan, T., 2003, Calcium and apoptosis: facts and hypotheses, *Oncogene* **22:**8619-27.
15. Mayer, B., and Oberbauer, R., 2003, Mitochondrial regulation of apoptosis, *News Physiol. Sci.* **18:**89-94.
16. Rao, R.V., Ellerby, H.M., and Bredesen, D.E., 2004, Coupling endoplasmic reticulum to the cell death program, *Cell Death Differ.* **11:**372-380.
17. Antonsson, B., and Martinou, J-C., 2000, The Bcl-2 protein family, *Exp. Cell Res.* **256:**50-57.
18. Wang, H.G., Pathan, N., Ethell, I.M., Krajewski, S., Yamaguchi, Y., Shibasaki, F., McKeon, F., Bobo, T., Franke, T.F., and Reed, J.C., 1999, $Ca^{2+}$-induced apoptosis through calcineurin dephosphorylation of BAD, *Science* **284:**339-343.
19. Rao, A., Luo, C., and Hogan, P.G., 1997, Transcription factors of the NFAT family: regulation and function. *Annu. Rev. Immunol.* **15:**707-47.
20. Robertson, J.D., Orrenius, S., and Zhivotovsky, B., 2000, Review: nuclear events in apoptosis, *J. Struct. Biol.* **129:**346-358.
21. Altznauer, F., Conus, S., Cavalli, A., Folkers, G., and Simon, H.U., 2004, Calpain-1 regulates Bax and subsequent Smac-dependent caspase-3 activation in neutrophil apoptosis, *J. Biol. Chem.* **279:**5947-5957.
22. Nakagawa, T., Zhu, H., Morishima, N., Li, E., Xu, J., Yankner, B.A., and Yuan, J., 2000, Caspase-12 mediates endoplasmic-reticulum-specific apoptosis and cytotoxicity by amyloid-beta, *Nature* **403:**98-103.
23. Morishima, N., Nakanishi, K., Takenouchi, H., Shibata, T., and Yasuhiko, Y., 2002, An endoplasmic reticulum stress-specific caspase cascade in apoptosis. Cytochrome c-independent activation of caspase-9 by caspase-12, *J. Biol. Chem.* **277:**34287-34294.
24. Schinzel, A., Kaufmann, T., and Borner, C., 2004, Bcl-2 family members: intracellular targeting, membrane-insertion, and changes in subcellular localization, *Biochim. Biophys. Acta* **1644:**95-105.
25. Annis, M.G., Yethon, J.A., Leber, B., and Andrews, D.W., 2004, There is more life and death than mitochondria: Bcl-2 proteins at the endoplasmic reticulum, *Biochim. Biophys. Acta* **1644:**115-123.
26. Distelhorst, C.W., and Shore, G.C., 2004, Bcl-2 and calcium: controversy beneath the surface, *Oncogene* **23:**2875-2880.
27. Kyprianou, N., English, H.F., and Isaacs, J.T., 1990, Programmed cell death during regression of PC-82 human prostate cancer following androgen ablation, *Cancer Res.* **50:**3748-3753.
28. Westin, P., and Bergh, A., 1998, Apoptosis and other mechanisms in androgen ablation treatment and androgen independent progression of prostate cancer: a review, *Cancer Detect. Prev.* **22:**476-844.
29. Martikainen, P., and Isaacs, J., 1990, Role of calcium in the programmed death of rat prostatic glandular cells, *Prostate* **17:**175-187.
30. Kyprianou, N., English, H.F., and Isaacs, J.T., 1988, Activation of a $Ca^{2+}$-$Mg^{2+}$-dependent endonuclease as an early event in castration-induced prostatic cell death, *Prostate* **13:**103-117.

31. Connor, J., Sawczuk, I.S., Benson MC., Tomashefsky, P., O'Toole, K.M., Olsson, C.A., and Buttyan, R., 1988, Calcium channel antagonists delay regression of androgen-dependent tissues and suppress gene activity associated with cell death, *Prostate* **13:**119-130.
32. Gong, Y., Blok, L.J., Perry, J.E., Lindzey, J.K., and Tindall, D.J., 1995, Calcium regulation of androgen receptor expression in the human prostate cancer cell line LNCaP, *Endocrinology* **136:**2172-2178.
33. Wang, Z., Tufts, R., Haleem, R., and Cai, X., 1997, Genes regulated by androgen in the rat ventral prostate, *Proc. Natl. Acad. Sci. USA* **94:**12999-13004.
34. Zhu, N., Pewitt, E.B., Cai, X., Cohn, E.B., Lang, S., Chen, R., and Wang, Z., 1998, Calreticulin: an intracellular $Ca^{2+}$-binding protein abundantly expressed and regulated by androgen in prostatic epithelial cells, *Endocrinology* **139:**4337-4344.
35. Pinton, P., Ferrari, D., Rapizzi, E., Di Virgilio, F., Pozzan, T., and Rizzuto, R., 2001, The $Ca^{2+}$ concentration of the endoplasmic reticulum is a key determinant of ceramide-induced apoptosis: significance for the molecular mechanism of Bcl-2 action, *EMBO J.* **20:**2690-2701.
36. Furuya, Y., Lundmo, P., Short, A.D., Gill, D.L., and Isaacs, J.T., 1994, The role of calcium, pH, and cell proliferation in the programmed (apoptotic) death of androgen-independent prostatic cancer cells induced by thapsigargin, *Cancer Res.* **54:**6167-6175.
37. Tombal, B., Weeraratna, A.T., Denmeade, S.R., and Isaacs, J.T., 2000, Thapsigargin induces a calmodulin/calcineurin-dependent apoptotic cascade responsible for the death of prostatic cancer cells, *Prostate* **43:**303-317.
38. Wertz, I.E., and Dixit, V.M., 2000, Characterization of calcium release-activated apoptosis of LNCaP prostate cancer cells, *J. Biol. Chem.* **275:**11470-11477.
39. Skryma, R., Mariot, P., Bourhis, X.L., Coppenolle, F.V., Shuba, Y., Vanden Abeele, F., Legrand, G., Humez, S., Boilly, B., and Prevarskaya, N., 2000, Store depletion and store-operated $Ca^{2+}$ current in human prostate cancer LNCaP cells: involvement in apoptosis, *J. Physiol.* **527:**71-83.
40. Mariot, P., Prevarskaya, N., Roudbaraki MM., Le Bourhis, X., Van Coppenolle, F., Vanoverberghe, K., and Skryma, R., 2000, Evidence of functional ryanodine receptor involved in apoptosis of prostate cancer (LNCaP) cells, *Prostate* **43:**205-214.
41. Humez, S., Legrand, G., Vanden-Abeele, F., Monet, M., Marchetti, P., Lepage, G., Crepin, A., Dewailly, E., Wuytack, F., and Prevarskaya, N., 2004, Role of endoplasmic reticulum calcium content in prostate cancer cell growth regulation by IGF and TNFalpha, *J. Cell Physiol.* **201:**201-213.
42. Nutt, L.K., Chandra, J., Pataer, A., Fang, B., Roth, J.A., Swisher, S.G., O'Neil, R.G., and McConkey, D.J., 2002, Bax-mediated $Ca^{2+}$ mobilization promotes cytochrome c release during apoptosis, *J. Biol. Chem.* **277:**20301-20308.
43. Tombal, B., Denmeade, S.R., Gillis, J.M., and Isaacs, J.T., 2002, A supramicromolar elevation of intracellular free calcium ($[Ca^{2+}]_i$) is consistently required to induce the execution phase of apoptosis, *Cell Death Differ.* **9:**561-573.
44. Gizatullina, Z.Z., Grapengiesser, E., Shabalina, I.G., Nedergaard, J., Heldin, C.H., and Aspenstrom, P., 2003, Effect of transforming growth factor-beta on calcium homeostasis in prostate carcinoma cells, *Biochem. Biophys. Res. Commun.* **304:**643-649.
45. McConkey, D.J., Lin, Y., Nutt, L.K., Ozel, H.Z., and Newman, R.A., 2000, Cardiac glycosides stimulate $Ca^{2+}$ increases and apoptosis in androgen-independent, metastatic human prostate adenocarcinoma cells, *Cancer Res.* **60:**3807-3812.
46. Huang, Y.T., Chueh, S.C., Teng, C.M., and Guh, J.H., 2004, Investigation of ouabain-induced anticancer effect in human androgen-independent prostate cancer PC-3 cells, *Biochem. Pharmacol.* **67:**727-733.

47. Venkatachalam, K., van Rossum, D.B., Patterson, R.L., Ma, H.T., and Gill, D.L., 2002, The cellular and molecular basis of store-operated calcium entry, *Nat. Cell Biol.* **4:**E263-E272.
1. 48 Hoth, M., and Penner, R., 1992, Depletion of intracellular calcium stores activates a calcium current in mast cells, *Nature* **355:**353-356.
48. Clapham, D.E., Runnels, L.W., and Strubing, C., 2001, The TRP ion channel family, *Nat. Rev. Neurosci.* **2:**387-396.
49. Minke, B., and Cook, B., 2002, TRP channel proteins and signal transduction, *Physiol. Rev.* **82:**429-472.
50. Voets, T., and Nilius, B., 2003, TRPs make sense, *J. Membr. Biol.* **192:**1-8.
51. Vanden Abeele, F., Shuba, Y., Roudbaraki, M., Lemonnier, L., Vanoverberghe, K., Mariot, P., Skryma, R., and Prevarskaya, N., 2003, Store-operated $Ca^{2+}$ channels in prostate cancer epithelial cells: function, regulation, and role in carcinogenesis, *Cell Calcium* **33:**357-373.
52. Vanden Abeele, F., Lemonnier, L., Thebault, S., Lepage, G., Parys, J., Shuba, Y., Skryma, R., and Prevarskaya, N., 2004, Two types of store-operated $Ca^{2+}$ channels with different activation modes and molecular origin in LNCaP human prostate cancer epithelial cells, *J. Biol. Chem.* **279:**30326-30337.
53. Vanden Abeele, F., Roudbaraki, M., Shuba, Y., Skryma, R., and Prevarskaya, N., 2003, Store-operated $Ca^{2+}$ current in prostate cancer epithelial cells. Role of endogenous $Ca^{2+}$ transporter type 1, *J. Biol. Chem.* **278:**15381-15389.
54. Peng, J.B., Chen, X.Z., Berger, U.V., Vassilev, P.M., Tsukaguchi, H., Brown, E.M., and Hediger, M.A., 1999, Molecular cloning and characterization of a channel-like transporter mediating intestinal calcium absorption, *J. Biol. Chem.* **274:**22739-22746.
55. Peng, J.B., Zhuang, L., Berge,r U.V., Adam, R.M., Williams, B.J., Brown, E.M., Hediger, M.A., and Freeman, M.R., 2001, CaT1 expression correlates with tumor grade in prostate cancer. *Biochem. Biophys. Res. Commun.* **282:**729-734.
56. Wissenbach, U., Niemeyer, B.A., Fixeme,r T., Schneidewind, A., Trost, C., Cavalie, A., Reus, K., Meese, E., Bonkhoff, H., and Flockerz,i V., 2001, Expression of CaT-like, a novel calcium-selective channel, correlates with the malignancy of prostate cancer. *J. Biol. Chem.* **276:**19461-19468.
57. Fixemer, T., Wissenbach, U., Flockerzi, V., and Bonkhoff, H., 2003, Expression of the $Ca^{2+}$-selective cation channel TRPV6 in human prostate cancer: a novel prognostic marker for tumor progression, *Oncogene* **22:**7858-7861.
58. Bodding, M., Fecher-Trost, C., and Flockerzi, V., 2003, Store-operated $Ca^{2+}$ current and TRPV6 channels in lymph node prostate cancer cells. *J. Biol. Chem.* **278:**50872-50879.
59. Vermassen, E., Parys, J.B., and Mauger, J.-P., 2004, Subcellular distribution of the inositol 1,4,5-trisphosphate receptors: functional relevance and molecular determinants, *Biol. Cell* **96:**3-17.
60. Vanoverberghe, K., Mariot, P., Vanden Abeele, F., Delcourt, P., Parys, J.B., and Prevarskaya, N., 2003, Mechanisms of ATP-induced calcium signaling and growth arrest in human prostate cancer cells, *Cell Calcium* **34:**75-85.
61. Birnbaumer, L., Boulay, G., Brown, D., Jiang, M., Dietrich, A., Mikoshiba, K., Zhu, X., and Qin, N., 2000, Mechanism of capacitative $Ca^{2+}$ entry (CCE): interaction between IP3 receptor and TRP links the internal calcium storage compartment to plasma membrane CCE channels, *Recent Prog. Horm. Res.* **55:**127-161.
62. Jayaraman, T., and Marks, A.R., 1997, T cells deficient in inositol 1,4,5-trisphosphate receptor are resistant to apoptosis, *Mol. Cell Biol.* **17:**3005-3012.

63. Li, C., Fox, C.J., Master, S.R., Bindokas, V.P., Chodosh, L.A., and Thompson, C.B., 2002, Bcl-$X_L$ affects $Ca^{2+}$ homeostasis by altering expression of inositol 1,4,5-trisphosphate receptors, *Proc. Natl. Acad. Sci. USA* **99:**9830-9835.
64. Boehning, D., Patterson, R.L., Sedaghat, L., Glebova, N.O., Kurosaki, T., and Snyder, S.H., 2003, Cytochrome c binds to inositol (1,4,5) trisphosphate receptors, amplifying calcium-dependent apoptosis, *Nat. Cell Biol.* **5:**1051-1061.
65. Camello, C., Lomax, R., Petersen, O.H., and Tepikin, A.V., 2002, Calcium leak from intracellular stores – the enigma of calcium signalling, *Cell Calcium* **32:**355-361.
66. Van Coppenolle, F., Vanden Abeele, F., Slomianny, C., Flourakis, M., Hesketh, J., Dewailly, E., and Prevarskaya, N., 2004, Ribosome-translocon complex mediates calcium leak from the endoplasmic reticulum stores, *J. Cell Sci.* **117:**4135-4142.
67. Wuytack, F., Raeymaekers, L., and Missiaen, L., 2002, Molecular physiology of the SERCA and SPCA pumps, *Cell Calcium* **32:**279-305.
68. Legrand, G., Humez, S., Slomianny, C., Dewailly, E., Vanden Abeele, F., Mariot, P., Wuytack, F., and Prevarskaya, N., 2001, $Ca^{2+}$ pools and cell growth. Evidence for sarcoendoplasmic $Ca^{2+}$-ATPases 2B involvement in human prostate cancer cell growth control, *J. Biol. Chem.* **276:**47608-47614.
69. Vanden Abeele, F., Skryma, R., Shuba, Y., Van Coppenolle, F., Slomianny, C., Roudbaraki, M., Mauroy, B., Wuytack, F., and Prevarskaya, N., 2002, Bcl-2-dependent modulation of $Ca^{2+}$ homeostasis and store-operated channels in prostate cancer cells, *Cancer Cell* **1:**169-179.
70. Michalak, M., Mariani, P., and Opas, M., 1998, Calreticulin, a multifunctional $Ca^{2+}$ binding chaperone of the endoplasmic reticulum, *Biochem. Cell Biol.* **76:**779-85.
71. Rizzuto, R., Bernardi, P., and Pozzan, T., 2000, Mitochondria as all-round players of the calcium game, *J. Physiol.* **529:**37-47.
72. Kirichok, Y., Krapivinsky, G., and Clapham, D.E., 2004, The mitochondrial calcium uniporter is a highly selective ion channel, *Nature* **427:**360-364.
73. Catz, S.D., and Johnson, J.L., 2003, BCL-2 in prostate cancer: a minireview, *Apoptosis* **8:**29-37.
74. McDonnell, T.J., Troncoso, P., Brisbay, S.M., Logothetis, C., Chung, L.W., Hsieh, J.T., Tu, S.M., and Campbell, M.L., 1992, Expression of the protooncogene bcl-2 in the prostate and its association with emergence of androgen-independent prostate cancer, *Cancer Res.* **52:**6940-6944.
75. Colombel, M., Symmans, F., Gil, S., O'Toole, K.M., Chopin, D., Benson, M., Olsson, C.A., Korsmeyer, S., and Buttyan, R., 1993, Detection of the apoptosis-suppressing oncoprotein bcl-2 in hormone-refractory human prostate cancers, *Am. J. Pathol.* **143:**390-400.
76. Furuya, Y., Krajewski, S., Epstein, J.I., Reed, J.C., and Isaacs, J.T., 1996, Expression of bcl-2 and the progression of human and rodent prostatic cancers, *Clin. Cancer Res.* **2:**389-398.
77. Krajewska, M., Krajewski, S., Epstein, J.I., Shabaik, A., Sauvageot, J., Song, K., Kitada, S., and Reed, J.C., 1996, Immunohistochemical analysis of bcl-2, bax, bcl-X, and mcl-1 expression in prostate cancers, *Am. J. Pathol.* **148:**1567-1576.
78. di Sant'Agnese, P.A., 1998, Neuroendocrine cells of the prostate and neuroendocrine differentiation in prostatic carcinoma: a review of morphologic aspects, *Urology* **51:**121-124.
79. Abrahamsson, P.A., 1999, Neuroendocrine cells in tumour growth of the prostate, *Endocr. Relat. Cancer* **6:**503-519.
80. Bonkhoff, H., 2001, Neuroendocrine differentiation in human prostate cancer. Morphogenesis, proliferation and androgen receptor status, *Ann. Oncol.* **12:**S141-S144.

81. Xue, Y., Verhofstad, A., Lange, W., Smedts, F., Debruyne, F., de la Rosette, J., and Schalken, J., 1997, Prostatic neuroendocrine cells have a unique keratin expression pattern and do not express Bcl-2: cell kinetic features of neuroendocrine cells in the human prostate, *Am. J. Pathol.* **151:**1759-1765.
82. Xing, N., Qian, J., Bostwick, D., Bergstralh, E., and Young, C.Y., 2001, Neuroendocrine cells in human prostate over-express the anti-apoptosis protein survivin, *Prostate* **48:**7-15.
83. July, L.V., Akbari, M., Zellweger, T., Jones, E.C., Goldenberg, S.L., and Gleave, M.E., 2002, Clusterin expression is significantly enhanced in prostate cancer cells following androgen withdrawal therapy, *Prostate* **50:**179-188.
84. Zelivianski, S., Verni, M., Moore, C., Kondrikov, D., Taylor, R., and Lin, M.F., 2001, Multipathways for transdifferentiation of human prostate cancer cells into neuroendocrine-like phenotype, *Biochim. Biophys. Acta* **1539:**28-43.
85. Mariot, P., Vanoverberghe, K., Lalevee, N., Rossier, M.F., and Prevarskaya, N., 2002, Overexpression of an alpha 1H (Cav3.2) T-type calcium channel during neuroendocrine differentiation of human prostate cancer cells, *J. Biol. Chem.* **277:**10824-10833.
86. Vanoverberghe, K., Vanden Abeele, F., Mariot, P., Lepage, G., Roudbaraki, M., Bonnal, J.L., Mauroy, B., Shuba, Y., Skryma, R., and Prevarskaya, N., 2004, $Ca^{2+}$ homeostasis and apoptotic resistance of neuroendocrine-differentiated prostate cancer cells, *Cell Death Differ.* **11:**321-330.
87. Zhu, N., and Wang, Z., 1999, Calreticulin expression is associated with androgen regulation of the sensitivity to calcium ionophore-induced apoptosis in LNCaP prostate cancer cells, *Cancer Res.* **59:**1896-1902.
88. Schendel, S.L., Montal, M., and Reed, J.C., 1998, Bcl-2 family proteins as ion-channels, *Cell Death Differ.* **5:**372-380.
89. Bouillon, R., Van Cromphaut, S., and Carmeliet, G., 2003, Intestinal calcium absorption: Molecular vitamin D mediated mechanisms, *J. Cell Biochem.* **88:**332-339.
90. Guzey, M., Kitada, S., and Reed, J.C., 2002, Apoptosis induction by 1alpha,25-dihydroxyvitamin D3 in prostate cancer, *Mol. Cancer Ther.* **1:**667-677.
91. Kim, J.H., Shin, S.Y., Yun, S.S., Kim, T.J., Oh, S.J., Kim, K.M., Chung, Y.S., Hong, E.K., Uhm, D.Y., and Kim, S.J., 2003, Voltage-dependent ion channel currents in putative neuroendocrine cells dissociated from the ventral prostate of rat, *Pflugers Arch.* **446:**88-99.
92. Jackisch, C., Hahm, H.A., Tombal, B., McCloskey, D., Butash, K., Davidson, N.E., and Denmeade, S.R., 2000, Delayed micromolar elevation in intracellular calcium precedes induction of apoptosis in thapsigargin-treated breast cancer cells, *Clin. Cancer Res.* **6:**2844-2850.
93. Sokolova, I.A., Cowan, K.H., and Schneider, E., 1995, $Ca^{2+}/Mg^{2+}$-dependent endonuclease activation is an early event in VP-16-induced apoptosis of human breast cancer MCF7 cells in vitro, *Biochim. Biophys. Acta*, **1266:**135-142.
94. Mathiasen, I.S., Sergeev, I.N., Bastholm, L., Elling, F., Norman, A.W., and Jaattela, M., 2002, Calcium and calpain as key mediators of apoptosis-like death induced by vitamin D compounds in breast cancer cells, *J. Biol. Chem.* **277:**30738-39745.
95. Hashimoto, T., Shibata, M.A., Ito, Y., Nakao, K.I., Sasaki, S., and Otsuki, Y., 2003, Elevated levels of intracellular Ca2+ and apoptosis in human lung cancer cells given heat-shock. *Int. J. Hyperthermia* **19:**178-192.
96. Aranha, O., Zhu, L., Alhasan, S., Wood, D.P Jr., Kuo, T.H., and Sarkar, F.H., 2002, Role of mitochondria in ciprofloxacin induced apoptosis in bladder cancer cells, *J. Urol.* **167:**1288-1294.

97. Zhang, T.C., Cao, E.H., Li, J.F., Ma, W., and Qin, J.F., 1999, Induction of apoptosis and inhibition of human gastric cancer MGC-803 cell growth by arsenic trioxide, *Eur. J. Cancer.* **35:**1258-1263.
98. Glinsky, G.V., Glinsky, V.V., Ivanova, A.B., and Hueser, C.J., 1997, Apoptosis and metastasis: increased apoptosis resistance of metastatic cancer cells is associated with the profound deficiency of apoptosis execution mechanisms, *Cancer Lett.* **115:**185-193.

Chapter 3

# CELL CLEARANCE AND CANCER

Bengt Fadeel
*Division of Molecular Toxicology, Institute of Environmental Medicine, Nobels väg 13, Karolinska Institutet, 171 77 Stockholm, Sweden*

Abstract: Elimination of cells through apoptosis is crucial for tissue homeostasis. The process of cell removal is regulated by the expression of recognition signals on the dying cell and corresponding phagocytosis receptors on the engulfing cell; in addition, chemotactic factors emitted by apoptotic cell corpses serve to attract macrophages to the site of cell attrition. Cell clearance is thus an active, programmed process and as such is thought to play an important role in the resolution of inflammation and the prevention of autoimmune diseases. Moreover, recent studies suggest that engulfment of apoptotic cell remnants may result in lateral transfer of genetic information within a tumour cell population, and thus contribute to cancer progression. In addition, several recent studies have established a link between malignant transformation and autophagy, an alternative form of programmed cell death. Further elucidation of the mechanism of cell disposal (occurring either through apoptosis or autophagy/self-digestion) is expected to aid in our understanding of cancer development and may unveil novel targets of therapeutic intervention.

Key words: apoptosis, autophagy, cancer, horizontal gene transfer, phagocytosis

## 1. INTRODUCTION

Apoptosis is a process of cellular suicide that is essential for the sculpting of organs during embryogenesis and for the maintenance of homeostasis in adult tissues (Wyllie et al., 1980; Jacobson et al., 1997). Characteristic morphological features of apoptosis (a term derived from the Greek word describing the falling off of petals from a flower or leaves from a tree) are nuclear and cytoplasmic condensation with concomitant blebbing (zeiosis)



*M. Sluyser (ed.), Application of Apoptosis to Cancer Treatment,* 51-84.

of the plasma membrane, and fragmentation of the cell into so-called apoptotic bodies, containing structurally intact organelles as well as portions of the nucleus (Kerr et al., 1972). Studies in recent decades have provided a detailed characterization of the biochemical features of apoptosis, including the fragmentation of nuclear DNA, activation of cysteine proteases known as caspases, and externalization of recognition signals such as phosphatidyserine (PS) on the cell surface (for an excellent review, see Danial and Korsmeyer, 2004). Importantly, the caspase-driven dismantling of the cell ultimately results in the recognition, ingestion, and degradation of apoptotic bodies and cell remnants by neighboring phagocytes. The removal of dying cells prior to their necrotic disintegration requires phagocytosis receptors and recognition signals expressed on phagocyte and prey, respectively, and is thought to preclude tissue scarring and inflammation (Lauber et al., 2004; Fadeel, 2004).

An emerging theme in recent years is that tissue homeostasis is dependent not only on the degree of mitosis (cell proliferation) and apoptosis, but also on the balance between cell death and cell clearance. Consequently, a mismatch between apoptosis and cell disposal may contribute to disease pathogenesis (Savill and Fadok, 2000). For instance, although vast numbers of immature thymocytes die *in situ*, minimal evidence of apoptosis is seen in the thymus (Surh and Sprent, 1994), thus reflecting the efficient clearance of dying cells in this tissue under normal conditions. In contrast, mice treated with agonistic anti-Fas antibodies display fulminant liver destruction (Ogasawara et al., 1993), due most likely to massive death of Fas-expressing hepatocytes such that the phagocytic capacity of the liver is overwhelmed. Apoptosis can thus be viewed essentially as a mechanism of cell clearance; indeed, the term "programmed cell clearance" was recently introduced to underscore this notion and to emphasize that distinct molecular events govern the removal of apoptotic cell corpses (Fadeel, 2003). The current review aims to summarize the mechanisms of clearance of mammalian cells, as well as to discuss the functional significance of (defective) clearance of apoptotic cells in the context of inflammation, autoimmune disease, and cancer. Autophagy, a mode of cell-autonomous clearance that is distinct from apoptosis (Levine and Klionsky, 2004), and its role in the development of cancer, will also be considered.

# 2. MECHANISMS OF CELL CLEARANCE

## 2.1 Recognition signals

Cells undergoing apoptosis express recognition or "eat me" signals, including lipids, proteins, and modified sugar moieties, that facilitate recognition and ingestion by macrophages or neighboring cells (Savill et al., 2002; Lauber et al., 2004). The best-known "eat me" signal is the anionic phospholipid PS that translocates from the inner to the outer leaflet of the plasma membrane in a caspase-dependent manner (Fadok et al., 1992; Martin et al., 1995; Vanags et al., 1996). Early studies suggested that the simultaneous activation of a calcium-dependent phospholipid scramblase and inhibition of an ATP-dependent aminophospholipid translocase, which under normal conditions maintains the asymmetric distribution of PS in the plasma membrane, is required for PS externalization (Verhoven et al., 1995; Bratton et al., 1997). There is also evidence for the involvement of the ATP-binding cassette transporter ABC1 in the redistribution of PS, although its specific role during apoptosis remains to be determined (Hamon et al., 2000). More recent studies suggest that PS exposure is modulated by the level of intracellular ATP and transpires downstream of Bcl-2-regulated mitochondrial events, irrespective of the expression of the phospholipid scramblase (Gleiss et al., 2002; Uthaisang et al., 2003). In support of this notion, Blom and colleagues (2003) have shown that regional loss of mitochondrial membrane potential in hepatocytes is rapidly followed by externalization of PS at that specific site during apoptosis. These investigators proposed that a spatio-temporal relationship may exist between the drop in mitochondrial ATP production and inhibition of aminophospholipid translocation, thus yielding localized externalization of PS. Furthermore, Ricci et al. (2004) have demonstrated that disruption of mitochondrial function during apoptosis is mediated by caspase cleavage of the p75 subunit of complex I of the electron transport chain; importantly, the disruption of mitochondrial function and subsequent dissipation of intracellular levels of ATP was coupled with plasma membrane externalization of PS in this model. In addition, microinjection of the so-called apoptosis-inducing factor (AIF) into the cytoplasm of intact cells induces caspase-independent dissipation of the mitochondrial membrane potential and concomitant PS externalization (Susin et al., 1999). Taken together, the egress of PS during apoptosis appears to be a mitochondria-dependent event that is linked to a drop in ATP production within the cell. Furthermore, while PS externalization can be dissociated from other features of the apoptotic program (Zhuang et al., 1998; Uthaisang et al., 2003), we and others have shown that this event is nonetheless essential for phagocytosis of apoptotic cells (Fadok et al., 2001; Kagan et al., 2002).

Evidence of selective oxidation of PS has been provided in several models of oxidative stress-induced apoptosis (for a review, see Kagan et al., 2003). Moreover, recent studies of non-oxidant-triggered apoptosis have yielded further evidence that oxidation of PS is selective and precedes its externalization (Kagan et al., 2002; Matsura et al., 2002; Koty et al., 2002). Moreover, we have shown, in collaboration with Kagan and his associates, that the expression of oxidized PS (PS-OX) in conjunction with its non-oxidized counterpart on the surface of apoptotic cells serves as a critical recognition signal for macrophages (Kagan et al., 2002; Arroyo et al., 2002). These findings were supported by the demonstration that enrichment of the plasma membrane of non-apoptotic cells with exogenous PS and/or PS-OX resulted in phagocytic removal of these cells. Moreover, antibodies to oxidized low-density lipoprotein (LDL) have been shown to bind to apoptotic cells and prevent their uptake by macrophages, thus providing further evidence for oxidation-specific epitopes on the surface of apoptotic cells (Chang et al., 1999). Indeed, C-reactive protein (CRP) was recently found to bind both oxidized LDL and "late" apoptotic cells (i.e. apoptotic cells that have undergone secondary necrosis due to prolonged *in vitro* culture) through recognition of a common ligand, oxidized phosphatidylcholine (PC-OX) (Chang et al., 2002). The mechanism of PS (and PC) oxidation remains to be clarified; however, recent data suggest that cytochrome *c* released from mitochondria into the cytosol may act as a catalyst that utilizes reactive oxygen species generated by disrupted mitochondrial electron transport for selective PS oxidation (Jiang et al., 2003; Fadeel et al., 2004a).

Recent proteomics studies have identified annexin I (also known as lipocortin I) as a novel recognition signal (Arur et al., 2003). Annexin I is thus recruited from the cytosol to the plasma membrane during apoptosis, and is required for efficient uptake of apoptotic cells by human umbilical vein endothelial cells, i.e. non-professional phagocytes. Moreover, annexin I, a PS-binding protein (Schlaepfer and Haigler, 1987), co-localizes with PS on the surface of apoptotic cells leading to the clustering of PS receptors on the phagocytic cell surface. Similarly, Fan et al. (2004) have reported that apoptotic Jurkat cells and primary T cells, but not apoptotic thymocytes, express annexin I on their cell surface. These investigators were also able to demonstrate constitutive annexin I and II expression on macrophages, and could show that antibodies to these annexins suppressed macrophage engulfment of apoptotic target cells. Hence, annexins may potentially serve as both ligand and receptor and thereby promote cell clearance, alone or in conjunction with PS or PS-OX, albeit in a cell type-specific manner.

Interestingly, macrophages derived from annexin I-deficient mice display anomalies in phagocytosis of zymosan particles (Hannon et al., 2003); whether annexin I-null cells also exhibit defects in engulfment of apoptotic targets remains to be determined.

Recent studies have identified detachment signals that contribute to the protection of viable cells from accidental clearance. Hence, CD47 (also known as integrin-associated protein) was shown to serve as a marker of "self" on viable erythrocytes; red blood cells that lack CD47 are rapidly cleared from the bloodstream by splenic red pulp macrophages (Oldenborg et al., 2000). Similarly, non-apoptotic neutrophils were reported to express CD31 (platelet-endothelial cell adhesion molecule 1, PECAM-1) (Brown et al., 2002). Apoptosis of neutrophils apparently disables CD31-mediated cell detachment from phagocytes, thereby promoting binding and engulfment of target cells. In line with the latter observations, a recent report has provided evidence that CD31 is required for efficient engulfment of apoptotic B lineage cells by bone marrow-derived macrophages (Dogusan et al., 2004). In addition, Tada et al. (2003) have shown that CD47 may act as a tethering molecule during apoptosis; concomitant externalization of PS, however, was essential for engulfment. The molecular mechanism underlying the switch from CD47- and CD31-mediated repulsion to adhesion remains an important area of investigation. Further studies are also required to assess whether an influx of calcium following CD31 engagement (O'Brien et al., 2001) can promote the calcium-dependent externalization of annexin I, as suggested by Orrenius et al. (2003).

## 2.2 Phagocytosis receptors

Macrophages utilize an array of phagocytosis receptors for engulfment of cell corpses. These include the so-called PS receptor (PSR), the class A scavenger receptor (SRA), CD36 (a class B scavenger receptor), CD68 (macrosialin; a class D scavenger receptor), the integrin receptors $\alpha_V\beta_3$ and $\alpha_V\beta_5$, the bacterial lipopolysaccharide (LPS) receptor CD14, and the calreticulin-CD91 receptor complex (Savill et al., 2002, and references therein). Tissue- and cell type-specific differences in receptor usage provides a partial explanation for this broad repertoire of phagocytosis receptors. For example, monocyte-derived macrophages and dendritic cells utilize the $\alpha_V\beta_3$ and $\alpha_V\beta_5$ integrin receptor, respectively, for the uptake of cell corpses (Savill et al., 1990; Albert et al., 1998a). In addition, the engulfment process may require the serial engagement of distinct receptors, some of which are involved in the initial tethering of apoptotic cells and others in the subsequent stage of cytoskeletal rearrangement that is needed for ingestion of cells. Indeed, Hoffmann et al. (2001) have shown, using a surrogate system consisting of erythrocytes coated with various protein ligands, anti-receptor antibodies, or phospholipids, that the majority of "eat me" signals mediate tethering (attachment) of erythrocytes to macrophages. Internalization of target cells, on the other hand, required the presence of PS as well as the PS-specific receptor, PSR. PSR was originally described to be a cell surface receptor that interacts in a stereospecific manner with externalized PS on apoptotic cells (Fadok et al., 2000); however, more recent findings call into question the membrane localization of this molecule (Cui et al., 2004; Cikala et al., 2004). Notwithstanding the issue of nuclear versus cell surface localization of PSR, a role for this molecule in the ingestion of cell corpses has been demonstrated in several mammalian model systems (Fadok et al., 2000; Todt et al., 2002; Hisatomi et al., 2003). Moreover, a PSR homolog was recently identified in C. elegans and shown to be an upstream receptor for the signaling pathway containing CED-2, CED-5, CED-10, and CED-12 (CED, cell death abnormal) proteins, and to play an important role in PS recognition during corpse clearance in the nematode (Wang et al., 2003). The contact site between the apoptotic cell and the phagocyte may thus be viewed as an "engulfment synapse", akin to the neural or immunological synapse, in which complex interactions between numerous ligands and receptors take place (Henson et al., 2001; Somersan and Bhardwaj, 2001). This model makes it feasible for signaling to proceed despite the low avidity of PS-PSR interactions (Fadok et al., 2000). In addition, an engulfment synapse allows for regulation and specificity in the uptake of dying cells. Indeed, because some cell types can transiently express PS upon activation (Dillon et al., 2000; Martin et al., 2000), the two-step process of tethering and engulfment may also serve to protect against accidental clearance of viable cells.

Several of the aforementioned receptors (not only the PSR) bind PS on apoptotic cells, either directly or indirectly (via bridging molecules) (Savill et al., 2002; and see below). Importantly, macrophage scavenger receptors were originally identified based on their ability to bind chemically modified structures, such as acetylated or oxidized LDL, but not their unmodified counterparts (Steinbrecher, 1999). A common feature of these proteins is their ability to recognize a wide range of structurally unrelated ligands, including oxidized LDL, bacterial LPS and the anionic phospholipid PS, and this lack of specificity is consistent with the idea that scavenger receptors act as receptors for apoptotic cells (Platt et al., 1998). For instance, the class B scavenger receptor, CD36, is required for phagocytosis of apoptotic cells by various classes of macrophages and non-professional phagocytes (Fadok et al., 1998a; Shiratsuchi et al., 1999). Similarly, croquemort ("catcher of death"), a Drosophila melanogaster homolog of CD36, is required for the engulfment of apoptotic cells in the fruitfly (Franc et al., 1999), and CED-1, a scavenger receptor-like molecule in C. elegans, was suggested to recognize a phospholipid ligand on the surface of apoptotic cell corpses in the nematode (Zhou et al., 2001). As discussed above, apoptotic cells express oxidation-specific epitopes, including PS-OX and PC-OX, on their cell surface, and it is tempting to suggest that scavenger receptors are specifically involved in the recognition of such oxidized phospolipid species on apoptotic cells. Indeed, we have shown that anti-PSR antibodies, but not anti-CD36 antibodies, are able to inhibit phagocytosis of Jurkat cells with exogenous PS integrated into their plasma membrane (Kagan et al., 2003). In contrast, both anti-PSR antibodies and anti-CD36 antibodies were effective in suppressing phagocytosis of target cells enriched with a combination of PS and PS-OX. These data, which imply that CD36 and PSR cooperate to recognize PS and its oxidized counterpart on the surface of apoptotic cells, are thus in concordance with the engulfment synapse model outlined above. Indeed, Fadok and colleagues (1998a) have previously demonstrated that CD36 contributes not only to $\alpha_V\beta_3$ integrin-dependent recognition, but also serves as a co-factor during PS-dependent clearance.

Recent studies have shown that members of the collectin family, such as the lung surfactant proteins A and D (SP-A and SP-D), mannose-binding lectin (MBL), and the collectin-like complement protein, C1q, facilitate the attachment of apoptotic cells to phagocytes (for a recent review, see Roos et al., 2004). Hence, C1q and MBL engagement of calreticulin and CD91 on the macrophage cell surface was shown to initiate macropinocytosis and uptake of apoptotic cells in vitro (Ogden et al., 2001). Similarly, SP-A and SP-D bind to apoptotic cells and drive apoptotic cell ingestion by phagocytes through a mechanism dependent on calreticulin and CD91 (Vandivier et al., 2002). Of note, several studies suggest that these opsonins, and related molecules such as the pentraxin, PTX3, recognize primarily "late" apoptotic cells and/or apoptotic debris (Rovere et al., 2000; Gaipl et al., 2001; Nauta et al., 2003); therefore, other mechanisms of cell disposal may prevail in normal clearance of early apoptotic cells (Roos et al., 2004). Furthermore, it is important to consider that collectins, pentraxins, and other molecules of the innate immune system, could trigger undesirable pro-inflammatory responses, thus precluding the "meaning" of cell death (Savill and Fadok, 2000). For comparison, CD14 induces pro-inflammatory responses upon recognition of bacterial LPS, yet mediates clearance of apoptotic cells without inciting inflammation (Gregory, 2000). Interestingly, SP-A and SP-D were recently shown to act in a dual manner, enhancing or suppressing inflammatory cytokine production depending on their orientation and on their specific binding partners (Gardai et al., 2003). Further studies are warranted to elucidate how opsonization of apoptotic cells, in the context of a putative engulfment synapse between phagocyte and apoptotic prey, can facilitate clearance without triggering a deleterious inflammatory response.

## 2.3 Bridging molecules

As mentioned in the preceding sections, recognition signals such as PS may not bind directly to phagocytosis receptors on the macrophage, but rather indirectly via membrane-bound co-factors (such as annexin I) or via soluble bridging molecules (such as collectins or pentraxins). Indeed, the basic principle of PS recognition via bridging molecules holds true for a number of phagocytosis receptors, such as the $\alpha_V\beta_3$ integrin receptor and the receptor tyrosine kinase, Mer (Lauber et al., 2004). Hence, it was shown some 10 years ago that macrophages are able to recognize $\alpha_V\beta_3$ together with the soluble bridging protein, thrombospondin, and the co-receptor, CD36 (Savill et al., 1992). More recent studies show that milk fat globule epidermal growth factor 8 (MFG-E8), a protein that is secreted by certain classes of macrophages and DCs, serves as a molecular bridge between PS on the apoptotic cell and $\alpha_V\beta_3$ or $\alpha_V\beta_5$ on the phagocyte surface (Hanayama et al., 2002; Akakura et al., 2004; Miyasaka et al., 2004). Indeed, our recent *in vitro* data demonstrate that PS-OX may serve as a preferential ligand for MFG-E8 binding (Borisenko et al., 2004), suggesting that exposure of PS-OX during apoptosis could promote clearance of cell corpses at least in part through interaction with MFG-E8. Further studies are needed to determine whether other bridging molecules such as Del-1, protein S, and Gas6 also bind preferentially to PS-OX, as opposed to non-oxidized PS.

Hanayama and colleagues (2004a) have shown that developmental endothelial locus-1 (Del-1) is functionally homologous to MFG-E8, insofar as it binds to PS on apoptotic cells and $\alpha_V\beta_3$ on engulfing cells. Importantly, thioglycolate-elicited macrophages expressed MFG-E8, but not Del-1, whereas fetal liver and thymic (resident) macrophages expressed Del-1, but not MFG-E8. These findings suggest that these two bridging molecules may play non-redundant roles in the clearance of cell corpses. Furthermore, β2-glycoprotein I, a soluble serum protein, also binds PS and enhances phagocytosis of apoptotic targets through recognition of a phagocyte receptor that is distinct from CD36, CD68, and CD14 (Balasubramanian et al., 1997; Balasubramanian and Schroit, 1998). Of note, β2-glycoprotein I has also been reported to bind to annexin II on the surface of endothelial cells (Ma et al., 2000). The latter data suggest that constitutive expression of annexin II on the macrophage surface (discussed above) may link the engulfing cell to PS on apoptotic cells via the bridging molecule, β2-glycoprotein I; future experiments should address this possibility.

Mer is another engulfment receptor that binds PS on apoptotic cells via a soluble bridging molecule, Gas6 (growth arrest-specific gene 6) (Nakano et al., 1997; Scott et al., 2001). Furthermore, protein S, a serum protein that is related to Gas6, was shown to facilitate clearance of cell corpses through binding of PS (Anderson et al., 2003). Interestingly, protein S also mediates binding of the complement-regulating protein, C4BP (C4b-binding protein), to apoptotic cells, and this C4BP-protein S complex was shown to inhibit phagocytosis of apoptotic cells (Webb et al., 2002; Kask et al., 2004). Although the C4BP-protein S complex and free protein S exert opposing effects on cell clearance, binding of the protein complex may nonetheless be beneficial to the cell since this will prevent further complement attack and subsequent necrosis (Kask et al., 2004). Similarly, binding of CRP to apoptotic cells protects these cells from assembly of terminal complement components (Gershov et al., 2000). Importantly, several studies have demonstrated that complement opsonization renders apoptotic cells more appetizing to macrophages; PS externalization has been shown to promote this process (Roos et al., 2004; Hart et al., 2004). Hence, both iC3b and C1q can bind to the surface of apoptotic cells, and facilitate macrophage engulfment of cell corpses (Takizawa et al., 1996; Korb and Ahearn, 1997; Mevorach et al., 1998; Ogden et al., 2001; Nauta et al., 2002). Opsonization of apoptotic cells with C3bi also facilitates clearance of dying cells by immature dendritic cells (DCs) (Verbovetski et al., 2002), and internalization of circulating apoptotic cells by splenic marginal zone DCs was shown to require complement receptors (Morelli et al., 2003). In addition, Taylor and colleagues (2000) have provided compelling *in vivo* evidence that complement plays a role in the phagocytic clearance of apoptotic cells by inflammatory macrophages; these investigators also found that defects in apoptotic cell clearance correlated with predisposition to autoimmune disease (discussed below).

## 2.4 Chemotactic factors

*C. elegans* lacks dedicated macrophages and dying cells are usually engulfed by neighboring cells (Robertson and Thomson, 1982). Engulfment by neighboring cells (i.e. non-professional phagocytes) also occurs in higher organisms, as evidenced in macrophage-less mice, null for the transcription factor, PU.1, in which the task of phagocytosis is taken over by mesenchymal neighbouring cells (Wood et al., 2000). However, such resident phagocytes are less mobile and have been shown to engulf with slower kinetics than professional phagocytes (Parnaik et al., 2000), perhaps because they fail to express the full repertoire of engulfment receptors and thus are unable to engage the "engulfment synapse". In some specific instances, cell clearance may occur through a cell-autonomous, non-apoptotic mechanism and not depend on phagocytosis, as suggested for chondrocytes that are embedded in extracellular matrix in cartilage tissue (Roach et al., 2004). Notwithstanding such alternative modes of cell clearance, it is evident that professional phagocytes (macrophages) are required for efficient disposal of dying cells in higher organisms. It is thus important to consider how (and where) apoptotic cells encounter macrophages. First, effete cells may traffic through the body until they reach the site of cell disposal; for instance, clearance of circulating neutrophils is effected by so-called Kupffer cells, resident (professional) macrophages of the liver (Shi et al., 2001). Similarly, chilling of blood platelets causes these cells to rapidly leave the circulation, and this was shown in an elegant series of experiments to occur through recognition by hepatic macrophage complement type 3 (CR3) receptors (Hoffmeister et al., 2003). However, since apoptotic cells in solid tissues are not always situated in close proximity to macrophages, it is reasonable to assume that the dying cell needs to emit soluble chemotactic signals in order to selectively recruit more phagocytes to the site of cell death. Indeed, recent studies have provided evidence for apoptosis-specific "come-and-get-me" signals (Grimsley and Ravichandran, 2003; Lauber et al., 2004).

Horino and colleagues (1998) have provided evidence that a cross-linked homodimer of S19 ribosomal protein can function as a chemotactic factor in the recruitment of monocytes from the circulation to apoptotic lesions. Subsequent studies revealed that intracellular transglutaminase activity was required for the generation of the monocyte chemotactic properties (Nishimura et al., 2001). Moreover, membranous vesicles (blebs) derived from apoptotic germinal center B cells were shown to be chemotactic for monocytes *in vitro*, and it was hypothesized that a gradient of apoptotic blebs released from dying B cells may attract macrophages *in vivo* (Segundo et al., 1999). Of note, recent studies indicate that membrane vesicles and apoptotic blebs contain biologically active oxidized phospholipids (Huber et al., 2002; and V. Kagan, personal communication), suggesting that membrane-bound PS-OX and/or other modified phospholipid species could play a role not only as recognition signals, but also as chemotactic agents. Moreover, in addition to its role as a bridging molecule, thrombospondin derived from apoptotic cells may also act as a signal to recruit macrophages (Moodley et al., 2003).

Using an *in vitro* transmigration system, it was recently demonstrated that apoptotic cells secrete a chemotactic signal, in a caspase-3-dependent fashion (Lauber et al., 2003). This factor was identified as lysophosphatidylcholine (LPC), a phospholipid previously shown to be a chemoattractant for monocytes and T cells (Hoffman et al., 1982; McMurray et al., 1993). Apoptotic vesicles could be excluded in this model system since neither filtration nor ultracentrifugation could abrogate the chemotactic activity of apoptotic cell supernatants (Lauber et al., 2003). Interestingly, the release of LPC during apoptosis was linked to caspase-3-mediated activation of the calcium-independent phospholipase $A_2$ ($iPLA_2$). Of note, Kim et al. (2002) have previously reported that $iPLA_2$ activation during apoptosis can promote cell surface exposure of LPC leading to the binding of natural IgM antibodies and subsequent complement opsonization of the dying cell. Taken together, these novel findings, which further emphasize the importance of phospholipid-dependent signaling in cell clearance (Fadeel, 2004), raise several interesting questions. For instance, does LPC act alone, or in concert with other co-factors or binding partners, such as vitamin D-binding protein (Homma et al., 1993), and is metabolism of LPC required for chemotaxis to occur (McMurray et al., 1993)? Moreover, are the chemotactic effects of secreted LPC receptor-mediated, and if so, which receptors (on which classes of macrophages) are involved; indeed, do DCs also respond to chemotactic signals emitted by apoptotic cell corpses?

# 3. CELL CLEARANCE AND DISEASE

## 3.1 Inflammation

Neutrophils are short-lived cells of the innate immune system that contain proteolytic enzymes, reactive oxygen species (ROS), and numerous other bactericidal factors. These cells therefore need to be removed prior to their disintegration as noxious contents would otherwise expel into the extracellular space and prolong inflammation (Savill, 1997; Fadeel and Kagan, 2003). Neutrophils cultured *ex vivo* undergo constitutive death within 24 hours, with typical hallmarks of apoptosis, and senescent neutrophils are believed to undergo apoptosis *in vivo* prior to their removal by macrophages. Indeed, Savill and associates (1989) have shown in pioneering studies that macrophages from acutely inflamed joints preferentially ingest apoptotic neutrophils, and histological evidence was presented for the *in situ* occurrence of this process. The recognition and disposal of intact senescent neutrophils may thus serve as a means to limit the degree of tissue injury and prevent chronic inflammation. Importantly, the process of cell clearance is not a passive event, but is thought to play an active role in the resolution of inflammation, through macrophage production of anti-inflammatory cytokines such as TGF-β and downregulation of pro-inflammatory mediators such as TNF-α (Voll et al., 1997; Fadok et al., 1998b; Byrne and Reen,

2002; Huynh et al., 2002). Furthermore, injection of apoptotic cells into the knee joint and uptake of these cells by synovial lining macrophages was recently shown to inhibit the onset of experimental arthritis in mice, suggesting an active role in the resolution of joint inflammation (van Lent et al., 2001).

Chronic granulomatous disease (CGD) is a rare hereditary condition characterized by severe recurrent bacterial and fungal infections, and an inability of neutrophils and other phagocytes to generate ROS; the underlying genetic defect is a mutation in the NADPH oxidase (Roos et al., 1996). We have previously shown that ROS-dependent externalization of PS is defective in CGD neutrophils (Fadeel et al., 1998). We surmise that the absence of this crucial recognition signal (PS and/or PS-OX) on the surface of neutrophils may disrupt the clearance of cells *in vivo*, thus contributing to the formation of inflammatory granulomas and tissue destruction evidenced in these patients. Indeed, an increased accumulation of neutrophils was observed in peritoneal exudates of NADPH oxidase-defective mice injected with heat-inactivated bacteria, indicative of a clearance defect in this model of CGD (Hampton et al., 2002). In addition, macrophages derived from CGD patients are compromised in their ability to produce anti-inflammatory mediators such as TGF-β upon ingestion of apoptotic targets (Brown et al., 2003). Early studies suggested that constitutive and Fas-triggered apoptosis of CGD neutrophils was impaired (Kasahara et al., 1997). However, subsequent examinations of CGD patients in several laboratories have shown that the execution of apoptosis is normal in CGD patient-derived neutrophils and monocytes (Fadeel et al., 1998; Yamamoto et al., 2002; Bernuth et al., 2004). The latter findings do not, however, rule out a role for oxidative stress in the modulation of apoptosis in neutrophils (Hampton et al., 1998; Fadeel and Kagan, 2003). Nevertheless, these data indicate that while NADPH oxidase-derived ROS are required for PS-mediated signaling during the disposal phase of cell death, ROS may not be crucial for the demolition phase of apoptosis in these cells.

## 3.2 Autoimmune disease

Apoptotic cells may serve as potential reservoirs of "self" antigens that might initiate and drive autoimmune responses (Rosen and Casciola-Rosen, 1999). Indeed, caspase-driven dismantling of cells may be potentially harmful as this generates neoantigens that become accessible on surface structures of apoptotic cells (Casciola-Rosen et al., 1994; Casciola-Rosen et al., 1995; Casiano et al., 1996). These findings suggest that autoimmune responses could result from an impairment of cell clearance. Several recent *in vivo* studies have provided evidence in support of this notion. Hence, mice deficient for the complement component (and opsonizing factor) C1q have high titers of autoantibodies and systemic lupus erythematosus (SLE)-like glomerulonephritis with evidence of numerous unengulfed apoptotic bodies (Botto et al., 1998). A similar lupus-like syndrome has been described in mice with defects in the tyrosine kinase Mer, a phagocytosis receptor (Scott et al., 2001; Cohen et al., 2002). Interestingly, macrophages isolated from these animals were defective for ingestion of apoptotic targets, while tethering of target cells remained intact, thus providing further support for the two-step model of tethering and engulfment of cell corpses. Recent studies have shown that deletion of the bridging molecule, MFG-E8, also results in autoimmune disease and impaired uptake of apoptotic cells (Hanayama et al., 2004b). These mice developed enlargement of the spleen, with formation of numerous germinal centers, and suffered from glomerulonephritis as a result of autoantibody production. It will be of interest to learn whether deletion of the related bridging molecule, Del-1, elicits a similar phenotype.

Additional lessons come from two recent studies of PSR-deficient mice. In the first study, Li and colleagues (2003) report a dramatic phenotype in PSR-null animals, with fatal neonatal respiratory failure associated with a reduction in the number of airways and an accumulation of non-engulfed cells and cellular debris in the developing lung. These findings suggest that PSR may be of particular importance for clearance in the lung as opposed to other tissues; for comparison, C1q-deficient mice exhibit impaired clearance of cell corpses in the peritoneum and kidney, but not in the skin (Pickering et al., 2001). In addition, these investigators found that a proportion of PSR-defective mice also exhibited hyperplastic brain malformations due to an overproduction of cells within the brain. The latter phenotype might suggest the loss of a non-phagocytic function of PSR; alternatively, PSR-expressing macrophages in normal mice could be responsible for killing of other cell types. In this context, previous studies in *C. elegans* have shown that engulfment may, indeed, promote the execution phase of cell death (Reddien et al., 2001; Hoeppner et al., 2001). In fact, cell death-related nuclease (*crn*) genes that regulate DNA degradation in *C. elegans* have also been implicated in cell clearance (Parrish and Xue, 2003). In a second study, Kunisaki et al. (2004) generated a strain of PSR-null mice that also exhibited perinatal lethality, albeit with a severe block in definitive erythropoiesis and thymocyte development. This phenotype is reminiscent of that seen in animals deficient for DNase II, an endonuclease expressed in macrophages (Kawane et al., 2001; Kawane et al., 2003). In addition, clearance of apoptotic cells by macrophages was impaired in both liver and thymus of PSR-deficient embryos. Again, these studies suggest that cell clearance may be linked to the execution of cell death, as shown in the nematode.

In contrast to PSR-null mice, mice defective for tissue transglutaminase 2 (TGase2), a protein-crosslinking enzyme, show no major developmental abnormalities (De Laurenzi and Melino, 2001). However, Szondy and colleagues (2003) have recently uncovered a stress-inducible phagocytosis defect in these animals. Hence, clearance of apoptotic cells was found to be defective during the involution of the thymus elicited by dexamethasone, anti-CD3 antibody, or γ-irradiation, and in the liver after $PbNO_3$-induced hyperplasia. The TGase2 deficiency was associated with the development of splenomegaly, autoantibody production, and glomerulonephritis. Interestingly, the lack of TGase2 prevented the production of TGF-β by macrophages exposed to apoptotic cells, which, in turn, was required for the upregulation of TGase2 in the thymus and for efficient clearance of apoptotic bodies. These studies thus suggest extensive cross-talk between apoptotic target cells and macrophages, and further emphasize the importance of corpse clearance in the prevention of autoimmunity.

Defects in macrophage clearance of apoptotic β-cells could contribute to the initiation of insulin-dependent (autoimmune) diabetes. Peritoneal and bone marrow-derived macrophages from diabetes-prone NOD mice are defective for engulfment of apoptotic cell corpses (O'Brien et al., 2002). In addition, NOD macrophages react aberrantly to both necrotic and apoptotic cells, with secretion of inappropriately high amounts of pro-inflammatory cytokines, including TNF-α (Stoffels et al., 2004). However, treatment of NOD mice with a superoxide dismutase (SOD) mimetic results in a reversal of macrophage defects in phagocytosis and cytokine production (Haskins et al., 2003). The latter findings are of considerable interest and suggest that defects in corpse clearance are amenable to pharmacological intervention *in vivo*. Finally, recent studies have disclosed that macrophages from human SLE patients display an impairment in phagocytosis of apoptotic cell; it was suggested that persistently circulating apoptotic "waste" (debris) may serve as immunogen for the induction of autoreactive responses in these individuals (Herrmann et al., 1998).

## 3.3 Cancer

### 3.3.1 Tumor-host cell interactions

Tumor-associated macrophages (TAMs) represent a major component of host cell infiltrates in malignant tumors *in vivo* (Mantovani et al., 1992; Opdenakker and Van Damme, 1992; Klein and Mantovani, 1993). These cells exert pleiotropic functions and have a complex (sometimes symbiotic) relationship to the neoplastic cells of the tumor; hence, TAMs have been suggested to affect tumor growth by influencing the proliferation of cancer cells, and promoting vascularization, but they can also kill cancer cells directly and/or elicit tumor-destructive reactions in collaboration with other components of the immune system. Early studies on the macrophage content of tumors suggested that infiltration of macrophages is associated with an effective immune response of the host to the tumor (Eccles and Alexander, 1974). However, other investigators have provided evidence that the macrophage content of tumors is regulated by tumor-derived chemotactic factors (Bottazzi et al., 1983; Graves et al., 1989). In addition, a recent study has shown that the infiltration rate of macrophages in malignant lymphoma depends on the rate of proliferation of the tumor; of interest, the degree of phagocytosis of apoptotic cells was reduced in malignant lymphoma as compared to normal lymphoid tissue (Hermann et al., 1998). These investigators proposed that the lack of efficient clearance of apoptotic tumor cells in certain lymphoma sub-entities might be due to a defective expression of recognition ("eat me") signals, rather that an over-burdening of phagocytes by a high cellular turnover in these tumors.

A classic example of macrophage infiltration in tumors is the "starry sky" appearance of Burkitt lymphoma (BL), representing scattered macrophages that have engaged in phagocytosis of cell debris among proliferating lymphoma cells. Fujita et al. (2004) have provided evidence that Epstein-Barr virus (EBV)-infected BL cells in the so-called lytic viral cycle, which eventually lapse into cell death, are phagocytosed prior to their rupture by macrophages that have migrated into the parenchyma of the tumor. This mechanism of cell clearance, which constitutes an example of a beneficial interaction between host and tumor cells, might serve to prevent EBV from spreading beyond infected lymphoma cells. In contrast, tumors may sometimes subvert the host response and take advantage of apoptotic cell effects on macrophages. Reiter and colleagues (1999) have found that exposure of macrophages to apoptotic tumor cells results in impairment of macrophage-mediated tumor defense *in vitro* and supports tumor cell growth. These findings could have implications for cancer treatment since chemotherapy-induced apoptosis of cancer cells (discussed below) might lead to a suppression of local anti-tumor reactions.

Another important class of tumor-infiltrating immune cells are the dendritic cells. DCs are antigen-presenting cells whose primary function is to monitor the environment for "danger" signals and transduce these signals to T cells. DCs are also capable of engulfing apoptotic cells, and can present antigen derived from ingested cell corpses in an MHC class I-restricted manner (Albert et al., 1998b). The question of whether recognition of apoptotic "self" induces a tolerogenic or immunogenic response remains controversial (Rovere et al., 1998; Ronchetti et al., 1999; Sauter et al., 2000;

Steinman et al., 2000; Scheffer et al., 2003). Nevertheless, the outcome of apoptotic cell engulfment by DCs has significant implications for vaccine strategies and immunotherapeutic approaches to cancer (Savill et al., 2002). In a recent *in vivo* study, Goldszmid et al. (2003) showed that DCs loaded with cells undergoing apoptosis are able to prime melanoma-specific helper and cytotoxic T cells and provide long-term protection against a poorly immunogenic tumor in mice. In addition, enforced recruitment of intratumoral DCs has been shown to suppress tumor growth, thus providing additional evidence that DCs can be manipulated *in vivo* to boost anti-tumor immunity (Fushimi et al., 2000). Indeed, recent results obtained in two different murine models support the notion that the number of tumor-associated DCs as well as the tumor milieu determines the ability of tumor-bearing hosts to mount an effective immune response (Furumoto et al., 2004).

### 3.3.2 Horizontal gene transfer

Horizontal gene transfer has been described in bacteria and fungi and is thought to play an important role in the generation of resistance to antibiotics as well as to the adaptation of microorganisms to new environments (Jain et al., 2002). The range and frequencies of horizontal gene transfer in higher organisms are often constrained by selective barriers (Kurland et al., 2003). However, *in vitro* transfer of DNA from bacteria to somatic cells has been demonstrated (Darji et al., 1997). Moreover, recent studies from two laboratories have shown that tumor cells are able to engulf apoptotic bodies and re-utilize the salvaged DNA, suggesting that horizontal (or lateral) transfer of genetic information between somatic cells may, indeed, occur. Propagation of intact genes through phagocytosis of apoptotic cells is counterintuitive, given the fact that the DNA of apoptotic cells is degraded into oligonucleosomal fragments (Wyllie, 1980). Nevertheless, Holmgren et al. (1999) observed that co-cultivation of cell lines containing integrated copies of EBV resulted in rapid uptake and transfer of EBV DNA as well as genomic DNA to the nucleus of the engulfing cell. In a subsequent study, these investigators provided evidence for horizontal transfer of oncogenes and cellular transformation upon uptake of apoptotic bodies (Bergsmedh et al., 2001). Importantly, no transformation was detected in the engulfing cell when apoptotic bodies were cultured with recipient cells harboring intact p53, indicating that p53 may protect normal cells from incorporation of “foreign” DNA. Indeed, p53 activation of the *p21* (*Cip1/Waf1*) gene was shown to block the propagation of horizontally transferred DNA in normal cells (Bergsmedh et al., 2002). In contrast, feeding apoptotic bodies derived from a rat fibrosarcoma to p21-deficient murine embryonic fibroblasts resulted in focus formation *in vitro* and tumor growth *in vivo* in a SCID mouse model. In an independent study, de la Taille and colleagues (1999) demonstrated that prostate cancer cells exchange and propagate drug resistance genes *in vitro* through the engulfment of apoptotic bodies; these investigators termed this phenomenon "apoptotic conversion."

These data suggest that transfer of oncogenes through the engulfment of apoptotic cell remnants may transpire within a tumor cell population, thus providing a novel mechanism for the propagation of genetic instability and/or diversity in tumors. Such events may be of particular relevance in cases of high frequencies of apoptosis within a tumor, such as conditions of hypoxia or restricted angiogenic support (Holmgren et al., 1995; Graeber et al., 1996). In addition, the aforementioned experimental findings may have clinical implications, since treatment of tumors with chemotherapeutic agents or radiation therapy induces apoptosis of cancer cells, thus likely enhancing the horizontal gene transfer effects. Indeed, horizontal passage of genetic defects (eg. defects in the apoptosis machinery of a cell) may accelerate the acquisition of drug resistance in cancer cells. Horizontal gene transfer could thus be considered as a means of adaptation of the tumor to its environment, analogous to the adaptation of microorganisms to changes in growth conditions. Further studies are required to determine the *in vivo* role of horizontal transfer, or apoptotic conversion, of engulfed DNA, and the putative role of this event in chemoresistance.

### 3.3.3 The "buried alive" hypothesis

The observation that anti-neoplastic drugs may trigger apoptosis of cancer cells was first reported by Searle et al. (1975). Today, we know that a wide variety of chemotherapeutic agents induce apoptosis in tumor cells (Makin and Dive, 2001). However, classic chemotherapeutic agents are associated with tumor cell resistance, toxicity due to bystander effects, and occasionally secondary neoplasia. We have recently presented a provocative hypothesis that enforced phagocytosis in the absence of a death signal may serve as an efficient means of deleting cancer cells without the associated bystander effects observed during conventional treatment (Fadeel et al., 2004b). In other words, we propose that cancer cells may be "buried alive" upon exposition of appropriate macrophage recognition signals.

Schroit and colleagues (1985) have shown that red blood cells containing an exogenous PS analogue in their plasma membrane are rapidly cleared from the peripheral circulation of syngeneic mice, suggesting that viable cells can be engulfed if they express appropriate "eat me" signals. Moreover, our *in vitro* studies have provided further impetus for this proposal. For instance, *N*-ethylmaleimide (NEM) treatment of the BL cell line, Raji, not only triggers PS oxidation and externalization, but also results in the efficient engulfment of these cells by macrophages, despite the absence of other indices of apoptosis, such as caspase activation and DNA fragmentation (Kagan et al., 2002). Furthermore, enrichment of the plasma membrane of non-apoptotic tumor cells of lymphoid (Raji, Jurkat) and myeloid (HL-60) origin with exogenous PS and/or PS-OX suffices to induce macrophage engulfment of these cells. In line with these findings, earlier studies have shown that undifferentiated murine erythroleukemic cells and certain tumorigenic cell lines of human origin express increased amounts of PS on the cell surface while in a viable state; PS expression correlated with recognition and tethering by macrophages (Connor et al., 1989; Utsugi et al., 1991). Moreover, transfection of the phagocytosis receptor CD36 into human melanoma cells was reported to confer an increased capacity to ingest apoptotic cells, comparable to that exhibited by professional phagocytes (macrophages) (Ren et al., 1995). Taken together, these data support our hypothesis that the expression of recognition signals (such as PS and/or PS-OX) on target cells, as well as the enforced expression of their cognate receptors on phagocytic cells, could enhance clearance of "un-dead" cancer cells. Indeed, one could take this a step further and propose a model whereby the concomitant induction of phagocytic receptors and recognition signals in cancer cells might result in self-engulfment of the tumor, i.e. cancer clearance by the cancer itself (Fadeel et al., 2004b). Recent *in vitro* studies indicate that cancer cells are capable of devouring their homotypic, apoptotic neighbors; interestingly, the efficiency of engulfment was determined, in part, by the dying cells themselves (Simamura et al., 2001; Wiegand et al., 2001). Future studies are needed to determine whether the "buried alive" hypothesis also holds true for *in vivo* models of tumor clearance.

### 3.3.4 Autophagy (self-digestion)

A tremendous amount of information has accumulated in the last decades regarding the molecular mechanisms that govern apoptotic cell death. However, programmed cell death (PCD) does not always occur by apoptosis (Schwartz et al., 1993; Sperandio et al., 2000). Indeed, alternative types of PCD and cell clearance were classified some 30 years ago, based on their respective morphologies (Schweichel and Merker, 1973). Hence, type I PCD corresponds to "classical" apoptosis and is usually caspase-dependent; apoptotic cells are swiftly removed by neighboring cells or macrophages prior to the rupture of the plasma membrane, as discussed above. Type II PCD or autophagic cell death, on the other hand, refers to self-digestion of cellular components through the lysosomal system of the same cell; in other words, these cells are essentially "cannabilized" from inside. Type III PCD is defined as non-lysosomal cellular degradation, and will not be discussed further in the present review. Importantly, cells undergoing autophagy seem to contain the machinery that is needed both to activate cell death and to degrade the dying cell (a process that occurs largely in the engulfing cell during apoptosis). Moreover, while apoptosis typically affects scattered single cells, autophagy is often observed when groups of contiguous cells or entire tissues die (Wyllie et al., 1980; Baehrecke, 2002). However, both forms of PCD appear to play a role during normal development; furthermore, recent studies indicate that these two types of PCD may be intricately connected also at the molecular level (Cohen et al., 2002).

Recent data suggest a role for autophagy in cancer development (for a recent review, see Gozuacik and Kimchi, 2004). Liang et al. (1999) have shown that *beclin 1* (the mammalian homolog of the yeast autophagy gene, *Apg6*) triggers autophagy in human breast carcinoma cells; this autophagy-promoting activity of beclin 1 was associated with inhibition of cellular proliferation *in vitro*, and with tumorigenesis in nude mice. To investigate whether *beclin 1* acts as a tumor suppressor and whether loss of *beclin 1* would contribute to an increased incidence of cancer, these investigators, and another independent laboratory, recently generated *beclin 1*-deficient mice (Qu et al., 2003; Yue et al., 2003). These studies demonstrate that loss of *beclin 1* is associated with a reduction in autophagic vacuole formation, that beclin 1-mediated regulation of autophagy is required for normal development, and that animals with reduced levels of beclin 1 display a pronounced increase in hematopoietic and other malignancies. Hence, *beclin 1* acts as a tumor suppressor gene, and autophagic cell death (self-digestion) appears to be an important means to prevent cellular transformation (Edinger and Thompson, 2003).

Some cancer cells respond to chemotherapeutic agents and irradiation by undergoing autophagy, rather than apoptosis (Bursch et al., 2000; Kanzawa et al., 2003; Paglin et al., 2003). These data thus indicate the potential utility of autophagic cell death induction in the treatment of cancer (Okada and Mak, 2004). However, it is important to consider the recent observation that caspase inhibition can trigger beclin 1-dependent, autophagic cell death (Yu et al., 2004). These findings indicate that therapeutic caspase inhibition could have the untoward effect of exacerbating cell death by activating an alternative program of cell degradation. Further studies are thus needed to investigate the putative "molecular switches" between type I PCD and type II PCD, and the role of such molecules in cancer development. Moreover, the design of cancer therapies capable of targeting the machinery of both apoptotic and autophagic cell death may be particularly advantageous.

# 4. CONCLUDING REMARKS

An appreciation of the molecular mechanisms that regulate the attraction, recognition, and degradation of apoptotic cells has emerged in recent years (Figure 1). The importance of this programmed cell clearance for the resolution of inflammation and the prevention of autoimmune responses is also evident. However, there is a paucity of studies on the mechanisms and consequences of clearance of cancer cells; in particular, our understanding of cancer cell clearance *in vivo* remains limited. Exciting recent studies show that horizontal transfer of genetic information may occur through engulfment of apoptotic bodies within a tumor population (so-called apoptotic conversion). However, more mechanistic studies of cancer cell clearance are required, and may unveil novel targets for cancer treatment. Another emerging theme is that cell clearance is not a passive event. Macrophages play a crucial role in the remodeling of tissues, not only as scavengers of apoptotic debris, but also through the active induction of apoptosis in certain tissues (Lang and Bishop, 1993). Moreover, even the dying cell itself may contribute to tissue homeostasis by promoting compensatory proliferation in response to cell death (Huh et al., 2004). The latter findings thus suggest that dying cells may communicate cell fate or behaviour instructions to their neighbors. Finally, evidence has accrued for alternative, non-apoptotic forms of cell death, including autophagy (self-digestion), so-called paraptosis, and necrosis-like PCD (Leist and Jäättelä, 2001). Future studies should focus on the molecular regulation of these divergent modes of cell demise, and on the harnessing of these signaling pathways for therapeutic purposes in the treatment of cancer and other diseases.

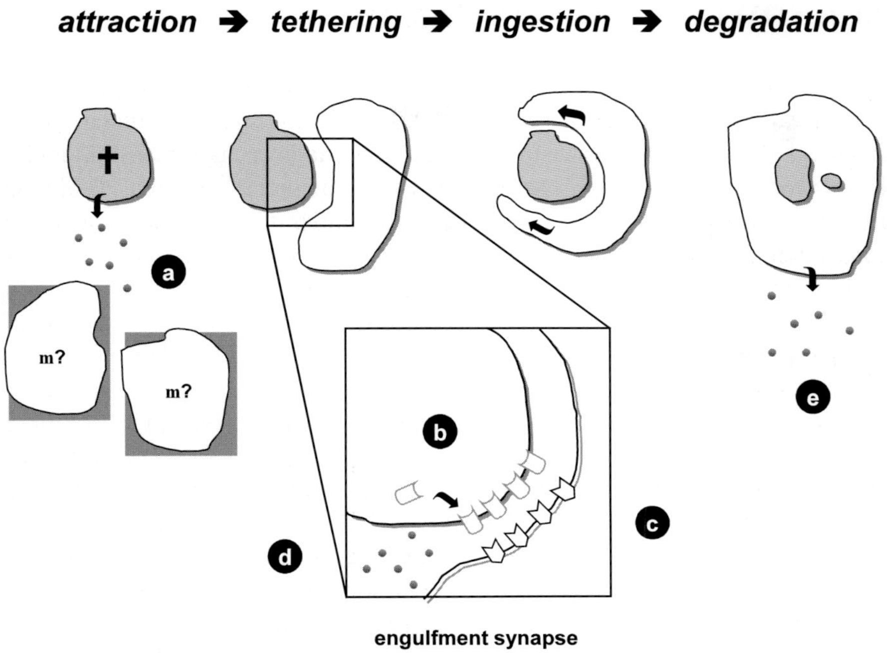


*Figure 1.* Programmed cell clearance.Mammalian cells dying by apoptosis emit chemotactic factors (a) that serve to recruit macrophages to the site of cell attrition. The engulfment of apoptotic cells by phagocytes is regulated by numerous recognition signals (b) on the apoptotic cell corpse and corresponding receptors (c) on the engulfing cell; soluble bridging molecules (d) serve to facilitate this recognition process. In addition, repulsion signals on the apoptotic cell are disabled, thus preventing the active rejection of phagocytes. Recent studies suggest that recognition involves an initial step of tethering (attachment) with subsequent ingestion of cell corpses; the engagement of multiple receptor-ligand pairs within the so-called engulfment synapse ensures the specific interaction between phagocyte and apoptotic prey. Following ingestion and degradation of apoptotic cells, macrophages produce cytokines (e) that contribute to the resolution of inflammation and thus precludes scarring of the surrounding tissue. Consult text for details.

*We call it a grain of sand,*
*but it calls itself neither grain nor sand.*
*It does fine without a name,*
*whether general, particular,*
*permanent, passing,*
*incorrect, or apt.*

*(Wislawa Szymborska)*

## ACKNOWLEDGEMENTS

I thank the Swedish Research Council, the Swedish Children's Cancer Foundation, the Swedish Medical Society, and the Jeansson Foundation for financial support. I also wish to thank Dr. Magnus Ingelman-Sundberg, Karolinska Institutet, for continuous encouragement, and Dr. Valerian Kagan, University of Pittsburgh, for illuminating discussions on cell clearance.

## REFERENCES

1. Akakura, S., Singh, S., Spataro, M., Akakura, R., Kim, J.I., Albert, M.L., and Birge, R.B. 2004. The opsonin MFG-E8 is a ligand for the $\alpha_v\beta_5$ integrin and triggers DOCK180-dependent Rac1 activation for the phagocytosis of apoptotic cells. Exp. Cell Res. **292**:403-416.
2. Albert, M.L., Pearce, S.F., Francisco, L.M., Sauter, B., Roy, P., Silverstein, R.L., and Bhardwaj, N. 1998a. Immature dendritic cells phagocytose apoptotic cells via $\alpha_v\beta_5$ and CD36, and cross-present antigens to cytotoxic T lymphocytes. J. Exp. Med. **188**:1359-1368.
3. Albert, M.L., Sauter, B., and Bhardwaj, N. 1998b. Dendritic cells acquire antigen from apoptotic cells and induce class I-restricted CTLs. Nature **392**:86-89.
4. Anderson, H.A., Maylock, C.A., Williams, J.A., Paweletz, C.P., Shu, H., and Shacter, E. 2003. Serum-derived protein S binds to phosphatidylserine and stimulates the phagocytosis of apoptotic cells. Nat. Immunol. **4**:87-91.
5. Arroyo, A., Modriansky, M., Serinkan, F.B., Bello, R.I., Matsura, T., Jiang, J., Tyurin, V.A., Tyurina, Y.Y., Fadeel, B., and Kagan, V.E. 2002. NADPH oxidase-dependent oxidation and externalization of phosphatidylserine during apoptosis in $Me_2SO$-differentiated HL-60 cells. Role in phagocytic clearance. J. Biol. Chem. **277**:49965-49975.
6. Arur, S., Uche, U.E., Rezaul, K., Fong, M., Scranton, V., Cowan, A.E., Mohler, W., and Han, D.K. 2003. Annexin I is an endogenous ligand that mediates apoptotic cell engulfment. Dev. Cell **4**:587-598.
7. Baehrecke, E.H. 2002. How death shapes life during development. Nat. Rev. Mol. Cell Biol. **3**:779-787.
8. Balasubramanian, K., Chandra, J., and Schroit, A.J. 1997. Immune clearance of phosphatidylserine-expressing cells by phagocytes. The role of β2-glycoprotein I in macrophage recognition. J. Biol. Chem. **272**:31113-31117.
9. Balasubramanian, K., and Schroit, A.J. 1998. Characterization of phosphatidylserine-dependent β2-glycoprotein I macrophage interactions. Implications for apoptotic cell clearance by phagocytes. J. Biol. Chem. **273**:29272-29277.
10. Bergsmedh, A., Szeles, A., Henriksson, M., Bratt, A., Folkman, M.J., Spetz, A.L., and Holmgren, L. 2001. Horizontal transfer of oncogenes by uptake of apoptotic bodies. Proc. Natl. Acad. Sci. U. S. A. **98**:6407-6411.
11. Bergsmedh, A., Szeles, A., Spetz, A.L., and Holmgren, L. 2002. Loss of the p21(Cip1/Waf1) cyclin kinase inhibitor results in propagation of horizontally transferred DNA. Cancer Res. **62**:575-579.

12. Bernuth, H., Kulka, C., Roesler, J., Gahr, M., and Rosen-Wolff, A. 2004. NADPH oxidase is not required for spontaneous and Staphylococcus aureus-induced apoptosis of monocytes. Ann. Hematol. **83**:206-211.
13. Blom, W.M., de Bont, H.J., and Nagelkerke, J.F. 2003. Regional loss of the mitochondrial membrane potential in the hepatocyte is rapidly followed by externalization of phosphatidylserines at that specific site during apoptosis. J. Biol. Chem. **278**:12467-12474.
14. Borisenko, G.G., Iverson, S.L., Ahlberg, S., Kagan, V.E., and Fadeel, B. 2004. Milk fat globule epidermal growth factor 8 (MFG-E8) binds to oxidized phosphatidylserine: implications for macrophage clearance of apoptotic cells. Cell Death Differ. **11**:943-945.
15. Bottazzi, B., Polentarutti, N., Acero, R., Balsari, A., Boraschi, D., Ghezzi, P., Salmona, M., and Mantovani, A. 1983. Regulation of the macrophage content of neoplasms by chemoattractants. Science **220**:210-212.
16. Botto, M., Dell'Agnola, C., Bygrave, A.E., Thompson, E.M., Cook, H.T., Petry, F., Loos, M., Pandolfi, P.P., and Walport, M.J. 1998. Homozygous C1q deficiency causes glomerulonephritis associated with multiple apoptotic bodies. Nat. Genet. **19**:56-59.
17. Bratton, D.L., Fadok, V.A.,. Richter, D.A, Kailey, J.M., Guthrie, L.A., and Henson, P.M. 1997. Appearance of phosphatidylserine on apoptotic cells requires calcium-mediated nonspecific flip-flop and is enhanced by loss of the aminophospholipid translocase. J. Biol. Chem. **272**:26159-26165.
18. Brown, J.R., Goldblatt, D., Buddle, J., Morton, L., and Thrasher, A.J. 2003. Diminished production of anti-inflammatory mediators during neutrophil apoptosis and macrophage phagocytosis in chronic granulomatous disease (CGD). J. Leukoc. Biol. **73**:591-599.
19. Brown. S., Heinisch. I., Ross. E., Shaw. K., Buckley. C.D., and Savill, J. 2002. Apoptosis disables CD31-mediated cell detachment from phagocytes promoting binding and engulfment. Nature **418**:200-203.
20. Bursch, W., Hochegger, K., Torok, L., Marian, B., Ellinger, A., and Hermann, R.S. 2000. Autophagic and apoptotic types of programmed cell death exhibit different fates of cytoskeletal filaments. J. Cell Sci. **113**:1189-1198.
21. Byrne, A., and Reen, D.J. 2002. Lipopolysaccharide induces rapid production of IL-10 by monocytes in the presence of apoptotic neutrophils. J. Immunol. **168**:1968-1977.
22. Casciola-Rosen, L.A., Anhalt, G., and Rosen, A. 1994. Autoantigens targeted in systemic lupus erythematosus are clustered in two populations of surface structures on apoptotic keratinocytes. J. Exp. Med. **179**:1317-1330.
23. Casciola-Rosen, L.A., Anhalt, G.J., and Rosen, A. 1995. DNA-dependent protein kinase is one of a subset of autoantigens specifically cleaved early during apoptosis. J. Exp. Med. **182**:1625-1634.
24. Casiano, C.A., Martin, S.J., Greenm D.R., and Tan E.M. 1996. Selective cleavage of nuclear autoantigens during CD95 (Fas/APO-1)-mediated T cell apoptosis. J. Exp. Med. **184**:765-770.
25. Chang, M.K., Bergmark, C., Laurila, A., Horkko, S., Han, K.H., Friedman, P., Dennis, E.A., and Witztum, J.L. 1999. Monoclonal antibodies against oxidized low-density lipoprotein bind to apoptotic cells and inhibit their phagocytosis by elicited macrophages: evidence that oxidation-specific epitopes mediate macrophage recognition. Proc. Natl. Acad. Sci. U. S. A. **96**:6353-6358.
26. Chang, M.K., Binder, C.J., Torzewskim M., and Witztum, J.L. 2002. C-reactive protein binds to both oxidized LDL and apoptotic cells through recognition of a common ligand: phosphorylcholine of oxidized phospholipids. Proc. Natl. Acad. Sci. U. S. A. **99**:13043-13048.

27. Cikala, M., Alexandrova, O., David, C.N., Proschel, M., Stiening, B., Cramer, P., and Bottger, A. 2004. The phosphatidylserine receptor from Hydra is a nuclear protein with potential Fe(II) dependent oxygenase activity. BMC Cell Biol. **5**(1):26.
28. Cohen, P.L., Caricchio, R., Abraham, V., Camenisch, T.D., Jennette, J.C., Roubey, R.A., Earp, H.S., Matsushima, G., and Reap, E.A. 2002. Delayed apoptotic cell clearance and lupus-like autoimmunity in mice lacking the c-mer membrane tyrosine kinase. J. Exp. Med. **196**:135-140.
29. Cohen, I., Castedo, M., and Kroemer, G. 2002. Tantalizing Thanatos: unexpected links in death pathways. Trends Cell Biol. **12**:293-295.
30. Connor, J., Bucana, C., Fidler, I.J., and Schroit, A.J. 1989. Differentiation-dependent expression of phosphatidylserine in mammalian plasma membranes: quantitative assessment of outer-leaflet lipid by prothrombinase complex formation. Proc. Natl. Acad. Sci. U. S. A. **86**:3184-3188.
31. Cui, P., Qin, B., Liu, N., Pan, G., Pei, D. 2004. Nuclear localization of the phosphatidylserine receptor protein via multiple nuclear localization signals. Exp. Cell Res. **293**:154-163.
32. Danial, N.N., and Korsmeyer, S.J. 2004. Cell death: critical control points. Cell **116**:205-219.
33. Darji, A., Guzman, C.A., Gerstel, B., Wachholz, P., Timmis, K.N., Wehland, J., Chakraborty, T., and Weiss, S. 1997. Oral somatic transgene vaccination using attenuated S. typhimurium. Cell **91**:765-775.
34. de la Taille, A., Chen, M.W., Burchardt, M., Chopin, D.K., and Buttyan, R. 1999. Apoptotic conversion: evidence for exchange of genetic information between prostate cancer cells mediated by apoptosis. Cancer Res. **59**:5461-5463.
35. De Laurenzi, V., and Melino, G. 2001. Gene disruption of tissue transglutaminase. Mol. Cell Biol. **21**:148-155.
36. Dillon, S.R., Mancini, M., Rosen, A., and Schlissel, M.S. 2000. Annexin V binds to viable B cells and colocalizes with a marker of lipid rafts upon B cell receptor activation. J. Immunol. **164**:1322-1332.
37. Dogusan, Z., Montecino-Rodriguez, E., and Dorshkind, K. 2004. Macrophages and stromal cells phagocytose apoptotic bone marrow-derived B lineage cells. J. Immunol. **172**:4717-4723.
38. Eccles, S.A., and Alexander, P. 1974. Macrophage content of tumours in relation to metastatic spread and host immune reaction. Nature **250**:667-669.
39. Edinger, A.L., and Thompson, C.B. 2003. Defective autophagy leads to cancer. Cancer Cell **4**:422-424.
40. Fadeel, B. 2003. Programmed cell clearance. Cell. Mol. Life Sci. **60**:2575-2585.
41. Fadeel, B. 2004. Plasma membrane alterations during apoptosis: role in corpse clearance. Antioxid. Redox Signal. **6**:269-275.
42. Fadeel, B., and Kagan, V.E. 2003. Apoptosis and macrophage clearance of neutrophils: regulation by reactive oxygen species. Redox Report **8**:143-150.
43. Fadeel, B., Karpova, M.B., Enoksson, M., and Orrenius, S. 2004a. Phosphatidylserine externalization in cardiolipin-deficient cells. Blood (in press).
44. Fadeel, B., Orrenius, S., and Pervaiz, S. 2004b. Buried alive: a novel approach to cancer treatment. FASEB J. **18**:1-4.
45. Fadeel, B., Åhlin, A., Henter, J-I., Orrenius, S., and Hampton, M.B. 1998. Involvement of caspases in neutrophil apoptosis: regulation by reactive oxygen species. Blood **92**:4808-4818.
46. Fadok. V.A., Bratton D.L., Konowal, A., Freed, P.W., Westcott, J.Y., and Henson, P.M. 1998b. Macrophages that have ingested apoptotic cells in vitro inhibit proinflammatory

cytokine production through autocrine/paracrine mechanisms involving TGF-β, PGE2, and PAF. J. Clin. Invest. **101**:890-898.

47. Fadok, V.A., Warner, M.L., Bratton, D.L., and Henson, P.M. 1998a. CD36 is required for phagocytosis of apoptotic cells by human macrophages that use either a phosphatidylserine receptor or the vitronectin receptor ($\alpha_v\beta_3$). J. Immunol. **161**:6250-6257.
48. Fadok, V.A., Bratton, D.L., Rose, D.M., Pearson, A., Ezekewitz, R.A., and Henson, P.M. 2000. A receptor for phosphatidylserine-specific clearance of apoptotic cells. Nature **405**:85-90.
49. Fadok, V.A., de Cathelineau, A., Daleke, D.L., Henson, P.M., and Bratton, D.L. 2001. Loss of phospholipid asymmetry and surface exposure of phosphatidylserine is required for phagocytosis of apoptotic cells by macrophages and fibroblasts. J. Biol. Chem. **276**:1071-1077.
50. Fadok, V.A., Voelker, D.R., Campbell, P.A., Cohenm J.J., Bratton, D.L., and Henson P.M. 1992. Exposure of phosphatidylserine on the surface of apoptotic lymphocytes triggers specific recognition and removal by macrophages. J. Immunol. **148**:2207-2216.
51. Fan, X., Krahling, S., Smith, D., Williamson, P., and Schlegel, R.A. 2004. Macrophage surface expression of annexins I and II in the phagocytosis of apoptotic lymphocytes. Mol Biol. Cell **15**:2863-2872.
52. Franc, N.C., Heitzler, P., Ezekowitz, R.A., and White, K. 1999. Requirement for croquemort in phagocytosis of apoptotic cells in Drosophila. Science **284**:1991-1994.
53. Fujita, S., Buziba, N., Kumatori, A., Senba, M., Yamaguchi, A., and Toriyama, K. 2004. Early stage of Epstein-Barr virus lytic infection leading to the "starry sky" pattern formation in endemic Burkitt lymphoma. Arch. Pathol. Lab. Med. **128**:549-552.
54. Furumoto, K., Soares, L., Engleman, E.G., and Merad, M. 2004. Induction of potent antitumor immunity by in situ targeting of intratumoral DCs. J. Clin. Invest. **113**:774-783.
55. Fushimi, T., Kojima, A., Moore, M.A., and Crystal, R.G. 2000. Macrophage inflammatory protein 3α transgene attracts dendritic cells to established murine tumors and suppresses tumor growth. J. Clin. Invest. **105**:1383-1393.
56. Gaipl, U.S., Kuenkele, S., Voll, R.E., Beyer, T.D., Kolowos, W., Heyder, P., Kalden, J.R., and Herrmann, M. 2001. Complement binding is an early feature of necrotic and a rather late event during apoptotic cell death. Cell Death Differ. **8**:327-334.
57. Gardai, S.J., Xiao, Y.Q., Dickinson, M., Nick, J.A., Voelker, D.R., Greene, K.E., and Henson, P.M. 2003. By binding SIRPα or calreticulin/CD91, lung collectins act as dual function surveillance molecules to suppress or enhance inflammation. Cell **115**:13-23.
58. Gershov, D., Kim, S., Brot, N., and Elkon, K.B. 2000. C-Reactive protein binds to apoptotic cells, protects the cells from assembly of the terminal complement components, and sustains an antiinflammatory innate immune response: implications for systemic autoimmunity. J. Exp. Med. **192**:1353-1364.
59. Gleiss, B., Gogvadze, V., Orrenius, S., and Fadeel, B. 2002. Fas-triggered phosphatidylserine exposure is modulated by intracellular ATP. FEBS Lett. **519**:153-158.
60. Goldszmid, R.S., Idoyaga, J., Bravo, A.I., Steinman, R., Mordoh, J., and Wainstok, R. 2003. Dendritic cells charged with apoptotic tumor cells induce long-lived protective $CD4^+$ and $CD8^+$ T cell immunity against B16 melanoma. J. Immunol. **171**:5940-5947.
61. Gozuacik, D., and Kimchi, A. 2004. Autophagy as a cell death and tumor suppressor mechanism. Oncogene **23**:2891-2906.

62. Graeber, T.G., Osmanian, C., Jacks, T., Housman, D.E., Koch, C.J., Lowe, S.W., and Giaccia, A.J. 1996. Hypoxia-mediated selection of cells with diminished apoptotic potential in solid tumours. Nature **379**:88-91.
63. Graves, D.T., Jiang, Y.L., Williamson, M.J., Valente, A.J. 1989. Identification of monocyte chemotactic activity produced by malignant cells. Science **245**:1490-1493.
64. Gregory, C.D. 2000. CD14-dependent clearance of apoptotic cells: relevance to the immune system. Curr. Opin. Immunol. **12**: 27-34.
65. Grimsley, C., and Ravichandran, K.S. 2003. Cues for apoptotic cell engulfment: eat-me, don't eat-me and come-get-me signals. Trends Cell Biol. **13**:648-656.
66. Hamon, Y., Broccardo, C., Chambenoit, O., Luciani, M.F., Toti, F., Chaslin, S., Freyssinet, J.M., Devaux, P.F., McNeish, J., Marguet, D., and Chimini, G. 2000. ABC1 promotes engulfment of apoptotic cells and transbilayer redistribution of phosphatidylserine. Nat. Cell Biol. **2**:399-406.
67. Hampton, M.B., Kettle, A.J., and Winterbourn, C.C. 1998. Inside the neutrophil phagosome: oxidants, myeloperoxidase, and bacterial killing. Blood **92**:3007-3017.
68. Hampton, M.B., Vissers, M.C., Keenan, J.I., and Winterbourn, C.C. 2002. Oxidant-mediated phosphatidylserine exposure and macrophage uptake of activated neutrophils: possible impairment in chronic granulomatous disease. J. Leukoc. Biol. **71**:775-781.
69. Hanayama. R., Tanaka. M., Miwa, K., Shinohara, A., Iwamatsu, A., and Nagata, S. 2002. Identification of a factor that links apoptotic cells to phagocytes. Nature **417**:182-187.
70. Hanayama, R., Tanaka, M., Miwa, K., and Nagata, S. 2004a. Expression of developmental endothelial locus-1 in a subset of macrophages for engulfment of apoptotic cells. J. Immunol. **172**:3876-3882.
71. Hanayama, R., Tanaka, M., Miyasaka, K., Aozasa, K., Koike, M., Uchiyama, Y., and Nagata, S. 2004b. Autoimmune disease and impaired uptake of apoptotic cells in MFG-E8-deficient mice. Science **304**:1147-1150.
72. Hannon, R., Croxtall, J.D., Getting, S.J., Roviezzo, F., Yona, S., Paul-Clark, M.J., Gavins, F.N., Perretti, M., Morris, J.F., Buckingham, J.C., and Flower, R.J. 2003. Aberrant inflammation and resistance to glucocorticoids in annexin $1^{-/-}$ mouse. FASEB J. **17**:253-255.
73. Hart, S.P., Smith, J.R., and Dransfield, I. 2004. Phagocytosis of opsonized apoptotic cells: roles for "old-fashioned" receptors for antibody and complement. Clin. Exp. Immunol. **135**:181-185.
74. Haskins, K., Bradley, B., Powers, K., Fadok, V., Flores, S., Ling, X., Pugazhenthi, S., Reusch, J., and Kench, J. 2003. Oxidative stress in type 1 diabetes. Ann. N. Y. Acad. Sci. **1005**:43-54.
75. Henson, P.M., Bratton, D.L., and Fadok, V.A. 2001. The phosphatidylserine receptor: a crucial molecular switch? Nat. Rev. Mol. Cell. Biol. **2**:627-633.
76. Hermann, M., Niemitz, C., Marafioti, T., and Schriever, F. 1998. Reduced phagocytosis of apoptotic cells in malignant lymphoma. Int. J. Cancer **75**:675-679.
77. Herrmann, M., Voll, R.E., Zoller, O.M., Hagenhofer, M., Ponner, B.B., and Kalden, J.R. 1998. Impaired phagocytosis of apoptotic cell material by monocyte-derived macrophages from patients with systemic lupus erythematosus. Arthritis Rheum. **41**:1241-1250.
78. Hisatomi, T., Sakamoto, T., Sonoda, K.H., Tsutsumi, C., Qiao, H., Enaida, H., Yamanaka, I., Kubota, T., Ishibashi, T., Kura, S., Susin, S.A., and Kroemer, G. 2003. Clearance of apoptotic photoreceptors: elimination of apoptotic debris into the subretinal space and macrophage-mediated phagocytosis via phosphatidylserine receptor and integrin $\alpha_v\beta_3$. Am. J. Pathol. **162**:1869-1879.

79. Hoeppner, D.J., Hengartner, M.O., and Schnabel, R. 2001. Engulfment genes cooperate with ced-3 to promote cell death in Caenorhabditis elegans. Nature **412**:202-206.
80. Hoffman, R.D., Kligerman, M., Sundt, T.M., Anderson, N.D., and Shin, H.S. 1982. Stereospecific chemoattraction of lymphoblastic cells by gradients of lysophosphatidylcholine. Proc. Natl. Acad. Sci. U. S. A. **79**:3285-3289.
81. Hoffmann, P.R., de Cathelineau, A.M., Ogden, C.A., Leverrier, Y., Bratton, D.L., Daleke, D.L., Ridley, A.J., Fadok, V.A., and Henson, P.M. 2001. Phosphatidylserine (PS) induces PS receptor-mediated macropinocytosis and promotes clearance of apoptotic cells. J. Cell Biol. **155**:649-659.
82. Hoffmeister, K.M., Felbinger, T.W., Falet, H., Denis, C.V., Bergmeier, W., Mayadas, T.N., von Andrian, U.H., Wagner, D.D., Stossel, T.P., Hartwig, J.H. 2003. The clearance mechanism of chilled blood platelets. Cell **112**:87-97.
83. Holmgren, L., O'Reilly, M.S., and Folkman, J. 1995. Dormancy of micrometastases: balanced proliferation and apoptosis in the presence of angiogenesis suppression. Nat. Med. **1**:149-153.
84. Holmgren, L., Szeles, A., Rajnavolgyi, E., Folkman, J., Klein, G., Ernberg, I., Falk, K.I. 1999. Horizontal transfer of DNA by the uptake of apoptotic bodies. Blood **93**:3956-3963.
85. Homma, S., Yamamoto, M., and Yamamoto, N. 1993. Vitamin D-binding protein (group-specific component) is the sole serum protein required for macrophage activation after treatment of peritoneal cells with lysophosphatidylcholine. Immunol. Cell. Biol. **71**:249-257.
86. Horino, K., Nishiura, H., Ohsako, T., Shibuya, Y., Hiraoka, T., Kitamura, N., and Yamamoto, T. 1998. A monocyte chemotactic factor, S19 ribosomal protein dimer, in phagocytic clearance of apoptotic cells. Lab. Invest. **78**:603-617.
87. Huber, J., Vales, A., Mitulovic, G., Blumer, M., Schmid, R., Witztum, J.L., Binder, B.R., and Leitinger, N. 2002. Oxidized membrane vesicles and blebs from apoptotic cells contain biologically active oxidized phospholipids that induce monocyte-endothelial interactions. Arterioscler. Thromb. Vasc. Biol. **22**:101-107.
88. Huh, J.R., Guo, M., and Hay, B.A. 2004. Compensatory proliferation induced by cell death in the Drosophila wing disc requires activity of the apical cell death caspase dronc in a nonapoptotic role. Curr. Biol. **14**:1262-1266.
89. Huynh, M.L., Fadok, V.A., and Henson P.M. 2002. Phosphatidylserine-dependent ingestion of apoptotic cells promotes TGF-β1 secretion and the resolution of inflammation. J. Clin. Invest. **109**:41-50.
90. Jacobson, M.D., Weil, M., and Raff M.C. 1997. Programmed cell death in animal development. Cell **88**:347-354.
91. Jain, R., Rivera, M.C., Moore, J.E., and Lake, J.A. 2002. Horizontal gene transfer in microbial genome evolution. Theor. Popul. Biol. **61**:489-495.
92. Jiang, J., Serinkan, B.F., Tyurina, Y.Y., Borisenko, G.G., Mi, Z., Robbins, P.D., Schroit, A.J., and Kagan, V.E. 2003. Peroxidation and externalization of phosphatidylserine associated with release of cytochrome c from mitochondria. Free Radic. Biol. Med. **35**:814-825.
93. Kagan V.E., Borisenko, G.G., Serinkan, B.F., Tyurina, Y.Y., Tyurin, V.A., Jiang, J., Liu, S.X., Shvedova, A.A., Fabisiak, J.P., Uthaisang, W., and Fadeel, B. 2003. Appetizing rancidity of apoptotic cells for macrophages: oxidation, externalization, and recognition of phosphatidylserine. Am. J. Physiol. (Lung Cell. Mol. Physiol.) **285**:1-17.
94. Kagan, V.E., Gleiss, B., Tyurina, Y.Y., Tyurin, V.A., Elenström-Magnusson, C., Liu, S.X., Serinkan, F.B., Arroyo, A., Chandra, J., Orrenius, S., and Fadeel, B. 2002. A role for oxidative stress in apoptosis: oxidation and externalization of phosphatidylserine is

required for macrophage clearance of cells undergoing Fas-mediated apoptosis. J. Immunol. **169**:487-499.

95. Kanzawa, T., Kondo, Y., Ito, H., Kondo, S., and Germano, I. 2003. Induction of autophagic cell death in malignant glioma cells by arsenic trioxide. Cancer Res. **63**:2103-2108.
96. Kasahara, Y., Iwai, K., Yachie, A., Ohta, K., Konno, A., Seki, H., Miyawaki, T., and Taniguchi, N. 1997. Involvement of reactive oxygen intermediates in spontaneous and CD95 (Fas/APO-1)-mediated apoptosis of neutrophils. Blood **89**:1748-1753.
97. Kask, L., Trouw, L.A., Dahlbäck, B., Blom, A.M. 2004. The C4b-binding protein-protein S complex inhibits the phagocytosis of apoptotic cells. J. Biol. Chem. **279**:23869-23873.
98. Kawane, K., Fukuyama, H., Kondoh, G., Takeda, J., Ohsawa, Y., Uchiyama, Y., and Nagata, S. 2001. Requirement of DNase II for definitive erythropoiesis in the mouse fetal liver. Science **292**:1546-1549.
99. Kawane, K., Fukuyama, H., Yoshida, H., Nagase, H., Ohsawa, Y., Uchiyama, Y., Okada, K., Iida, T., and Nagata, S. 2003. Impaired thymic development in mouse embryos deficient in apoptotic DNA degradation. Nat. Immunol. **4**:138-144.
100. Kerr, J.F.R., Wyllie, A.H., and Currie, A.R. 1972. Apoptosis: a basic biological phenomenon with wide-ranging implications in tissue kinetics. Br. J. Cancer **26**:239-257.
101. Kim, S.J, Gershov, D., Ma, X., Brot, N., and Elkon, K.B. 2002. I-$PLA_2$ activation during apoptosis promotes the exposure of membrane lysophosphatidylcholine leading to binding by natural immunoglobulin M antibodies and complement activation. J. Exp. Med. **196**:655-665.
102. Klein, E., and Mantovani, A. 1993. Action of natural killer cells and macrophages in cancer. Curr. Opin. Immunol. **5**:714-718.
103. Korb, L.C., and Ahearn, J.M. 1997. C1q binds directly and specifically to surface blebs of apoptotic human keratinocytes: complement deficiency and systemic lupus erythematosus revisited. J. Immunol. **158**:4525-4528.
104. Koty, P.P., Tyurina, Y.Y., Tyurin, V.A., Li, S.X., and Kagan, V.E. 2002. Depletion of Bcl-2 by an antisense oligonucleotide induces apoptosis accompanied by oxidation and externalization of phosphatidylserine in NCI-H226 lung carcinoma cells. Mol. Cell. Biochem. **234-235**:125-133.
105. Kunisaki, Y., Masuko, S., Noda, M., Inayoshi, A., Sanui, T., Harada, M., Sasazuki, T., and Fukui, Y. 2004. Defective fetal liver erythropoiesis and T lymphopoiesis in mice lacking the phosphatidylserine receptor. Blood **103**:3362-3364.
106. Kurland, C.G., Canback, B., and Berg, O.G. 2003. Horizontal gene transfer: a critical view. Proc. Natl. Acad. Sci. U. S. A. **100**:9658-9662.
107. Lang, R.A., and Bishop, J.M. 1993. Macrophages are required for cell death and tissue remodeling in the developing mouse eye. Cell **74**:453-462.
108. Lauber, K., Blumenthal, S.G., Waibel, M., and Wesselborg, S. 2004. Clearance of apoptotic cells: getting rid of the corpses. Mol. Cell. **14**:277-287.
109. Lauber, K., Bohn, E., Krober, S.M., Xiao, Y.J., Blumenthal, S.G., Lindemann, R.K., Marini, P., Wiedig, C., Zobywalski, A., Baksh, S., Xu, Y., Autenrieth, I.B., Schulze-Osthoff, K., Belka, C., Stuhler, G., and Wesselborg, S. 2003. Apoptotic cells induce migration of phagocytes via caspase-3-mediated release of a lipid attraction signal. Cell **113**:717-730.
110. Leist, M., and Jäättelä, M. 2001. Four deaths and a funeral: from caspases to alternative mechanisms. Nat. Rev. Mol. Cell. Biol. **2**:589-598.

111. Levine, B., and Klionsky, D.J. 2004. Development by self-digestion: molecular mechanisms and biological functions of autophagy. Dev. Cell **6**:463-477.
112. Li, M.O., Sarkisian, M.R., Mehal, W.Z., Rakic, P., and Flavell, R.A. 2003. Phosphatidylserine receptor is required for clearance of apoptotic cells. Science **302**:1560-1563.
113. Liang, X.H., Jackson, S., Seaman, M., Brown, K., Kempkes, B., Hibshoosh, H., and Levine, B. 1999. Induction of autophagy and inhibition of tumorigenesis by beclin 1. Nature **402**:672-676.
114. Ma, K., Simantov, R., Zhang, J.C., Silverstein, R., Hajjar, K.A., and McCrae, K.R. 2000. High affinity binding of β2-glycoprotein I to human endothelial cells is mediated by annexin II. J. Biol. Chem. **275**:15541-15548.
115. Makin, G., and Dive, C. 2001. Apoptosis and cancer chemotherapy. Trends Cell Biol. **11**:22-26.
116. Mantovani, A., Bottazzi, B., Colotta, F., Sozzani, S., and Ruco, L. 1992. The origin and function of tumor-associated macrophages. Immunol. Today **13**:265-270.
117. Martin, S., Pombo, I., Poncet, P., David, B., Arock, M., and Blank, U. 2000. Immunologic stimulation of mast cells leads to the reversible exposure of phosphatidylserine in the absence of apoptosis. Int. Arch. Allergy Immunol. **123**:249-258.
118. Martin, S.J., Reutelingsperger, C.P., McGahon, A.J., Rader, J.A., van Schie, R. C., LaFace, D.M., and Green, D.R. 1995. Early redistribution of plasma membrane phosphatidylserine is a general feature of apoptosis regardless of the initiating stimulus: inhibition by overexpression of Bcl-2 and Abl. J. Exp. Med. **182**:1545-1556.
119. Matsura, T., Serinkan, B.F., Jiang, J., and Kagan, V.E. 2002. Phosphatidylserine peroxidation/externalization during staurosporine-induced apoptosis in HL-60 cells. FEBS Lett. **524**:25-30.
120. McMurray, H.F., Parthasarathy, S., and Steinberg, D. 1993. Oxidatively modified low density lipoprotein is a chemoattractant for human T lymphocytes. J. Clin. Invest. **92**:1004-1008.
121. Mevorach, D., Mascarenhas, J.O., Gershov, D., and Elkon K.B. 1998. Complement-dependent clearance of apoptotic cells by human macrophages. J. Exp. Med. **188**:2313-2320.
122. Miyasaka, K., Hanayama, R., Tanaka, M., and Nagata, S. 2004. Expression of milk fat globule epidermal growth factor 8 in immature dendritic cells for engulfment of apoptotic cells. Eur. J. Immunol. **34**:1414-1422.
123. Moodley, Y., Rigby, P., Bundell, C., Bunt, S., Hayashi, H., Misso, N., McAnulty, R., Laurent, G., Scaffidi, A., Thompson, P., and Knight, D. 2003. Macrophage recognition and phagocytosis of apoptotic fibroblasts is critically dependent on fibroblast-derived thrombospondin 1 and CD36. Am. J. Pathol. **162**:771-779.
124. Morelli, A.E., Larregina, A.T., Shufesky, W.J., Zahorchak, A.F., Logar, A.J., Papworth, G.D., Wang, Z., Watkins, S.C., Falo, L.D., and Thomson, A.W. 2003. Internalization of circulating apoptotic cells by splenic marginal zone dendritic cells: dependence on complement receptors and effect on cytokine production. Blood **101**:611-620.
125. Nakano, T., Ishimoto, Y., Kishino, J., Umeda, M., Inoue, K., Nagata, K., Ohashi, K., Mizuno, K., and Arita, H. 1997. Cell adhesion to phosphatidylserine mediated by a product of growth arrest-specific gene 6. J. Biol. Chem. **272**:29411-29414.
126. Nauta, A.J., Raaschou-Jensen, N., Roos, A., Daha, M.R., Madsen, H.O., Borrias-Essers, M.C., Ryder, L.P., Koch, C., and Garred, P. 2003. Mannose-binding lectin engagement with late apoptotic and necrotic cells. Eur. J. Immunol. **33**:2853-2863.

127. Nauta, A.J., Trouw, L.A., Daha, M.R., Tijsma, O., Nieuwland, R., Schwaeble, W.J., Gingras, A.R., Mantovani, A., Hack, E.C., and Roos, A. 2002. Direct binding of C1q to apoptotic cells and cell blebs induces complement activation. Eur. J. Immunol. **32**:1726-1736.
128. Nishimura, T., Horino, K., Nishiura, H., Shibuya, Y., Hiraoka, T., Tanase, S., and Yamamoto, T. 2001. Apoptotic cells of an epithelial cell line, AsPC-1, release monocyte chemotactic S19 ribosomal protein dimer. J. Biochem. (Tokyo) **129**:445-454.
129. O'Brien, B.A. Huang, Y., Geng, X., Dutz, J.P., and Finegood, D.T. 2002. Phagocytosis of apoptotic cells by macrophages from NOD mice is reduced. Diabetes **51**:2481-2488.
130. O'Brien, C.D., Ji, G., Wang, Y.X., Sun, J., Krymskaya, V.P., Ruberg, F.L., Kotlikoff, M.I., and Albelda, S.M. 2001. PECAM-1 (CD31) engagement activates a phosphoinositide-independent, nonspecific cation channel in endothelial cells. FASEB J. **15**:1257-1260.
131. Ogasawara, J., Watanabe-Fukunaga, R., Adachi, M., Matsuzawa, A., Kasugai, T., Kitamura, Y., Itoh, N., Suda, T., and Nagata, S. 1993. Lethal effect of the anti-Fas antibody in mice. Nature **364**:806-809.
132. Ogden, C.A., de Cathelineau, A., Hoffmann, P.R., Bratton, D., Ghebrehiwet, B., Fadok, V.A., and Henson, P.M. 2001. C1q and mannose binding lectin engagement of cell surface calreticulin and CD91 initiates macropinocytosis and uptake of apoptotic cells. J. Exp. Med. **194**:781-795.
133. Okada, H., and Mak, T.W. 2004. Pathways of apoptotic and non-apoptotic death in tumour cells. Nat. Rev. Cancer **4**:592-603.
134. Oldenborg, P.A., Zheleznyak, A., Fang, Y.F., Lagenaur, C. F., Gresham, H.D., and Lindberg, F.P. 2000. Role of CD47 as a marker of self on red blood cells. Science **288**:2051-2054.
135. Opdenakker, G., and Van Damme, J. 1992. Chemotactic factors, passive invasion and metastasis of cancer cells. Immunol. Today **13**:463-464.
136. Orrenius, S., Zhivotovsky, B., and Nicotera, P. 2003. Regulation of cell death: the calcium-apoptosis link. Nat. Rev. Mol. Cell. Biol. **4**:552-565.
137. Paglin, S., Hollister, T., Delohery, T., Hackett, N., McMahill, M., Sphicas, E., Domingo, D., and Yahalom, J. 2001. A novel response of cancer cells to radiation involves autophagy and formation of acidic vesicles. Cancer Res. **61**:439-444.
138. Parnaik, R., Raff, M.C., and Scholes, J. 2000. Differences between the clearance of apoptotic cells by professional and non-professional phagocytes. Curr. Biol. **10**:857-860.
139. Parrish, J.Z., and Xue, D. 2003. Functional genomic analysis of apoptotic DNA degradation in C. elegans. Mol. Cell **11**:987-996.
140. Pickering, M.C., Fischer, S., Lewis, M.R., Walport, M.J., Botto, M., and Cook, H.T. 2001. Ultraviolet-radiation-induced keratinocyte apoptosis in C1q-deficient mice. J. Invest. Dermatol. **117**:52-58.
141. Platt, N., da Silva, R.P., and Gordon, S. 1998. Recognizing death: the phagocytosis of apoptotic cells. Trends Cell Biol. **8**:365-372.
142. Qu, X., Yu, J., Bhagat, G., Furuya, N., Hibshoosh, H., Troxel, A., Rosen, J., Eskelinen, E.L., Mizushima, N., Ohsumi, Y., Cattoretti, G., and Levine, B. 2003. Promotion of tumorigenesis by heterozygous disruption of the beclin 1 autophagy gene. J. Clin. Invest. **112**:1809-1820.
143. Reddien, P.W., Cameron, S., and Horvitz, H.R. 2001. Phagocytosis promotes programmed cell death in C. elegans. Nature **412**:198-202.

144. Reiter, I., Krammer, B., and Schwamberger, G. 1999. Differential effect of apoptotic versus necrotic tumor cells on macrophage antitumor activities. J. Immunol. **163**:1730-1732.
145. Ren, Y., Silverstein, R.L., Allen, J., and Savill, J. 1995. CD36 gene transfer confers capacity for phagocytosis of cells undergoing apoptosis. J. Exp. Med. **181**:1857-1862.
146. Ricci, J.E., Munoz-Pinedo, C., Fitzgerald, P., Bailly-Maitre, B., Perkins, G.A., Yadava, N., Scheffler, I.E., Ellisman, M.H., and Green, D.R. 2004. Disruption of mitochondrial function during apoptosis is mediated by caspase cleavage of the p75 subunit of complex I of the electron transport chain. Cell **117**:773-786.
147. Roach, H.I., Aigner, T., and Kouri, J.B. 2004. Chondroptosis: a variant of apoptotic cell death in chondrocytes? Apoptosis **9**:265-277.
148. Robertson, A.M.G., and Thomson, J.N. 1982. Morphology of programmed cell death in the ventral nerve cord of Caenorhabditis elegans larvae. J. Embryol. Exp. Morphol. **67**: 89-100.
149. Ronchetti, A., Rovere, P., Iezzi, G., Galati, G., Heltai, S., Protti, M.P., Garancini, M.P., Manfredi, A.A., Rugarli, C., and Bellone, M. 1999. Immunogenicity of apoptotic cells in vivo: role of antigen load, antigen-presenting cells, and cytokines. J. Immunol. **163**:130-136.
150. Roos, D., de Boer, M., Kuribayashi, F., Meischl, C., Weening, R.S., Segal, A.W., Åhlin, A., Nemet, K., Hossle, J.P., Bernatowska-Matuszkiewicz, E., and Middleton-Price, H. 1996. Mutations in the X-linked and autosomal recessive forms of chronic granulomatous disease. Blood **87**:1663-1681.
151. Roos, A., Xu, W., Castellano, G., Nauta, A.J., Garred, P., Daha, M.R., and van Kooten, C. 2004. A pivotal role for innate immunity in the clearance of apoptotic cells. Eur. J. Immunol. **34**:921-929.
152. Rosen, A., and Casciola-Rosen, L. 1999. Autoantigens as substrates for apoptotic proteases: implications for the pathogenesis of systemic autoimmune disease. Cell Death Differ. **6**:6-12.
153. Rovere, P., Peri, G., Fazzini, F., Bottazzi, B., Doni, A., Bondanza, A., Zimmermann, V.S., Garlanda, C., Fascio, U., Sabbadini, M.G., Rugarli, C., Mantovani, A., and Manfredi, A.A. 2000. The long pentraxin PTX3 binds to apoptotic cells and regulates their clearance by antigen-presenting dendritic cells. Blood **96**:4300-4306.
154. Rovere, P., Vallinoto, C., Bondanza, A., Crosti, M.C., Rescigno, M., Ricciardi-Castagnoli, P., Rugarli, C, and Manfredi, A.A. 1998. Bystander apoptosis triggers dendritic cell maturation and antigen-presenting function. J. Immunol. **161**:4467-4471.
155. Rubartelli, A., Poggi, A., and Zocchi, M.R. 1997. The selective engulfment of apoptotic bodies by dendritic cells is mediated by the $\alpha_V\beta_3$ integrin and requires intracellular and extracellular calcium. Eur. J. Immunol. **27**:1893-1900.
156. Sauter, B., Albert, M.L., Francisco, L., Larsson, M., Somersan, S., and Bhardwaj, N. 2000. Consequences of cell death: exposure to necrotic tumor cells, but not primary tissue cells or apoptotic cells, induces the maturation of immunostimulatory dendritic cells. J. Exp. Med. **191**:423-434.
157. Savill, J. 1997. Apoptosis in resolution of inflammation. J. Leukoc. Biol. **61**:375-380.
158. Savill, J., Dransfield, I., Gregory, C., and Haslett, C. 2002. A blast from the past: clearance of apoptotic cells regulates immune responses. Nat. Rev. Immunol. **2**:965-975.
159. Savill, J., Dransfield, I., Hogg, N., and Haslett, C. 1990. Vitronectin receptor-mediated phagocytosis of cells undergoing apoptosis. Nature 343:170-173.
160. Savill, J., and Fadok V. 2000. Corpse clearance defines the meaning of cell death. Nature **407**:784-788.

161. Savill, J., Hogg, N., Ren, Y., and Haslett, C. 1992. Thrombospondin cooperates with CD36 and the vitronectin receptor in macrophage recognition of neutrophils undergoing apoptosis. J. Clin. Invest. **90**:1513-1522.
162. Savill, J.S., Wyllie, A.H., Henson, J.E., Walport, M.J., Henson, P.M., and Haslett, C. 1989. Macrophage phagocytosis of aging neutrophils in inflammation. Programmed cell death in the neutrophil leads to its recognition by macrophages. J. Clin. Invest. **83**:865-875.
163. Scott, R.S., McMahon, E.J., Pop, S.M., Reap, E.A., Caricchio, R., Cohen, P.L., Earp, H.S., Matsushima, G.K. 2001. Phagocytosis and clearance of apoptotic cells is mediated by MER. Nature **411**: 207-211.
164. Scheffer, S.R., Nave, H., Korangy, F., Schlote, K., Pabst, R., Jaffee, E.M., Manns, M.P., and Greten, T.F. 2003. Apoptotic, but not necrotic, tumor cell vaccines induce a potent immune response in vivo. Int. J. Cancer **103**:205-211.
165. Shi, J., Gilbert, G.E., Kokubo, Y., and Ohashi, T. 2001. Role of the liver in regulating numbers of circulating neutrophils. Blood **98**:1226-1230.
166. Schlaepfer, D.D., and Haigler, H.T. 1987. Characterization of $Ca^{2+}$-dependent phospholipid binding and phosphorylation of lipocortin I. J. Biol. Chem. **262**:6931-6937.
167. Schroit, A.J., Madsen, J.W., and Tanaka, Y. 1985. In vivo recognition and clearance of red blood cells containing phosphatidylserine in their plasma membranes. J. Biol. Chem. **260**:5131-5138.
168. Schweichel, J.U., and Merker, H.J. 1973. The morphology of various types of cell death in prenatal tissues. Teratology **7**:253-266.
169. Schwartz, L.M., Smith, S.W., Jones, M.E., and Osborne, B.A. 1993. Do all programmed cell deaths occur via apoptosis? Proc. Natl. Acad. Sci. U. S. A. **90**:980-984.
170. Scott, R.S., McMahon, E.J., Pop, S.M., Reap, E.A., Caricchio, R., Cohen, P.L., Earp, H.S., and Matsushima, G.K. 2001. Phagocytosis and clearance of apoptotic cells is mediated by MER. **Nature 411**:207-211.
171. Searle, J., Lawson, T.A., Abbott, P.J., Harmon, B., and Kerr, J.F. 1975. An electron-microscope study of the mode of cell death induced by cancer-chemotherapeutic agents in populations of proliferating normal and neoplastic cells. J. Pathol. **116**:129-138.
172. Segundo, C., Medina, F., Rodriguez, C., Martinez-Palencia, R., Leyva-Cobian, F., and Brieva, J.A. 1999. Surface molecule loss and bleb formation by human germinal center B cells undergoing apoptosis: role of apoptotic blebs in monocyte chemotaxis. Blood **94**:1012-1020.
173. Shiratsuchi, A., Kawasaki, Y., Ikemoto, M., Arai, H., and Nakanishi, Y. 1999. Role of class B scavenger receptor type I in phagocytosis of apoptotic rat spermatogenic cells by Sertoli cells. J. Biol. Chem. **274**:5901-5908.
174. Segundo, C., Medina, F., Rodriguez, C., Martinez-Palencia, R., Leyva-Cobian, F., and Brieva, J.A. 1999. Surface molecule loss and bleb formation by human germinal center B cells undergoing apoptosis: role of apoptotic blebs in monocyte chemotaxis. Blood **94**:1012-1020.
175. Simamura, E., Hirai, K.I., Shimada, H., and Koyama, J. 2001. Apoptosis and epithelial phagocytosis in mitomycin C-treated human pulmonary adenocarcinoma A549 cells. Tissue Cell. **33**:161-168.
176. Somersan, S., and Bhardwaj, N. 2001. Tethering and tickling: a new role for the phosphatidylserine receptor. J. Cell Biol. **155**:501-504.
177. Sperandio, S., de Belle, I., and Bredesen, D.E. 2000. An alternative, nonapoptotic form of programmed cell death. Proc. Natl. Acad. Sci. U. S. A. **97**:14376-14381.

178. Steinbrecher, U.P. 1999. Receptors for oxidized low density lipoprotein. Biochim. Biophys. Acta **1436**:279-298.
179. Steinman, R.M., Turley, S., Mellman, I., and Inaba, K. 2000. The induction of tolerance by dendritic cells that have captured apoptotic cells. J. Exp. Med. **191**:411-416.
180. Stoffels, K., Overbergh, L., Giulietti, A., Kasran, A., Bouillon, R., Gysemans, C., and Mathieu, C. 2004. NOD macrophages produce high levels of inflammatory cytokines upon encounter of apoptotic or necrotic cells. J. Autoimmun. **23**:9-15.
181. Surh, C.D., and Sprent J. 1994. T cell apoptosis detected in situ during positive and negative selection in the thymus. Nature **372**:100-103.
182. Susin, S.A., Lorenzo, H.K., Zamzami, N., Marzo, I., Snow, B.E., Brothers, G.M., Mangion, J., Jacotot, E., Costantini, P., Loeffler, M., Larochette, N., Goodlett, D.R., Aebersold, R., Siderovski, D.P., Penninger, J.M., and Kroemer, G. 1999. Molecular characterization of mitochondrial apoptosis-inducing factor. Nature **397**:441-446.
183. Szondy, Z., Sarang, Z., Molnar, P., Nemeth, T., Piacentini, M., Mastroberardino, P.G., Falasca, L., Aeschlimann, D., Kovacs, J., Kiss, I., Szegezdi, E., Lakos, G., Rajnavolgyi. E., Birckbichler. P.J., Melino, G., and Fesus, L. 2003. Transglutaminase $2^{-/-}$ mice reveal a phagocytosis-associated crosstalk between macrophages and apoptotic cells. Proc. Natl. Acad. Sci. U. S. A. **100**:7812-7817.
184. Tada, K., Tanaka, M., Hanayama, R., Miwa, K., Shinohara, A., Iwamatsu, A., and Nagata, S. 2003. Tethering of apoptotic cells to phagocytes through binding of CD47 to Src homology 2 domain-bearing protein tyrosine phosphatase substrate-1. J. Immunol. **171**:5718-5726.
185. Takizawa, F., Tsuji, S., and Nagasawa, S. 1996. Enhancement of macrophage phagocytosis upon iC3b deposition on apoptotic cells. FEBS Lett. **397**:269-272.
186. Taylor, P.R., Carugati, A., Fadok, V.A., Cook, H.T., Andrews, M., Carroll, M.C., Savill, J.S., Henson, P.M., Botto, M., and Walport, M.J. 2000. A hierarchical role for classical pathway complement proteins in the clearance of apoptotic cells in vivo. J. Exp. Med. **192**:359-366.
187. Todt, J.C., Hu, B., Punturieri, A., Sonstein, J., Polak, T., and Curtis, J.L. 2002. Activation of protein kinase C βII by the stereo-specific phosphatidylserine receptor is required for phagocytosis of apoptotic thymocytes by resident murine tissue macrophages. J. Biol. Chem. **277**:35906-35914.
188. Uthaisang, W., Nutt, L.K., Orrenius, S., and Fadeel, B. 2003. Phosphatidylserine exposure in Fas type I cells is mitochondria-dependent. FEBS Lett. **545**:110-114.
189. Utsugi, T., Schroit, A.J., Connor, J., Bucana, C.D., and Fidler, I.J. 1991. Elevated expression of phosphatidylserine in the outer membrane leaflet of human tumor cells and recognition by activated human blood monocytes. Cancer Res. **51**:3062-3066.
190. Vanags, D.M., Pörn-Ares, M.I., Coppola, S., Burgess, D.H., and Orrenius, S. a996. Protease involvement in fodrin cleavage and phosphatidylserine exposure in apoptosis. J. Biol. Chem. **271**:31075-31085.
191. Vandivier, R.W., Ogden, C.A., Fadok, V.A., Hoffmann, P.R., Brown, K.K., Botto, M., Walport, M.J., Fisher, J.H., Henson, P.M., and Greene, K.E. 2002. Role of surfactant proteins A, D, and C1q in the clearance of apoptotic cells in vivo and in vitro: calreticulin and CD91 as a common collectin receptor complex. J. Immunol. **169**:3978-3986.
192. van Lent, P.L., Licht, R., Dijkman, H., Holthuysen, A.E., Berden, J.H., and van den Berg, W.B. 2001. Uptake of apoptotic leukocytes by synovial lining macrophages inhibits immune complex-mediated arthritis. J. Leukoc. Biol. **70**:708-714.
193. Verbovetski, I., Bychkov, H., Trahtemberg, U., Shapira, I., Hareuveni, M., Ben-Tal, O., Kutikov, I., Gill, O., and Mevorach, D. 2002. Opsonization of apoptotic cells by

autologous iC3b facilitates clearance by immature dendritic cells, down-regulates DR and CD86, and up-regulates CC chemokine receptor 7. J. Exp. Med. **196**:1553-15561.

194. Verhoven, B., Schlegel, R.A., and Williamson P. 1995. Mechanisms of phosphatidylserine exposure, a phagocyte recognition signal, in apoptotic T lymphocytes. J. Exp. Med. **182**:1597-1601.
195. Voll, R.E., Herrmann, M., Roth, E.A., Stach, C., Kalden, J.R., and Girkontaite, I. 1997. Immunosuppressive effects of apoptotic cells. Nature **390**:350-351.
196. Wang, X., Wu, Y.C., Fadok, V.A., Lee, M.C., Gengyo-Ando, K., Cheng, L.C., Ledwich, D., Hsu, P.K., Chen, J.Y., Chou, B.K., Henson, P., Mitani, S., and Xue, D. 2003. Cell corpse engulfment mediated by C. elegans phosphatidylserine receptor through CED-5 and CED-12. Science **302**:1563-1566.
197. Webb, J.H., Blom, A.M., Dahlbäck, B. 2002. Vitamin K-dependent protein S localizing complement regulator C4b-binding protein to the surface of apoptotic cells. J. Immunol. **169**:2580-2586.
198. Wood, W., Turmaine, M., Weber, R., Camp, V., Maki, R.A., McKercher, S.R., and Martin, P. 2000. Mesenchymal cells engulf and clear apoptotic footplate cells in macrophageless PU.1 null mouse embryos. Development **127**:5245-5252.
199. Wiegand, U.K., Corbach, S., Prescott, A.R., Savill, J., and Spruce, B.A. 2001. The trigger to cell death determines the efficiency with which dying cells are cleared by neighbours. Cell Death Differ. **8**:734-746.
200. Wyllie, A.H. 1980. Glucocorticoid-induced thymocyte apoptosis is associated with endogenous endonuclease activation. Nature **284**:555-556.
201. Wyllie, A.H., Kerr, J.F., and Currie A.R. 1980. Cell death: the significance of apoptosis. Int. Rev. Cytol. **68**:251-306.
202. Yamamoto, A., Taniuchi, S., Tsuji, S., Hasui, M., and Kobayashi, Y. 2002. Role of reactive oxygen species in neutrophil apoptosis following ingestion of heat-killed Staphylococcus aureus. Clin. Exp. Immunol. **129**:479-484.
203. Yu, L., Alva, A., Su, H., Dutt, P., Freundt, E., Welsh, S., Baehrecke, E.H., and Lenardo, M.J. 2004. Regulation of an ATG7-beclin 1 program of autophagic cell death by caspase-8. Science **304**:1500-1502.
204. Yue, Z., Jin, S., Yang, C., Levine, A.J., and Heintz, N. 2003. Beclin 1, an autophagy gene essential for early embryonic development, is a haploinsufficient tumor suppressor. Proc. Natl. Acad. Sci. U. S. A. **100**:15077-15082.
205. Zhou, Z., Hartweig, E., and Horvitz, H.R. 2001. CED-1 is a transmembrane receptor that mediates cell corpse engulfment in C. elegans. Cell **104**:43-56.
206. Zhuang, J., Ren, Y., Snowden, R.T., Zhu, H., Gogvadze, V., Savill, J.S., and Cohen, G.M. 1998. Dissociation of phagocyte recognition of cells undergoing apoptosis from other features of the apoptotic program. J. Biol. Chem. **273**:15628-15632.

Chapter 4

# SPECIFIC, PRO-POPTOTIC CELL-SIGNALING: DESIGN OF NOVEL, RECOMBINANT TARGETED ANTITUMOR AGENTS OPERATING EXCLUSIVELY THROUGH MODULATION OF CELLULAR APOPTOTIC EVENTS

*Novel, Targeted Pro-Apoptotic Therapeutic Agents*

Yuying Liu, Lawrence H. Cheung and Michael G. Rosenblum
*Immunopharmacology and Targeted Therapy Section, Department of Experimental Therapeutics, M.D. Anderson Cancer Center, 1515 Holcombe Blvd., Unit 044, Houston, TX 77030*

Abstract: Normal cellular homeostasis employs a delicate balance between pro-apoptotic (cell death) and anti-apoptotic (growth-stimulatory) forces. Unrestricted/unregulated cell growth, which occurs in neoplasia frequently, accompanies an imbalance in these forces and frequently the apoptotic pathways are disrupted leading to increased metastatic spread, resistance to chemotherapy and radiotherapy. Pro-apoptotic enzymes such as the serine protease Granzyme B can activate the pro-apoptotic cascade within tumor cells through multiple, independent mechanisms. We have generated novel cell-targeting fusion constructs containing enzymatically active GrB and have demonstrated that these constructs are highly cytotoxic to target cells. Apoptotic cascades can be maximally activated within 4 hr of treatment and these agents are synergistic with chemotherapeutic agents. We propose that these types of signaling agents, which operate exclusively through activation of pro-apoptotic mechanisms, may describe a new class of targeted therapeutic agents with unique biological properties.

Key words: apoptosis, caspases, fusion protein, Granzyme B, targeted therapy

Cellular homeostasis requires a balance between cell proliferation and programmed cell death (apoptosis). Apoptosis is defined as genetically programmed autonomous cell death, and it occurs in normal healthy tissues



*M. Sluyser (ed.), Application of Apoptosis to Cancer Treatment,* 85-115.

such as breast cells at varying rates during the estrus cycle in response to changes in hormone levels [1]. Apoptosis is also regulated by non-hormonal signals. Changes in the genetics of apoptotic regulatory mechanisms may result in an increase in cell numbers, as well as the preservation of genetically altered cells, which begins the process of tumorigenesis [2]. Studies on apoptosis have led to the identification of a central tripartite death switch i.e. apoptosome consisting of Apaf-1, Apaf-2 and Apaf-3. The caspases, a family of cysteine-dependent aspartate directed-proteases, constitute the central executioners of apoptosis. Much of the attention on programmed cell death is focused on caspases, however, apoptosis can still occur even when the caspase cascade is blocked, revealing the existence of caspase-independent alternative pathway(s). The mitochondrial release of cytochrome c following a programmed cell death inducing stimulus in both plants and animals suggests the evolutionary conservation of death pathways [3]. The major apoptosis pathways are summarized in Figure 1. There are at least three recognized points at which apical proteases are activated to initiate apoptosis. The extrinsic pathways include delivery of Granzyme B (GrB) to the cell as well as receptor interaction with certain ligands. The mitochondrial pathway is an intrinsic pathway triggering apoptosis and is apparently subordinate to the dominant, direct pathway in most cell types.

In addition, many of the survival pathways including growth factors or their receptors such as ErbB-2 [4;5] and oncogenes such as Bcl-2 that drive tumor cells into unregulated growth and directly inhibit cellular apoptosis pathways in multiple ways. Bcl-2 is an anti-apoptotic and anti-proliferative protein over-expressed in several different human cancers including breast. Gain of Bcl-2 function in mammary epithelial cells was superimposed on the WAP-TAg transgenic mouse model of breast cancer progression to determine its effect on epithelial cell survival and proliferation at three key stages in oncogenesis: the initial proliferative process, hyperplasia, and cancer. During the initial proliferative process, Bcl-2 strongly inhibits both apoptosis and mitotic activity. However as tumorigenesis progresses to hyperplasia and adenocarcinoma, the inhibitory effects on mitotic activity were lost. In contrast, anti-apoptotic activity persisted in both hyperplasias and adenocarcinomas. These results demonstrate that the inhibitory effect of Bcl-2 on epithelial cell proliferation and apoptosis can separate during cancer progression. In this model, retention of anti-apoptotic activity with loss of anti-proliferative action resulted in earlier tumor presentation [6;7].

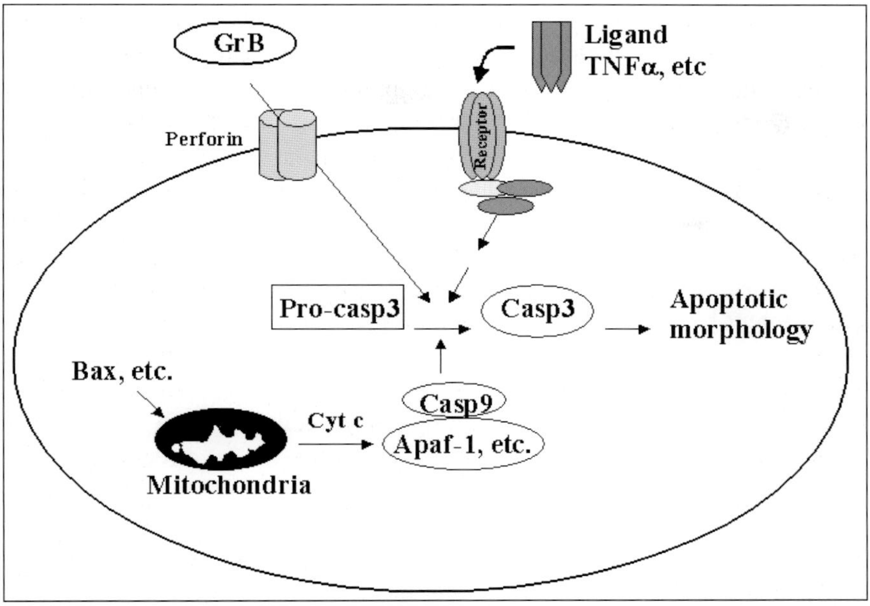


*Figure 1.* Major cellular apoptotic pathways: the extrinsic pathways include delivery of Granzyme B (GrB) to the cell as well as receptor interaction with certain ligands. The mitochondrial pathway is an intrinsic pathway triggering apoptosis.

Traditional chemotherapeutic approaches for treatment of neoplastic disease have generally relied upon the targeting of rapidly proliferating cells by inhibiting DNA replication or cell division. Although this strategy has been effective, its innate lack of selectivity for tumor cells has resulted in diminishing returns, approaching the limits of acceptable toxicity. Apoptosis also contributes to cell death in tumors treated with various anticancer agents [8]. Development of resistance of cancer cells to chemotherapeutic agents has been clearly shown to be correlated with a blockade of apoptotic signaling in resistant cells [9;10]. Metastatic spread of breast cancer also appears to directly involve the apoptotic pathway. Some studies have shown that high metastatic potential is strictly associated with increased resistance to apoptosis [11-14]. Suppression of drug-induced apoptosis by over-expression of oncogenes such as bcl-2, growth factors or their receptors [15-18] may be an important cause of both metastases and intrinsic chemo- and radiation- resistance.

Recent clinical studies [19] in patients with breast cancer suggest a direct link between tumor apoptosis, labeling index (Ki-67) and patient response to chemotherapeutic regimens. The studies assess the in vivo relationship of apoptosis to proliferation and Bcl-2 protein in human breast tumors both prior to chemotherapy and in the residual resistant cell population at the completion of treatment by evaluation of apoptotic index (AI), Ki67 and Bcl-2 protein expression in the tissue of patients with operable breast cancer immediately before ECF [6 cycles of epirubicin (50 mg/m2, iv, Day 1), cisplatin (50 mg /m2, iv, Day 1) and continuous infusional 5 Fu (200 mg/m2 /24 hrs)] preoperative chemotherapy. There was a significant positive association between AI and Ki67 both before and after chemotherapy. In the residual specimens AI and Ki67 were significantly reduced compared with pre-treatment biopsies, while Bcl-2 expression showed a significant increase. These data suggest that apoptosis and proliferation are closely related in vivo. It is possible that the phenotype of reduced apoptosis and increased Bcl-2 may be associated with breast cancer cells resistant to cytotoxic chemotherapy [20]. Several studies have suggested a direct correlation between response to chemotherapy and activation of apoptotic pathways [21-24]. Chemo-resistance appears to frequently accompany clinical progression of breast cancers from a hormone-dependent, non-metastatic, antiestrogen-sensitive phenotype to a hormone-independent, invasive, metastatic, antiestrogen-resistant phenotype [25].

A growing understanding of the molecular events that mediate tumor growth and metastases has led to the development of rationally designed targeted therapeutics that offer the dual hope of maximizing efficacy and minimizing toxicity to normal tissue [26;27]. In recent years, a strategy in cancer therapy in general, and breast cancer in particular, has been the use of maximum tolerated doses of toxic non-specific agents as well as the investigation of a range of new agents that specifically target tumor-related molecules through a variety of biological pathways. Approaches to directly modulate of apoptosis pathways are of particular interest in terms of new drug development in breast cancer. The activation of apoptotic mechanisms in breast tumor cells has been directly implicated in the response of patient tumors to chemotherapy, response to radiotherapy and propensity to metastasize (Figure 2).

Therapeutic intervention strategies focusing on inactivation of anti-apoptotic pathways have employed strategies such as antisense molecules to down-regulate constitutively over-expressed Bcl-2 [28]. Treatment with antisense molecules has been shown to restore sensitivity to chemotherapeutic agents in breast tumors. In addition, dendritic cells [29-31] have been shown to initiate apoptosis in breast tumor cells (in vitro and in vivo) which are resistant to chemotherapeutic agents. As expected, dendritic cells appear to synergize with chemotherapeutic agents in generating cytotoxic effects on resistant breast tumor cells.

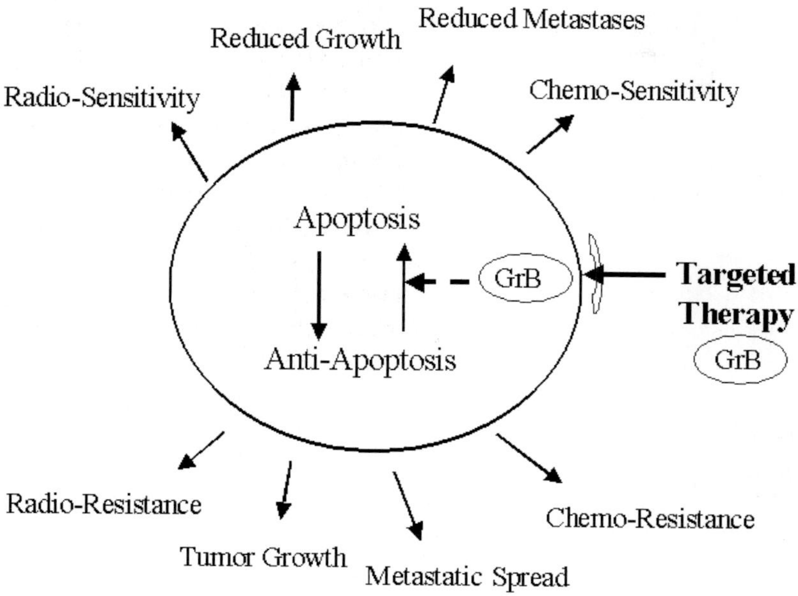


*Figure 2.* Apoptosis and anti-apoptotic events directly control ability of tumor cells to respond to external factors and control response to chemotherapeutic agents, radiation therapy and metastatic spread.

Cytotoxic T-lymphocytes (CTL) and Natural Killer (NK) cell types destroy tumor cells by direct physical delivery of the serine protease granzyme B (GrB) thereby inducing apoptosis and this is one of the major mechanisms in the cellular immune-response [32;33]. Cytotoxic T-lymphocyte granules contain perforin (PFN), a pore-forming protein, and a family of serine proteases termed granzymes. In CTL-mediated cytolysis, PFN is initially released and it inserts into the target cell membranes where it polymerizes to form transmembrane pores [34;35], which facilitates access of NK or CTL-released GrB to the target cell cytoplasm. Alternatively, other authors have suggested that after release from CTLs, GrB may be internalized into target cells by receptor-mediated endocytosis and that the role of PFN is to mediate release of GrB from endocytic vesicles. These studies suggest that PFN can be replaced by other vesicle-disrupting factors such as those produced by adenovirus [36-38].

Once delivered to the cytoplasm, GrB induces apoptosis (Figure 3) by directly activating caspases and inducing rapid DNA fragmentation [39;40].

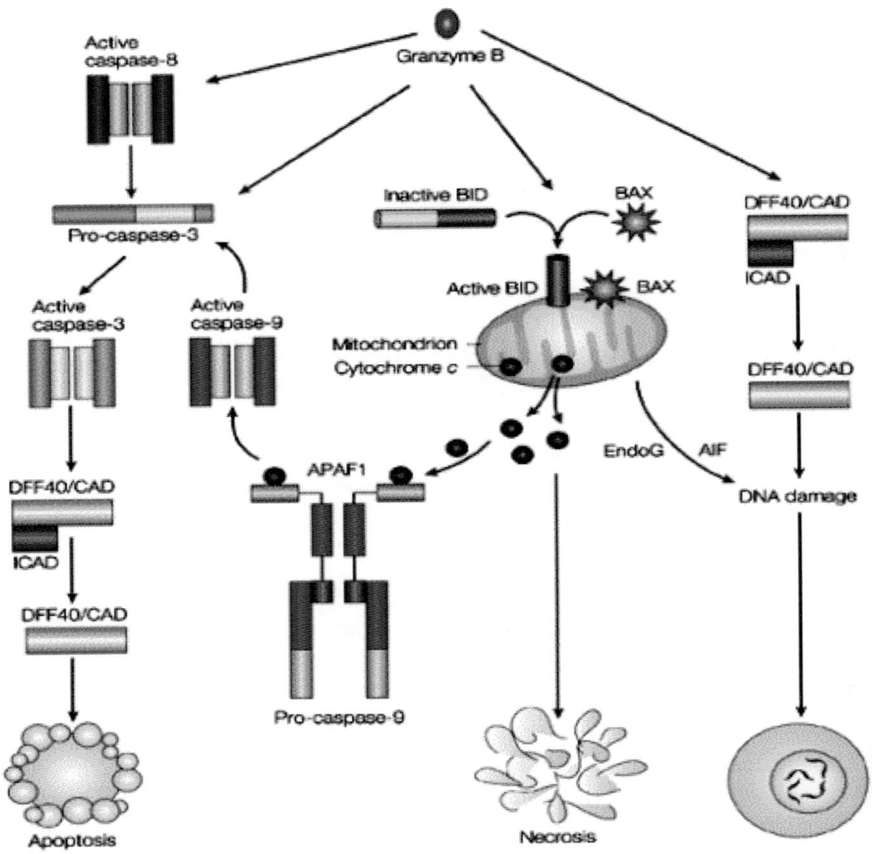


*Figure 3.* Pathways to apoptotic cell death initiated by GrB. Once released into the cytoplasm, GrB can initiate apoptotic cell death through both caspase-dependent and caspase-independent pathways. (From M. Barry and R. C. Bleackley. "Cytotoxic T Lymphocytes: All Roads Lead to Death". Nature Reviews-Immunology. 2002, 2: 401-409).

GrB can cleave many procaspases in vitro, and has been an important tool in analyzing the maturation of caspase-3 [41], caspase-7 [42;43], caspase-6 [44], caspase-8 [45], caspase-9 [46], and caspase-10a/b [47;48]. Although many procaspases are efficiently cleaved in vitro, GrB-induced caspase activation occurs in a hierarchical manner in intact cells, commencing at the level of "executioner caspases" such as caspase-3, followed by caspase-7 [49]. Some studies have shown that GrB activated cell death pathways through cleavage of Bid and activation of the mitochondrial death pathway in intact cells [50;51]. In addition to the caspase-mediated cytotoxic events, GrB can also rapidly translocate to the nucleus and cleave poly (ADP-ribose) polymerase and nuclear matrix antigen, utilizing different cleavage sites than those preferred by caspases [52;53]. In addition, some studies have demonstrated that GrB can directly damage to non-nuclear structures such as mitochondria, subsequently induce cell death through caspase-independent pathway [54;55].

Since almost all cells contain mechanisms responsible for mediating cell death (apoptosis) we propose that targeted delivery of GrB protein to the interior of cells will result in cell death through apoptotic mechanisms assuming that sufficient quantities of active enzyme can be successfully delivered to the appropriate subcellular compartment.

Recombinant single-chain Fv antibody (scFv)-based agents have been used in pre-clinical studies for cell-targeted delivery of cytokines [56] and intracellular delivery of highly cytotoxic n-glycosidases such as recombinant gelonin (rGel) toxin [57-59]. The smaller size of these antibody fragments may allow better penetration into tumor tissue, improved pharmacokinetics, and a reduction in the immunogenicity observed with intravenously administered murine antibodies. Initially, to target melanoma cells, we chose a recombinant single-chain antibody designated scFvMEL which recognizes the high-molecular-weight glycoprotein gp240, found on a majority (80 %) of melanoma cell lines and fresh tumor samples [60]. Our group and others have demonstrated that this antibody possesses high specificity for melanoma and is minimally reactive with a variety of normal tissues, making it a promising candidate for further study [61-63].

Recently, studies have demonstrated that the gp240 antigen is present not only on melanoma but also on 66 % of lobular breast cancer [64]. Lobular breast cancer is an important subtype classification of breast cancer which defines a population of women with tumors which are relatively radiation resistant and resistant to chemotherapeutic regimens [65;66]. Data suggests that cellular apoptotic mechanisms in this population of tumors may be inhibited thus suppressing response to therapeutic agents although this is speculative.

Antibodies designated ZME-018 or 225.28 S targeting the gp240 antigen have been extensively studied in melanoma patients and have demonstrated an impressive ability to localize in metastatic tumors after systemic administration [64;67-69] (Figure 4).

The recombinant scFvMEL single-chain antibody has been used extensively in our laboratory to target gp240 bearing cells in vitro and using xenograft models [70-77]. This antibody binds to target cells and is efficiently internalized making this an excellent carrier to deliver toxins or other therapeutic payloads.

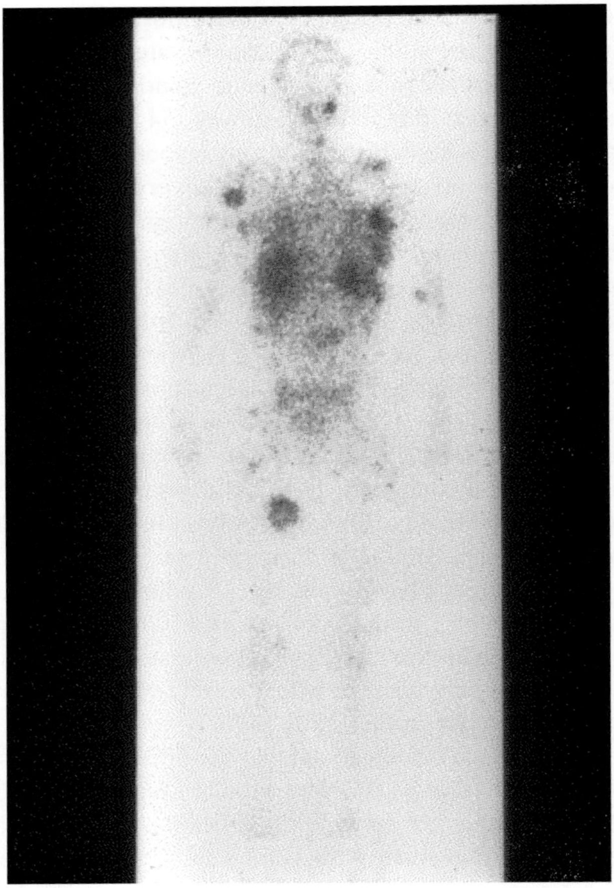

*Figure 4.* Tumor uptake of $^{111}$In monoclonal antibody ZME-018 in a patient with metastatic melanoma after systemic administration.

More importantly, the gp240 antigen is not expressed on normal cells thus making this an interesting target for therapeutic intervention. In this study, we used scFvMEL as a tumor cell-targeting carrier and designed a novel recombinant fusion construct designated GrB/scFvMEL, containing human pro-apoptotic enzyme GrB. The purpose of these studies was to determine whether we could deliver sufficient quantities of active GrB enzyme to drive cellular apoptotic events specifically in breast cancer target cells. Moreover, we propose to examine if apoptotic modulation by delivery of GrB into breast tumor is able to re-establish sensitivity of breast cancer tumor cells resistant to chemotherapy or to specifically sensitize tumor cells to radiation therapy.

Delivery of human pro-apoptotic proteins to tumor cells is a novel concept. We have recently demonstrated that nascent cellular apoptotic cascades which are normally held in check can be specifically activated using a directed therapeutic approach in which activated signal transduction cascade proteins are delivered to cell cytoplasm using recombinant fusion constructs containing antibodies or growth factors capable of binding to a cell surface antigen and internalizing a payload.

The concept of delivering human pro-apoptotic proteins as functional components of targeted therapeutics we believe is a novel and previously untried approach. Our recent publications on constructs containing GrB are the first reported such constructs found in the literature [78;79]. The development of gp240-targeted therapeutic agents containing GrB offers an unprecedented opportunity to utilize such agents to probe the criticality of apoptotic processes during tumor growth, metastatic spread and response to therapeutic agents. In addition, information regarding these studies will also be used to further develop the GrB/scFvMEL fusion construct as a potential therapeutic agent.

## 1. CAN WE GENERATE A RECOMBINANT CONSTRUCT CONTAINING A TARGETING ANTIBODY AND ACTIVE GRB?

We successfully generated a recombinant construct GrB/scFvMEL containing active GrB and single-chain Fv fragment targeting the gp240 antigen present on 80% of melanoma and 66% of lobular breast cancer. First, we successfully obtained the human pre-mature GrB gene composed of mature GrB and a signal sequence from human cutaneous T-cell lymphoma (Hut-78) cell RNA utilizing reverse transcription-PCR (Figure 5).

In the premature GrB protein, the first 20 amino acids at as a signal sequence. In cytotoxic T cells, active GrB is nominally generated by dipeptidyl peptidase I (DPPI)-mediated proteolysis [80] which removes the two-residue (Gly Glu) propeptide and exposes a terminal Ile $^{21}$ residue. The NH2-terminal Ile-Ile-Gly-Gly sequence of GrB is necessary for enzymatically active GrB.

In our PCR engineering design and construction of the final molecule, this enzymatic requirement dictated that the GrB protein leads the molecule followed by a flexible linker and the targeting antibody. In addition, we insured that the EK cleavage site (DDDDK) for removal of the purification tag was immediately adjacent to Ile $^{21}$.

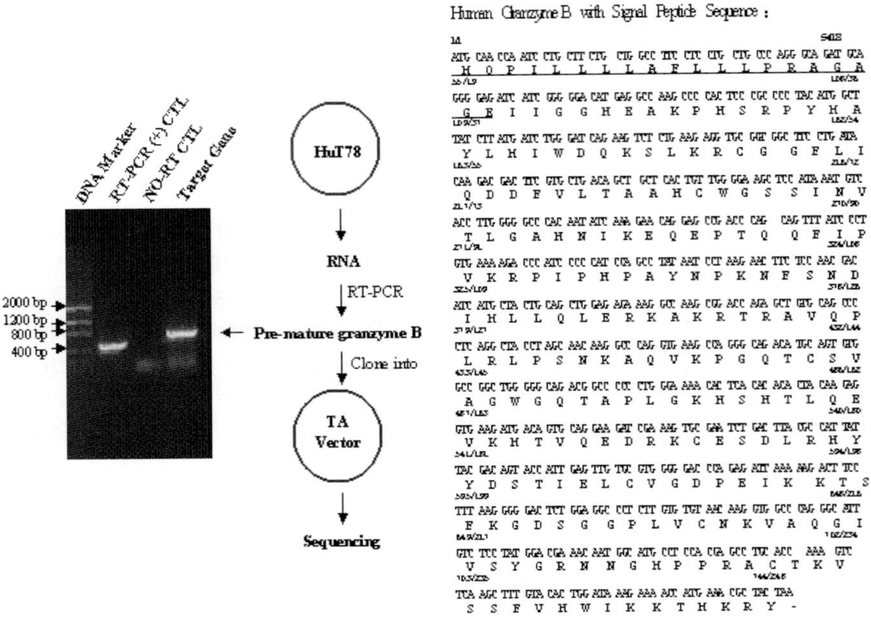


*Figure 5.* Cloning the human GrB gene from HuT-78 cells. RNA was isolated, and premature GrB cDNA (~ 800 bp) was amplified by reverse transcription –PCR and cloned into the PCR2.1 TA vector. The human GrB sequence with a 20- amino acid signal sequence was confirmed and designated as pre-mature GrB. Once the signal peptide was removed, the mature amino-terminal Ile-Ile-Gly-Gly sequence of active GrB was generated.

The fusion gene was then introduced into the pET32a (+) bacterial expression vector to form pET32GrB/scFvMEL (Figure 6). The recombinant protein GrB/scFvMEL was expressed in bacterial expression system (Figure 7) as a polyhistidine-tagged protein (A, lane 2) and then purified by nickel-NTA metal affinity chromatography (A, lane 3). The his-tag was cleaved by addition of recombinant enterokinase (rEK) to produce active form of GrB fusion and then Q-sepharose ion exchange resin was used for final purified GrB/scFvMEL (A, lane 4). Specificity of the cleaved fusion protein was confirmed by Western blot using either mouse anti-GrB or rabbit anti-scFvMEL antibody (B). To assess the biological activity of the GrB component of the fusion construct, the ability of the enzyme to cleave a BAADT substrate was assessed and compared to native GrB [81]. The fusion construct GrB/scFvMEL was shown to have intact GrB enzymatic activity with ΔmA/min = 68.6 and a Specific Activity (SA) = $2.6 \times 10^5$ units/ μmole. This activity was comparable to that of native GrB with ΔmA/min = 48.2 and a SA = $4.8 \times 10^5$ units/ μmole. As expected, the GrB/scFvMEL construct containing the purification tag (before rEK digestion) was shown to be unable to cause hydrolysis of the BAADT substrate (ΔmA/min ≤ 5).

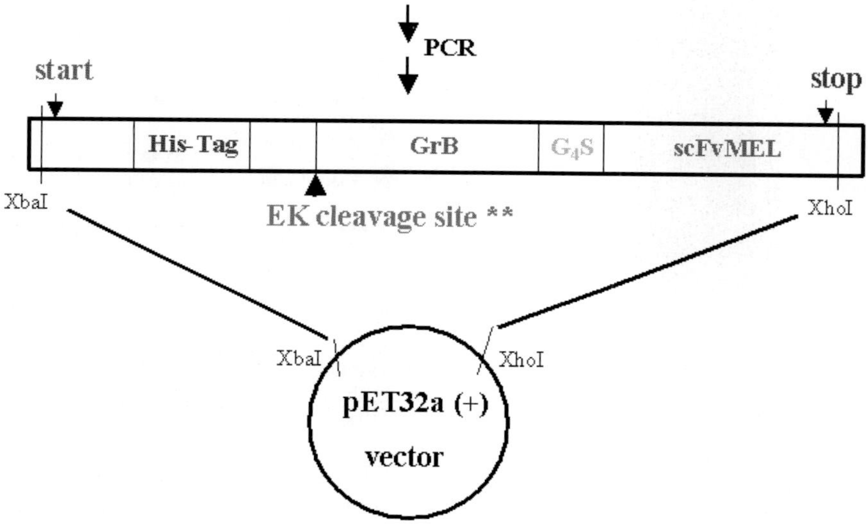


*Figure 6.* Construction of the GrB/scFvMEL fusion toxin by PCR and insertion into pET32a (+) vector. Mature GrB was attached to the recombinant antibody scFvMEL via a flexible tether ($G_4S$). A cleavage site for EK (DDDDK) was inserted upstream of the first amino acid (Ile) of mature GrB. The fused gene was then introduced into pET32a (+) vector to form the expression vector pET-32GrB/scFvMEL. Once the protein tag was removed by rEK digestion, the first reside (Ile) of mature GrB was exposed, thereby activating the GrB moiety of the fusion construct.

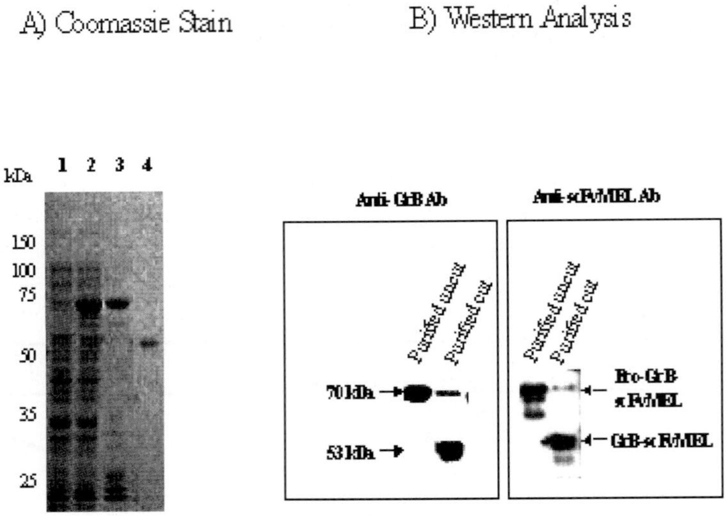


**Enzymatic Activity of GrB Moiety of Fusion Protein Compared with Native GrB ***

| Samples | Δ mOD/ min | Units (U) | U/μg | MW (kDa) | Specific Activity (U/μM) |
|---|---|---|---|---|---|
| Native GrB | 48.2 | 1.0 | 19.2 | 25 | $4.8 \times 10^5$ |
| GrB/scFvMEL (Un-rEK cut) | 2.0** | - | - | 70 | - |
| GrB/scFvMEL (rEK-cut) | 68.6 | 1.42 | 4.7 | 53 | $2.6 \times 10^5$ |

* BAADT Assay
** The rate of non-enzymatic hydrolysis of BAADT at 0.2 nM, in 0.3 nM Ellman's Buffer at 25 °C is ≤ 5 Δ m OD/min

*Figure 7.* SDS-PAGE and Western analysis of GrB/scFvMEL expression in E. coli. A. 10 % SDS-PAGE gel with coomassie blue staining under reducing conditions showed that the GrB/scFvMEL construct was expressed as a 70- kDa molecule (53 + 17 kDa purification tag). The size of final purified GrB/scFvMEL was ~ 53 kDa. Lane1, non-induced bacterial cell lysate; lane 2, induced cell lysate; lane 3, pro-GrB/scFvMEL (+ tag) after purification by nickel-NTA metal affinity chromatography; lane 4, final purified GrB/scFvMEL. B. Western blotting confirmed that the fusion protein reacted with both mouse anti-GrB and rabbit anti-scFvMEL antibodies.

## 2. DOES THIS RECOMBINANT GRB/SCFVMEL INTERNALIZE EFFICIENTLY INTO GP240 POSITIVE CELLS?

The GrB moiety of the fusion construct was efficiently delivered into the cytosol of gp240 positive A375-M melanoma cells after treatment with GrB/scFvMEL for 1 h (D) or 6 h (E) as assessed by confocal microscope imaging detected using goat anti-GrB antibody (Figure 8). Pretreatment of cells with the original anti-gp240 antibody ZME-018 followed by incubation with the GrB/scFvMEL was shown to completely suppress internalization of this agent (B, C). This effectively demonstrates that binding of the construct to gp240 on the tumor cell surface is responsible for internalization of the construct.

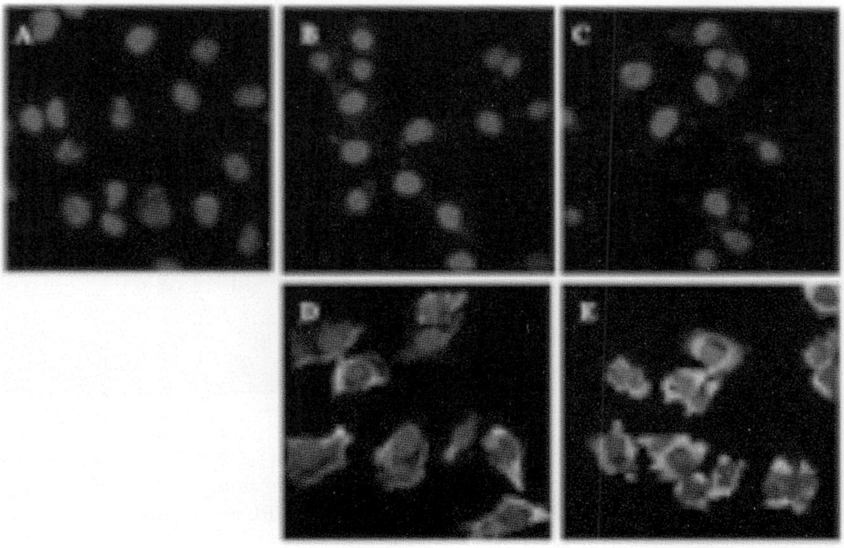


*Figure 8.* Internalization of GrB/scFvMEL into A375-M cells assessed by confocal microscopy. A375-M cells were treated with 40 nM GrB/scFvMEL for 1 or 6 h. As indicated, some cells were pre-treated with ZME018 to assess antigen-mediated internalization. Molecules bound to the cell surface were removed by brief treatment with glycine buffer (pH 2.5). Cells were then fixed, permeabilized and incubated with goat anti-GrB antibody, followed by FITC-coupled anti-goat IgG and propidium iodide (PI). The slides were mounted and analyzed by Zeiss LSM 510 confocal laser scanning microscopy. A, no treatment control. B, pretreatment with ZME-018, then GrB/scFvMEL treatment for 1 h. C, pretreatment with ZME-018, then GrB/scFvMEL treatment for 6 h. D, GrB/scFvMEL treatment for 1 h. E, GrB/scFvMEL treatment for 6 h.

# 3. DOES THE GRB/SCFVMEL CONSTRUCT GENERATE APOPTOSIS IN TARGET CELLS?

Both antigen-positive and antigen-negative cells were treated with an $IC_{50}$ concentration of the GrB/scFvMEL fusion construct. Apoptotic cells were observed at 8 h after treatment when assessed by TUNEL assay. In contrast, there were no apoptotic cells in non-target cells treated with identical doses of the fusion construct, thereby demonstrating cell specificity of the construct (Figure 9).

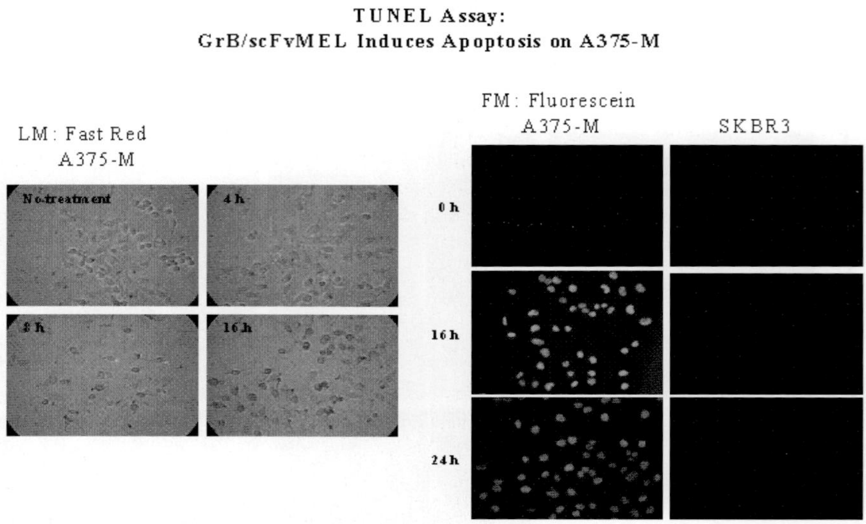


*Figure 9.* Treatment with GrB/scFvMEL induces apoptosis on antigen-positive A375-M cells but not on antigen-negative SKBR3 cells as assessed by TUNEL assay.

Procaspase-3 was cleaved into one fragment (~ 20 kDa) at 4 h, and further cleaved into smaller fragments after treatment for 8 h. We were not able to demonstrate caspase-3 cleavage on antigen-negative SKBR3 cells treated with GrB/scFvMEL (Figure 10). Treatment of GrB/scFvMEL was shown to result in cytochrome c translocation from the mitochondrial compartment into cytosol by 4 h after treatment of A375-M cells but not observed on SKBR3 cells (Figure 11). These studies showed that the cytotoxic /apoptotic events observed after treatment of target cells with the fusion construct are the result of both caspase-dependent and caspase-independent mechanisms.

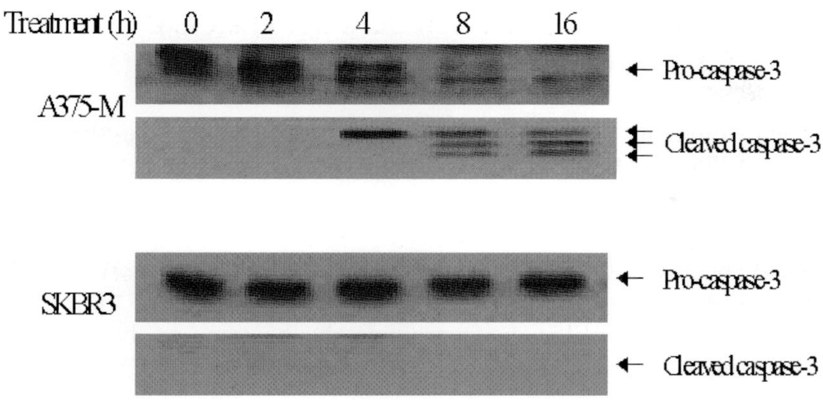


*Figure 10.* GrB/scFvMEL induced caspase-3 cleavage on antigen-positive A375-M cells. A375-M and SKBR3 cells were treated with GrB/scFvMEL at 50 nM for various times. Whole cell lysates (30 μg) were analyzed by 12 % SDS-PAGE and followed by immunoblotting to detect caspase-3 or cleaved caspase-3. Pro-caspase-3 was cleaved into an active fragment at 4 h and further cleaved into smaller fragments after treatment for 8 h by GrB/scFvMEL on A375-M cells. We found no caspase-3 cleavage on SKBR3 cells treated with GrB/scFvMEL.

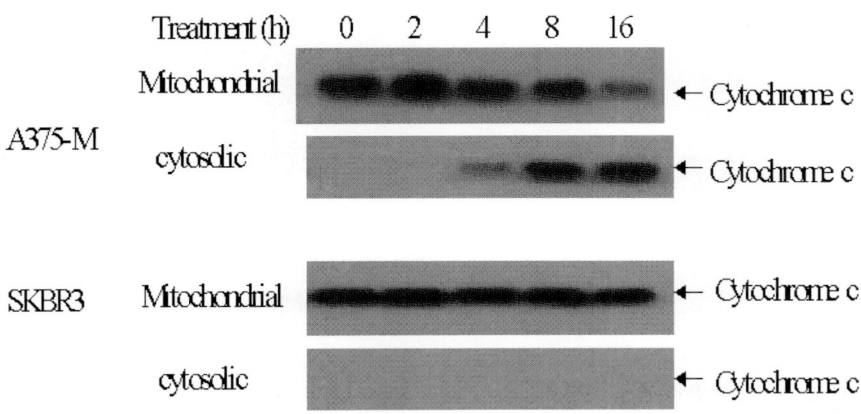


*Figure 11.* Cytochrome c was released from mitochondria fractions into cytosol by GrB/scFvMEL on A375-M cells. Cells were treated with GrB/scFvMEL at 50 nM for various times. Cells were collected and the cytosolic and mitochondrial fractions were isolated. Fractions (30 μg) from nontreated and treated cells were analyzed by 15 % SDS-PAGE followed by immunoblotting and detection with an anti-cytochrome c antibody. As shown, cytochrome c was found to be released from mitochondria into the cytosol on A375-M cells but not on SKBR3 cells within 4 h after treatment by GrB/scFvMEL.

# 4. IS THE GRB/SCFVMEL CONSTRUCT CYTOTOXIC TO TARGET CELLS?

The cytotoxicity of GrB/scFvMEL was assessed against log-phase gp240 positive A375-M, MDA-MB435 and negative SKBR3 cells in culture. A 50 % growth inhibitory effect was found at a concentration of ~ 20 nM on A375-M cells and MDA-MB435 cells.

However, no cytotoxic effects were found on antigen-negative SKBR3 cells at doses of up to 1 μM (Figure 12). By comparison, the cytotoxic effects of GrB/scFvMEL were approximately the same as that of another fusion toxin, MELsFv/rGel [82] on A375-M cells.

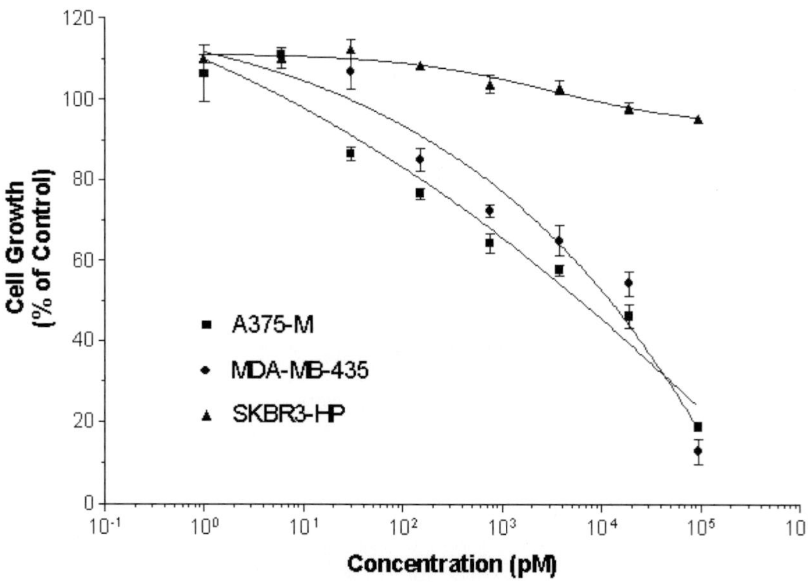


*Figure 12.* Cytotoxicity of the GrB/scFvMEL fusion toxin on A375-M, MDA-MB435 and SKBR3-HP cells. Log-phase cells were plated into 96-well plates at a density of 2.5 x $10^3$ cells per well and allowed to attach for 24 h. The medium was replaced with medium containing various concentrations of GrB/scFvMEL. After 72 h, the effect of fusion toxin on the growth of cells in culture was determined using crystal violet staining. The IC $_{50}$ of GrB/scFvMEL was ~ 20 nM on A375-M and MDA-MB435 cells. In contrast, no cytotoxicity was observed on SKBR3 cells.

When A375-M cells were pretreated with a full-length anti-gp240 antibody (ZME-018) for 6 h and then treated with GrB/scFvMEL for 72 h, the cytotoxicity of GrB/scFvMEL was abolished, thereby demonstrating a requirement for antigen recognition in the cytotoxic effect of the GrB/scFvMEL fusion construct (Figure 13). These data confirm the confocal imaging data that demonstrate that pre-treatment of cells with native ZME-018 can abolish internalization of this agent. In addition, the uncut GrB/scFvMEL construct or rEK showed no cytotoxicity to target cells as expected (Figure 13).

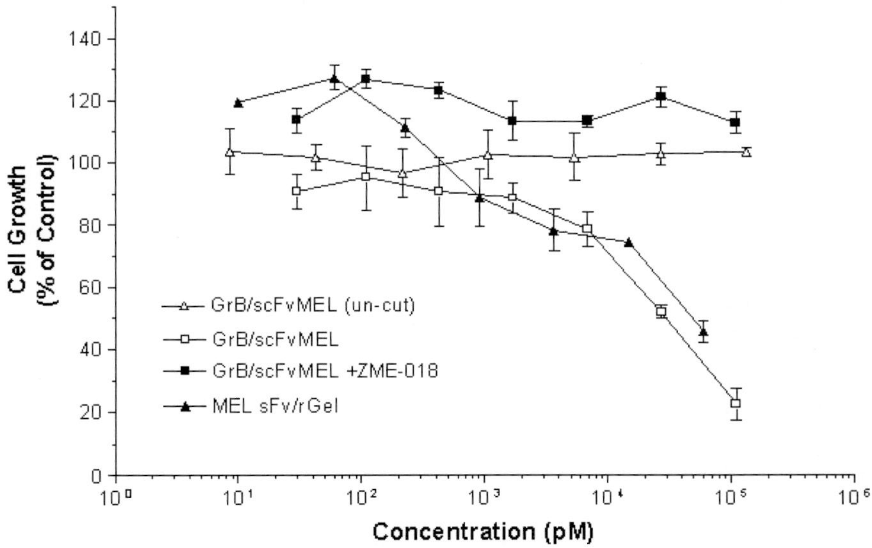


*Figure 13.* Comparative cytotoxicity of GrB/scFvMEL and MEL sFv/rGel and effect of addition of ZME-018 on cytotoxicity of GrB/scFvMEL against A375-M cells. Log-phase cells were plated into 96-well plates at a density of 2.5 x $10^3$ cells per well and allowed to attach for 24 h. Cells were treated with different concentrations of GrB/scFvMEL or MEL sFv/rGel and were also pretreated with ZME-018 (40 mg/ml) for 6 h and then co-treated with various concentrations of GrB/scFvMEL. After 72 h, the cells were stained with crystal violet. The $IC_{50}$ of GrB/scFvMEL was approximately identical to that of MEL sFv/rGel on A375-M. ZME-018 pretreatment inhibited the cytotoxicity of GrB/scFvMEL on A375-M cells.

## 5. CAN GRB/SCFVMEL FUSION PROTEIN SENSITIZE CELLS TO CONVENTIONAL CHEMOTHERAPEUTIC AGENTS?

Cells in exponential growth phase were plated into 96-well plates. After 24 hr, the cells were treated with drug-containing medium. At the end of the indicated incubation period, growth inhibition was assessed by crystal violet staining. In order to determine the effects of sequencing, cells were treated with the different sequences.

Sequence I (C1): cells were pretreated with chemotherapeutic agent for 6 h, and then co-administered with chemotherapeutic agent and GrB/scFvMEL for 72 h.

Sequence II (C2): cells were pretreated with GrB/scFvMEL for 6 h, and then co-administered with chemotherapeutic agent and GrB/scFvMEL for 72 h.

Sequence III: cells were pretreated with GrB/scFvMEL for 6 h, followed by treatment with chemotherapeutic agents for 72 h.

Sequence IV: cells were treated with various chemotherapeutic agents for 72 h without GrB/scFvMEL pretreatment.

Chemotherapeutic agents include doxorubicin (DOX), vincristine (VCR), etoposide (VP-16), cisplantin (CDDP), cytarabine (Ara C) and 5 –FU.

There were synergistically cytotoxic effects on A375-M cells when the GrB/scFvMEL fusion construct in combination with the chemotherapeutic agents such as DOX, VCR and CDDP. In addition, this new class of agents has additive interactions with VP-16, Ara-C and 5-Fu. The cytotoxicities significantly increased when A375-M cells were pretreated with GrB/scFvMEL for 6 h followed by co-treatment with chemotherapeutic agents (sequence II-C2) compared to that the cells were pretreated with chemotherapeutic agents for 6 h followed by combination treatment (sequence I-C1) (Figure 14). Our data also demonstrated that the cytotoxicities of chemotherapeutic agents significantly increased when cells were pretreated with the fusion construct for 6 h followed by various chemotherapeutic agents for 72 h (sequence III) compared to without GrB/scFvMEL pretreatment (sequence IV) ($p < 0.01$). The results indicated that the effects of conventional chemotherapeutic agents could be sensitized by pretreatment with GrB/scFvMEL fusion protein for 6 h on gp240 positive targeted cells.

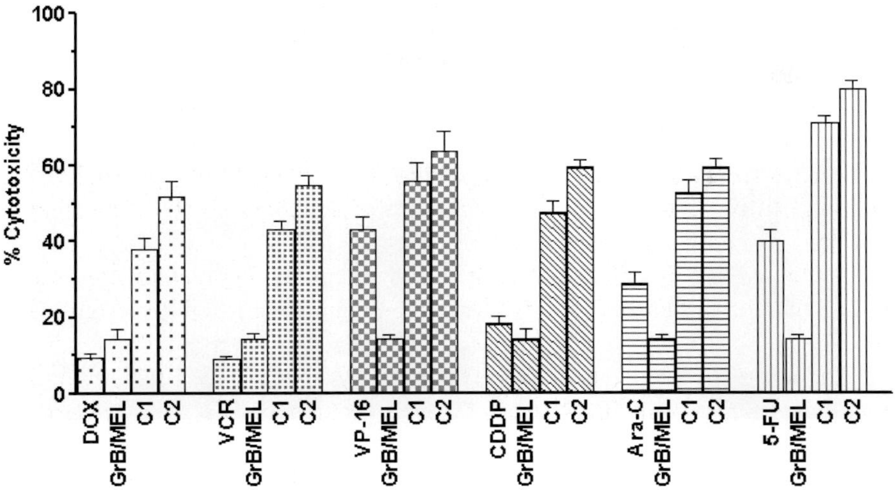


*Figure 14.* Studies of GrB/scFvMEL in combination with various chemotherapeutic agents. Antigen-positive(A375) cells were pretreated (at $IC_{25}$ doses) with various chemotherapeutic agents for 6 h followed by addition of GrB/scFvMEL($IC_{25}$). The cells were then incubated for a total of 72 h (sequence C1). Alternatively, cells were first treated with GrB/scFvMEL for 6 h, and then various chemotherapeutic agents were added for 72 h (sequence C2). Chemotherapeutic agents include doxorubicin (DOX), vincristine (VCR), etoposide (VP-16), cisplantin (CDDP), cytarabine (Ara C) and 5-Fu.

## 6. DOES GRB/SCFVMEL LOCALIZE IN TUMOR XENOGRAFT CELLS AND INDUCE TUMOR APOPTOSIS AFTER SYSTEMIC ADMINISTRATION?

Mice bearing A375-M xenograft tumors were administered GrB/scFvMEL (75 μg in 0.1 ml saline). Twenty-four hours later, animals were sacrificed and representative tissue sections were removed and formalin fixed and stained (H&E and TUNEL). Tumor tissue displayed apoptotic neuclei in treatment group (Figure 15).

# Treatment of Mice with GrB/scFvMEL Increases Apoptosis (TUNEL) in Tumors

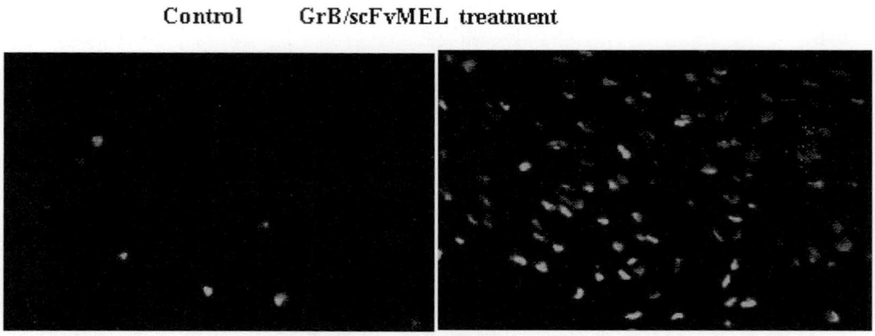


*Figure 15.* Mice bearing A375-M xenograft tumors were administered GrB/scFvMEL (37.5 mg/kg total dose, iv, tail vein in divided doses at the schedule indicated). Twenty-four hours after the last dose, a group of amimals were sacrificed and tumor tissues were removed and fixed. Tumor tissue sections were stained by TUNEL and analyzed under Nikon Eclipse TS 100 fluorescent microscope. Tumor tissue displayed highly apoptotic neuclei in the treatment group compared to the control group.

Immunohistochemical stains for GrB/scFvMEL detected by either anti-GrB or anti-scFvMEL antibody were performed. Localization or internalization of GrB/scFvMEL was observed in tumor tissue (Figure 16).

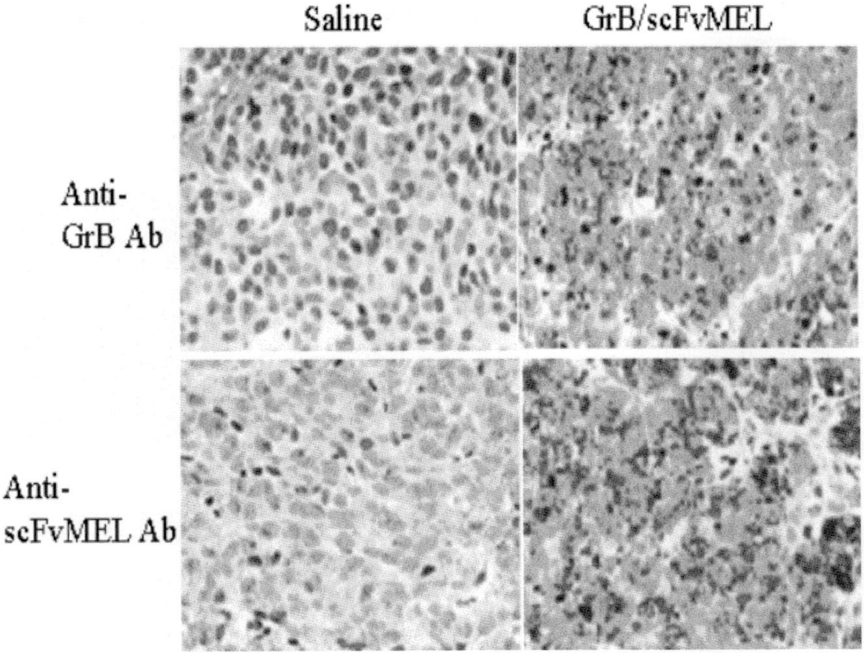


*Figure 16.* Mice bearing A375-M xenograft tumors were administered GrB/scFvMEL (37.5 mg/kg). Twenty-four hours after the last dose, animals were sacrificed and tumor tissues were removed, fixed and stained by immunohistochemical staining for GrB/scFvMEL detected by either anti-GrB or anti-scFvMEL antibody. Localization and internalization of GrB/scFvMEL was observed in tumor tissue in the treatment group but not in the control group.

# 7. DOES GRB/SCFVMEL HAVE IN VIVO ANTITUMOR EFFECTS?

Preliminary studies to examine the in vivo anti-tumor effects of GrB/scFvMEL were performed on A375-M human melanoma tumor xenografts. Mice bearing the tumors were treated (iv tail vein) 5 × every other day with either GrB/scFvMEL or saline. Tumor volumes were measured for 42 days. The saline-treated control tumors increased from 50 $mm^3$ to 1200 $mm^3$ over this period. Tumors treated with GrB/scFvMEL (37.5 mg/kg) increased from 50 $mm^3$ to 200 $mm^3$ (Figure 17).

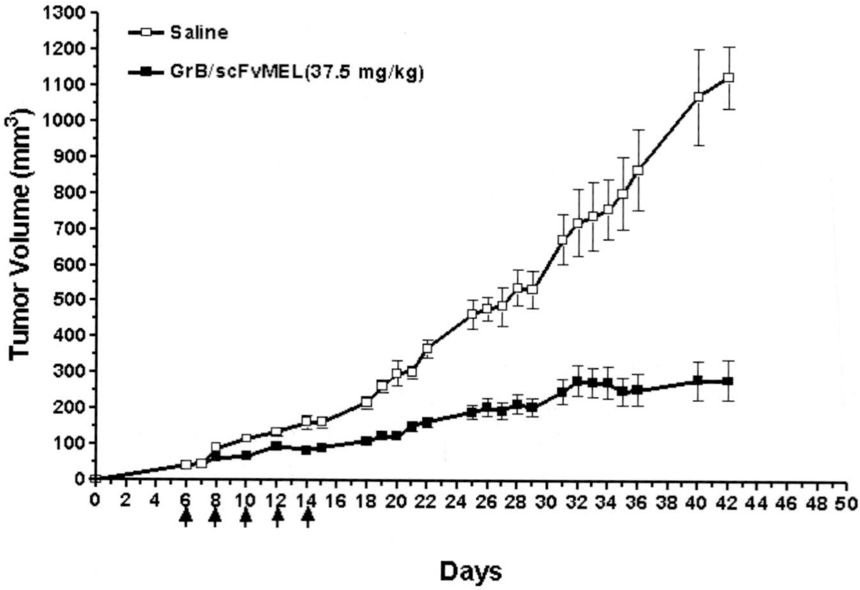


*Figure 17.* Athymic (nu/nu) mice, female, 6-8 weeks of age, were injected subcutaneously, right flank with 3 x 10 $^6$ log-phase A375-M cells and tumors were allowed to establish. Once tumors reached measurable size (~ 30 – 50 mm$^3$), animals were treated via i. v. tail vein with either saline (control) or GrB/scFvMEL fusion construct (37.5 mg/kg total dose) for 5 times every other day. Animals were monitored and tumors were measured for an additional 28 days. The saline-treated control tumors increased from 50 mm$^3$ to 1200 mm$^3$ over this period. In contrast, tumors treated with GrB/scFvMEL (37.5 mg/kg) increased from 50 mm$^3$ to 200 mm$^3$.

This study clearly demonstrates the biological activity of a new class of tumor-targeted enzymes that are cytotoxic to target cells because they are capable of direct activation of the nascent cellular pro-apoptotic pathway. Although several groups have generated antibody-enzyme chemical conjugates and fusion constructs, the purpose of the majority of these targeted enzymes have been to locally convert inactive pro-drugs to active therapeutic agents [83]. The primary types of directly cytotoxic enzymes commonly delivered by cell- targeting proteins such as antibodies and growth factors usually fall into the class of ribosome-inhibiting proteins (RIPs). Toxins such as pseudomonas exotoxin (PE) and gelonin (rGel) have been successfully utilized because only a few molecules are needed to irretrievably intoxicate a target cell [84;85]. Recently, Newton et al. described a new class of immunoconjugates containing human RNase which has in vitro and in vivo cytotoxic activity against human tumor cell lines and xenografts [86] primarily through degradation of RNA. The current construct, to our knowledge, is one of the first descriptions of a targeted enzymatic agent that operates primarily through activation of the pro-apoptotic cascade process. We were encouraged to note that the GrB/scFvMEL fusion construct demonstrated equivalent cytotoxic effect on target cells compared to a fusion toxin containing a highly potent plant n-glycosidase such as recombinant gelonin. Studies in our laboratory demonstrate that the scFvMEL/rGel fusion construct apparently cells through a necrotic rather than an apoptotic process. These comparative studies demonstrate that the robust cytotoxic effects of the rGel toxin can be matched by that of the GrB apoptotic effects.

On interaction of a cytotoxic T lymphocyte (CTL) with a target cell, there is a directed exocytosis of the CTL granules into the extracellular space between the two cells. The original view of entry for GrB into the target cell was that perforin polymerized to form a pore in the target cell membrane through which GrB could pass [87]. More recently, the discovery of a receptor for GrB that is the mannose-6-phosphate receptor has indicated that GrB might be taken up by receptor-mediated endocytosis [88;89] and that perforin may act to release endosomal GrB into the cytosol of the target cell [90].

The current study clearly demonstrates that an antibody delivery vehicle can provide the cellular entry access for the enzyme specifically on antigen-positive cells and is capable of delivering the enzyme to cytoplasm without the need for perforins. Moreover, the delivery vehicle can provide enzyme concentrations in the cytoplasm which are apparently sufficient to drive apoptosis. The anti-gp240 scFv antibody used has been well-studied in our laboratory and is capable of internalizing and delivering attached proteins such as toxins into the cytoplasm of target cells. The exact mechanism by which this particular antibody internalizes has not been fully elucidated. However, our studies suggest that there is no need for perforins or vesicle-disrupting factors in the antibody-mediated delivery of GrB to the cytoplasm. In addition, since the recombinant GrB protein is unglycosylated, it will not be taken up through the mannose-6-phosphate receptors, thereby reducing the likelihood of uptake by antigen-negative cells.

We have also described other constructs containing GrB as the cytotoxic moiety. Studies describing GrB fused to $VEGF_{121}$ for directed apoptosis targeting tumor vasculature indicated that the vasculature of many solid tumors were susceptible to the cytotoxic properties of the fusion construct [91]. In addition, as in the current study, the $I.C._{50}$ values for the GrB/$VEGF_{121}$ were similar to that of the $VEGF_{121}$/rGel construct.

The current study of GrB/scFvMEL extends these observations. These studies demonstrate that delivery of GrB to the cytosol of human tumor cells in culture can induce apoptosis. Recently studies showed that apoptosis induced by GrB is characteristically a rapid process taking approximately 1 h to result in the hallmarks of apoptosis such as DNA fragmentation, caspase cleavage and cytochrome c release[92]. Our studies indicated that apoptotic event happened within 8 h. This is because the antibody internalization process requires several hours for optimal delivery and the current process requires delivery of GrB to the cytoplasm in quantities sufficient to overcome normal homeostasis and to drive cells into apoptosis. The apoptosis induced by antibody delivery of GrB suggests that tumor cells, like normal endothelial cells, contain sufficient pro-apoptotic substrates capable of activation by GrB and that there appear to be no apparent downstream control-point mechanisms nominally in place to prevent pharmacologically directed apoptosis using this particular mechanistic approach. Mechanistic studies demonstrate release of cytochrome c from the mitochondria to the cytosol induced by the fusion protein GrB/scFvMEL as well as clear activation of caspase-3 specifically in target cells. These studies suggest that the cytotoxic/apoptotic events observed after treatment of target cells with the fusion construct are the result of both caspase-dependent and caspase-independent mechanisms. It remains to be determined whether there are cellular resistance mechanisms capable of protecting cells against GrB-containing targeted therapeutic agents. In addition, still to be determined is whether there are specific cell types inherently or iatrogenically resistant to this cytotoxic approach.

This appears to be the first description of the use of the proapoptotic, serine protease GrB in targeted therapeutic constructs. This agent is the culmination of our search for a human therapeutic protein that is both highly active and potentially universal. In addition, as a critical consideration, we demonstrated that the GrB enzyme was effective even while it remained covalently attached to an assembled targeting vehicle. This advance avoids problems with linker technology that can result in premature release of the active component prior to target internalization or loss of biological activity of the active component due to permanent attachment to the targeting molecule. Importantly, the demonstration of biological activity with a GrB-containing fusion protein does not appear to be confined to the current approach using $VEGF_{121}$ as a vehicle to target tumor endothelium. Recent studies in our laboratory have demonstrated impressive cytotoxic activity of GrB-containing fusion constructs composed of single-chain antibodies targeting numerous human tumor cell types. The unique mechanism of action of GrB- based fusion targeted therapeutic agents may also provide for novel interactions with other types of therapeutic agents or with other modalities such as hyperthermia or radiation therapy. Further in vitro and animal studies with this agent are clearly warranted.

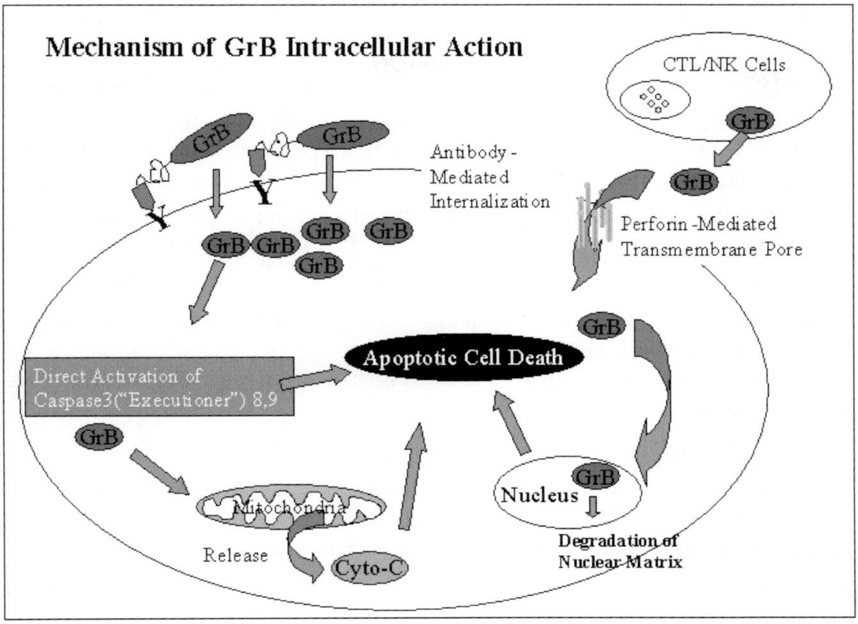


*Figure 18.* GrB intracellular activity can be produced by both perforin- and antibody-mediated internalization.

## AKNOWLEDGEMENTS

The authors wish to express their sincere thanks to Ms. Michelle McCall for her assistance in editing and formatting this article. In addition, the authors wish to thank the Clayton Foundation for Research for their support and encouragement.

## REFERENCES

1. J. Wu, Apoptosis and angiogenesis: two promising tumor markers in breast cancer (review), *Anticancer Res.* 16, pp. 2233-2239 (1996).
2. P. A. Furth, U. Bar-Peled, M. Li, A. Lewis, R. Laucirica, R. Jager, H. Weiher and R. G. Russell, Loss of anti-mitotic effects of Bcl-2 with retention of anti-apoptotic activity during tumor progression in a mouse model, *Oncogene* 18, pp. 6589-6596 (1999).
3. B. Katoch, S. Sebastian, S. Sahdev, H. Padh, S. E. Hasnain and R. Begum, Programmed cell death and its clinical implications, *Indian J. Exp. Biol.* 40, pp. 513-524 (2002).

4. S. Kato, Y. Masuhiro, M. Watanabe, Y. Kobayashi, K. I. Takeyama, H. Endoh and J. Yanagisawa, Molecular mechanism of a cross-talk between oestrogen and growth factor signalling pathways, *Genes Cells* 5, pp. 593-601 (2000).
5. D. Karunagaran, E. Tzahar, R. R. Beerli, X. Chen, D. Graus-Porta, B. J. Ratzkin, R. Seger, N. E. Hynes and Y. Yarden, ErbB-2 is a common auxiliary subunit of NDF and EGF receptors: implications for breast cancer, *EMBO J.* 15, pp. 254-264 (1996).
6. P. A. Furth, U. Bar-Peled, M. Li, A. Lewis, R. Laucirica, R. Jager, H. Weiher and R. G. Russell, Loss of anti-mitotic effects of Bcl-2 with retention of anti-apoptotic activity during tumor progression in a mouse model, *Oncogene* 18, pp. 6589-6596 (1999).
7. K. Schorr, M. Li, S. Krajewski, J. C. Reed and P. A. Furth, Bcl-2 gene family and related proteins in mammary gland involution and breast cancer, *J. Mammary. Gland. Biol. Neoplasia.* 4, pp. 153-164 (1999).
8. K. M. Debatin, Activation of apoptosis pathways by anticancer drugs, *Adv. Exp. Med. Biol.* 457, pp. 237-244 (1999).
9. L. Serrone and P. Hersey, The chemoresistance of human malignant melanoma: an update, *Melanoma Res.* 9, pp. 51-58 (1999).
10. S. Inoue, A. E. Salah-Eldin and K. Omoteyama, Apoptosis and anticancer drug resistance, *Hum. Cell* 14, pp. 211-221 (2001).
11. T. Lin, L. Zhang, J. Davis, J. Gu, M. Nishizaki, L. Ji, J. A. Roth, M. Xiong and B. Fang, Combination of TRAIL gene therapy and chemotherapy enhances antitumor and antimetastasis effects in chemosensitive and chemoresistant breast cancers, *Mol. Ther.* 8, pp. 441-448 (2003).
12. B. Bucci, E. Carico, A. Rinaldi, F. Froio, Y. M. Puce, I. D'Agnano, A. Vecchione and E. Brunetti, Biological indicators of aggressiveness in T1 ductal invasive breast cancer, *Anticancer Res.* 21, pp. 2949-2955 (2001).
13. A. Sierra, X. Castellsague, A. Escobedo, B. Lloveras, M. Garcia-Ramirez, A. Moreno and A. Fabra, Bcl-2 with loss of apoptosis allows accumulation of genetic alterations: a pathway to metastatic progression in human breast cancer, *Int. J. Cancer* 89, pp. 142-147 (2000).
14. P. Lipponen, Apoptosis in breast cancer: relationship with other pathological parameters, *Endocr. Relat Cancer* 6, pp. 13-16 (1999).
15. W. Wu, S. Chaudhuri, D. R. Brickley, D. Pang, T. Karrison and S. D. Conzen, Microarray analysis reveals glucocorticoid-regulated survival genes that are associated with inhibition of apoptosis in breast epithelial cells, *Cancer Res.* 64, pp. 1757-1764 (2004).
16. C. Botti, S. Buglioni, M. Benevolo, D. Giannarelli, P. Papaldo, F. Cognetti, P. Vici, F. Di Filippo, F. Del Nonno, F. M. Venanzi, P. G. Natali and M. Mottolese, Altered expression of FAS system is related to adverse clinical outcome in stage I-II breast cancer patients treated with adjuvant anthracycline-based chemotherapy, *Clin. Cancer Res.* 10, pp. 1360-1365 (2004).
17. S. Abramovitch and H. Werner, Functional and physical interactions between BRCA1 and p53 in transcriptional regulation of the IGF-IR gene, *Horm. Metab Res.* 35, pp. 758-762 (2003).
18. E. K. Rowinsky, Signal events: Cell signal transduction and its inhibition in cancer, *Oncologist.* 8 Suppl 3, pp. 5-17 (2003).
19. F. Martinez-Arribas, M. J. Nunez-Villar, A. R. Lucas, J. Sanchez, A. Tejerina and J. Schneider, Immunofluorometric study of Bcl-2 and Bax expression in clinical fresh tumor samples from breast cancer patients, *Anticancer Res.* 23, pp. 565-568 (2003).
20. P. A. Ellis, I. E. Smith, S. Detre, S. A. Burton, J. Salter, R. A'Hern, G. Walsh, S. R. Johnston and M. Dowsett, Reduced apoptosis and proliferation and increased Bcl-2 in

residual breast cancer following preoperative chemotherapy, *Breast Cancer Res. Treat.* 48, pp. 107-116 (1998).
21. S. J. Park, C. H. Wu and A. R. Safa, A P-glycoprotein- and MRP1-independent doxorubicin-resistant variant of the MCF-7 breast cancer cell line with defects in caspase-6, -7, -8, -9 and -10 activation pathways, *Anticancer Res.* 24, pp. 123-131 (2004).
22. C. D. Archer, M. Parton, I. E. Smith, P. A. Ellis, J. Salter, S. Ashley, G. Gui, N. Sacks, S. R. Ebbs, W. Allum, N. Nasiri and M. Dowsett, Early changes in apoptosis and proliferation following primary chemotherapy for breast cancer, *Br. J. Cancer* 89, pp. 1035-1041 (2003).
23. A. P. Simoes-Wust, T. Schurpf, J. Hall, R. A. Stahel and U. Zangemeister-Wittke, Bcl-2/bcl-xL bispecific antisense treatment sensitizes breast carcinoma cells to doxorubicin, paclitaxel and cyclophosphamide, *Breast Cancer Res. Treat.* 76, pp. 157-166 (2002).
24. M. J. Serrano, P. Sanchez-Rovira, I. Algarra, A. Jaen, A. Lozano and J. J. Gaforio, Evaluation of a gemcitabine-doxorubicin-paclitaxel combination schedule through flow cytometry assessment of apoptosis extent induced in human breast cancer cell lines, *Jpn. J. Cancer Res.* 93, pp. 559-566 (2002).
25. R. A. Campbell, P. Bhat-Nakshatri, N. M. Patel, D. Constantinidou, S. Ali and H. Nakshatri, Phosphatidylinositol 3-kinase/AKT-mediated activation of estrogen receptor alpha: a new model for anti-estrogen resistance, *J. Biol. Chem.* 276, pp. 9817-9824 (2001).
26. S. Syed and E. Rowinsky, The new generation of targeted therapies for breast cancer, *Oncology (Huntingt)* 17, pp. 1339-1351 (2003).
27. A. Awada, F. Cardoso, G. Atalay, R. Giuliani, M. Mano and M. J. Piccart, The pipeline of new anticancer agents for breast cancer treatment in 2003, *Crit Rev. Oncol. Hematol.* 48, pp. 45-63 (2003).
28. Y. Gutierrez-Puente, P. Zapata-Benavides, A. M. Tari and G. Lopez-Berestein, Bcl-2-related antisense therapy, *Semin. Oncol.* 29, pp. 71-76 (2002).
29. B. Yu, S. Kusmartsev, F. Cheng, M. Paolini, Y. Nefedova, E. Sotomayor and D. Gabrilovich, Effective combination of chemotherapy and dendritic cell administration for the treatment of advanced-stage experimental breast cancer, *Clin. Cancer Res.* 9, pp. 285-294 (2003).
30. P. P. Manna and T. Mohanakumar, Human dendritic cell mediated cytotoxicity against breast carcinoma cells in vitro, *J. Leukoc. Biol.* 72, pp. 312-320 (2002).
31. K. A. Candido, K. Shimizu, J. C. McLaughlin, R. Kunkel, J. A. Fuller, B. G. Redman, E. K. Thomas, B. J. Nickoloff and J. J. Mule, Local administration of dendritic cells inhibits established breast tumor growth: implications for apoptosis-inducing agents, *Cancer Res.* 61, pp. 228-236 (2001).
32. P. A. Henkart, Mechanism of lymphocyte-mediated cytotoxicity, *Annu. Rev. Immunol.* 3, pp. 31-58 (1985).
33. M. J. Smyth and J. A. Trapani, Granzymes: exogenous proteinases that induce target cell apoptosis, *Immunol. Today* 16, pp. 202-206 (1995).
34. E. R. Podack, Perforin: structure, function, and regulation, *Curr. Top. Microbiol. Immunol.* 178, pp. 175-184 (1992).
35. H. Yagita, M. Nakata, A. Kawasaki, Y. Shinkai and K. Okumura, Role of perforin in lymphocyte-mediated cytolysis, *Adv. Immunol.* 51, pp. 215-242 (1992).
36. K. A. Browne, E. Blink, V. R. Sutton, C. J. Froelich, D. A. Jans and J. A. Trapani, Cytosolic delivery of granzyme B by bacterial toxins: evidence that endosomal disruption, in addition to transmembrane pore formation, is an important function of perforin, *Mol. Cell Biol.* 19, pp. 8604-8615 (1999).

37. C. J. Froelich, K. Orth, J. Turbov, P. Seth, R. Gottlieb, B. Babior, G. M. Shah, R. C. Bleackley, V. M. Dixit and W. Hanna, New paradigm for lymphocyte granule-mediated cytotoxicity. Target cells bind and internalize granzyme B, but an endosomolytic agent is necessary for cytosolic delivery and subsequent apoptosis, *J. Biol. Chem.* 271, pp. 29073-29079 (1996).
38. M. J. Pinkoski, M. Hobman, J. A. Heibein, K. Tomaselli, F. Li, P. Seth, C. J. Froelich and R. C. Bleackley, Entry and trafficking of granzyme B in target cells during granzyme B-perforin-mediated apoptosis, *Blood* 92, pp. 1044-1054 (1998).
39. L. Shi, R. P. Kraut, R. Aebersold and A. H. Greenberg, A natural killer cell granule protein that induces DNA fragmentation and apoptosis, *J. Exp. Med.* 175, pp. 553-566 (1992).
40. M. Barry and R. C. Bleackley, Cytotoxic T lymphocytes: all roads lead to death, *Nat. Rev. Immunol.* 2, pp. 401-409 (2002).
41. A. J. Darmon, D. W. Nicholson and R. C. Bleackley, Activation of the apoptotic protease CPP32 by cytotoxic T-cell-derived granzyme B, *Nature* 377, pp. 446-448 (1995).
42. A. J. Darmon, D. W. Nicholson and R. C. Bleackley, Activation of the apoptotic protease CPP32 by cytotoxic T-cell-derived granzyme B, *Nature* 377, pp. 446-448 (1995).
43. Y. Gu, C. Sarnecki, M. A. Fleming, J. A. Lippke, R. C. Bleackley and M. S. Su, Processing and activation of CMH-1 by granzyme B, *J. Biol. Chem.* 271, pp. 10816-10820 (1996).
44. K. Orth, A. M. Chinnaiyan, M. Garg, C. J. Froelich and V. M. Dixit, The CED-3/ICE-like protease Mch2 is activated during apoptosis and cleaves the death substrate lamin A, *J. Biol. Chem.* 271, pp. 16443-16446 (1996).
45. A. J. Darmon, D. W. Nicholson and R. C. Bleackley, Activation of the apoptotic protease CPP32 by cytotoxic T-cell-derived granzyme B, *Nature* 377, pp. 446-448 (1995).
46. H. Duan, K. Orth, A. M. Chinnaiyan, G. G. Poirier, C. J. Froelich, W. W. He and V. M. Dixit, ICE-LAP6, a novel member of the ICE/Ced-3 gene family, is activated by the cytotoxic T cell protease granzyme B, *J. Biol. Chem.* 271, pp. 16720-16724 (1996).
47. T. Fernandes-Alnemri, R. C. Armstrong, J. Krebs, S. M. Srinivasula, L. Wang, F. Bullrich, L. C. Fritz, J. A. Trapani, K. J. Tomaselli, G. Litwack and E. S. Alnemri, In vitro activation of CPP32 and Mch3 by Mch4, a novel human apoptotic cysteine protease containing two FADD-like domains, *Proc. Natl. Acad. Sci. U. S. A* 93, pp. 7464-7469 (1996).
48. C. Vincenz and V. M. Dixit, Fas-associated death domain protein interleukin-1beta-converting enzyme 2 (FLICE2), an ICE/Ced-3 homologue, is proximally involved in, *J. Biol. Chem.* 272, pp. 6578-6583 (1997).
49. X. Yang, H. R. Stennicke, B. Wang, D. R. Green, R. U. Janicke, A. Srinivasan, P. Seth, G. S. Salvesen and C. J. Froelich, Granzyme B mimics apical caspases. Description of a unified pathway for trans-activation of executioner caspase-3 and -7, *J. Biol. Chem.* 273, pp. 34278-34283 (1998).
50. M. Barry and R. C. Bleackley, Cytotoxic T lymphocytes: all roads lead to death, *Nat. Rev. Immunol.* 2, pp. 401-409 (2002).
51. V. R. Sutton, J. E. Davis, M. Cancilla, R. W. Johnstone, A. A. Ruefli, K. Sedelies, K. A. Browne and J. A. Trapani, Initiation of apoptosis by granzyme B requires direct cleavage of bid, but not direct granzyme B-mediated caspase activation, *J. Exp. Med.* 192, pp. 1403-1414 (2000).

52. D. A. Jans, P. Jans, L. J. Briggs, V. Sutton and J. A. Trapani, Nuclear transport of granzyme B (fragmentin-2). Dependence of perforin in vivo and cytosolic factors in vitro, *J. Biol. Chem.* 271, pp. 30781-30789 (1996).
53. J. A. Trapani, K. A. Browne, M. J. Smyth and D. A. Jans, Localization of granzyme B in the nucleus. A putative role in the mechanism of cytotoxic lymphocyte-mediated apoptosis, *J. Biol. Chem.* 271, pp. 4127-4133 (1996).
54. J. A. Heibein, M. Barry, B. Motyka and R. C. Bleackley, Granzyme B-induced loss of mitochondrial inner membrane potential (Delta Psi m) and cytochrome c release are caspase independent, *J. Immunol.* 163, pp. 4683-4693 (1999).
55. J. A. Trapani, D. A. Jans, P. J. Jans, M. J. Smyth, K. A. Browne and V. R. Sutton, Efficient nuclear targeting of granzyme B and the nuclear consequences of apoptosis induced by granzyme B and perforin are caspase-dependent, but cell death is caspase-independent, *J. Biol. Chem.* 273, pp. 27934-27938 (1998).
56. Y. Liu, L. H. Cheung, J. W. Marks and M. G. Rosenblum, Recombinant single-chain antibody fusion construct targeting human melanoma cells and containing tumor necrosis factor, *Int. J. Cancer* 108, pp. 549-557 (2004).
57. J. A. Heibein, M. Barry, B. Motyka and R. C. Bleackley, Granzyme B-induced loss of mitochondrial inner membrane potential (Delta Psi m) and cytochrome c release are caspase independent, *J. Immunol.* 163, pp. 4683-4693 (1999).
58. M. G. Rosenblum, L. H. Cheung, Y. Liu and J. W. Marks, III, Design, expression, purification, and characterization, in vitro and in vivo, of an antimelanoma single-chain Fv antibody fused to the toxin gelonin, *Cancer Res.* 63, pp. 3995-4002 (2003).
59. J. A. Trapani, D. A. Jans, P. J. Jans, M. J. Smyth, K. A. Browne and V. R. Sutton, Efficient nuclear targeting of granzyme B and the nuclear consequences of apoptosis induced by granzyme B and perforin are caspase-dependent, but cell death is caspase-independent, *J. Biol. Chem.* 273, pp. 27934-27938 (1998).
60. R. R. Kantor, A. K. Ng, P. Giacomini and S. Ferrone, Analysis of the NIH workshop monoclonal antibodies to human melanoma antigens, *Hybridoma* 1, pp. 473-482 (1982).
61. K. Mujoo, L. Cheung, J. L. Murray and M. G. Rosenblum, Pharmacokinetics, tissue distribution, and in vivo antitumor effects of the antimelanoma immunotoxin ZME-gelonin, *Cancer Immunol. Immunother.* 40, pp. 339-345 (1995).
62. M. G. Rosenblum, L. Cheung, K. Mujoo and J. L. Murray, An antimelanoma immunotoxin containing recombinant human tumor necrosis factor: tissue disposition, pharmacokinetic, and therapeutic studies in xenograft models, *Cancer Immunol. Immunother.* 40, pp. 322-328 (1995).
63. M. G. Rosenblum, J. L. Murray, L. Cheung, R. Rifkin, S. Salmon and R. Bartholomew, A specific and potent immunotoxin composed of antibody ZME-018 and the plant toxin gelonin, *Mol. Biother.* 3, pp. 6-13 (1991).
64. D. J. Macey, S. J. Denardo, G. L. Denardo, J. K. Goodnight and M. W. Unger, Uptake of indium-111-labeled monoclonal antibody ZME-018 as a function of tumor size in a patient with melanoma, *Am. J. Physiol Imaging* 3, pp. 1-6 (1988).
65. N. Cordes, M. A. Blaese, L. Plasswilm, H. P. Rodemann and D. Van Beuningen, Fibronectin and laminin increase resistance to ionizing radiation and the cytotoxic drug Ukrain in human tumour and normal cells in vitro, *Int. J. Radiat. Biol.* 79, pp. 709-720 (2003).
66. S. Lehnert, D. Greene and G. Batist, Radiation response of drug-resistant variants of a human breast cancer cell line: the effect of glutathione depletion, *Radiat. Res.* 124, pp. 208-215 (1990).

67. R. R. Kantor, A. P. Albino, A. K. Ng and S. Ferrone, Biosynthesis and intracellular processing of four human melanoma associated antigens, *Cancer Res.* 46, pp. 5223-5228 (1986).
68. M. G. Rosenblum, J. L. Murray, L. Cheung, R. Rifkin, S. Salmon and R. Bartholomew, A specific and potent immunotoxin composed of antibody ZME-018 and the plant toxin gelonin, *Mol. Biother.* 3, pp. 6-13 (1991).
69. M. G. Rosenblum, B. Levin, M. Roh, D. Hohn, R. McCabe, L. Thompson, L. Cheung and J. L. Murray, Clinical pharmacology and tissue disposition studies of 131I-labeled anticolorectal carcinoma human monoclonal antibody LiCO 16.88, *Cancer Immunol. Immunother.* 39, pp. 397-400 (1994).
70. Y. Liu, L. H. Cheung, W. N. Hittelman and M. G. Rosenblum, Targeted delivery of human pro-apoptotic enzymes to tumor cells: In vitro studies describing a novel class of recombinant highly cytotoxic agents, *Mol. Cancer Ther.* 2, pp. 1341-1350 (2003).
71. M. G. Rosenblum, J. L. Murray, L. Cheung, R. Rifkin, S. Salmon and R. Bartholomew, A specific and potent immunotoxin composed of antibody ZME-018 and the plant toxin gelonin, *Mol. Biother.* 3, pp. 6-13 (1991).
72. M. G. Rosenblum, L. Cheung, J. L. Murray and R. Bartholomew, Antibody-mediated delivery of tumor necrosis factor (TNF-alpha): improvement of cytotoxicity and reduction of cellular resistance, *Cancer Commun.* 3, pp. 21-27 (1991).
73. M. G. Rosenblum, B. Levin, M. Roh, D. Hohn, R. McCabe, L. Thompson, L. Cheung and J. L. Murray, Clinical pharmacology and tissue disposition studies of 131I-labeled anticolorectal carcinoma human monoclonal antibody LiCO 16.88, *Cancer Immunol. Immunother.* 39, pp. 397-400 (1994).
74. M. G. Rosenblum, L. Cheung, K. Mujoo and J. L. Murray, An antimelanoma immunotoxin containing recombinant human tumor necrosis factor: tissue disposition, pharmacokinetic, and therapeutic studies in xenograft models, *Cancer Immunol. Immunother.* 40, pp. 322-328 (1995).
75. M. G. Rosenblum, L. Cheung, S. K. Kim, K. Mujoo, N. J. Donato and J. L. Murray, Cellular resistance to the antimelanoma immunotoxin ZME-gelonin and strategies to target resistant cells, *Cancer Immunol. Immunother.* 42, pp. 115-121 (1996).
76. M. G. Rosenblum, J. W. Marks and L. H. Cheung, Comparative cytotoxicity and pharmacokinetics of antimelanoma immunotoxins containing either natural or recombinant gelonin, *Cancer Chemother. Pharmacol.* 44, pp. 343-348 (1999).
77. M. G. Rosenblum, L. H. Cheung, Y. Liu and J. W. Marks, III, Design, expression, purification, and characterization, in vitro and in vivo, of an antimelanoma single-chain Fv antibody fused to the toxin gelonin, *Cancer Res.* 63, pp. 3995-4002 (2003).
78. Y. Liu, L. H. Cheung, W. N. Hittelman and M. G. Rosenblum, Targeted delivery of human pro-apoptotic enzymes to tumor cells: In vitro studies describing a novel class of recombinant highly cytotoxic agents, *Mol. Cancer Ther.* 2, pp. 1341-1350 (2003).
79. Y. Liu, L. H. Cheung, P. Thorpe and M. G. Rosenblum, Mechanistic studies of a novel human fusion toxin composed of vascular endothelial growth factor (VEGF)121 and the serine protease granzyme B: directed apoptotic events in vascular endothelial cells, *Mol. Cancer Ther.* 2, pp. 949-959 (2003).
80. M. J. Smyth, M. J. McGuire and K. Y. Thia, Expression of recombinant human granzyme B. A processing and activation role for dipeptidyl peptidase I, *J. Immunol.* 154, pp. 6299-6305 (1995).
81. M. Poe, J. T. Blake, D. A. Boulton, M. Gammon, N. H. Sigal, J. K. Wu and H. J. Zweerink, Human cytotoxic lymphocyte granzyme B. Its purification from granules and the characterization of substrate and inhibitor specificity, *J. Biol. Chem.* 266, pp. 98-103 (1991).

82. M. G. Rosenblum, L. H. Cheung, Y. Liu and J. W. Marks, III, Design, expression, purification, and characterization, in vitro and in vivo, of an antimelanoma single-chain Fv antibody fused to the toxin gelonin, *Cancer Res.* 63, pp. 3995-4002 (2003).
83. M. S. Tallman, Monoclonal antibody therapies in leukemias, *Semin. Hematol.* 39, pp. 12-19 (2002).
84. M. G. Rosenblum, L. H. Cheung, Y. Liu and J. W. Marks, III, Design, expression, purification, and characterization, in vitro and in vivo, of an antimelanoma single-chain Fv antibody fused to the toxin gelonin, *Cancer Res.* 63, pp. 3995-4002 (2003).
85. L. M. Veenendaal, H. Jin, S. Ran, L. Cheung, N. Navone, J. W. Marks, J. Waltenberger, P. Thorpe and M. G. Rosenblum, In vitro and in vivo studies of a VEGF121/rGelonin chimeric fusion toxin targeting the neovasculature of solid tumors, *Proc. Natl. Acad. Sci. U. S. A* 99, pp. 7866-7871 (2002).
86. D. L. Newton, H. J. Hansen, H. Liu, D. Ruby, M. S. Iordanov, B. E. Magun, D. M. Goldenberg and S. M. Rybak, Specifically targeting the CD22 receptor of human B-cell lymphomas with RNA damaging agents, *Crit Rev. Oncol. Hematol.* 39, pp. 79-86 (2001).
87. P. A. Henkart, Mechanism of lymphocyte-mediated cytotoxicity, *Annu. Rev. Immunol.* 3, pp. 31-58 (1985).
88. C. J. Froelich, K. Orth, J. Turbov, P. Seth, R. Gottlieb, B. Babior, G. M. Shah, R. C. Bleackley, V. M. Dixit and W. Hanna, New paradigm for lymphocyte granule-mediated cytotoxicity. Target cells bind and internalize granzyme B, but an endosomolytic agent is necessary for cytosolic delivery and subsequent apoptosis, *J. Biol. Chem.* 271, pp. 29073-29079 (1996).
89. B. Motyka, G. Korbutt, M. J. Pinkoski, J. A. Heibein, A. Caputo, M. Hobman, M. Barry, I. Shostak, T. Sawchuk, C. F. Holmes, J. Gauldie and R. C. Bleackley, Mannose 6-phosphate/insulin-like growth factor II receptor is a death receptor for granzyme B during cytotoxic T cell-induced apoptosis, *Cell* 103, pp. 491-500 (2000).
90. M. J. Pinkoski, M. Hobman, J. A. Heibein, K. Tomaselli, F. Li, P. Seth, C. J. Froelich and R. C. Bleackley, Entry and trafficking of granzyme B in target cells during granzyme B-perforin-mediated apoptosis, *Blood* 92, pp. 1044-1054 (1998).
91. Y. Liu, L. H. Cheung, P. Thorpe and M. G. Rosenblum, Mechanistic studies of a novel human fusion toxin composed of vascular endothelial growth factor (VEGF)121 and the serine protease granzyme B: directed apoptotic events in vascular endothelial cells, *Mol. Cancer Ther.* 2, pp. 949-959 (2003).
92. Y. Liu, L. H. Cheung, W. N. Hittelman and M. G. Rosenblum, Targeted delivery of human pro-apoptotic enzymes to tumor cells: In vitro studies describing a novel class of recombinant highly cytotoxic agents, *Mol. Cancer Ther.* 2, pp. 1341-1350 (2003).

# Chapter 5

# TARGETING ONCOGENES WHICH REGULATE APOPTOSIS

Andrew C. Phillips
*Institute of Molecular Medicine and Genetics, Medical College of Georgia, CB2803, 1120 15th Street, Augusta, Georgia 30912.*

Abstract: Characterization of the genetic changes driving the malignant phenotype has revealed that mutations promoting cell survival are common in human cancer. The increased understanding of cancer at the molecular level has lead to the development of therapeutics aimed at inhibiting the function of the genes responsible for the tumor phenotype. Inhibition of the activity of the oncogenes promoting tumor cell survival is an attractive approach for the development of cancer therapeutics that has already yielded some success and promises more.

Key words: Oncogene, tyrosine kinase inhibitor, Bcl-2, IAP, BCR-ABL, EGFR, RAS, AKT, p53, NF-kB

## 1. ACTIVATION OF ONCOGENES CAN PROMOTE CELL SURVIVAL WITHOUT INCREASING PROLIFERATION

Proto-oncogenes are normal cellular genes that when mutated, or expressed in appropriately contribute to tumor formation. The first oncogenes discovered were mutated forms of genes involved in the regulation of cell growth. As such, activation of these genes by mutation promotes proliferation in a cell autonomous fashion, regardless of external growth factor signaling. The first oncogene described that contributed to tumor formation by inhibition of cell death was identified by characterization of B-cell lymphoma 2 gene (*bcl2*). Cloning of a reciprocal



*M. Sluyser (ed.), Application of Apoptosis to Cancer Treatment,* 117-144.

translocation site associated with Follicular lymphoma determined that *bcl2* is brought under the regulation of an immunoglobulin locus, deregulating its expression. This over-expression inhibits the normal apoptosis of B-cells, contributing to tumorigenesis by promoting survival rather than enhancing proliferation. Transgenic studies have demonstrated that proliferative oncogenes such as c-Myc can cooperate with Bcl2 in tumor formation underlining the importance of controlling both cell proliferation and death in restricting tumor development (1).

Subsequently many other oncogenes have been shown to promote cell survival and function in diverse manners. In this review we will discuss a variety of oncogenes that promote tumor cell survival and are currently the subject investigation as potential targets for the development or evaluation of therapeutics that inhibit their function.

## 2. INHIBITION OF ONCOGENE ACTIVITY AS A VIABLE THERAPEUTIC STRATEGY

Since the identification of oncogenes and tumor suppressors as the driving force behind malignant transformation, much effort has been expended identifying and characterizing their function in both normal and tumor cells. This increased understanding has allowed the rational design of therapeutics aimed at inhibiting oncogene function. In developing inhibitors of these oncogenes as therapeutic agents for cancer treatment we make the assumption that in addition to being necessary for the initiation of the tumor these oncogenes are required for its maintenance. Furthermore given the plasticity of the cancer genome, is targeting a specific oncogene likely to have a positive clinical outcome? The problem of toxicity to normal tissue is also a consideration. As oncogenes exist in non-transformed cells as proto-oncogenes, pharmacological inhibition of oncogenes could have profound implications for certain normal cells function.

The role of oncogenes in tumor maintenance has been investigated using transgenic animals and provided compelling evidence that the inactivation of a single oncogene can lead to dramatic regression of tumor. A transgenic animal was generated by targeting a tetracycline inducible allele of *H-RasV12G* to melanocytes(2). RAS is a signaling molecule that is activated by direct mutation in many human malignancies including melanoma. Mice were crossed into a INK4a null background to generate a model of melanoma. *INK4a* is a tumor suppressor that is deleted or mutated in both familial and sporadic melanoma, and encodes a cyclin-dependent kinase inhibitor that restricts cell proliferation(3). In the presence of tetracycline 25% of the mice developed melanomas with an average latency of 60 days.

The removal of tetracycline from the animal's drinking water resulted in rapid tumor regression with marked apoptosis. However, in some of these animals tumors can re-emerge that do not express *H-RasV12G*, indicating that alternative pathways for tumor maintenance can evolve.

A similar approach targeted mutant *K-Ras* to type II pneumocytes, resulting in adenocarcinomas in the lung(4). Switching off expression of mutant *K-Ras* resulted in dramatically reduced tumor burden after three days and undectable disease after a month. The tumor regression was accompanied by dramatic apoptosis(4). To investigate the mechanism of apoptotic regression the experiments were repeated in *INK4a/ARF* and *p53* null backgrounds. These critical tumor suppressors function by eliminating potentially malignant cells by inducing growth arrest or apoptosis(3). In this genetic background tumor formation was accelerated, but the tumor remained dependent on mutant *K-Ras* expression, with rapid regression seen when mutant *K-Ras* was switched off(4). This indicates that the apoptotic regression is not a consequence of activation of a p53 or INK4a tumor suppressor pathways, an important finding since these are frequently deleted in human lung cancer (and many other malignancies). In this case the apoptosis was seen only in the tumor cells suggesting a cell autonomous effect. In contrast, apoptosis was seen in both tumor and host endothelial cells in the melanoma model, indicating a role for RAS in maintaining poorly understood tumor-host cell interactions. These results and others indicate that inhibition of RAS activity would be a viable treatment protocol, even in cancer with mutations in critical tumor suppressors.

Similar work has indicated that inhibition of other oncogenes including *Myc*(5, 6), *Bcr-abl*(7), *HER2*(8) and *Met*(9) can lead to rapid tumor regression even in the case of complex metastatic disease presumably driven by a number of oncogenic mutations in addition to the regulated oncogene(8). The mechanism of regression seems to differ depending on the oncogene and tumor type, with both apoptosis and differentiation of the tumor cells playing a role.

Whilst these results indicate that certain experimental tumors in mice require sustained oncogene expression, in treatment of human disease, pharmacological inhibition of some oncogenes for long periods of time may not be possible due to the role of these genes in normal cells. In this regard it is interesting that even a short inactivation of Myc can lead to a sustained loss of neoplastic phenotype(10). *Myc* is an oncogene that can drive proliferation, but when over-expressed in the absence of compensating mutation can trigger apoptosis. In a transgenic model with features similar to human osteosarcoma, including frequent metastasis, inhibition of *Myc* expression leads to regression, but in this case the mechanism seems to be differentiation rather than apoptosis. Interestingly reactivation of *Myc* leads not to re-emergence of the tumor, but apoptosis. This shows the complexity of the tumor phenotype and the difficulty of predicting the effect of inhibition of an oncogene *in vivo*, never the less it further indicates that the inhibition of oncogene function is a viable therapeutic strategy.

Despite significant regression in response to oncogene inactivation, in some animal models the tumors can re-emerge, indicating that it is no longer dependent on the targeted oncogene (4, 6). Particular genetic changes can predispose tumor to relapse despite dramatic initial regression of tumor after inhibition of oncogene expression. In mouse models of mammary tumors, activation of RAS(11) or loss of p53 (12) increased the chance of relapse despite inactivation of the *Myc* or *Wnt* oncogenes respectively. Given the plasticity of the cancer genome this is not surprising. Good animal models may allow the identification of the pathways activated to over-come the dependence on the oncogene. As such a second line of therapy targeting these pathways would be the goal.

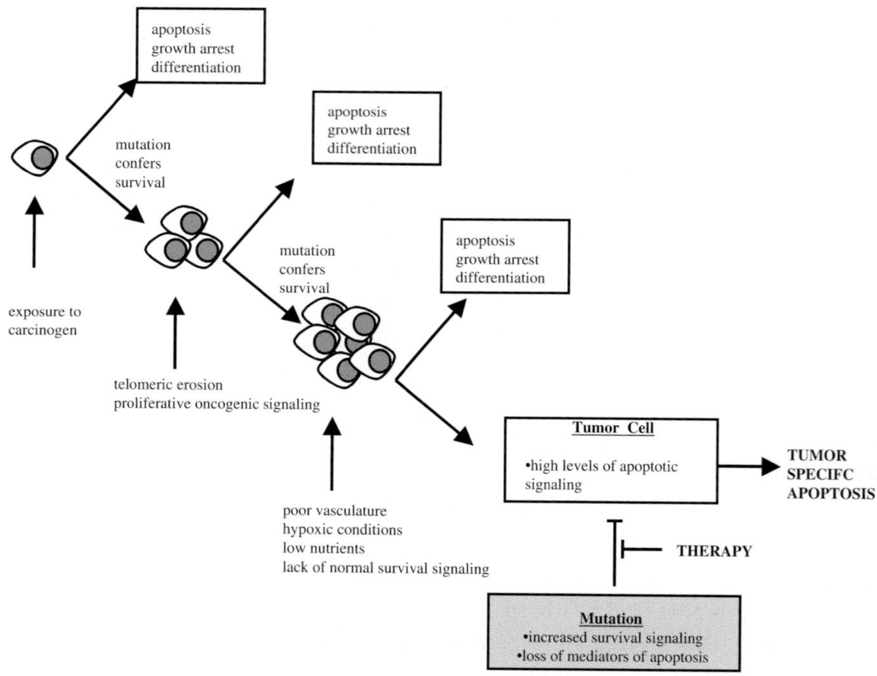


*Figure 1.* Mutations that inhibit apoptosis allow tumor cells to tolerate apoptotic signals. The developing tumor cell is subject to variety of stresses that can lead to the elimination of the cell before full malignant transformation can occur such as DNA damage, hypoxia, low nutrients, telomeric erosion etc, and survive only as a result of compensating mutation. Therapy aimed at abrogating the activity of the oncogenes promoting tumor cell survival may have a greater effect on tumor than normal tissue.

A pre-requisite for therapies that would inhibit this signaling survival driven by oncogenes is that the tumor cells are more dependent on this signaling than normal cells. The developing tumor is subject to a variety of stresses generating selection pressure for cells that have mutation increasing resistance to cell death, for example amplification of the pro-survival kinase Akt(13). The consequent increased resistance to apoptosis allows the cells to tolerate additional stresses that may be associated with the developing tumor such as increased DNA damage, low oxygen conditions, low nutrients, telomeric erosion and lack of survival signaling (figure 4-1). As such the developing tumor cell is exposed to more stress than its normal counter-part and survives as a consequence of mutations blocking apoptosis. Furthermore, activation of oncogenes that drive proliferation such as *Myc* or *E2F* can also trigger apoptosis(14). These oncoproteins are normally transiently activated by external growth factor signaling to promote proliferation, and these growth factors concomitantly activate survival-signaling pathways suppressing the apoptosis stimulated by expression of these oncogenes. This 'oncogene-induced apoptosis' is thought to be a defense mechanism against malignant transformation whereby a single genetic event stimulating proliferation does not lead to cancer as the cell is eliminated by apoptosis. For tumor cells to survive sustained proliferative oncogene expression, compensating mutations that inhibit apoptosis are required. Hence tumor cells may carry a high apoptotic burden compared to their normal counterparts that is tolerated as a consequence of acquisition of mutations promoting survival. Inhibition of oncogenes promoting survival may have a greater effect on tumor than normal tissues and this is supported by the development of therapeutics against specific oncogenes that are relatively well tolerated despite triggering apoptosis of tumor cells (figure 4-1).

## 3. DIRECT MODULATORS OF THE APOPTOTIC MACHINERY AS MOLECULAR TARGETS

Understanding how apoptosis is controlled and mediated at the molecular level is an area of intense research, and has identified a number of targets for the development of therapeutics to enhance apoptosis(15). For the treatment of cancer, a greater effect on the tumor compared to the normal tissue is required. If a tumor commonly contains an amplification of an oncogene that can inhibit apoptosis, it is reasonable too predict that the tumor may be more dependent on the activity of this gene than normal tissue. As such common genetic lesions in a particular tumor type may point to the critical molecules regulating tumor cell survival, and targeting these oncogenes may be a fruitful approach to develop cancer therapeutics. A limited number of genes directly involved in the control of apoptosis are known to function as oncogenes and as such are the subject of investigation as molecular targets for cancer therapy.

## 3.1 Bcl2 antagonists

The intrinsic pathway of apoptosis is activated in response to many stimuli, including radiation and chemotherapeutic drugs. The mitochondria play a central role in the mediation of this apoptotic pathway sequestering a number of pro-apoptotic molecules(16). In response to signals activating this pathway, mitochondrial outer membrane permeabilization (MOMP) is triggered, and pro-apoptotic molecules are released into the cytosol initiating the caspase proteolytic cascade as well as caspase independent cell death pathways(16, 17). Whilst the precise mechanism of action is unclear, Bcl2 inhibits this pathway, maintaining mitochondrial membrane integrity(18).

There is a large family of Bcl2 related proteins that contain at least one of the four Bcl2 homology regions (BH domains). These include proteins such as $BclX_L$, Bcl-w and Mcl-1 that inhibit apoptosis(1). Other family members including Bax, Bim, Bok, Puma and Noxa promote apoptosis. A subset of these proteins contain only BH3 domains, and are otherwise unrelated. This domain mediates protein-protein interactions and is necessary for the promotion of apoptosis. These BH3-only proteins respond to a number of signals to initiate the intrinsic apoptotic pathways and can interact with Bcl2 like proteins inhibiting their function. For example death receptor activation mediates the activation of caspase 8, this protease cleaves the BH3-only protein BID to generate a truncated form (tBID). tBID translocates to the mitochondria triggering MOMP, coupling death receptor signaling to the activation of the intrinsic pathway of apoptosis(19). In contrast the BH3-only proteins PUMA and Noxa are transcriptionally activated by p53 and related family members, and are critical mediators of p53 and chemotherapy induced apoptosis(20-23).

*Bcl2* was originally identified by characterization of a translocation in follicular lymphoma, but activation of this oncogene is not restricted to this leukemia. Over-expression of Bcl2 and the related $BclX_L$ has been reported in a wide variety of human malignancies, including prostate, colorectal, lung and breast cancers(24). Not surprisingly, the search for therapies that can inhibit the activity of the Bcl2 like proteins is being pursued. The identification of small molecule inhibitors that block the interaction between Bcl2/$BclX_L$ and BH3-only proteins is one such approach. Generally inhibition of protein-protein interactions with small molecule inhibitors is extremely difficult, however, although most interaction surfaces are relative large and flat, structural studies indicate that $BclX_L$ and Bcl2 contain a deep hydrophobic groove, that serves as the site of interaction with BH3 domains(25, 26). This structural feature has spurred the search for inhibitors of these protein-proteins interactions. The first small molecule inhibitor of Bcl2 was discovered by an *in silico* screen, identifying HA14-1 as molecule that can bind to Bcl2(27). This molecule can displace a Bak peptide containing the BH3 domain from Bcl2 at micro-molar concentrations. HA14-1 is able to kill tumor cells, and there is a correlation between the level of expression of Bcl2 and the effectiveness of the drug. Comparison of the sensitivity of leukemic blast and normal cells, revealed that low doses of HA14-1 in combination with chemotherapeutic agents triggers apoptosis specifically in the leukemic blasts.

Testing known inhibitors of mitochondrial respiration for the ability to induce apoptosis identified another inhibitor of Bcl2/$BclX_L$. Antimycin A can inhibit Bcl2/$BclX_L$, and a derivative molecule 2-methoxy antimycin A, can bind Bcl2/$BclX_L$ and trigger apoptosis without the general effect on mitochondrial function of antimycin A(27). Another natural inhibitor of Bcl2 functions is tetrocarcin A, although the mechanism is unknown(28). These inhibitors are promising lead molecules that might allow translation of a successful small molecule antagonist of Bcl2 related proteins in the clinic.

An antisense approach to the inhibition of Bcl2 has been successfully employed in pre-clinical models and is currently in clinical trials. Antisense technology utilizes short synthetic stretches of DNA that are complementary to specific regions of the target mRNA. Binding of the DNA to mRNA results in inhibition of gene expression as a result of blocking translation of the mRNA or targeting the mRNA for RNAseH cleavage. Genasense is a phosphorothioate oligo nucletide complementary to the first 6 codons of Bcl2. This modification renders the oligo more stable *in vitro*, and can efficiently down regulate Bcl2 expression. In xenograph experiments Genasense was able to inhibit tumor formation, and the effect was enhanced by combination with chemotherapeutic drugs(29). This treatment is currently in phase III clinical trials for a number of malignancies including chronic lymphocytic leukemia, multiple myeloma and melanoma after positive clinical responses in phase I/II trial for chronic lymphocytic leukemia(30).

Successful antagonists of Bcl2 family members may have broad applicability in cancer treatment. In addition to mutations that increase Bcl2 or $BclX_L$ in tumors, radiation and chemotherapy trigger apoptosis by the mitochondrial pathways, consequently inhibition of their function may increase the response to standard therapies.

## 3.2 Inhibitor of Apoptosis Protein (IAP) antagonists

The Inhibitors of apoptosis (IAP) are able to suppress apoptosis, at least in part, by inhibiting caspase activity. IAPs are defined by the presence of a Baculovirus IAP repeat domain (BIR) and this domain is necessary for the inhibition of apoptosis and at least for certain IAPs is responsible for the binding and inhibition of caspase activity(31). Several IAPs also contain a RING domain and can function as ubiquitin ligases involved in the proteolytic degradation of themselves and interacting molecules including caspase 3(32-34).

Interacting proteins play a critical role in the regulation of IAP function. Smac/DIABLO (human/mouse) are proteins that in non-apoptotic cells are located in the inter-membrane space of the mitochondria and during translocation the amino-terminus is removed revealing an protein interaction domain termed the IAP-binding motif (IBM). In response to apoptotic signals these proteins are released into the cytoplasm(35, 36), bind IAPs, displacing caspases thus promoting apoptosis. Omi/HtrA2 is a second mitochondrial-sequestered protein containing an IBM, but is also a serine protease that can trigger cell death in an IAP independent manner(37-39).

Since IAPs primarily function downstream of caspase activity (figure 4-2), whether modulating their activity would be a good therapy is unclear, but in the least could serve as adjunct treatments with other agents. However there is accumulating evidence that IAPs may have a role in the development of cancer, suggesting they may contribute to tumor cell survival, as such are potential targets for therapy. Chromosomal translocation of c-IAP2 causes aberrant expression of a chimeric IAP-MALT1 gene in mucosal associated lymphoma (40). Amplification of c-IAP1 is implicated in esophageal squamous cell carcinoma(41), and ML-IAP is over-expressed in melanomas(42). Over-expression of survivin is common in a variety of cancers, although whether this is a causative in the sense of inhibiting apoptosis or is a consequence of the enhanced proliferation of tumor tissues in unclear(43). XIAP expression is elevated in a number of cancers, and the expression level of this gene has been correlated with poor prognosis(44). In addition the XIAP interacting protein Xaf1 has been suggested to be a tumor suppressor gene, with low expression in tumor lines (45, 46), and gastric tumor samples(47).

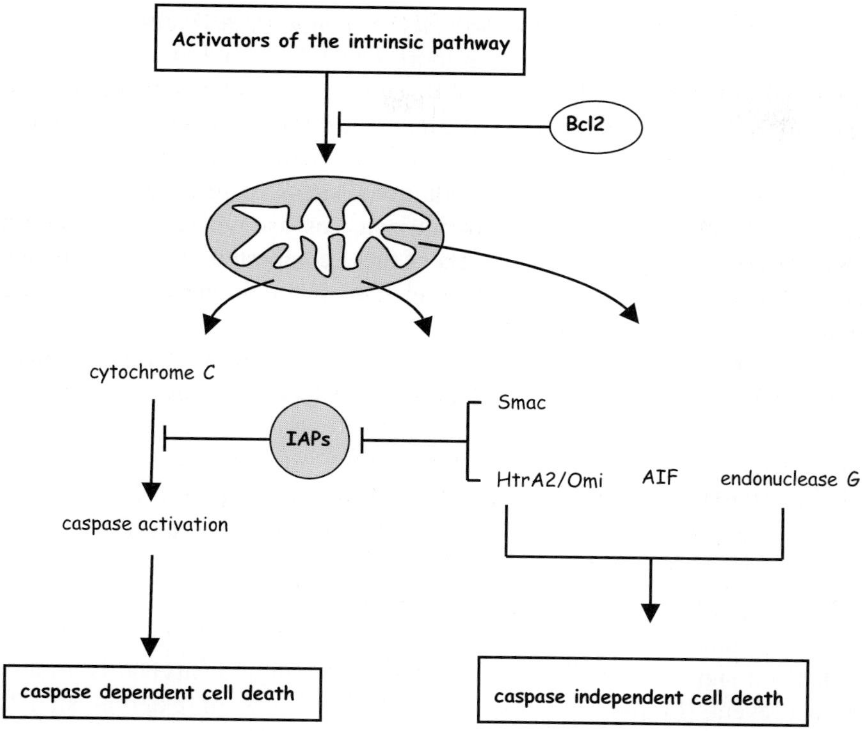


*Figure 2.* Bcl2 and IAPs regulate apoptosis at different points in the apoptotic cascade. Bcl2 inhibits loss of mitochondrial outer membrane potential, and IAPs inhibit caspase activity.

Studies in tumor cell lines have demonstrated that inhibition of IAPs function can sensitize cells to apoptosis. Moreover small cell permeable peptides of the IAP binding protein SMAC that inhibit IAP function are able to sensitize cells to a variety of agents including cisplatin and the death receptor ligand Apo2L/TRAIL, as can specific inhibition of XIAP itself providing evidence that inhibition of IAP activity may be a viable therapeutic strategy(48-53). Small molecule inhibitors of XIAP have recently been identified by high through-put screening of combinatorial libraries of polyphenylurea compounds for the ability to relieve XIAP repression of caspase activity(54). This screen identified molecules that are able to induce apoptosis in tumor cells and inhibit tumor growth in xenograft models, whilst exhibiting little toxicity to normal tissues. In addition the drugs were able to cooperate with a broad range of cytotoxic agents, including radiation, chemotherapeutic agents and the death receptor ligand TRAIL. These inhibitors may have potential for broad applicability as chemo-sensitizers for agents that trigger activation of caspases.

The success of the IAP inhibition as a therapeutic strategy depends on the assumption that the tumor cells are more dependent on IAP activity than normal cells. Supporting this is the absence of dramatic phenotypes for any of the IAP null mice generated to date, including XIAP, indicating that in the absence of stress they are dispensable or redundant. The ability of XIAP inhibitors to kill tumor cells without additional stress suggests that the tumor cells may have a greater amount of intrinsic caspase activity than normal cells, if so this may explain the success of these compounds. It is unknown if these compounds inhibit other IAPs, and if so if this is important for their tumor killing properties or if there are targets other than IAPs. Never the less these inhibitors are promising lead compounds for to develop cancer therapeutics and validate targeting IAPs as a therapeutic strategy.

## 4. TARGETING MOLECULES MEDIATING SURVIVAL SIGNALING IN TUMOR CELLS

Direct mutation of downstream effectors of the apoptotic cascade is not the most common mechanism tumor cells use to evade apoptosis. Mutation in upstream signaling pathways that regulate growth and survival is a more frequent event. In multi-cellular organisms, a series of external signals control growth differentiation and apoptosis. Without the appropriate signaling, cells will not survive. As such components of the signal transduction pathway are frequently mutated in human cancer promoting tumor cell survival, and have been investigated as potential molecular targets for cancer therapy.

### 4.1 BCR-ABL

Chronic Myeloid Leukemia is characterized by the presence of the Philadelphia chromosome. This reciprocal translocation results in the generation a chimeric protein termed BCR-ABL, leading to constitutive and inappropriate activation of the ABL tyrosine kinase driving tumor formation. An inhibitor of the ABL tyrosine kinase activity was designed (Gleevec/Imatinib) and is able to block the growth and induce apoptosis of BCR-ABL containing myeloid cell lines or cells from CML patients and is in use in the clinic as a highly effective agent for this disease(55). This drug has high selectively at sub μιχρο–molar levels, affecting the proliferation and survival of only BCR-ABL+ cells. In clinical trials comparing effectiveness with the standard treatment of interferon A and cytarabine for chronic phase CML, complete cytological responses were seen in 76% of the patients compared with 15% for those treated with interferon/cytarabine(56).

In addition Gleevec caused fewer side effects and reduced the probability of progression from chronic to the blast crisis phase of the disease. Patients in blast crisis respond to this drug, but normally resistance evolves. This resistance is most commonly the result of an increase of BCR-ABL levels or mutation affecting the ABL kinase domain(57). That the resistant tumor cells normally directly mutate the *bcr-abl* gene indicates that the leukemic cells remain dependent on continued ABL activity for survival, although in some cases mutations downstream of ABL signaling have also been shown to mediate resistance. A series of different small molecular inhibitors of ABL have been developed and are showing promise in pre-clinical studies(58, 59), and these may be useful as alternative treatments to counter Gleevec resistance.

In addition to inhibiting BCR-ABL, Gleevec can also inhibit v-Abl, c-Abl, the PDFR, c-Kit and ARG kinases and as such its applicability may be extended to other treatment of tumors. Currently this drug has been approved for the treatment of gastrointestinal stromal tumors (GIST) after successful clinical trials(60). In this tumor type gain of function mutations in the c-kit oncogene are common. Several trials are underway to assess the effect of this inhibitor on different tumor types.

## 4.2 EGFR family

The epidermal growth factor receptor (EGFR) and the human EGFR related receptor (HER2/neu) are receptor tyrosine kinases (RTK) that contribute to the malignant phenotype of a variety of human cancers(61). Ligand binding results in homo or hetero-dimerization and auto-phosphorylation stimulating the tyrosine kinase activity. Subsequently adaptor and signaling molecules recognize the C-terminal phospho-tyrosines and trigger activation of a variety of important signal transduction pathways including RAS and phosphatidylinositol 3-kinase (PI3K) mediated activation of Akt, promoting a variety of cellular responses including proliferation and cell survival(61).

Evidence for a role of these RTKs in cancer was first provided by studies of the transforming properties of the Avian Erthyroblastosis Virus identifying a viral oncogene *v-erbB* that is homologous to the human *EGFR*, but encodes a truncated version with constitutive tyrosine kinase activity. In human malignancies the *EGFR* family of RTK are frequently over-expressed in a variety of tumor types including stomach, lung, breast and ovarian cancer. In addition to over-expression, mutation of the *EGFR* can lead to elevated catalytic tyrosine kinase activity irrespective of external signaling, most commonly deletion of exons 2-7 by gene rearrangement or alternative splicing(62-64). The importance of these RTK in a variety of human cancers has lead to the development of a number of therapeutics designed to inhibit their function. Quinazolines were identified as compounds capable of inhibition of EGFR tyrosine kinase activity(65) and lead to the development of the metabolically stable and specific EGFR tyrosine kinase inhibitor Gefitinib (ZD1839/Iressa)(66). Preclinical studies established this inhibitor as an effective agent for blocking the growth and survival of a variety of human tumor cell lines and xenografts, regardless of the level of EGFR expression. After successful clinical trials in Japan, Europe and the USA Gefitinib was approved for the second line treatment of advanced non-small cell lung carcinoma (NSCLC) in Japan and of third line treatment in the USA(67). The original rationale of testing this agent in NSCLC was that the EGFR is often over-expressed and current chemotherapy is not particularly effective for this cancer(68, 69). However in contrast to the ABL-specific tyrosine kinase inhibitor where the majority of CML patients treated with Gleevec show a dramatic response, only a subset of the NSCLC patients treated with Gefitinib respond, although in these patients the response can be dramatic and rapid suggesting EGFR activity is necessary for the survival of these tumor cells. No correlation with the level of the EGFR expression and response to tumor therapy could be found(67), however, recent studies suggest that patients with gain of function mutations in the EGFR responded to the treatment with Gefitinib(70, 71). This work correlated mutations in the tyrosine kinase domain to positive response to treatment. Functional analysis of two of these mutations revealed an enhanced and extended response to ligand treatment. Furthermore these mutant EGFRs were 10 fold more sensitive to Gefitinib than their wild type counterparts(71). A variety of other EGFR inhibitors have been successful in preclinical models including CI-1033 (Pfizer) and EKB-569 (Wyeth), that in contrast to Gefitinib bind non-competitively to EGFR family of RTK and are curently in Phase II trials for ovarian and NSCLC(72-74).

In addition to small molecule inhibitors of RTK activity, monoclonal antibodies have been developed to inhibit the activity of these receptors. HER2 is amplified in breast and ovarian cancer and Genetech developed a humanized antibody that targets HER2 termed trastuzumab (Herceptin). This antibody binds to HER2 inducing receptor internalization leading to growth arrest and apoptosis of tumor cells, and recruitment of immune effector cells. HER2 is over-expressed in approximately 25% of breast cancers and this is correlated with poor prognosis. Trastuzumab is approved by the FDA for first line treatment of metastatic breast cancer over-expressing HER2 in combination with chemotherapy. Trastuzumab treatment significantly improves the response to therapy, reducing disease progression and increasing survival of patients when compared to chemotherapy alone(75). A second antibody against HER2 that inhibits receptor signaling by blocking dimerization with other family members is currently in phase II trials(75). An anti-EGFR monoclonal antibody Cetuximab is also approved for the treatment of metastatic colorectal cancer in patients that do not respond to standard irinotecan chemotherapy. Cetuximab reduced tumors in a number of patients, increasing survival and delaying disease progression both alone and in combination with irinotecan of patients with irinotecan refractory colo-rectal cancer (76).

## 4.3 RAS as a molecular target

The RAS proteins play critical roles in regulating a number of signaling pathways controlling cell proliferation and survival(77). The membrane associated RAS proteins are active when bound to GTP and inactive when bound to GDP. RAS has intrinsic GTPase activity but this process can be regulated *in vivo* by the guanidine activating proteins (GAPs) that down regulate RAS activity, and Guanidine nucleotide exchange factors (GEF) that replace GDP with GTP. A large family of proteins mediates these reactions *in vivo*, allowing RAS activity to be modulated by multiple signaling pathways in the cell(77). These include the receptor tyrosine kinases such as the EGFR, which activate RAS by recruitment of GEFs to the plasma membrane where they promote the formation of GTP-bound RAS. GTP bound RAS can bind and activate effector enzymes, the most studied of which is RAF, (leading to activation of the mitogen activated protein kinase (MAPK) cascade promoting proliferation) and the catalytic subunit of PI3K resulting the activation of a large number of downstream modulators including Akt, a powerful inhibitor of apoptosis(78).

Analysis of the viral oncogenes in the Harvey and Kirsten sarcoma viruses, and the isolation of their cellular counterparts as oncogenes in transformation assays indicated a role for the *Ras* genes in cancer. Both the viral and the transforming *Ras* genes derived from tumors are constitutively active mutants, and the importance of these genes in human cancer is indicated by direct mutation in around 20% of tumors overall(79). The RAS proteins can contribute to a number of different properties of the malignant phenotype including promoting proliferation, survival, invasion, metastasis and angiogenesis(77). In addition even in tumors without direct mutation of *Ras*, the RAS pathway contributes to the tumor phenotype as a consequence of its activation by growth factor signaling or mutation of other oncogenes such as RTK or deletion of GAP proteins such as the tumor suppressor neurofibromin (*NF1*).

The high frequency of *Ras* mutation in human cancers and experimental evidence that sustained RAS activity was necessary for tumor maintenance suggests inhibition of RAS, or the RAS pathway may be a viable therapeutic approach(80). A series of Farnesyltransferase inhibitors (FTI) have been designed to inhibit the activity of RAS proteins. The rational of this approach is that the RAS proteins need to be targeted to the plasma membrane to be active and *in vivo* farnesylation of the c-terminus is the first step in the processing of the RAS proteins to achieve membrane localization. A variety of effective inhibitors of farnesylation were developed by a number of pharmaceutical companies, but there initial promise has not been realized(81, 82). Although these inhibitors can block farnesylation of H-RAS in cell culture and reduce tumor formation in *H-ras* transgenic mice, results in clinical trials have been disappointing(83). The disappointing results may reflect alternative modifications that can target RAS proteins to the membrane. Geranylgeranyltransferase can add a related isoprenoid group to K-RAS and N-RAS when farnesylation is blocked, allowing these proteins to be biologically active. Since mutations in *K-Ras* and *N-Ras* account for the majority of mutations in human cancers, this may explain the disappointing results in clinical trials, and efforts to combine FTI and GTI have been hampered due to toxicity(84). Whether this indicates that inhibition of all RAS activity is toxic to normal tissues, or other effects of these agents is unclear. The FTI may yet become a tool in the in treatment of certain cancers since positive anti-tumor effects have been seen in certain leukemia patients, although at this stage the critical molecular targets of these agents are unclear(85).

In addition to the FTI, anti-sense oligo-nucleotides targeting *H-Ras* and *K-Ras* have been developed that are stabilized by chemical modification and are able to reduce the expression these proteins in tumor cell lines and reduce tumor growth in nude mice. The *H-Ras* anti-sense therapy is currently in phase II trials. A problem of anti-sense therapy in general is the ability to deliver large enough quantities of these molecules to the tumor cells. In addition the specificity of the agents, targeting only a single family member may be ineffective. Agents effective against multiple family members are likely to be more effective treatments as long as this does not result in increased toxicity to normal tissues.

Targeting pathways that are activated by RAS is an area of intense effort. The RAF-MAPK pathway itself can be active in tumor cells as a result of mutation of *Ras*, mutations upstream (e.g. RTK such as *EGFR*) loss of negative regulators of RAS (the GAP proteins) or mutation of kinases such as *Raf*(86). An inhibitor of RAF has shown promise in pre-clinical work and is currently in clinical trials. This inhibitor BAY43-9600 targets the ATP-binding site efficiently inhibiting kinase activity(87). This molecule also inhibits the related BRAF protein, which may make the drug more effective since *c-Raf1* mutation is not common in human malignancies, in contrast in mutation of *BRAF* is found in around 90% of melanomas(88).

The kinases MEK1 and MEK2 are RAF substrates and inhibitors of these molecules are able to block proliferation and apoptosis of a wide range of tumor cell lines, as well as inhibiting the growth of tumors in nude mice(89, 90). An inhibitor of MEK, CI-1040 has undergone phase I trials and is well tolerated by patients and some partial responses were seen. Analysis of the peripheral blood mono-nucleocytes demonstrated that these doses of drug were able to inhibit the activation of ERK (a MEK substrate) indicating that the drug is an effective inhibitor of MEK in humans(91). Given the prevalence of mutations that activate this pathway in human cancer these drugs have the potential of broad specificity. Phase II trials are underway and will determine their effectiveness.

## 4.4 Akt

The Akt kinases are activated by PI3K, a downstream target of active RAS, and Akt plays an important role in mediating the survival signal generated by RAS(92). PI3K phosphorylates the lipid phosphatidylinositol-4,5-bisphosphate (PtdIns(4,5)$P_2$) to generate phosphatidylinositol-3,4,5-triphosphate (PtdIns(3,4,5)$P_{3)}$. This lipid second messenger binds to Akt activating it serine/threonine kinase activity. Akt has a number of substrates involved in the regulation of apoptosis including BAD, a BH3-only protein that can trigger the release of pro-apoptotic molecules from the mitochondria. Akt-mediated phosphorylation of this protein inhibits its activity. When activated this kinase is a powerful inhibitor of apoptosis and plays a role in the development of a number of human malignancies(78). These kinases can function as oncogenes and *Akt* genes are amplified in a number of malignancies including Ovarian and Breast cancers(93). A more common mutation leading to increased Akt activity is loss of the tumor suppressor PTEN (Phosphatase and Tensin homologue). This protein is a lipid phosphatase removing phosphates from (PtdIns(3,4,5)$P_3$) to generate (PtdIns(4,5)$P_2$) thus reducing Akt activation. PTEN loss is extremely common in a variety of human malignancies(94).

Inhibitors of both PI3K and Akt are currently being tested in preclinical and early clinical trials. Unc-01 was originally identified as a protein kinase C inhibitor but seems to have broader specificity being able to block the PI3K mediated activation of Akt inducing apoptosis of tumor cells and is currently in clinical trials(95). Development of Akt inhibitors is at an early stage, but targeting downstream mediators of receptors signaling such as Akt is an attractive approach as therapeutic intervention at earlier points in the signal transduction cascade would not be predicted to affect tumors containing mutations downstream of this points in the signal transduction cascade.

## 5. REACTIVATING P53 IN TUMORS. HDM2 AS A THERAPEUTIC TARGET

The tumor suppressor *p53* is the most frequently mutated gene in human cancer, and encodes a transcription factor that activates a number of biological programs that can prevent malignant transformation, most notably growth arrest and apoptosis(96). Under normal conditions the levels of p53 are maintained by a negative feedback loop involving Hdm2. p53 induces expression of Hdm2, which functions as a ubiquitin ligase, binding to p53 and mediating its ubiquitin dependent degradation. In response to a variety of stresses associated with the initiation or progression of tumorigenesis, this negative feedback loop is interrupted leading to the rapid stabilization of p53 (figure 4-3) (97). Amplification of *Hdm2* is associated with a number of malignancies, and inhibition of Hdm2 activity is a potential therapeutic strategy(98). Inhibition of the Hdm2-p53 interaction as a potential therapeutic approach was first investigated with small peptides, leading to p53 stabilization growth arrest and apoptosis(99). Peptides as therapeutics raise the considerable difficulty of efficient delivery to tumor cells, and ideally classical cell permeable small molecule drugs are desirable. Using a high through-put screen for small molecules that can inhibit the p53-Hdm2 interaction, Vassilev *et. al.,* identified a series of synthetic cis-imidazole analogues they termed Nutlins capable of blocking p53-Hdm2 interaction(100). Whilst historically it has been difficult to inhibit non-enzymatic protein-protein interactions with small molecule inhibitors, crystal structure analysis revealed that the p53 binding site of Hdm2 is a deep hydrophobic pocket, and the inhibitors bind in this same pocket(99, 100).

As would be predicted these inhibitors are capable of stabilizing p53 and inducing apoptosis and growth arrest of p53 positive tumor cell lines, with the greatest effect seen on a cell line with amplified *Hdm2*. In contrast although proliferation of non-transformed human and mouse fibroblasts was inhibited when treated with one of the inhibitors (Nutlin-3), the cells remain viable. Oral administration of this inhibitor was able to inhibit growth of a xenograft tumor with amplified *Hdm2* by 90% (similar to the effect of intravenous administration of the chemotherapeutic agent doxorubicin) without obvious detrimental effects on the mouse. Although the research is at an early stage, the development of these inhibitors opens the promise of reactivating the tumor suppressive effects of p53 in tumors with amplified *Hdm2*. In principle targeting Hdm2 as a strategy to elevate p53 may not be restricted only to tumors that contain amplified *Hdm2*, but any tumor retaining wild type p53. The success in these preliminary experiments may reflect the preference of p53 to induce apoptosis in tumor cells, whilst only triggering growth arrest in normal cells(101).

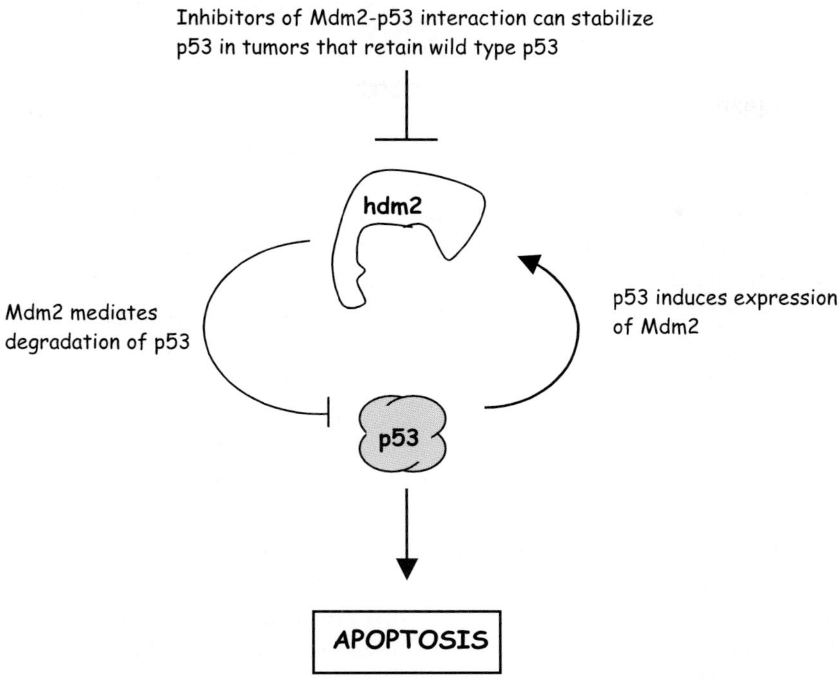


*Figure 3.* Inhibition of Hdm2 as a therapeutic strategy. The tumor suppressor p53 is regulated by a negative feedback loop; p53 induces expression of hdm2, and Hdm2 degrades p53. Blocking the activity of Hdm2 in tumors that retain wild type p53 can lead to apoptosis.

## 6. NF-KB AS A THERAPEUTIC TARGET

The NF-kB family consists of transcription factors that are defined by the presence of a Rel domain, that specifies DNA binding properties and interaction with the Ik-B proteins(102). In the absence of an activating signal NF-kB is normally in the cytoplasm bound to the repressors Ik-Bα and Ik-Bβ. In response to an activating signal such as TNFα treatment, Ik-B kinase (IKK) phosphorylates Ik-B targeting this protein for ubiquitin-mediated degradation in the proteosome(103). This allows NF-kB to translocate to the nucleus and activate genes typically involved in immune and immflamatory responses, promotion of cell growth and the inhibition of apoptosis.

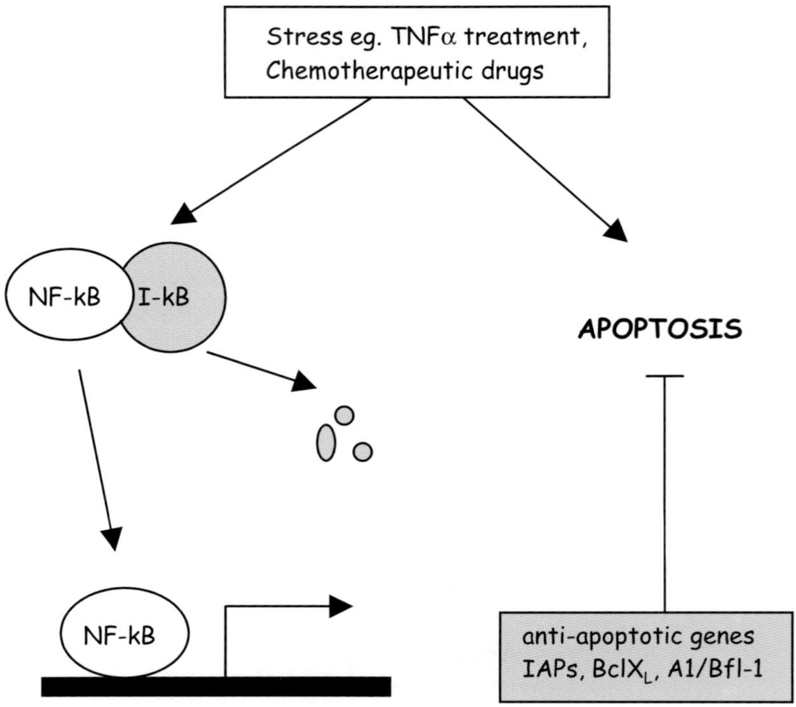


*Figure 4.* NF-kB inhibits apoptosis. Under most conditions, NF-kB is bound in the cytoplasm by an inhibitor I-kB. In response to activating signals, I-kB is degraded allowing NF-kB to translocate to the nucleus activating genes that inhibit apoptosis. Inhibition of NF-kB can enhance chemotherapeutic and radiation responses in animal models.

NF-kB has been demonstrated to play a role in oncogenesis with *v-Rel* the viral homolog of *c-Rel* causing aggressive tumors in chickens. In human maligancy direct activation by amplification or translocation has been seen for *c-Rel*, *NF-kB2* and increased NF-kB activity is often found in tumor lines(104). In addition a number of oncogenes have been shown to require NF-kB activity for the transforming activity. Expression of a ' Ik-B super-repressor' that blocks activation of NF-kB, prevents foci formation by oncogenic *Ras* in immortalized mouse fibroblasts and BCR-ABL driven tumorigenesis(105).Whilst it is clear that NF-kB is involved in certain human malignancies, what role it is playing is not as clear for this plieotropic factor. Depending on the context, inhibition of apoptosis or growth promotion may be the critical property, although roles in metastasis and angiogenesis have also been postulated. Multiple studies have demonstrated that NF-kB is anti-apoptotic, and can activate potent inhibitors of apoptosis including IAPs and Bcl2 homologues(106, 107).

An additional property of NF-kB that makes it an interesting molecular target for cancer treatment is that it is activated in response to a variety of stress signals, including radiation and chemotherapeutic agents (figure 4-4). Inhibition of NF-kB by the use of the super-repressor enhances the effect of radiation and chemotherapy in tumor cell lines(106, 107). Experiments in xenograft models have confirmed this and have been extended to show that an inhibitor of the proteosome (PS-341) can be substituted for infection with the super-repressor(105, 108, 109). The logic here being that this agent blocks the normal degradation of Ik-B and activation of NF-kB. Clearly a caveat for these experiments is that the observed effects may not be the result of modulation of NF-kB activation, as inhibition of proteosomal degradation will lead to changes in the abundance of many proteins within a cell.The proteosome inhibitor PS-341 has been utilized in a variety of clinical trials and its use was stimulated in part by its ability to inhibit activation of NF-kB(103). PS-341 (Velcade) has been successful in clinical trials for multiple myeloma and has received accelerated FDA approval for use a third line of therapy for multiple myeloma after phase II trial showed that 28-38% of patients responsed to treatment(110). What importance the targeting of NF-kB plays in the effectiveness of this treatment is unknown, but a compound (PS-1145) that blocks the phosphorylation of IKK can also inhibit the growth and survival of multiple myeloma cells, although less efficiently than PS-341(111).

Other agents that can inhibit NF-kB are currently in use as cancer therapeutics are arsenic trioxide ($As_2O_3$) and Thalidomide. $As_2O_3$ is effective in treating acute promyelocytic leukemia and is FDA approved for treated patients with relapse, with complete responses of 85% in trials (although the reduction in expression of a driving chimeric oncogene is likely to play a role in the response of this leukemia to treatment)(112). It is currently being evaluated in multiple myeloma in combination therapies(111). Thalidomide is also capable of blocking NF-kB activation and has shown promise in clinical trials(113-115). Caution however for all of these agents is required in ascribing the effect to an inhibition of NF-kB as they all have diverse effects on cell function.

NF-kB is generally thought of as a promoter of growth or survival, however in certain contexts it can be growth inhibitory or pro-apoptotic. In the epidermis inhibition of NF-kB leads to hyper-proliferation, cooperating with RAS in the induction of squamous cell carcinomas (SCC), and cytoplasmic sequestration of NF-kB has been reported in SCC of the skin(116-118). In addition an obligatory role for NF-kB in p53-induced apoptosis has also been demonstrated suggesting that in certain tumors inhibition of NF-kB could potentially be counter productive(119). Furthermore, although NF-kB can protect against radiation and certain chemotherapeutic drugs, NF-kB has been shown to be necessary for Taxol induced apoptosis(120). Despite these limitations, inhibition of NF-kB remains a promising approach for tumors dependent on NF-kB activity, or as an adjunct treatment for radiation or chemotherapy.

## 7. FUTURE PERSPECTIVES

Animal models have demonstrated that oncogenes that drive the tumorigenic process are also required for tumor maintenance and in some cases, loss of expression of the oncogene is sufficient to eliminate the tumor(48-53). This indicates that targeting oncogenes is a viable strategy for the treatment of human malignancies. The success of Gleevec in the treatment of chronic myeloid leukemia indicates that rational design of drugs based on a good molecular understanding of the disease can be successful. It is of note that the re-emergence of disease in some patients has indicated that these leukemic cells retain a dependence on BCR-ABL signaling for survival, since mutations within BCR-ABL conferring resistance to Gleevec are common(121). The development of related inhibitors may allow a cocktail approach that should be a more effective treatment of the disease. The success of the treatment is due to the ability to match an effective treatment targeted to a specific oncogenic lesion - in this case presence of the Philadelphia Chromosome. The ability to identify patients with amplified HER2 in breast cancer allows identification of the patients likely to respond to Trastuzumab treatment(75). Work with the EGFR tyrosine kinase inhibitors further illustrate that even when a drug is effective it may be effective in only a subset of patients, presumably with specific constellation of oncogenic changes conferring dependence on EGFR signaling. In the case of NSCLC described above, it seems that mutants of the EGFR with enhanced activity predicts treatment efficacy(70, 71). This illustrates the need for better molecular classification of many solid tumor types, to be able to identify the underlying oncogenes that are driving the malignant phenotype and allowing the use of therapeutics targeted to particular oncogenes. Micro array and proteomic technologies offer the potential to more accurately define tumor subtypes, as well matching expression sets that are characteristic of positive responses to different targeted therapies. Hopefully the explosion in the understanding of the molecular basis of tumor formation and maintenance will begin to be successfully translated in better more specific therapies for many human cancers.

# REFERENCES


1. Cory, S. and Adams, J. M. The Bcl2 family: regulators of the cellular life-or-death switch. Nat Rev Cancer, *2:* 647-656, 2002.
2. Chin, L., Tam, A., Pomerantz, J., Wong, M., Holash, J., Bardeesy, N., Shen, Q., O'Hagan, R., Pantginis, J., Zhou, H., Horner, J. W., 2nd, Cordon-Cardo, C., Yancopoulos, G. D., and DePinho, R. A. Essential role for oncogenic Ras in tumour maintenance. Nature, *400:* 468-472, 1999.
3. Sherr, C. J. The INK4a/ARF network in tumour suppression. Nat Rev Mol Cell Biol, *2:* 731-737, 2001.
4. Fisher, G. H., Wellen, S. L., Klimstra, D., Lenczowski, J. M., Tichelaar, J. W., Lizak, M. J., Whitsett, J. A., Koretsky, A., and Varmus, H. E. Induction and apoptotic regression of lung adenocarcinomas by regulation of a K-Ras transgene in the presence and absence of tumor suppressor genes. Genes Dev, *15:* 3249-3262, 2001.
5. Pelengaris, S., Khan, M., and Evan, G. I. Suppression of Myc-induced apoptosis in beta cells exposes multiple oncogenic properties of Myc and triggers carcinogenic progression. Cell, *109:* 321-334, 2002.
6. Giuriato, S., Rabin, K., Fan, A. C., Shachaf, C. M., and Felsher, D. W. Conditional animal models: a strategy to define when oncogenes will be effective targets to treat cancer. Semin Cancer Biol, *14:* 3-11, 2004.
7. Huettner, C. S., Zhang, P., Van Etten, R. A., and Tenen, D. G. Reversibility of acute B-cell leukaemia induced by BCR-ABL1. Nat Genet, *24:* 57-60, 2000.
8. Moody, S. E., Sarkisian, C. J., Hahn, K. T., Gunther, E. J., Pickup, S., Dugan, K. D., Innocent, N., Cardiff, R. D., Schnall, M. D., and Chodosh, L. A. Conditional activation of Neu in the mammary epithelium of transgenic mice results in reversible pulmonary metastasis. Cancer Cell, *2:* 451-461, 2002.
9. Wang, R., Ferrell, L. D., Faouzi, S., Maher, J. J., and Bishop, J. M. Activation of the Met receptor by cell attachment induces and sustains hepatocellular carcinomas in transgenic mice. J Cell Biol, *153:* 1023-1034, 2001.
10. Jain, M., Arvanitis, C., Chu, K., Dewey, W., Leonhardt, E., Trinh, M., Sundberg, C. D., Bishop, J. M., and Felsher, D. W. Sustained loss of a neoplastic phenotype by brief inactivation of MYC. Science, *297:* 102-104, 2002.
11. D'Cruz, C. M., Gunther, E. J., Boxer, R. B., Hartman, J. L., Sintasath, L., Moody, S. E., Cox, J. D., Ha, S. I., Belka, G. K., Golant, A., Cardiff, R. D., and Chodosh, L. A. c-MYC induces mammary tumorigenesis by means of a preferred pathway involving spontaneous Kras2 mutations. Nat Med, *7:* 235-239, 2001.
12. Gunther, E. J., Moody, S. E., Belka, G. K., Hahn, K. T., Innocent, N., Dugan, K. D., Cardiff, R. D., and Chodosh, L. A. Impact of p53 loss on reversal and recurrence of conditional Wnt-induced tumorigenesis. Genes Dev, *17:* 488-501, 2003.
13. Evan, G. I. and Vousden, K. H. Proliferation, cell cycle and apoptosis in cancer. Nature, *411:* 342-348, 2001.
14. Labazi, M. and Phillips, A. C. Oncogenes as regulators of apoptosis. Essays Biochem, *39:* 89-104, 2003.
15. Danial, N. N. and Korsmeyer, S. J. Cell death: critical control points. Cell, *116:* 205-219, 2004.
16. Brenner, C. and Kroemer, G. Apoptosis. Mitochondria--the death signal integrators. Science, *289:* 1150-1151, 2000.
17. Green, D. R. and Kroemer, G. The pathophysiology of mitochondrial cell death. Science, *305:* 626-629, 2004.

18. Sharpe, J. C., Arnoult, D., and Youle, R. J. Control of mitochondrial permeability by Bcl-2 family members. Biochim Biophys Acta, *1644:* 107-113, 2004.
19. Sprick, M. R. and Walczak, H. The interplay between the Bcl-2 family and death receptor-mediated apoptosis. Biochim Biophys Acta, *1644:* 125-132, 2004.
20. Nakano, K. and Vousden, K. H. PUMA, a novel proapoptotic gene, is induced by p53. Molecular Cell, *7:* 683-694, 2001.
21. Oda, E., Ohki, R., Murasawa, H., Nemoto, J., Shibue, T., Yamashita, T., Tokino, T., Taniguchi, T., and Tanaka, N. Noxa, a BH3-only member of the Bcl-2 family and candidate mediator of p53-induced apoptosis. Science, *288:* 1053-1058, 2000.
22. Yu, J., Zhang, L., Hwang, P. M., Kinzler, K. W., and Vogelstein, B. PUMA induces the rapid apoptosis of colorectal cancer cells. Mol Cell, *7:* 673-682, 2001.
23. Jeffers, J. R., Parganas, E., Lee, Y., Yang, C., Wang, J., Brennan, J., MacLean, K. H., Han, J., Chittenden, T., Ihle, J. N., McKinnon, P. J., Cleveland, J. L., and Zambetti, G. P. Puma is an essential mediator of p53-dependent and -independent apoptotic pathways. Cancer Cell, *4:* 321-328, 2003.
24. Kirkin, V., Joos, S., and Zornig, M. The role of Bcl-2 family members in tumorigenesis. Biochim Biophys Acta, *1644:* 229-249, 2004.
25. Petros, A. M., Medek, A., Nettesheim, D. G., Kim, D. H., Yoon, H. S., Swift, K., Matayoshi, E. D., Oltersdorf, T., and Fesik, S. W. Solution structure of the antiapoptotic protein bcl-2. Proc Natl Acad Sci U S A, *98:* 3012-3017, 2001.
26. Sattler, M., Liang, H., Nettesheim, D., Meadows, R. P., Harlan, J. E., Eberstadt, M., Yoon, H. S., Shuker, S. B., Chang, B. S., Minn, A. J., Thompson, C. B., and Fesik, S. W. Structure of Bcl-xL-Bak peptide complex: recognition between regulators of apoptosis. Science, *275:* 983-986, 1997.
27. Wang, J. L., Liu, D., Zhang, Z. J., Shan, S., Han, X., Srinivasula, S. M., Croce, C. M., Alnemri, E. S., and Huang, Z. Structure-based discovery of an organic compound that binds Bcl-2 protein and induces apoptosis of tumor cells. Proc Natl Acad Sci U S A, *97:* 7124-7129, 2000.
28. Nakashima, T., Miura, M., and Hara, M. Tetrocarcin A inhibits mitochondrial functions of Bcl-2 and suppresses its anti-apoptotic activity. Cancer Res, *60:* 1229-1235, 2000.
29. Cotter, F. E. Antisense oligonucleotides for haematological malignancies. Haematologica, *84 Suppl EHA-4:* 19-22, 1999.
30. Marcucci, G., Byrd, J. C., Dai, G., Klisovic, M. I., Kourlas, P. J., Young, D. C., Cataland, S. R., Fisher, D. B., Lucas, D., Chan, K. K., Porcu, P., Lin, Z. P., Farag, S. F., Frankel, S. R., Zwiebel, J. A., Kraut, E. H., Balcerzak, S. P., Bloomfield, C. D., Grever, M. R., and Caligiuri, M. A. Phase 1 and pharmacodynamic studies of G3139, a Bcl-2 antisense oligonucleotide, in combination with chemotherapy in refractory or relapsed acute leukemia. Blood, *101:* 425-432, 2003.
31. Duckett, C. S., Li, F., Wang, Y., Tomaselli, K. J., Thompson, C. B., and Armstrong, R. C. Human IAP-like protein regulates programmed cell death downstream of Bcl-x(L) and cytochrome c. Molecular and Cellular Biology, *18:* 608-615, 1998.
32. Huang, H., Joazeiro, C. A., Bonfoco, E., Kamada, S., Leverson, J. D., and Hunter, T. The inhibitor of apoptosis, cIAP2, functions as a ubiquitin-protein ligase and promotes in vitro monoubiquitination of caspases 3 and 7. J Biol Chem, *275:* 26661-26664, 2000.
33. Yang, Y., Fang, S., Jensen, J. P., Weissman, A. M., and Ashwell, J. D. Ubiquitin protein ligase activity of IAPs and their degradation in proteasomes in response to apoptotic stimuli. Science, *288:* 874-877, 2000.
34. Li, X., Yang, Y., and Ashwell, J. D. TNF-RII and c-IAP1 mediate ubiquitination and degradation of TRAF2. Nature, *416:* 345-347, 2002.

35. Du, C. Y., Fang, M., Li, Y. C., Li, L., and Wang, X. D. Smac, a mitochondrial protein that promotes cytochrome c- dependent caspase activation by eliminating IAP inhibition. Cell, *102:* 33-42, 2000.
36. Verhagen, A. M., Ekert, P. G., Pakusch, M., Silke, J., Connolly, L. M., Reid, G. E., Moritz, R. L., Simpson, R. J., and Vaux, D. L. Identification of DIABLO, a mammalian protein that promotes apoptosis by binding to and antagonizing IAP proteins. Cell, *102:* 43-53, 2000.
37. Verhagen, A. M., Silke, J., Ekert, P. G., Pakusch, M., Kaufmann, H., Connolly, L. M., Day, C. L., Tikoo, A., Burke, R., Wrobel, C., Moritz, R. L., Simpson, R. J., and Vaux, D. L. HtrA2 promotes cell death through its serine protease activity and its ability to antagonize inhibitor of apoptosis proteins. Journal of Biological Chemistry, *277:* 445-454, 2002.
38. Hegde, R., Srinivasula, S. M., Zhang, Z. J., Wassell, R., Mukattash, R., Cilenti, L., DuBois, G., Lazebnik, Y., Zervos, A. S., Fernandes-Alnemri, T., and Alnemri, E. S. Identification of Omi/HtrA-2 as a mitochondrial apoptotic serine protease that disrupts inhibitor of apoptosis protein- caspase interaction. Journal of Biological Chemistry, *277:* 432-438, 2002.
39. Suzuki, Y., Imai, Y., Nakayama, H., Takahashi, K., Takio, K., and Takahashi, R. A serine protease, HtrA2, is released from the mitochondria and interacts with XIAP, inducing cell death. Molecular Cell, *8:* 613-621, 2001.
40. Dierlamm, J., Baens, M., Wlodarska, I., Stefanova-Ouzounova, M., Hernandez, J. M., Hossfeld, D. K., De Wolf-Peeters, C., Hagemeijer, A., Van den Berghe, H., and Marynen, P. The apoptosis inhibitor gene API2 and a novel 18q gene, MLT, are recurrently rearranged in the t(11;18)(q21;q21)p6ssociated with mucosa-associated lymphoid tissue lymphomas. Blood, *93:* 3601-3609, 1999.
41. Imoto, I., Yang, Z. Q., Pimkhaokham, A., Tsuda, H., Shimada, Y., Imamura, M., Ohki, M., and Inazawa, J. Identification of cIAP1 as a candidate target gene within an amplicon at 11q22 in esophageal squamous cell carcinomas. Cancer Res, *61:* 6629-6634, 2001.
42. Vucic, D., Stennicke, H. R., Pisabarro, M. T., Salvesen, G. S., and Dixit, V. M. ML-IAP, a novel inhibitor of apoptosis that is preferentially expressed in human melanomas. Current Biology, *10:* 1359-1366, 2000.
43. Altieri, D. C. Survivin, versatile modulation of cell division and apoptosis in cancer. Oncogene, *22:* 8581-8589, 2003.
44. Tamm, I., Kornblau, S. M., Segall, H., Krajewski, S., Welsh, K., Kitada, S., Scudiero, D. A., Tudor, G., Qui, Y. H., Monks, A., Andreeff, M., and Reed, J. C. Expression and prognostic significance of IAP-family genes in human cancers and myeloid leukemias. Clin Cancer Res, *6:* 1796-1803, 2000.
45. Fong, W. G., Liston, P., Rajcan-Separovic, E., St Jean, M., Craig, C., and Korneluk, R. G. Expression and genetic analysis of XIAP-associated factor 1 (XAF1) in cancer cell lines. Genomics, *70:* 113-122, 2000.
46. Liston, P., Young, S. S., Mackenzie, A. E., and Korneluk, R. G. Life and death decisions: the role of the IAPs in modulating programmed cell death. Apoptosis, *2:* 423-441, 1997.
47. Byun, D. S., Cho, K., Ryu, B. K., Lee, M. G., Kang, M. J., Kim, H. R., and Chi, S. G. Hypermethylation of XIAP-associated factor 1, a putative tumor suppressor gene from the 17p13.2 locus, in human gastric adenocarcinomas. Cancer Res, *63:* 7068-7075, 2003.
48. Yang, L. L., Mashima, T., Sato, S., Mochizuki, M., Sakamoto, H., Yamori, T., Oh-hara, T., and Tsuruo, T. Predominant suppression of apoptosome by inhibitor of apoptosis

protein in non-small cell lung cancer H460 cells: Therapeutic effect of a novel polyarginine-conjugated Smac peptide. Cancer Research, *63:* 831-837, 2003.

49. Usuda, J., Chiu, S. M., Azizuddin, K., Xue, L. Y., Lam, M., Nieminen, A. L., and Oleinick, N. L. Promotion of photodynamic therapy-induced apoptosis by the mitochondrial protein Smac/DIABLO: Dependence on Bax. Photochemistry and Photobiology, *76:* 217-223, 2002.
50. MacKenzie, A. and LaCasse, E. Inhibition of IAP's protection by Diablo/Smac: new therapeutic opportunities? Cell Death and Differentiation, *7:* 866-867, 2000.
51. Jia, L., Patwari, Y., Kelsey, S. M., Srinivasula, S. M., Agrawal, S. G., Alnemri, E. S., and Newland, A. C. Role of Smac in human leukaemic cell apoptosis and proliferation. Oncogene, *22:* 1589-1599, 2003.
52. Chauhan, D., Hideshima, T., Rosen, S., Reed, J. C., Kharbanda, S., and Anderson, K. C. Apaf-1/cytochrome c-independent and Smac-dependent induction of apoptosis in multiple myeloma (MM) cells. Journal of Biological Chemistry, *276:* 24453-24456, 2001.
53. Arnt, C. R., Chiorean, M. V., Heldebrant, M. V., Gores, G. J., and Kaufmann, S. H. Synthetic Smac/DIABLO peptides enhance the effects of chemotherapeutic agents by binding XIAP and cIAP1 in situ. Journal of Biological Chemistry, *277:* 44236-44243, 2002.
54. Aron D. Schimmer 15, K. W., Clemencia Pinilla 25, Zhiliang Wang 1, Maryla Krajewska 1, Marie-Josee Bonneau 1, Irene M. Pedersen 1, Shinichi Kitada 1, Fiona L. Scott 1, Beatrice Bailly-Maitre 1, Gennadi Glinsky 3, Dominick Scudiero 4, Edward Sausville 4, Guy Salvesen 1, Adel Nefzi 2, John M. Ostresh 2, Richard A. Houghten 2, and John C. Reed 1 Small-molecule antagonists of apoptosis suppressor XIAP exhibit broad antitumor activity. Cancer Cell, *5:* 25-35, 2004.
55. Demetri, G. D. Identification and treatment of chemoresistant inoperable or metastatic GIST: experience with the selective tyrosine kinase inhibitor imatinib mesylate (STI571). Eur J Cancer, *38 Suppl 5:* S52-59, 2002.
56. O'Brien, S. G., Guilhot, F., Larson, R. A., Gathmann, I., Baccarani, M., Cervantes, F., Cornelissen, J. J., Fischer, T., Hochhaus, A., Hughes, T., Lechner, K., Nielsen, J. L., Rousselot, P., Reiffers, J., Saglio, G., Shepherd, J., Simonsson, B., Gratwohl, A., Goldman, J. M., Kantarjian, H., Taylor, K., Verhoef, G., Bolton, A. E., Capdeville, R., and Druker, B. J. Imatinib compared with interferon and low-dose cytarabine for newly diagnosed chronic-phase chronic myeloid leukemia. N Engl J Med, *348:* 994-1004, 2003.
57. Tauchi, T. and Ohyashiki, K. Molecular mechanisms of resistance of leukemia to imatinib mesylate. Leuk Res, *28 Suppl 1:* S39-45, 2004.
58. Wisniewski, D., Lambek, C. L., Liu, C., Strife, A., Veach, D. R., Nagar, B., Young, M. A., Schindler, T., Bornmann, W. G., Bertino, J. R., Kuriyan, J., and Clarkson, B. Characterization of potent inhibitors of the Bcr-Abl and the c-kit receptor tyrosine kinases. Cancer Res, *62:* 4244-4255, 2002.
59. Sun, X., Layton, J. E., Elefanty, A., and Lieschke, G. J. Comparison of effects of the tyrosine kinase inhibitors AG957, AG490, and STI571 on BCR-ABL--expressing cells, demonstrating synergy between AG490 and STI571. Blood, *97:* 2008-2015, 2001.
60. Dagher, R., Cohen, M., Williams, G., Rothmann, M., Gobburu, J., Robbie, G., Rahman, A., Chen, G., Staten, A., Griebel, D., and Pazdur, R. Approval summary: imatinib mesylate in the treatment of metastatic and/or unresectable malignant gastrointestinal stromal tumors. Clin Cancer Res, *8:* 3034-3038, 2002.
61. Roskoski, R., Jr. The ErbB/HER receptor protein-tyrosine kinases and cancer. Biochem Biophys Res Commun, *319:* 1-11, 2004.

62. Ekstrand, A. J., Sugawa, N., James, C. D., and Collins, V. P. Amplified and rearranged epidermal growth factor receptor genes in human glioblastomas reveal deletions of sequences encoding portions of the N- and/or C-terminal tails. Proc Natl Acad Sci U S A, *89:* 4309-4313, 1992.
63. Wong, A. J., Ruppert, J. M., Bigner, S. H., Grzeschik, C. H., Humphrey, P. A., Bigner, D. S., and Vogelstein, B. Structural alterations of the epidermal growth factor receptor gene in human gliomas. Proc Natl Acad Sci U S A, *89:* 2965-2969, 1992.
64. Sugawa, N., Ekstrand, A. J., James, C. D., and Collins, V. P. Identical splicing of aberrant epidermal growth factor receptor transcripts from amplified rearranged genes in human glioblastomas. Proc Natl Acad Sci U S A, *87:* 8602-8606, 1990.
65. Ward, W. H., Cook, P. N., Slater, A. M., Davies, D. H., Holdgate, G. A., and Green, L. R. Epidermal growth factor receptor tyrosine kinase. Investigation of catalytic mechanism, structure-based searching and discovery of a potent inhibitor. Biochem Pharmacol, *48:* 659-666, 1994.
66. Barker, A. J., Gibson, K. H., Grundy, W., Godfrey, A. A., Barlow, J. J., Healy, M. P., Woodburn, J. R., Ashton, S. E., Curry, B. J., Scarlett, L., Henthorn, L., and Richards, L. Studies leading to the identification of ZD1839 (IRESSA): an orally active, selective epidermal growth factor receptor tyrosine kinase inhibitor targeted to the treatment of cancer. Bioorg Med Chem Lett, *11:* 1911-1914, 2001.
67. Pao, W., Miller, V. A., and Kris, M. G. 'Targeting' the epidermal growth factor receptor tyrosine kinase with gefitinib (Iressa) in non-small cell lung cancer (NSCLC). Semin Cancer Biol, *14:* 33-40, 2004.
68. Okamoto, I., Kenyon, L. C., Emlet, D. R., Mori, T., Sasaki, J., Hirosako, S., Ichikawa, Y., Kishi, H., Godwin, A. K., Yoshioka, M., Suga, M., Matsumoto, M., and Wong, A. J. Expression of constitutively activated EGFRvIII in non-small cell lung cancer. Cancer Sci, *94:* 50-56, 2003.
69. Hirsch, F. R., Varella-Garcia, M., Bunn, P. A., Jr., Di Maria, M. V., Veve, R., Bremmes, R. M., Baron, A. E., Zeng, C., and Franklin, W. A. Epidermal growth factor receptor in non-small-cell lung carcinomas: correlation between gene copy number and protein expression and impact on prognosis. J Clin Oncol, *21:* 3798-3807, 2003.
70. Paez, J. G., Janne, P. A., Lee, J. C., Tracy, S., Greulich, H., Gabriel, S., Herman, P., Kaye, F. J., Lindeman, N., Boggon, T. J., Naoki, K., Sasaki, H., Fujii, Y., Eck, M. J., Sellers, W. R., Johnson, B. E., and Meyerson, M. EGFR mutations in lung cancer: correlation with clinical response to gefitinib therapy. Science, *304:* 1497-1500, 2004.
71. Lynch, T. J., Bell, D. W., Sordella, R., Gurubhagavatula, S., Okimoto, R. A., Brannigan, B. W., Harris, P. L., Haserlat, S. M., Supko, J. G., Haluska, F. G., Louis, D. N., Christiani, D. C., Settleman, J., and Haber, D. A. Activating mutations in the epidermal growth factor receptor underlying responsiveness of non-small-cell lung cancer to gefitinib. N Engl J Med, *350:* 2129-2139, 2004.
72. Allen, L. F., Lenehan, P. F., Eiseman, I. A., Elliott, W. L., and Fry, D. W. Potential benefits of the irreversible pan-erbB inhibitor, CI-1033, in the treatment of breast cancer. Semin Oncol, *29:* 11-21, 2002.
73. Gieseg, M. A., de Bock, C., Ferguson, L. R., and Denny, W. A. Evidence for epidermal growth factor receptor-enhanced chemosensitivity in combinations of cisplatin and the new irreversible tyrosine kinase inhibitor CI-1033. Anticancer Drugs, *12:* 683-690, 2001.
74. Gschwind, A., Fischer, O. M., and Ullrich, A. The discovery of receptor tyrosine kinases: targets for cancer therapy. Nat Rev Cancer, *4:* 361-370, 2004.
75. Slamon, D. J., Leyland-Jones, B., Shak, S., Fuchs, H., Paton, V., Bajamonde, A., Fleming, T., Eiermann, W., Wolter, J., Pegram, M., Baselga, J., and Norton, L. Use of

chemotherapy plus a monoclonal antibody against HER2 for metastatic breast cancer that overexpresses HER2. N Engl J Med, *344:* 783-792, 2001.

76. Cunningham, D., Humblet, Y., Siena, S., Khayat, D., Bleiberg, H., Santoro, A., Bets, D., Mueser, M., Harstrick, A., Verslype, C., Chau, I., and Van Cutsem, E. Cetuximab monotherapy and cetuximab plus irinotecan in irinotecan-refractory metastatic colorectal cancer. N Engl J Med, *351:* 337-345, 2004.
77. Shields, J. M., Pruitt, K., McFall, A., Shaub, A., and Der, C. J. Understanding Ras: 'it ain't over 'til it's over'. Trends Cell Biol, *10:* 147-154, 2000.
78. Datta, S. R., Brunet, A., and Greenberg, M. E. Cellular survival: a play in three Akts. Genes Dev, *13:* 2905-2927, 1999.
79. Bos, J. L. ras oncogenes in human cancer: a review. Cancer Res, *49:* 4682-4689, 1989.
80. Downward, J. Targeting RAS signalling pathways in cancer therapy. Nat Rev Cancer, *3:* 11-22, 2003.
81. Sebti, S. M. and Hamilton, A. D. Farnesyltransferase and geranylgeranyltransferase I inhibitors and cancer therapy: lessons from mechanism and bench-to-bedside translational studies. Oncogene, *19:* 6584-6593, 2000.
82. Cox, A. D. and Der, C. J. Farnesyltransferase inhibitors: promises and realities. Curr Opin Pharmacol, *2:* 388-393, 2002.
83. Kohl, N. E., Omer, C. A., Conner, M. W., Anthony, N. J., Davide, J. P., deSolms, S. J., Giuliani, E. A., Gomez, R. P., Graham, S. L., Hamilton, K., and et al. Inhibition of farnesyltransferase induces regression of mammary and salivary carcinomas in ras transgenic mice. Nat Med, *1:* 792-797, 1995.
84. Lobell, R. B., Omer, C. A., Abrams, M. T., Bhimnathwala, H. G., Brucker, M. J., Buser, C. A., Davide, J. P., deSolms, S. J., Dinsmore, C. J., Ellis-Hutchings, M. S., Kral, A. M., Liu, D., Lumma, W. C., Machotka, S. V., Rands, E., Williams, T. M., Graham, S. L., Hartman, G. D., Oliff, A. I., Heimbrook, D. C., and Kohl, N. E. Evaluation of farnesyl:protein transferase and geranylgeranyl:protein transferase inhibitor combinations in preclinical models. Cancer Res, *61:* 8758-8768, 2001.
85. Sebti, S. M. and Der, C. J. Opinion: Searching for the elusive targets of farnesyltransferase inhibitors. Nat Rev Cancer, *3:* 945-951, 2003.
86. Hoshino, R., Chatani, Y., Yamori, T., Tsuruo, T., Oka, H., Yoshida, O., Shimada, Y., Ari-i, S., Wada, H., Fujimoto, J., and Kohno, M. Constitutive activation of the 41-/43-kDa mitogen-activated protein kinase signaling pathway in human tumors. Oncogene, *18:* 813-822, 1999.
87. Lyons, J. F., Wilhelm, S., Hibner, B., and Bollag, G. Discovery of a novel Raf kinase inhibitor. Endocr Relat Cancer, *8:* 219-225, 2001.
88. Davies, H., Bignell, G. R., Cox, C., Stephens, P., Edkins, S., Clegg, S., Teague, J., Woffendin, H., Garnett, M. J., Bottomley, W., Davis, N., Dicks, E., Ewing, R., Floyd, Y., Gray, K., Hall, S., Hawes, R., Hughes, J., Kosmidou, V., Menzies, A., Mould, C., Parker, A., Stevens, C., Watt, S., Hooper, S., Wilson, R., Jayatilake, H., Gusterson, B. A., Cooper, C., Shipley, J., Hargrave, D., Pritchard-Jones, K., Maitland, N., Chenevix-Trench, G., Riggins, G. J., Bigner, D. D., Palmieri, G., Cossu, A., Flanagan, A., Nicholson, A., Ho, J. W., Leung, S. Y., Yuen, S. T., Weber, B. L., Seigler, H. F., Darrow, T. L., Paterson, H., Marais, R., Marshall, C. J., Wooster, R., Stratton, M. R., and Futreal, P. A. Mutations of the BRAF gene in human cancer. Nature, *417:* 949-954, 2002.
89. Sebolt-Leopold, J. S. Development of anticancer drugs targeting the MAP kinase pathway. Oncogene, *19:* 6594-6599, 2000.
90. Sebolt-Leopold, J. S., Dudley, D. T., Herrera, R., Van Becelaere, K., Wiland, A., Gowan, R. C., Tecle, H., Barrett, S. D., Bridges, A., Przybranowski, S., Leopold, W. R.,

and Saltiel, A. R. Blockade of the MAP kinase pathway suppresses growth of colon tumors in vivo. Nat Med, *5:* 810-816, 1999.

91. Allen, L. F., Sebolt-Leopold, J., and Meyer, M. B. CI-1040 (PD184352), a targeted signal transduction inhibitor of MEK (MAPKK). Semin Oncol, *30:* 105-116, 2003.
92. Khwaja, A., Rodriguez-Viciana, P., Wennstrom, S., Warne, P. H., and Downward, J. Matrix adhesion and Ras transformation both activate a phosphoinositide 3-OH kinase and protein kinase B/Akt cellular survival pathway. Embo J, *16:* 2783-2793, 1997.
93. Bellacosa, A., de Feo, D., Godwin, A. K., Bell, D. W., Cheng, J. Q., Altomare, D. A., Wan, M., Dubeau, L., Scambia, G., Masciullo, V., and et al. Molecular alterations of the AKT2 oncogene in ovarian and breast carcinomas. Int J Cancer, *64:* 280-285, 1995.
94. Parsons, R. and Simpson, L. PTEN and cancer. Methods Mol Biol, *222:* 147-166, 2003.
95. Sato, S., Fujita, N., and Tsuruo, T. Interference with PDK1-Akt survival signaling pathway by UCN-01 (7-hydroxystaurosporine). Oncogene, *21:* 1727-1738, 2002.
96. Vogelstein, B., Lane, D., and Levine, A. J. Surfing the p53 network. Nature, *408:* 307-310, 2000.
97. Ryan, K. M., Phillips, A. C., and Vousden, K. H. Regulation and function of the p53 tumor suppressor protein. Current Opinion in Cell Biology, *13:* 332-337, 2001.
98. Momand, J., Jung, D., Wilczynski, S., and Niland, J. The MDM2 gene amplification database. Nucleic Acids Res, *26:* 3453-3459, 1998.
99. Chene, P. Inhibiting the p53-MDM2 interaction: an important target for cancer therapy. Nat Rev Cancer, *3:* 102-109, 2003.
100. Vassilev, L. T., Vu, B. T., Graves, B., Carvajal, D., Podlaski, F., Filipovic, Z., Kong, N., Kammlott, U., Lukacs, C., Klein, C., Fotouhi, N., and Liu, E. A. In vivo activation of the p53 pathway by small-molecule antagonists of MDM2. Science, *303:* 844-848, 2004.
101. Vousden, K. H. and Lu, X. Live or let die: The cell's response to p53. Nature Reviews Cancer, *2:* 594-604, 2002.
102. Ghosh, S., May, M. J., and Kopp, E. B. NF-kappa B and Rel proteins: evolutionarily conserved mediators of immune responses. Annu Rev Immunol, *16:* 225-260, 1998.
103. Karin, M. and Ben-Neriah, Y. Phosphorylation meets ubiquitination: the control of NF-[kappa]B activity. Annu Rev Immunol, *18:* 621-663, 2000.
104. Rayet, B. and Gelinas, C. Aberrant rel/nfkb genes and activity in human cancer. Oncogene, *18:* 6938-6947, 1999.
105. Baldwin, A. S. Control of oncogenesis and cancer therapy resistance by the transcription factor NF-kappaB. J Clin Invest, *107:* 241-246, 2001.
106. Wang, C. Y., Mayo, M. W., Korneluk, R. G., Goeddel, D. V., and Baldwin, A. S., Jr. NF-kappaB antiapoptosis: induction of TRAF1 and TRAF2 and c-IAP1 and c-IAP2 to suppress caspase-8 activation. Science, *281:* 1680-1683, 1998.
107. Wang, C. Y., Guttridge, D. C., Mayo, M. W., and Baldwin, A. S., Jr. NF-kappaB induces expression of the Bcl-2 homologue A1/Bfl-1 to preferentially suppress chemotherapy-induced apoptosis. Mol Cell Biol, *19:* 5923-5929, 1999.
108. Cusack, J. C., Jr., Liu, R., Houston, M., Abendroth, K., Elliott, P. J., Adams, J., and Baldwin, A. S., Jr. Enhanced chemosensitivity to CPT-11 with proteasome inhibitor PS-341: implications for systemic nuclear factor-kappaB inhibition. Cancer Res, *61:* 3535-3540, 2001.
109. Russo, S. M., Tepper, J. E., Baldwin, A. S., Jr., Liu, R., Adams, J., Elliott, P., and Cusack, J. C., Jr. Enhancement of radiosensitivity by proteasome inhibition: implications for a role of NF-kappaB. Int J Radiat Oncol Biol Phys, *50:* 183-193, 2001.

110. Richardson, P. G., Hideshima, T., and Anderson, K. C. Bortezomib (PS-341): a novel, first-in-class proteasome inhibitor for the treatment of multiple myeloma and other cancers. Cancer Control, *10:* 361-369, 2003.
111. Hideshima, T., Chauhan, D., Richardson, P., Mitsiades, C., Mitsiades, N., Hayashi, T., Munshi, N., Dang, L., Castro, A., Palombella, V., Adams, J., and Anderson, K. C. NF-kappa B as a therapeutic target in multiple myeloma. J Biol Chem, *277:* 16639-16647, 2002.
112. Niu, C., Yan, H., Yu, T., Sun, H. P., Liu, J. X., Li, X. S., Wu, W., Zhang, F. Q., Chen, Y., Zhou, L., Li, J. M., Zeng, X. Y., Yang, R. R., Yuan, M. M., Ren, M. Y., Gu, F. Y., Cao, Q., Gu, B. W., Su, X. Y., Chen, G. Q., Xiong, S. M., Zhang, T., Waxman, S., Wang, Z. Y., Chen, S. J., and et al. Studies on treatment of acute promyelocytic leukemia with arsenic trioxide: remission induction, follow-up, and molecular monitoring in 11 newly diagnosed and 47 relapsed acute promyelocytic leukemia patients. Blood, *94:* 3315-3324, 1999.
113. Singhal, S., Mehta, J., Desikan, R., Ayers, D., Roberson, P., Eddlemon, P., Munshi, N., Anaissie, E., Wilson, C., Dhodapkar, M., Zeddis, J., and Barlogie, B. Antitumor activity of thalidomide in refractory multiple myeloma. N Engl J Med, *341:* 1565-1571, 1999.
114. Singhal, S. and Mehta, J. Thalidomide in cancer. Biomed Pharmacother, *56:* 4-12, 2002.
115. Keifer, J. A., Guttridge, D. C., Ashburner, B. P., and Baldwin, A. S., Jr. Inhibition of NF-kappa B activity by thalidomide through suppression of IkappaB kinase activity. J Biol Chem, *276:* 22382-22387, 2001.
116. Seitz, C. S., Lin, Q., Deng, H., and Khavari, P. A. Alterations in NF-kappaB function in transgenic epithelial tissue demonstrate a growth inhibitory role for NF-kappaB. Proc Natl Acad Sci U S A, *95:* 2307-2312, 1998.
117. van Hogerlinden, M., Rozell, B. L., Ahrlund-Richter, L., and Toftgard, R. Squamous cell carcinomas and increased apoptosis in skin with inhibited Rel/nuclear factor-kappaB signaling. Cancer Res, *59:* 3299-3303, 1999.
118. Dajee, M., Lazarov, M., Zhang, J. Y., Cai, T., Green, C. L., Russell, A. J., Marinkovich, M. P., Tao, S., Lin, Q., Kubo, Y., and Khavari, P. A. NF-kappaB blockade and oncogenic Ras trigger invasive human epidermal neoplasia. Nature, *421:* 639-643, 2003.
119. Ryan, K. M., Ernst, M. K., Rice, N. R., and Vousden, K. H. Role of NF-kappa B in p53-mediated programmed cell death. Nature, *404:* 892-897, 2000.
120. Huang, Y., Johnson, K. R., Norris, J. S., and Fan, W. Nuclear factor-kappaB/IkappaB signaling pathway may contribute to the mediation of paclitaxel-induced apoptosis in solid tumor cells. Cancer Res, *60:* 4426-4432, 2000.
121. John, A. M., Thomas, N. S., Mufti, G. J., and Padua, R. A. Targeted therapies in myeloid leukemia. Semin Cancer Biol, *14:* 41-62, 2004.

# Chapter 6

# TARGETING CYCLINS TO CAUSE CANCER CELL APOPTOSIS

Suparna Mazumder[1], Dragos Plesca[1,2], and Alexandru Almasan[1,2,3]
*[1]Department of Cancer Biology, Lerner Research Institute, The Cleveland Clinic Foundation, Cleveland, OH 44195; [2]Department of Biological Sciences, Kent State University, Kent, 44242; [3]Department of Radiation Oncology, The Cleveland Clinic Foundation, Cleveland, OH 44195*

**Abstract**: Progression through each phase of the cell cycle is precisely controlled by the activity of distinct cyclin-dependent kinases (CDKs) and their regulatory subunits known as cyclins. Each cell cycle phase is characterized by specific cyclin-CDK combinations and it is tightly controlled by complex regulatory mechanisms that either allow or restrain its progression. Of these, cyclins play the key role in controlling cell cycle progression, with D-type cyclins being the sensors for environmental cues and their assembly into CDK4/6 complexes leading to sequestration of the CDK inhibitors that ultimately allow Cyclin E/CDK2 activation. Cyclin E is unique in its critical role in the G1/S transition, initiation of DNA replication, and centrosome duplication, functions that may provide the underpinnings of its role in tumor development. Some functions, such as its oncogenic function and its ability, following its caspase-mediated conversion to a truncated form, to regulate apoptosis and DNA repair, seem to be independent of its nuclear localization or its associated CDK2-kinase activity. Many cell cycle control genes, when deregulated, can cause cells that are not dividing to enter the cell cycle and begin to proliferate leading to cancer development. S-phase sensitization, by modulating levels of Cyclin E, A, or CDK2 and the associated kinase activity may provide a useful therapeutic approach. There is at present much optimism about the possibility of finding anticancer drug treatment strategies that modulate the activity of cyclins and their regulatory partners. This review summarizes what is known about the biological role of cyclins, their deregulation in cancer, and the opportunities they may provide as targets that may improve clinical therapy.

**Key words**: cyclin; CDK; cell cycle control; apoptosis; genotoxic stress; drug target; review.



*M. Sluyser (ed.), Application of Apoptosis to Cancer Treatment,* 145-177.

# 1. INTRODUCTION

The main families of regulatory proteins that play key roles in controlling cell cycle progression are the cyclins, the cyclin dependent kinases (CDKs) and, as their critical downstream substrates, the tumor suppressor gene products of the retinoblastoma (pRb) and p53 (Figure 1). The cell cycle is characterized by a highly conserved and ordered set of events, culminating in cell growth and division. The activity of CDKs oscillates leading to the cyclical changes in the activation of key proteins, which regulate the main cell cycle processes: DNA replication, mitosis, and cytokinesis. G1/S-cyclins commit cells to DNA replication, which is driven by the S-phase cyclins, whereas mitosis is governed by M-cyclins. Progression through the early G1-phase is regulated by signal transduction cascades activated by polypeptide growth factors and by extracellular matrix (ECM) components.

In higher eukaryotes, the S- and M- phases of the cell cycle are triggered by distinct CDKs. For example, in frog egg extracts, CDK1-Cyclin B catalyzes entry into mitosis but cannot trigger DNA replication. This observation can be explained in two ways: either CDK1-Cyclin B fails to recognize the key substrates of its S-phase-promoting counterparts, or somehow its activity is regulated to prevent it from activating DNA synthesis. A recent study has revealed that, surprisingly, the CDK1-cyclin B1 complex has cryptic S-phase-promoting abilities that can be exposed by its cytoplasmic to nuclear relocalization and also, by stimulating its activity. By grafting the nuclear localization signal from Cyclin E to Cyclin B, the chimeric protein entered the nucleus and induced DNA replication. This indicates that subcellular localization of vertebrate CDKs and control of their activity are critical factors that determine their specificity[1]. Therefore, much of the difference between the function of different cyclins may reflect where and when they are expressed rather than their direct effect on substrate specificity[2].

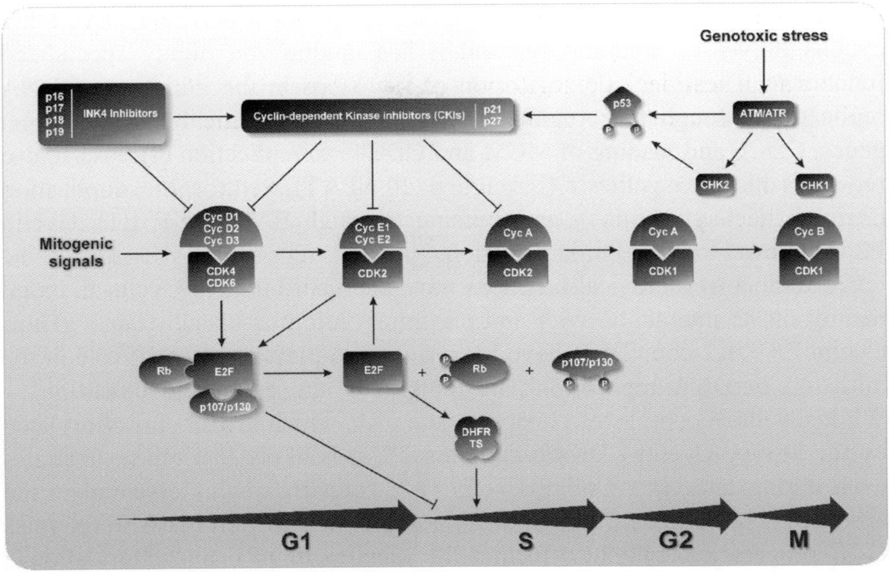


*Figure 1.* Cell cycle control in mammalian cells subjected to genotoxic stress. The mammalian cell cycle can be divided into four phases: G1 (gap phase 1), S (DNA synthesis), G2 (gap phase 2), and M (mitosis). The main families of regulatory proteins that play key roles in controlling cell cycle progression are the cyclins, the cyclin dependent kinases (CDKs), CDK kinase inhibitors (CKIs), the tumor suppressor gene product p53, which is regulated by checkpoint components such as ATM/ATR, CHK1/CHK2, pRb/pocket proteins (p107/p130) and E2F transcription factors, responsible for the transactivation of genes crirical for S-phase progression [e.g. Dihydrofolate reductase (DHFR) and Thymidylate synthase (TS)]. G1/S-cyclins commit cells to DNA replication, which is driven by the S-phase cyclins, whereas mitosis is governed by M-cyclins. Cyclin D1/D2/D3-CDK4/6 and Cyclin E1/E2-CDK2 are responsible for progression through the G1- into S-phase. Cyclin A-CDK2 is responsible for late S-phase progression, whereas Cyclin A-CDK1 and Cyclin B-CDK1 are critical for G2/M control.

## 2. CYCLIN E IS THE CRITICAL G1/S CYCLIN

The human G1 cyclins, the D- and E-type cyclins, were identified functionally by screening of human cDNA libraries for sequences that could complement G1 cyclin mutations in *S. cerevisiae*[3,4]. The human Cyclin E has several forms and splice variants, with the predominant species encoding a 395 amino acids protein with a molecular weight of 50-55 kDa[3,4]. Cyclin E is essential for progression through the G1-phase of the cell cycle and initiation of DNA replication by interacting with and activating its catalytic partner CDK2.

Cyclin E, unlike any other cyclin, can complement deficiencies of other cyclins in yeast[3,4] and mouse[5] and it has unique and highly specialized functions. These include regulation of key steps in the initiation of DNA replication through E2F regulation and thus of its critical S-phase target genes, CDC6 and loading of MCM and CDC45 to replication origins. It also regulates histone synthesis (through p220/NPAT), centrosome duplication (through nucleophosmins), and splicing (through SAP155/SF3B1). Cyclin E2 is believed to have similar functions[6,7].

A number of biological functions have been attributed to Cyclin E, based mainly on its interaction with and phosphorylation of its substrates. Thus, Cyclin E/CDK2 complexes have been shown to play an essential role in the initiation of DNA replication[8], in addition to the cell cycle transition[9,10]. While Cyclin E complexes phosphorylate pRb, which is also phosphorylated by the D-type Cyclin-CDK4/6 complexes[11], it is likely that other substrates exist during late G1. Unlike Cyclin D, Cyclin E remains essential in the absence of pRb, since: (i) its inducible expression in fibroblasts accelerates G1/S progression without affecting the kinetics of pRb phosphorylation[12]; (ii) unlike the D-type cyclins, Cyclin E is essential for cell-cycle progression in pRb-deficient cells[9]; (iii) ectopic expression of cyclin E bypasses pRb-mediated cell cycle arrest[13,14]; and (iv) Cyclin E is required for S-phase entry in *Drosophila*[15] and *Xenopus*[16]. These findings highlight a fundamental difference between the Cyclin D and Cyclin E complexes, strongly suggesting that other key rate-limiting substrates exist for Cyclin E/CDK2[13,14]. Constitutive activation of Cyclin E/CDK2 results in uncoupling of the initiation of centrosome and DNA duplication leading to unscheduled initiation of centrosome duplication prior to S-phase entry[17].

The Cyclin E/CDK2 complex was thought to be a critical regulator of the somatic cell cycle. Surprisingly, ablating these genes in the mouse had very limited consequences. Embryonic fibroblasts deficient in CDK2 had normal proliferation and become immortal after continuous passage in culture. Cdk2-/- mice are viable and can survive for up to two years, suggesting that CDK2 may not be necessary for proliferation and survival of most cell types. It was found, however, to be necessary for completion of prophase I during meiotic cell division in male and female germ cells[18,19]. Similarly, it was demonstrated that E-type cyclins are largely dispensable for mouse development[20]. Nevertheless, in the absence of Cyclin E, endoreplication of trophoblast giant cells and megakaryocytes is severely impaired. Cyclin E-deficient cells proliferate normally under conditions of continuous cell cycling but are not able to reenter the cell cycle from the quiescent G0 state. Molecular analyses revealed that cells lacking Cyclin E fail to normally incorporate MCM proteins into DNA replication origins during G0 to S-phase progression. These findings define a molecular function for E-type cyclins in cell cycle reentry and reveal a differential necessity for Cyclin E in normal versus oncogenic proliferation[20].

The dispensable function of E-type cyclins and CDK2 is puzzling. There are two main arguments used to try to explain it. First, acute and chronic inactivation may be quite different. The best example has been provided by conditional inactivation of an allele of the mouse Rb gene, to model sporadic cancers associated with inactivation of Rb in humans. It was shown that severe loss of Rb in primary quiescent cells is enough for cell cycle entry and has phenotypic consequences different from germline loss of Rb function. This difference may be explained partly by functional compensation by the Rb-related gene p107. In senescent cells, there is acute loss of Rb, which leads to reversal of the cellular senescence. Thus, the application of conditional knockout strategies might refine the understanding of gene function and will be helpful to model human cancer more precisely[21]. However, at least one study may not be reconciled with this interpretation, as removal of a conditional cdk2 allele in immortal cells may not have a striking effect on proliferation either[18,19]. Another interpretation for dispensable E-type cyclins and CDK2 is that cell division in early embryos is different than that of somatic cells as cells move directly after completing mitosis into DNA replication. During the somatic cell cycle, the G1 phase has emerged as a need to inactivate the cell cycle inhibitors and thus to drive the cell cycle[2].

## 2.1 Transcriptional Regulation of Cyclin E

There are a number of putative binding sites for E2F in the cyclin E promoter region, suggesting an E2F-dependent regulation[22]. A variant E2F binding site is a cyclin E repressor module responsible for the periodic down-regulation of the cyclin E promoter until the growing cells have reached the late G1-phase[22]. This site facilitates transcriptional repression by binding to a large complex, containing E2F4, DP1, and a pocket protein, which functions to delay the expression of cyclin E until late G1[23]. pRb forms a repressor complex with a histone deacetylase (HDAC) and the SNF2-like (BRG1 and hbrm) component of the mammalian hSW1/SNF nucleosome remodeling complex, starting from the end of S-phase until late G1[24]. The SW1-SNF complexes have an important role in transcriptional regulation, altering the chromatin structure by relieving the transcription from the nucleosome-mediated repression, thereby, opening access to the activators of transcription. The phosphorylation of pRb by Cyclin D/CDK4 abrogates its interaction with HDAC and transactivates cyclin E and, thereby, overcomes the G1 arrest[24]. Cyclin E/CDK2 can phosphorylate pRb or the hSW1/SNF component after Cyclin E reaches a certain level, when the interaction of pRb-hSW1/SNF is disrupted.

Deregulation of any of the components of this transcriptional complex could lead to the unscheduled expression of cyclin E, which is very common in cancer cells. We and others have previously shown that constitutive levels of several E2F1 target genes are increased in Rb-deficient fibroblasts[25,26]. Similarly, Cyclin E levels are increased constitutively when Rb has been inactivated by the HPV16-E7 oncogene expression in human foreskin fibroblasts[27].

The above reports suggest that the transcription factor responsible for cyclin E induction might be E2F. E2F1 has been reported to be upregulated in response to DNA damage in a manner analogous to p53, with its levels being unchanged following DNA damage of epithelial cells with mutated p53. In contrast, we found that cyclin E was upregulated in cells with mutated p53 expressed either endogenously or following its stable transfection[29]. Northern and run-on analyses have indicated that the cyclin E levels induced by radiation result mostly from transcriptional regulation. Cyclin E induction is not restricted to radiation and the hematopoietic cell lines we have investigated. Cyclin E levels were also increased following treatment with chemotherapeutic agents, such as the topoisomerase I inhibitor VP16[29]. In addition, we recently found that cyclin E is also upregulated in other cell types, such as those of prostate and epithelial cell types. Although further work is required to establish whether E2F plays a regulatory role in radiation-induced cyclin E expression, it is quite possible that cyclin E is regulated by different mechanism by mitogens and genotoxic stress such as radiation. Understanding how this regulation takes place may provide additional targets for therapy.

## 2.2 Post-transcriptional Regulation of Cyclins

Cyclin E periodicity is not only determined at the transcriptional level, but it is also maintained by post-translational regulation through ubiquitin-dependent proteolysis[30,31]. In normal proliferating cells, Cyclin E has a short half-life of less than 30 min, but this can be extended to more than 2 hrs by the addition of pharmacologic inhibitors of the proteasome. Turnover of Cyclin E by this proteasome pathway is regulated by the binding of Cyclin E to CDK2 and also by site-specific GSK3-mediated[32] and auto-phosphorylation of Cyclin E. The unbound Cyclin E is degraded by the proteasome, and binding to CDK2 protects it from degradation. Cyclin E/CDK2 activity reverses the stabilizing effect of complex assembly. By site-directed mutagenesis, it has been shown that the threonine (T380A) mutation of Cyclin E prevents it from ubiquitin-mediated proteolytic degradation[32].

Skp2 and Fbw7/CDC4 are responsible for ubiquitin-dependent proteolysis of Cyclin E. Skp2 was thought earlier to be the F-box protein responsible for Cyclin E degradation since skp2 gene inactivation by homologous recombination results in p27 and Cyclin E accumulation[33,34]. However, since p27 is the predominant substrate of Skp2, its accumulation leads indirectly to the inhibition of Cyclin E/CDK2 kinase activity and phosphorylation of Cyclin E on T380, preventing its recognition by the Cyclin E-specific F-box protein. Instead, Fbw7, specifically targets ubiquitin-mediated proteolysis of Cyclin E[35,36]. The interaction of Cyclin E with this F-box protein depends on Cyclin E phosphorylation by GSK3 on T380[32] and T62 autophosphorylation[30]. Some tumor cell lines that have high levels of Cyclin E also have mutations in the Fbw7 gene or express low levels of its mRNA, suggesting that Fbw7 may function as a tumor suppressor[37,38]. Cross-talk between the regulation of cyclin E transcription and that of Cyclin E protein stability results in fine tuning of Cyclin E levels, emphasizing its crucial role in cell cycle regulation and predicting deleterious effects of a constitutively high level of expression of Cyclin E, as seen in many cancer cells.

The oscillations in CDK activities are largely determined by the levels of their interacting partners, which in turn depend on their synthesis, through transcriptional regulation, and proteolytic degradation. Phosphorylation by Cyclin H/CDK7 (CAK) and Wee1/Myt 1 kinases and dephosphorylation by the KAP, Cdc25A, and PP2C phosphatase families play another important role in the regulation of the cell cycle machinery[7,39-44]. In normal cells, Cdc25A is negatively regulated by stress signals such as irradiation through checkpoint-mediated ubiquitination. In checkpoint-deficient cells, Cdc25A is overexpressed, leading to radioresistant DNA synthesis[45].

## 2.3 Cyclin E as a Critical Regulator of Apoptosis

Apoptosis is a universal genetic program of cell death in higher eukaryotes that represents a basic process involved in cellular development and differentiation[46,47]. The Bcl-2 family of proteins has been highly conserved during evolution, with its members (24 to date) being critical regulators of apoptosis[48]. Pro-apoptotic members of this family, such as Bax, promote apoptosis by causing release of cytochrome c from mitochondria into the cytosol, with anti-apoptotic proteins, such as Bcl-2 preventing it. Bax normally resides in the cytosol, but it is translocated to mitochondria to promote apoptosis[49]. Genotoxic stressors, such as IR, induce Bax expression and activation[50,51]. Recently, it was suggested that the N-terminus of Bax interacts with the C-terminus of Ku70 (aa 496-609 domain) that prevents its translocation to mitochondria and activation of apoptosis[52].

Caspases are synthesized as inactive precursors, which are activated by proteolytic cleavage to generate active enzymes. The activation of caspases is a common and critical regulator of the execution phase of apoptosis, triggered by many factors, including genotoxic agents, such as γ-irradiation or treatment with anti-cancer agents[51,53-57]. They further proteolytically cleave proteins critical for maintenance of cellular cytoskeleton, DNA repair, signal transduction, and cell cycle control. Once cells are committed to cell death, apoptogenic factors, the best known of which is cytochrome c, are released from mitochondria to initiate the caspase cascade[54,58]. Cytochrome c acts as a cofactor to stimulate the complexing of Apaf-1 with Caspase 9,[59] which then initiates activation of the caspase cascade[60]. There are over 300 *in vivo* caspase substrates, including transcription factors, kinases, enzymes involved in DNA repair, and cytoskeletal proteins (reviewed in[61]). Amongst them are proteins essential for cell cycle regulation, such as Cyclin E[62], pRb[63,64], MDM2[65], PITSLRE[66], $p21^{Cip1/Waf1}$ and $p27^{Kip1}$[67].

Induction of apoptosis by various stimuli has been shown to require activation of CDK2[68,69], whereas forced expression of CKIs in cultured cells[69], neurons[70], or during myocyte differentiation[71] prevents apoptosis. Importantly, similar observations were made in non-cycling developing thymocytes, in which CDK2 was activated; conversely, CDK2 inhibition abrogates apoptosis[72]. Apoptotic targets of CDKs are predicted, given reports of an interplay between the cell cycle control processes and apoptosis[73,74] and that the apoptosis regulatory proteins themselves can directly impact on the cell cycle machinery[75-78]. Inhibitors targeting CDKs have become of great interest in cancer therapy with more than 50 compounds being evaluated for antitumor activity, some of them already in preclinical trials[79,80].

Our studies have revealed that cyclin E plays an important role in apoptosis of hematopoietic cells, in addition to its reported key regulatory role in the control of the G1- to S-phase transition and the initiation of DNA replication. Based on our studies[29,62], a dual role for cyclin E in apoptosis of hematopoietic cells has emerged (Figure 2). Initially, we found a substantial induction of cyclin E mRNA, accompanied by increased production of Cyclin E protein and Cyclin E/CDK2 kinase activity in multiple myeloma and lymphoma cells following irradiation. This increase of Cyclin E levels might be implicated in the initiation phase of apoptosis. Consistent with a role of cyclin E in apoptosis, its overexpression in hematopoietic cells greatly sensitizes these cells to irradiation, while its inhibition by a dominant-negative CDK2 blocks cell death[29].

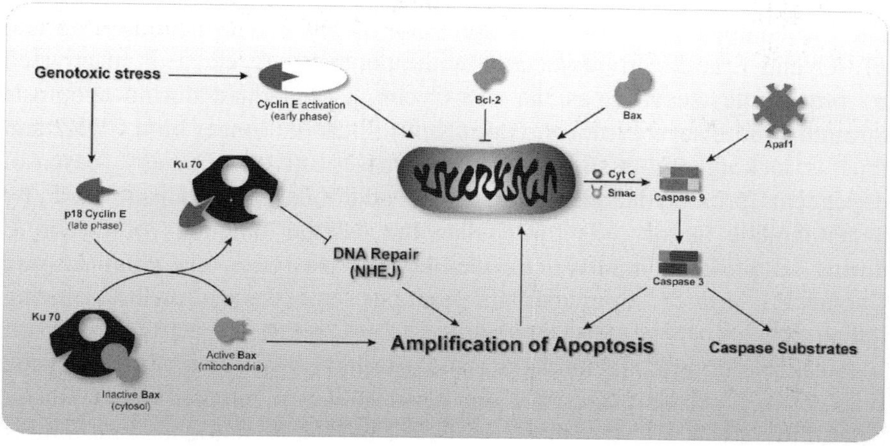


*Figure 2.* Cyclin E as a key determinant of apoptosis. Following genotoxic stress, activation of Cyclin E is critical for the early phase of apoptosis targeting the mitochondria which initiate the activation of Caspase-3. At the late stage of apoptosis, a Cyclin E derivative is generated through caspase-mediated proteolytic cleavage of Cyclin E1. The resulting p18-Cyclin E may trigger apoptosis by amplifying the mitochondrial death process. Also, it may either prevent the DNA repair process or it can trigger the activation of Bax by releasing it its from the Ku70 inhibitory complexes. Ku70 is a crucial component of Non-homologous end joining (NHEJ), the major pathway for DNA repair following genotoxic stress.

The consensus view has been that Cyclin E acts solely through its CDK2 associated kinase in the nucleus. However, important functions of Cyclin E, such as its centrosome regulatory function, require its presence in the cytoplasm. Genetic studies using mouse knock-outs have clearly indicated that the phenotype of cdk2 and cyclin E deficiencies are quite different and therefore cyclin E may have functions independent of cdk2[20]. Moreover, its oncogenic properties are not confined to CDK2 activation alone but rather rely on several other, distinct functions of the protein[81].

Recently, we found that the native p50-Cyclin E is proteolytically cleaved and thus converted to an 18-kD C-terminal fragment, which becomes the most abundant form of Cyclin E during the course of apoptosis induced by radiation and chemotherapeutic agents, such as VP16. Cyclin E cleavage results in abrogation of its binding to CDK2 and, therefore, inactivation of its associated kinase activity and cell cycle function[62]. The conversion of Cyclin E from a p50- to a p18-fragment may be a general process, as it is produced in all hematopoietic tumor cell lines we have examined and following treatment with multiple genotoxic stress agents which trigger apoptosis. The p18-Cyclin E, as it is generated at a later stage of apoptosis by caspase-mediated proteolytic cleavage, might participate, directly or indirectly, in the amplification of the apoptotic response initiated by caspase activation. Overexpression of Bcl-2 not only prevents apoptosis but also completely inhibits the expression of p18-Cyclin E, indicating that p18-Cyclin E is directly associated with apoptosis. Cyclin E is inactivated by proteolytic cleavage, as the p18-Cyclin E, generated during apoptosis through a more severe truncation of Cyclin E, can no longer bind CDK2 and thus is lacking any associated active kinase activity. Thereby, caspase-dependent proteolytic cleavage is an additional mechanism used by hematopoietic tumor cells to regulate the cellular functions of Cyclin E during apoptosis. Similarly, caspase-mediated cleavage of Cyclin A2 and Cyclin D1 induces apoptosis in Xenopus embryos following ionizing radiation[82].

Yeast two-hybrid and mass-spec analyses have only produced p18-Cyclin E interacting partners that have not been known to interact with Cylin E. One of these is Ku70, a critical component in Non-homologous End-joining Repair (NHEJ), which represents the major DNA-damage repair pathway following genotoxic stress, such as ionizing radiation. Cyclin E interaction with Ku70 inactivates the DNA repair machinery in cells undergoing apoptosis[83]. Recently, it was reported that in addition to its role in NHEJ repair, Ku70 has a cytoprotective function through regulation of the mitochondrial translocation of Bax by interacting with inactive Bax in the cytosol[52]. We find that the Ku70/Bax interaction is a general process that takes place not only in apoptosis–resistant cells, as previously suggested[52], but also in cells that readily undergo apoptosis, such as hematopoietic cells[84]. In all these cells, endogenously generated p18-Cyc E, or its ectopic expression induces the activation of Bax by releasing Bax from the Ku70 inhibitory complex to activate apoptosis. The interaction of Ku70 with p18-Cyclin E may provide the balance between cell death and survival in response to a genotoxic challenge. Augmenting apoptotic responses and minimizing DNA repair in tumor cells, while at the same time facilitating effective DNA repair in normal cells, is the ultimate goal of clinical therapy. Cyclin E may thus be unique in linking cell cycle control, DNA repair, and apoptosis during the cellular genotoxic stress response.

One other mechanism that may explain the sensitivity of cyclin E-overexpressing cells is that they display enhanced levels of the proapoptotic proteins Bad and Bax and a significant decrease in the expression of the antiapoptotic protein Bcl-2. As MCF7-cyclin E cells have an increased sensitivity to receptor ligand (Fas, Apo2L/TRAIL, and TNF$\alpha$) mediated cell death, this suggests a link between expression of cyclin E, deregulation of Bcl-2 protein and cytokine mediated cell death[85].

## 2.4 Deregulated G1/S Cyclin E Expression in Cancer

Normal cell proliferation is under strict regulation, governed by checkpoints located at distinct positions in the cell cycle. Cell cycle progression is controlled by two major checkpoints, one at the G1/S boundary, when cells commit to DNA replication, with the other one at the G2/M boundary during the commitment of cells to mitotic division[86]. The G1/S transition is regulated by the Restriction point (R point), which is a point of no return, such that, if the cell passes that point, it no longer needs mitogens and is committed to completion of DNA replication and the cell cycle[87]. Any type of deregulation of the G1/S transition, along with the disappearance of the R-point, is a hallmark of cancer, leading to uncontrolled cell proliferation. The periodic appearance of Cyclin E coincides precisely with the timing of the R point. In contrast to normal cells, mitotic cyclins appear prior to G1 cyclins in tumor cells[88]. Another view is that passage through the R point is a prerequisite for Cyclin E accumulation[89]. This study found that the postmitotic G1 cells that had not yet reached R were negative for Cyclin E accumulation, while cells that had passed R accumulated Cyclin E at variable times (1 to 8 h) after passage through R and 2 to 5 h before entry into S-phase.

As these cyclins and CDKs are important components of cell cycle control and cell proliferation, alterations or mutational changes in the expression of the corresponding genes leads to oncogenesis[87,90-92]. Cyclin E has been shown to be deregulated and overexpressed in several solid tumors, including breast, colon, and prostate carcinomas,[91,93-95,96] [as reviewed][7,97]. High levels of cyclin E expression have been associated with the progression of different other tumors, such as leukemias and lymphomas. Overexpression of Cyclin E results in accelerated G1 progression and chromosome instability[9,10,98]. Transgenic mouse models with constitutive overexpression of cyclin E develop malignant diseases indicating that Cyclin E is a dominant oncoprotein[7,99].

Although the protein is both synthesized and degraded in the cytoplasm, it is translocated to the nucleus, where it exerts its functions. In the nucleus, the nascent Cyclin E associates with its CDK2 partner, activating its serine-threonine kinase activity shortly before entry into S-phase[100,101]. Accumulation of Cyclin E in the cytoplasm either reflects increased synthesis, decreased degradation, or failure of transport to the nucleus. Loss or low p27 expression as well as overexpression of Cyclin E or CDK2 are significantly associated with malignancy in ovarian cancers[102].

Levels of Cyclin E and its low molecular weight derivatives in these tumor tissues correlate strongly with survival in these patients[95]. There can be significant changes of Cyclin E or CDK2 levels during carcinogenesis.

During progression from the primary to the lymph node-metastatic foci, the levels of Cyclin E protein remain the same, while CDK2 levels increase significantly[103]. However, during a similar transition from the primary to the liver-metastatic foci, Cyclin E levels are apparently reduced, and those of CDK2 diminish almost completely. Additionally, the decrease of Cyclin E is significantly associated with large tumor size and lymph nodal metastasis in primary carcinomas, with large tumor size and hepatic metastasis being strongly related to decreased CDK2 levels. Induced Cyclin E protein is related to increased CDK2, which is further associated with Ki-67 staining. Thus, CDK2 overexpression could facilitate lymph node metastasis and both Cyclin E and CDK2 overexpression may trigger the progression of early cancer.

Cyclin E is expressed at levels substantially higher that those physiological in many human cancers and with frequent amplification of the genomic locus at which the cyclin E gene is located (19q12-q13). In some tumors, cyclin E gene amplification and protein accumulation are late events, whereas in other neoplasms, the increase of cyclin E is observed during the early stages of malignancy. Increased protein expression does not necessarily reflect the mutations in the cyclin E gene. The question of whether Cyclin E is merely a link in the chain of events that leads to cell proliferation or it is the driving force for cell replication is difficult to ascertain, as it may be tumor dependent. Elevated transcript levels of cyclins E are found in breast, colorectal, lung, and ovary/uterus tumor samples as compared to levels in normal tissues. Its overexpression in breast carcinoma could pave the way for their genomic instability[104,105]. Expression levels of Cyclin E is most likely to be elevated in breast tumors that lacked the estrogen receptor as compared to breast tumors with the receptor and normal breast tissue[93,106].

The status of cyclin E expression is also important for other cancer types, such as sarcomas, NSCLC, leukemias, and lymphomas for prognosis and staging of tumors[107-109]. The finding that cyclin E is expressed at high levels in several types of leukemias, chronic lymphocytic leukemia, Hodgkin's and Non-Hodgkin's lymphoma[109,110], point to a role of Cyclin E in the development of these neoplastic diseases. Cyclin E and HPV expression seem to also correlate very well, making thus cyclin E a good biomarker for infection and early stage of preneoplastic lesions[27]. Cervical, prostate, breast[96], and renal cell carcinoma (65%)[111] as well as leiomyosarcomas[112] exhibit high expression of Cyclin E. Cyclin E may be also used as an independent marker in the prognosis of metastasis-free survival of lymph node-negative breast cancer[113] as well as in hepatocellular carcinomas[114] and testicular germ cell tumors[115].

## 3. CYCLIN D IN APOPTOSIS AND CANCER

Cyclin D-CDK4/6 complexes promote G1 cell cycle progression by both phosphorylating pRb and related negative cell cycle regulators as well as by sequestering the CDK inhibitors $p21^{Cip1/Waf1}$, $p27^{Kip1}$ and $p57^{Kip2}$ away from Cyclin E-CDK2. The Cip/Kip proteins also facilitate the assembly, stability and nuclear presence of the Cyclin D1-CDK4 complexes[116]. Therefore, the main function of cyclin D, by providing a link between the cell environment and the core cell cycle machinery, is to activate Cyclin E-CDK2. In fact, forced expression of Cyclin E induces cell proliferation[14] when Cyclin D-CDK4/6 complexes are inactive and, when expressed from the cyclin D1 promoter, restores many of the defects of cyclin D1-deficient mice[5]. In cells that are not dividing, CDK 4 is in low amounts allowing $p27^{Kip1}$ to bind and inhibit the Cyclin E-CDK2 complex. Cyclin D1 is induced following mitogen stimulation by c-Myc[117] and NF-κB[118], activates CDK4 and therefore sequesters $p27^{Kip1}$ ensuring the activation of the Cyclin E-CDK2 complex[119]. Both CDK4 and CDK2 phosphorylate pRb, which releases the E2F family of transcription factors that regulate genes important for DNA replication, as well as cyclin E and cyclin A. Members of the pocket protein family that includes in addition to pRb, p107 and p130[120], act as a brake upon the cell cycle by binding to the E2Fs and inhibiting their activity.[121] These suggest that the main role of cyclin D in cell cycle progression is to ensure cyclin E expression.[2]

Similarly to E-type cyclins and CDK2, it has now been established that D-type cyclins and their regulatory CDKs (4 and 6) are also dispensable for cell cycle control. Nevertheless, D-type cyclins and CDK6 are critical for expansion of hematopoietic stem cells.[122] Embryos defective for cdk4 and cdk6 die during the late stages of embryonic development due to severe anemia[123], similar to those deficient in all D-type cyclins[122].

As a sensor of environmental cues that activate cell cycle progression in response to mitogenic stimuli[124], the presence of cyclin D1 is critical for tumorigenesis as the overexpression of Ras or Her2/Neu in mammary epithelium lacking cyclin D1 failed to induce breast cancer[125]. Cyclin D1 is frequently found overexpressed in many human cancer types, particularly in lung cancer[91,126]. Its abnormal levels are often consequences of chromosomal rearrangements or gene amplification that take place in the early stages of tumor progression. About one third of breast cancer cases have either the cyclin D1 gene amplified or its encoded protein is overexpressed. Presence of the estrogen receptor, together with the constitutive expression of cyclin D1, render cells sensitive to retinoic acid treatment, evidenced by activation of Bax, cytochrome c release, and cleavage of Caspase-9[127]. Both cyclin D1 and D3 were found to be very frequently involved in IgH translocations in multiple myeloma. Since at least one member of the cyclin D family is deregulated in multiple myeloma, based on the expression pattern of cyclin D1, D2, and D3 a five-subtype myeloma classification has emerged, which may be effectively used in predicting the tumor progression and response to therapy[128]. Moreover, many studies have confirmed the correlation between cyclin D1 abnormal expression and chromosomal instability[129].

Cyclin D3 has been shown to be downregulated in T cell leukemias following activation of protein kinase C by PMA leading to a G1-phase arrest and induction of apoptosis[130]. Both cyclin D3 and cyclin E overexpression have been identified in Diffuse Large B-cell Lymphomas (DLBCL; 43% and 22%, respectively)[131]. Other types of cancers such as of breast, colorectal, prostate and non small cell lung cancer (NSCLC), have been also associated with dysregulated expression of Cyclin D and E[132].

Interestingly, cell cycle regulation was also proven to be involved in the apoptosis manifested in various neurodegenerative disorders such as Alzheimer's disease and neuronal trauma. Apparently, increased expression of cyclin D-CDK4/6 followed by uncontrolled activity of E2Fs may lead to neuronal apoptosis[133,134]. Cyclin D1 subcellular localization may play an important part in neuronal death and survival, the accumulation in the nucleus taking place only in proliferating cells whereas its ectopical expression being limited to the cytoplasm[135]. Furthermore, lowered radiosensitivity was reported in cases of Cyclin D overexpression[136]. Monitoring levels of Cyclin D1 may be used as a prognosis marker in patients with advanced hypopharyngeal carcinoma[137] or melanoma[138].

Many studies have shown that overexpression of cyclin D1 plays a pivotal role in tumor progression and lowering its levels by antisense may induce remission suggesting cyclin D1 as a target for therapy. Downregulation of cyclin D1 using an antisense approach may have therapeutic effect at least in lung and pancreatic cancers[139]. Several approaches in targeting cyclin D have been developed: modulation of mRNA translation by desferroxamine[140], eicosapentaenoic acid[141], post-translational modifications (through ubiquitination and proteasomal degradation)[142], or enzyme function (through selective CDK kinase inhibitors)[119,143].

## 4. CYCLIN A IN APOPTOSIS AND CANCER

Cancer onset and progression is frequently associated with disregulation not only of the E-type cyclin/CDK2-interacting partners but also of cyclin A. There are two A-type cyclins that are active in late S and early G2/M. While the role of the Cyclin A/CDK2 is best understood, recent studies suggest that

Cyclin A-CDK1 may also be involved in DNA replication because of its interaction with and phosphorylation of the Flap endonuclease 1 (Fen1), a key enzyme of the DNA replication machinery[144]. Absence of cyclin A1 leads to sterility in male mice due to cell cycle arrest in spermatocytes during meiosis, followed by apoptosis[145,146]. Myelocytic leukemias, particularly the ones at the promyelocyte and myeloblast stages of development overexpress Cyclin A1[147].

Cyclin A2 deficiency is embryonic lethal[148]. Its levels go up during S-phase and then drop following its proteasomal degradation before metaphase. During S-phase, it plays an important role in ensuring that DNA replication takes place only once per cell cycle. However, its role during mitosis has not been unequivocally established yet. While Cyclin A2 is predominantly localized in the nucleus, its cytosolic overexpression may lead to aneuploidy and transformation even in the absence of viral or Ha-ras influence[149]. On the other hand, Cyclin A-dependent kinase translocates to the cytoplasm following UV-induced apoptosis in proliferating mouse mesangial cells, a mechanism which is p53-independent[150].

One of the factors that lead to its deregulation is the transforming activity of viruses. For example, Epstein-Barr virus nuclear antigen 3C transforms primary B cells by binding to Cyclin A, enhancing the activity of its binding partner, CDK2, and inducing the phosphorylation of pRb.[151] Furthermore, integration of hepatitis virus B in the cyclin A gene intron and its disruption may contribute to transformation[152].

E2F1 in particular serves as the primary link between loss of pRb function and activation of p53-dependent apoptosis[153]. E2F1 plays a central role in signaling disturbances in the pRb growth control pathway by upregulation of Chk2 and p53 phosphorylation but which also require the DNA damage response proteins ATM and NBS1. E2F1 induced by DNA damage can bind to and promote the apoptotic function of p53 via the cyclin A binding site of E2F1. Cyclin A, but not Cyclin E, prevents E2F1 from interacting and cooperating with p53 to induce apoptosis. However, in response to DNA damage, cyclin A levels decrease, with a concomitant increase in E2F1-p53 complex formation[154]. Overexpression of cyclin A in breast cancer cells (MCF-7) interferes with the apoptotic function of p53[155].

Deregulated expression of cyclin A has been reported in several tumors. Cyclin A as well as Cdc2 plays a key role in the progression of thyroid[156] and hepatocellular carcinoma[157]. Prognosis of endometrial carcinoma[158], renal cell carcinoma[159], colorectal[160] as well as NSCLC carcinomas[161,162] and soft tissue sarcomas[163] may be assessed from the expression levels of cyclin A. Adhesion of cells to extracellular matrix plays an essential role in the regulation of cell proliferation. Cyclin A levels increase following engagement of CD44 surface matrix receptor by anti-CD44 antibodies followed by an apoptotic phenotype which was reverted by cyclin A antisense oligonucleotides, suggesting again the role of cyclin A in apoptosis[164].

Increased levels of cyclin A2 in colorectal cancer have been associated with progression from early dysplasia to adenoma and carcinoma, further metastasis to the liver correlating with lower expression levels[165]. Moreover, elevated levels of cyclin A correlate with poor prognosis and relapse of primary breast carcinomas[166] and superficial spreading melanomas[167].

## 5. CYCLIN B IN APOPTOSIS AND CANCER

Targeting the G2/M transition has been based on the observations that it represents the phase of the cell cycle when cells are most sensitive to therapeutic intervention with DNA-damaging agents, such as radiation. Also, in most tumor cells G2/M provides the main checkpoint by arresting the irradiated cells and its deregulation can further sensitize cells to therapy. Cyclin B plays a critical role in the cell cycle by regulating mitosis, and thus chromosome condensation, nuclear envelope breakdown, mitotic spindle assembly, and chromosome attachment to the microtubules. Intracellular localization of Cyclin B1 plays an important role in Cyclin B1-dependent apoptosis, as its nuclear expression is essential for DNA-damage-induced apoptosis whereas its nuclear export renders cells resistant[168]. Two types of cyclin B have been identified in mammals. Whereas cyclin B2-null mice undergo a normal development, cyclin B1 proved to be lethal[169]. The expression of cyclin B and its interacting partner CDK1 is essential for the progression through the M-phase of the cell cycle. However, its unscheduled expression in the early and late stages of G1 in MOLT-4 leukemia and breast cancer cells may play a role in apoptosis and pathogenesis of the disease[170].

Nucleoside analogues induce irreversible arrest at the G2/M checkpoint by increasing the levels of the CDC2-cyclin B inhibitor $p21^{Cip1/Waf1}$ [171]. Commonly used drugs in cancer therapy, taxol[172,173] and paclitaxel[174], cause accumulation of Cyclin B1 as well activation of CDK1, inducing cell-cycle arrest and eventually trigger apoptosis. Cyclin A and B1 have been found to be overexpressed in tumors such as those of colorectal cancers[175] and astrocytoma[176]. Moreover, Cyclin B1 may be associated with the progression from Helicobacter pylori-induced gastritis to mucosa-associated lymphoid tissue lymphoma[177] and in squamous non-small cell lung carcinoma[178]. Cyclin B-CDC2 activity has also a key role in the exit of cells from the G2/M arrest following ionizing radiation in an acidic environment[179]. The sensitivity of hematopoietic cells to radiation may depend on the levels of nuclear Cyclin B1 expression[168] in several mouse and human hematopoietic cells (Ramos, DP16, HL60, thymocytes) undergoing radiation-induced apoptosis. Its overexpression induces apoptosis, that can be suppressed by its antisense inhibition[180]. The nuclear retention of the Cyclin B1/CDK1 complex in response to genotoxic stress[181] or its activation may lead to apoptosis and to mitotic catastrophe after ionizing radiation in p53-deficient epithelial cells[182].

## 6. CELL CYCLE-BASED SENSITIZATION OF TUMORS

The deregulation of cell cycle checkpoints and the molecules associated with them may transform a normal cell into a cancer cell. This most often occurs by interfering with the basic cell cycle regulatory machinery to activate cell cycle entry that leads to deregulation of the cell cycle in dividing somatic cells. Finding anticancer drug treatment strategies tailored to modulate specific cell cycle components in particular tumors is of great interest. Candidate targets for such strategies include critical cell cycle molecules involved in the G1 to S phase or G2 to M phase transitions[183-185].

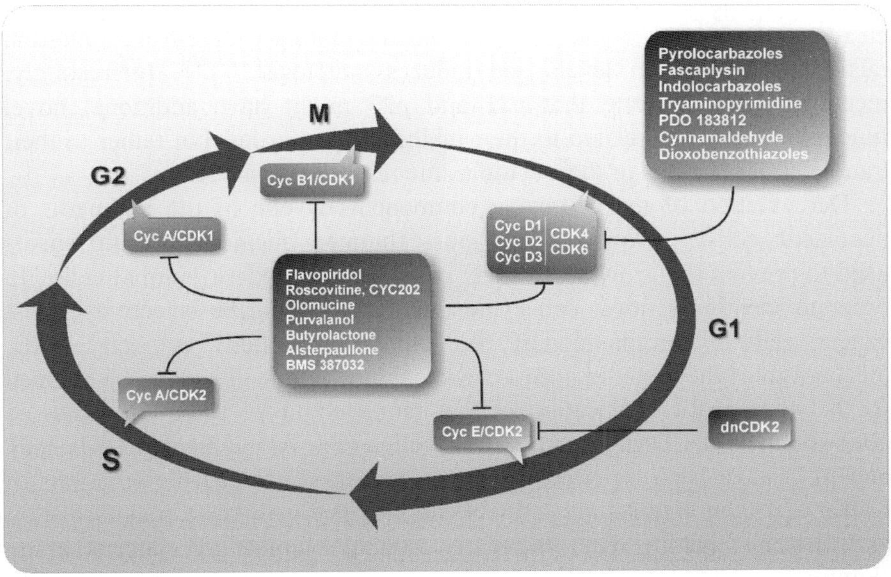


*Figure 3.* Therapeutic approaches for targeting Cyclin/CDKs in tumors. A number of pharmacologic inhibitors are generated targeting the cyclins/CDKs at different phases of the cell cycle and several of them are in preclinical trials. Most of the inhibitors prevent the activity of more than one Cyclin/CDK. The effect of Cyclin E can be blocked by the expression of a dominant–negative (dn) CDK2 which may constitute another therapeutic approach to target Cyclin E.

S-phase sensitization through G1/S Cyclin or E2F deregulation could be an effective therapeutic approach. Most tumor cells are characterized by aberrations in cell cycle controls primarily incurred through inactivation of pRb by the phosphorylation induced hyperactive CDKs or inactivating viral proteins[186]. Clearly, Rb-deficiency allows cells to progress into S-phase inappropriately, conditions under which they are vulnerable to chemo and radiotherapy[187]. Rb deficiency leads to increased expression of E2F target genes that include cyclin E. Increasing levels of cyclins using proteasome inhibitors and thus sensitize target cells has been also used effectively to target S-phase cyclin complexes as an effective therapeutic approach to induce apoptosis[188].

## 6.1 CDK Inhibition as a Therapeutic Approach

Uncontrolled CDK activity is often the cause of human cancer. The function of CDKs is tightly regulated by cell-cycle inhibitors such as $p21^{Cip1/Waf1}$ and $p27^{Kip1}$. Following DNA damage or anti-mitogenic signals, p21 and p27 associate with Cyclin-CDK complexes to inhibit their catalytic activity and induce cell cycle arrest. Loss of function of these genes are often observed in different types of human cancers[102,189]. Molecular targeting of the p21 and p27 inhibitors has been reported in epithelial cells. Thus, cleavage of $p21^{Cip1/Waf1}$ and $p27^{Kip1}$ releases cyclin A-CDK2 from inhibition raising its expression levels and induces apoptosis[67,190,191]. Interestingly, recent findings indicate that p21 and p27 might have additional novel functions that are unrelated to their inhibitory properties, but rather to their transcriptional targets or their cytoplasmic relocalization[192].

Deregulation of the cell-cycle components is one of the hallmarks of neoplastic cells[119]. However, the most common chemotherapeutic agents used to prevent the growth of cancer cells are also toxic to normal cells. In order to identify a novel cell cycle target that would be selective against cancer cells, determination of the expression pattern as well as the mechanism of the target in normal versus tumor cells will have to be pursued so that therapeutic application of that target could be achieved. Genetic abnormalities of CDKs (e.g.1, 2, 7), Cyclins (e.g. A and E) and CKIs (p21 and p27) have been reported in human cancers. Enforced perturbations of CDKs and their interacting partners can enhance neoplastic transformation and, thereby, making them extremely attractive targets for cancer therapy (Figure 3).

Structural studies of cyclins and CDKs have revealed that CDKs provide suitable drug targets given their more rigid structure and identifiable domains, such as that responsible for ATP-binding. In contrast, cyclins have a highly flexible structure that is not amenable for target development[193,194].

Inhibitors targeting CDKs have become of great interest in cancer therapy with more than 50 compounds being evaluated for antitumor activity, some of them already in preclinical trials[79,186]. These inhibitors either target the CDKs themselves, through their ATP-binding site, or their upstream regulators. The first group includes include flavopiridol and roscovitine, the second UCN-01, perifostine, and lovastatin[186].

Flavopiridol, which has now reached Phase III clinical trials, shows clinical efficacy when either used alone or in combination with standard and novel chemo- or radiotherapeutic modalities[80]. Leukemia and myeloma cells are particularly sensitive to flavopiridol, regardless of their origin, type, or resistance to agents that act by damaging DNA that include radiation[79]. Flavopiridol potently triggers apoptosis in small cell lung cancer[195]. It can be also combined effectively with PMA, deacetylase inhibitors[196] and Apo2L/TRAIL[197], a tumor-specific cell death ligand expected to go soon into clinical trials[198]. Synergism with conventional cytotoxic agents may be related to the agents used and the sequence of treatment. Nevertheless, pre-treatment with flavopiridol has been shown to recruit leukemia cells into a proliferative state thus priming them for S-phase related cytotoxicity of nucleoside analogues, such as ara-C[199]. A similar mechanism may be responsible for its combination with CPT-11 or its active metabolite, SN-38 known to be also active in S-phase cells. SCLC cells are particularly sensitive to flavopiridol in S-phase[195]. Flavopiridol may also influence the cellular radioresponse through inhibition of DNA repair[200], possibly through repression of gene transcription[201], including that of the DNA repair gene KU70[202] and antiapoptotic proteins[203]. It may also do so by suppression of phosphorylation of at least one of these proteins, surviving[204].

Indolinones, UCN-01, and paullones, are currently tested in clinical trials[79,186]. While demonstrating clinical activity, neither acts specifically only against CDK2. Other more specific CDK2 inhibitors are currently in preclinical development[205-207]. These include purine based CDK inhibitors, such as olomoucine, purvalanols, and roscovitine, that display greater selectivity for CDK1, 2 and probably 5. Novel derivatives, such as R-roscovitine (CYC202) show great promise[208]. The efficacy of these inhibitors may not be the result of their CDK-inhibitory activity alone but also their effect on transcription of CDK complex components, such as Cyclin D1 as well as apoptotic regulators[79].

A novel class of therapeutics comprises inhibitors of the proteasome (bortezomib) which lead to accumulation of Cyclin A and B, G2/M arrest and induces apoptosis in lung cancer cell lines[188]. Another new emerging compound with antitumoral activity is a fusion protein made up of a Cyclin A-CDK2 binding peptide and an F-box protein, which targets the complex to ubiquitination and proteasomal degradation. This leads to a decrease of

Cyclin A and CDK2 levels followed by apoptosis both in vivo and in vitro with no significant toxicity to normal cells. Apparently the down-regulation of Cyclin A is key for the cytotoxic activity. However, decreasing the expression of both Cyclin A and CDK2 causes the highest cytotoxic effect[209]. Previous studies have shown that such Cyclin A-CDK2-binding peptides induced apoptosis by further deregulating E2F expression[210]. Thus, an S-phase checkpoint seem to be activated that involves E2F, whose DNA binding function is in turn negatively regulated by Cyclin A[211-213]. A different type of CDK2-Cyclin A antagonists (which do not block the ATP site) bearing the RXL peptidic motif were found to selectively induce apoptosis in cells in which pRb and Cyclin D were deregulated[214].

## 7. CONCLUSIONS

A large body of evidence, only partly reviewed here support the case for the need for targeting cell cycle regulators, particularly those implicated in the pRb pathway for clinical therapy. Both small molecule inhibitors as well as molecular approaches for inhibition of cyclins and CDKs are being pursued. However, expression of not just a particular cyclin but of several if not may be important for some tumors, making the case that developing specific small molecule inhibitors may not be so critical. Thus, advanced malignancy in head and neck squamous cell carcinomas has been associated with elevated levels of Cyclin A, B1, and E expression[215]. Expression of dominant negative mutants of CDC2, CDK2, and CDK3 in HeLa cells inhibited apoptosis induced by TNF$\alpha$ and staurosporine[74]. Expression of cyclin A, B, and D1 have been downregulated by flavopiridol which impairs the CDK activity but also has a positive influence upon induction of apoptosis. Inhibition of cyclin A-dependent kinase leads to a persistent activity of E2F1 transcription factor which sensitizes cells to flavopiridol-induced apoptosis[216].

## ACKNOWLEDGMENTS

We would like to apologize to all our colleagues whose work has not, or has been only indirectly mentioned. This work was supported by research grants from the National Institutes of Health to AA (CA81504 and CA82858).

# REFERENCES


1. J. D. Moore, J. A. Kirk & T. Hunt. Unmasking the S-phase-promoting potential of cyclin B1. *Science* **300**, 987-90. (2003).
2. A. W. Murray. Recycling the cell cycle: cyclins revisited. *Cell* **116**, 221-34. (2004).
3. A. Koff, F. Cross, A. Fisher, J. Schumacher, K. Leguellec, M. Philippe & J. M. Roberts. Human cyclin E, a new cyclin that interacts with two members of the CDC2 gene family. *Cell* **66**, 1217-28 (1991).
4. D. J. Lew, V. Dulic & S. I. Reed. Isolation of three novel human cyclins by rescue of G1 cyclin (Cln) function in yeast. *Cell* **66**, 1197-206 (1991).
5. Y. Geng, W. Whoriskey, M. Y. Park, R. T. Bronson, R. H. Medema, T. Li, R. A. Weinberg & P. Sicinski. Rescue of cyclin D1 deficiency by knockin cyclin E. *Cell* **97**, 767-77. (1999).
6. S. Mazumder, E. L. DuPree & A. Almasan. A dual role of cyclin E in cell proliferation and apoptosis may provide a target for cancer therapy. *Curr Cancer Drug Targets* **4**, 65-75. (2004).
7. T. Moroy & C. Geisen. Cyclin E. *Int J Biochem Cell Biol* **36**, 1424-39. (2004).
8. P. K. Jackson, S. Chevalier, M. Philippe & M. W. Kirschner. Early events in DNA replication require cyclin E and are blocked by p21CIP1. *J Cell Biol* **130**, 755-69 (1995).
9. M. Ohtsubo, A. M. Theodoras, J. Schumacher, J. M. Roberts & M. Pagano. Human cyclin E, a nuclear protein essential for the G1-to-S phase transition. *Mol Cell Biol* **15**, 2612-24 (1995).
10. D. Resnitzky, M. Gossen, H. Bujard & S. I. Reed. Acceleration of the G1/S phase transition by expression of cyclins D1 and E with an inducible system. *Mol Cell Biol* **14**, 1669-79 (1994).
11. S. F. Dowdy, P. W. Hinds, K. Louie, S. I. Reed, A. Arnold & R. A. Weinberg. Physical interaction of the retinoblastoma protein with human D cyclins. *Cell* **73**, 499-511 (1993).
12. D. Resnitzky & S. I. Reed. Different roles for cyclins D1 and E in regulation of the G1-to-S transition. *Mol Cell Biol* **15**, 3463-9 (1995).
13. K. Alevizopoulos, J. Vlach, S. Hennecke & B. Amati. Cyclin E and c-Myc promote cell proliferation in the presence of p16INK4a and of hypophosphorylated retinoblastoma family proteins. *Embo J* **16**, 5322-33 (1997).
14. J. Lukas, T. Herzinger, K. Hansen, M. C. Moroni, D. Resnitzky, K. Helin, S. I. Reed & J. Bartek. Cyclin E-induced S phase without activation of the pRb/E2F pathway. *Genes Dev* **11**, 1479-92 (1997).
15. R. J. Duronio, A. Brook, N. Dyson & P. H. O'Farrell. E2F-induced S phase requires cyclin E. *Genes Dev* **10**, 2505-13 (1996).
16. X. H. Hua, H. Yan & J. Newport. A role for Cdk2 kinase in negatively regulating DNA replication during S phase of the cell cycle. *J Cell Biol* **137**, 183-92. (1997).
17. J. G. Mussman, H. F. Horn, P. E. Carroll, M. Okuda, P. Tarapore, L. A. Donehower & K. Fukasawa. Synergistic induction of centrosome hyperamplification by loss of p53 and cyclin E overexpression. *Oncogene* **19**, 1635-46 (2000).
18. S. Ortega, I. Prieto, J. Odajima, A. Martin, P. Dubus, R. Sotillo, J. L. Barbero, M. Malumbres & M. Barbacid. Cyclin-dependent kinase 2 is essential for meiosis but not for mitotic cell division in mice. *Nat Genet* **35**, 25-31. Epub 2003 Aug 17. (2003).
19. C. Berthet, E. Aleem, V. Coppola, L. Tessarollo & P. Kaldis. Cdk2 knockout mice are viable. *Curr Biol* **13**, 1775-85 (2003).

20. Y. Geng, Q. Yu, E. Sicinska, M. Das, J. E. Schneider, S. Bhattacharya, W. M. Rideout, R. T. Bronson, H. Gardner & P. Sicinski. Cyclin E ablation in the mouse. *Cell* **114**, 431-43. (2003).
21. J. Sage, A. L. Miller, P. A. Perez-Mancera, J. M. Wysocki & T. Jacks. Acute mutation of retinoblastoma gene function is sufficient for cell cycle re-entry. *Nature* **424**, 223-8. (2003).
22. L. Le Cam, J. Polanowska, E. Fabbrizio, M. Olivier, A. Philips, E. Ng Eaton, M. Classon, Y. Geng & C. Sardet. Timing of cyclin E gene expression depends on the regulated association of a bipartite repressor element with a novel E2F complex. *Embo J* **18**, 1878-90. (1999).
23. J. Polanowska, E. Fabbrizio, L. Le Cam, D. Trouche, S. Emiliani, R. Herrera & C. Sardet. The periodic down regulation of Cyclin E gene expression from exit of mitosis to end of G(1) is controlled by a deacetylase- and E2F- associated bipartite repressor element. *Oncogene* **20**, 4115-27. (2001).
24. H. S. Zhang, M. Gavin, A. Dahiya, A. A. Postigo, D. Ma, R. X. Luo, J. W. Harbour & D. C. Dean. Exit from G1 and S phase of the cell cycle is regulated by repressor complexes containing HDAC-Rb-hSWI/SNF and Rb-hSWI/SNF. *Cell* **101**, 79-89. (2000).
25. A. Almasan, Y. Yin, R. E. Kelly, E. Y. Lee, A. Bradley, W.-W. Li, J. R. Bertino & G. M. Wahl. Deficiency of retinoblastoma protein leads to inappropriate S-phase entry, activation of E2F-responsive genes, and apoptosis. *Proc. Natl. Acad. Sci. USA* **92**, 5436-5440. (1995).
26. R. E. Herrera, F. Chen & R. A. Weinberg. Increased histone H1 phosphorylation and relaxed chromatin structure in Rb-deficient fibroblasts. *Proc Natl Acad Sci U S A* **93**, 11510-5 (1996).
27. L. G. Martin, G. W. Demers & D. A. Galloway. Disruption of the G1/S transition in human papillomavirus type 16 E7-expressing human cells is associated with altered regulation of cyclin E. *J Virol* **72**, 975-85 (1998).
28. C. Blattner, A. Sparks & D. Lane. Transcription factor E2F-1 is upregulated in response to DNA damage in a manner analogous to that of p53. *Mol Cell Biol* **19**, 3704-13 (1999).
29. S. Mazumder, B. Gong & A. Almasan. Cyclin E induction by genotoxic stress leads to apoptosis of hematopoietic cells. *Oncogene* **19**, 2828-35. (2000).
30. K. A. Won & S. I. Reed. Activation of cyclin E/CDK2 is coupled to site-specific autophosphorylation and ubiquitin-dependent degradation of cyclin E. *Embo J* **15**, 4182-93 (1996).
31. B. E. Clurman, R. J. Sheaff, K. Thress, M. Groudine & J. M. Roberts. Turnover of cyclin E by the ubiquitin-proteasome pathway is regulated by cdk2 binding and cyclin phosphorylation. *Genes Dev* **10**, 1979-90 (1996).
32. M. Welcker, J. Singer, K. R. Loeb, J. Grim, A. Bloecher, M. Gurien-West, B. E. Clurman & J. M. Roberts. Multisite phosphorylation by Cdk2 and GSK3 controls cyclin E degradation. *Mol Cell* **12**, 381-92 (2003).
33. C. Bai, P. Sen, K. Hofmann, L. Ma, M. Goebl, J. W. Harper & S. J. Elledge. SKP1 connects cell cycle regulators to the ubiquitin proteolysis machinery through a novel motif, the F-box. *Cell* **86**, 263-74. (1996).
34. J. T. Winston, D. M. Koepp, C. Zhu, S. J. Elledge & J. W. Harper. A family of mammalian F-box proteins. *Curr Biol* **9**, 1180-2. (1999).
35. H. Strohmaier, C. H. Spruck, P. Kaiser, K. A. Won, O. Sangfelt & S. I. Reed. Human F-box protein hCdc4 targets cyclin E for proteolysis and is mutated in a breast cancer cell line. *Nature* **413**, 316-22. (2001).

36. D. M. Koepp, L. K. Schaefer, X. Ye, K. Keyomarsi, C. Chu, J. W. Harper & S. J. Elledge. Phosphorylation-dependent ubiquitination of cyclin E by the SCFFbw7 ubiquitin ligase. *Science* **294**, 173-7 (2001).
37. C. H. Spruck, H. Strohmaier, O. Sangfelt, H. M. Muller, M. Hubalek, E. Muller-Holzner, C. Marth, M. Widschwendter & S. I. Reed. hCDC4 gene mutations in endometrial cancer. *Cancer Res* **62**, 4535-9. (2002).
38. S. Ekholm-Reed, C. H. Spruck, O. Sangfelt, F. van Drogen, E. Mueller-Holzner, M. Widschwendter, A. Zetterberg & S. I. Reed. Mutation of hCDC4 leads to cell cycle deregulation of cyclin E in cancer. *Cancer Res* **64**, 795-800 (2004).
39. A. Cheng, P. Kaldis & M. J. Solomon. Dephosphorylation of human cyclin-dependent kinases by protein phosphatase type 2C alpha and beta 2 isoforms. *J Biol Chem* **275**, 34744-9 (2000).
40. M. Donzelli & G. F. Draetta. Regulating mammalian checkpoints through Cdc25 inactivation. *EMBO Rep* **4**, 671-7 (2003).
41. R. Y. Poon & T. Hunter. Dephosphorylation of Cdk2 Thr160 by the cyclin-dependent kinase-interacting phosphatase KAP in the absence of cyclin. *Science* **270**, 90-3 (1995).
42. L. L. Parker, P. J. Sylvestre, M. J. Byrnes, 3rd, F. Liu & H. Piwnica-Worms. Identification of a 95-kDa WEE1-like tyrosine kinase in HeLa cells. *Proc Natl Acad Sci U S A* **92**, 9638-42 (1995).
43. P. R. Mueller, T. R. Coleman, A. Kumagai & W. G. Dunphy. Myt1: a membrane-associated inhibitory kinase that phosphorylates Cdc2 on both threonine-14 and tyrosine-15. *Science* **270**, 86-90 (1995).
44. D. J. Lew & S. Kornbluth. Regulatory roles of cyclin dependent kinase phosphorylation in cell cycle control. *Curr Opin Cell Biol* **8**, 795-804 (1996).
45. J. Falck, N. Mailand, R. G. Syljuasen, J. Bartek & J. Lukas. The ATM-Chk2-Cdc25A checkpoint pathway guards against radioresistant DNA synthesis. *Nature* **410**, 842-7. (2001).
46. D. R. Green & G. I. Evan. A matter of life and death. *Cancer Cell* **1**, 19-30. (2002).
47. N. N. Danial & S. J. Korsmeyer. Cell death. Critical control points. *Cell* **116**, 205-19 (2004).
48. S. Cory & J. M. Adams. The bcl2 family: regulators of the cellular life-or-death switch. *Nat Rev Cancer* **2**, 647-56. (2002).
49. S. Ray & A. Almasan. Apoptosis induction in prostate cancer cells and xenografts by combined treatment with Apo2L/TRAIL and CPT-11. *Cancer Res* **63**, 4713-4723 (2003).
50. B. Gong & A. Almasan. Differential upregulation of p53-responsive genes by genotoxic stress in hematopoietic cells containing wild-type and mutant p53. *Gene Expr* **8**, 197-206 (1999).
51. B. Gong, Q. Chen, B. Endlich, S. Mazumder & A. Almasan. Ionizing radiation-induced, Bax-mediated cell death is dependent on activation of serine and cysteine proteases. *Cell Growth Diff.* **10**, 491-502 (1999).
52. M. Sawada, W. Sun, P. Hayes, K. Leskov, D. A. Boothman & S. Matsuyama. Ku70 suppresses the apoptotic translocation of Bax to mitochondria. *Nat Cell Biol* **5**, 320-9. (2003).
53. B. Gong & A. Almasan. Apo2 ligand/TNF-related apoptosis-inducing ligand and death receptor 5 mediate the apoptotic signaling induced by ionizing radiation in leukemic cells. *Cancer Res* **60**, 5754-60 (2000).
54. Q. Chen, B. Gong & A. Almasan. Distinct stages of cytochrome c release from mitochondria: evidence for a feedback amplification loop linking caspase activation to

mitochondrial dysfunction in genotoxic stress induced apoptosis. *Cell Death Differ* **7**, 227-33 (2000).
55. Q. Chen, M. Crosby & A. Almasan. Redox regulation of apoptosis before and after cytochrome c release. *Korean J Biol Sci* **7**, 1-9 (2003).
56. S. Ray, J. G. Hissong, M. Oancea & A. Almasan. *Expression and regulation of death receptors in multiple myeloma and prostate carcinoma* (ed. El-Deiry, W.) (Humana Press, 2004).
57. Q. Chen, Y.-C. Chai, S. Mazumder, C. Jiang, R. M. Macklis, G. M. Chisolm & A. Almasan. The late increase in intracellular free radical oxygen species during apoptosis is associated with cytochrome c release, caspase activation, and mitochondrial dysfunction. *Cell Death Differ* **10**, 323-334 (2003).
58. X. Liu, C. N. Kim, J. Yang, R. Jemmerson & X. Wang. Induction of apoptotic program in cell-free extracts: requirement for dATP and cytochrome c. *Cell* **86**, 147-57 (1996).
59. P. Li, D. Nijhawan, I. Budihardjo, S. M. Srinivasula, M. Ahmad, E. S. Alnemri & X. Wang. Cytochrome c and dATP-dependent formation of Apaf-1/caspase-9 complex initiates an apoptotic protease cascade. *Cell* **91**, 479-89 (1997).
60. X. Wang. The expanding role of mitochondria in apoptosis. *Genes Dev* **15**, 2922-33. (2001).
61. U. Fischer, R. U. Janicke & K. Schulze-Osthoff. Many cuts to ruin: a comprehensive update of caspase substrates. *Cell Death Differ* **10**, 76-100 (2003).
62. S. Mazumder, Q. Chen, B. Gong, J. A. Drazba, J. C. Buchsbaum & A. Almasan. Proteolytic cleavage of cyclin E leads to inactivation of associated kinase activity and amplification of apoptosis in hematopoietic cells. *Mol. Cell. Biol.* **22**, 2398-409. (2002).
63. R. U. Janicke, P. A. Walker, X. Y. Lin & A. G. Porter. Specific cleavage of the retinoblastoma protein by an ICE-like protease in apoptosis. *Embo J* **15**, 6969-78 (1996).
64. X. Tan, S. J. Martin, D. R. Green & J. Y. J. Wang. Degradation of retinoblastoma protein in tumor necrosis factor- and CD95-induced cell death. *J Biol Chem* **272**, 9613-6 (1997).
65. L. Chen, V. Marechal, J. Moreau, A. J. Levine & J. Chen. Proteolytic cleavage of the mdm2 oncoprotein during apoptosis. *J Biol Chem* **272**, 22966-73 (1997).
66. R. Beyaert, V. J. Kidd, S. Cornelis, M. Van de Craen, G. Denecker, J. M. Lahti, R. Gururajan, P. Vandenabeele & W. Fiers. Cleavage of PITSLRE kinases by ICE/CASP-1 and CPP32/CASP-3 during apoptosis induced by tumor necrosis factor. *J Biol Chem* **272**, 11694-7 (1997).
67. B. Levkau, H. Koyoma, E. W. raines, B. E. Clurman, B. Herren, K. Orth & J. M. Roberts. Cleavage of p21/cip1/waf1 and p27kip1 mediates apoptosis of endotelial cells through activation of Cdk2: Role of a caspase cascade. *Molec. Cell* **1**, 553-563 (1997).
68. J. Schlegel, I. Peters, S. Orrenius, D. K. Miller, N. A. Thornberry, T. T. Yamin & D. W. Nicholson. CPP32/apopain is a key interleukin 1 beta converting enzyme-like protease involved in Fas-mediated apoptosis. *J Biol Chem* **271**, 1841-4 (1996).
69. B. B. Zhou, H. Li, J. Yuan & M. W. Kirschner. Caspase-dependent activation of cyclin-dependent kinases during Fas-induced apoptosis in Jurkat cells. *Proc Natl Acad Sci U S A* **95**, 6785-90 (1998).
70. D. S. Park, E. J. Morris, J. Padmanabhan, M. L. Shelanski, H. M. Geller & L. A. Greene. Cyclin-dependent kinases participate in death of neurons evoked by DNA-damaging agents. *J Cell Biol* **143**, 457-67 (1998).
71. J. Wang & K. Walsh. Resistance to apoptosis conferred by Cdk inhibitors during myocyte differentiation. *Science* **273**, 359-61 (1996).

72. A. Hakem, T. Sasaki, I. Kozieradzki & J. M. Penninger. The cyclin-dependent kinase Cdk2 regulates thymocyte apoptosis. *J Exp Med* **189**, 957-68 (1999).
73. W. Meikrantz & R. Schlegel. Apoptosis and the cell cycle. *J Cell Biochem* **58**, 160-74 (1995).
74. W. Meikrantz & R. Schlegel. Suppression of apoptosis by dominant negative mutants of cyclin-dependent protein kinases. *J Biol Chem* **271**, 10205-9 (1996).
75. H. J. Brady, G. Gil-Gomez, J. Kirberg & A. J. Berns. Bax alpha perturbs T cell development and affects cell cycle entry of T cells. *Embo J* **15**, 6991-7001 (1996).
76. S. Mazel, D. Burtrum & H. T. Petrie. Regulation of cell division cycle progression by bcl-2 expression: a potential mechanism for inhibition of programmed cell death. *J Exp Med* **183**, 2219-26 (1996).
77. G. P. Linette, Y. Li, K. Roth & S. J. Korsmeyer. Cross talk between cell death and cell cycle progression: BCL-2 regulates NFAT-mediated activation. *Proc Natl Acad Sci U S A* **93**, 9545-52 (1996).
78. L. A. O'Reilly, D. C. Huang & A. Strasser. The cell death inhibitor Bcl-2 and its homologues influence control of cell cycle entry. *Embo J* **15**, 6979-90. (1996).
79. Y. Dai & S. Grant. Cyclin-dependent kinase inhibitors. *Curr Opin Pharmacol* **3**, 362-70. (2003).
80. A. M. Senderowicz. Novel small molecule cyclin-dependent kinases modulators in human clinical trials. *Cancer Biol Ther* **2**, S84-95 (2003).
81. C. Geisen & T. Moroy. The oncogenic activity of cyclin E is not confined to Cdk2 activation alone but relies on several other, distinct functions of the protein. *J Biol Chem* **277**, 39909-18. (2002).
82. C. V. Finkielstein, L. G. Chen & J. L. Maller. A role for G1/S cyclin-dependent protein kinases in the apoptotic response to ionizing radiation. *J Biol Chem* **277**, 38476-85 (2002).
83. S. Mazumder, D. Plesca, F. Chen, D. J. Chen & A. Almasan. unpublished (2005).
84. S. Mazumder, E. L. Dupree, M. Kintner & A. Almasan. unpublished (2004).
85. N. K. Dhillon & M. Mudryj. Cyclin E overexpression enhances cytokine-mediated apoptosis in MCF7 breast cancer cells. *Genes Immun* **4**, 336-42. (2003).
86. L. O'Connor, D. C. Huang, L. A. O'Reilly & A. Strasser. Apoptosis and cell division. *Curr Opin Cell Biol* **12**, 257-63. (2000).
87. A. B. Pardee. G1 events and regulation of cell proliferation. *Science* **246**, 603-8. (1989).
88. K. Keyomarsi & A. B. Pardee. Redundant cyclin overexpression and gene amplification in breast cancer cells. *Proc Natl Acad Sci U S A* **90**, 1112-6 (1993).
89. S. V. Ekholm, P. Zickert, S. I. Reed & A. Zetterberg. Accumulation of Cyclin E Is Not a Prerequisite for Passage through the Restriction Point. *Mol Cell Biol* **21**, 3256-65. (2001).
90. T. Hunter & J. Pines. Cyclins and cancer. II: Cyclin D and CDK inhibitors come of age. *Cell* **79**, 573-82 (1994).
91. C. J. Sherr. Cancer cell cycles. *Science* **274**, 1672-7 (1996).
92. G. I. Evan & K. H. Vousden. Proliferation, cell cycle and apoptosis in cancer. *Nature* **411**, 342-8. (2001).
93. K. Keyomarsi, N. O'Leary, G. Molnar, E. Lees, H. J. Fingert & A. B. Pardee. Cyclin E, a potential prognostic marker for breast cancer. *Cancer Res* **54**, 380-5 (1994).
94. K. Keyomarsi, D. Conte, Jr., W. Toyofuku & M. P. Fox. Deregulation of cyclin E in breast cancer. *Oncogene* **11**, 941-50 (1995).
95. K. Keyomarsi, S. L. Tucker, T. A. Buchholz, M. Callister, Y. Ding, G. N. Hortobagyi, I. Bedrosian, C. Knickerbocker, W. Toyofuku, M. Lowe, T. W. Herliczek & S. S.

Bacus. Cyclin E and survival in patients with breast cancer. *N Engl J Med* **347**, 1566-75. (2002).

96. F. Erlandsson, C. Wahlby, S. Ekholm-Reed, A. C. Hellstrom, E. Bengtsson & A. Zetterberg. Abnormal expression pattern of cyclin E in tumour cells. *Int J Cancer* **104**, 369-75. (2003).
97. S. Akli & K. Keyomarsi. Cyclin E and its Low Molecular Weight Forms in Human Cancer and as Targets for Cancer Therapy. *Cancer Biology & Therapy* **2**, S38-47. (2003).
98. M. Ohtsubo & J. M. Roberts. Cyclin-dependent regulation of G1 in mammalian fibroblasts. *Science* **259**, 1908-12 (1993).
99. H. Karsunky, C. Geisen, T. Schmidt, K. Haas, B. Zevnik, E. Gau & T. Moroy. Oncogenic potential of cyclin E in T-cell lymphomagenesis in transgenic mice: evidence for cooperation between cyclin E and Ras but not Myc. *Oncogene* **18**, 7816-24. (1999).
100. G. D'Urso, R. L. Marraccino, D. R. Marshak & J. M. Roberts. Cell cycle control of DNA replication by a homologue from human cells of the p34cdc2 protein kinase. *Science* **250**, 786-91. (1990).
101. A. Koff, A. Giordano, D. Desai, K. Yamashita, J. W. Harper, S. Elledge, T. Nishimoto, D. O. Morgan, B. R. Franza & J. M. Roberts. Formation and activation of a cyclin E-cdk2 complex during the G1 phase of the human cell cycle. *Science* **257**, 1689-94 (1992).
102. P. L. Porter, K. E. Malone, P. J. Heagerty, G. M. Alexander, L. A. Gatti, E. J. Firpo, J. R. Daling & J. M. Roberts. Expression of cell-cycle regulators p27Kip1 and cyclin E, alone and in combination, correlate with survival in young breast cancer patients. *Nat Med* **3**, 222-5. (1997).
103. J. Q. Li, H. Miki, M. Ohmori, F. Wu & Y. Funamoto. Expression of cyclin E and cyclin-dependent kinase 2 correlates with metastasis and prognosis in colorectal carcinoma. *Hum Pathol* **32**, 945-53. (2001).
104. W. L. Lingle, W. H. Lutz, J. N. Ingle, N. J. Maihle & J. L. Salisbury. Centrosome hypertrophy in human breast tumors: implications for genomic stability and cell polarity. *Proc Natl Acad Sci U S A* **95**, 2950-5. (1998).
105. P. E. Carroll, M. Okuda, H. F. Horn, P. Biddinger, P. J. Stambrook, L. L. Gleich, Y. Q. Li, P. Tarapore & K. Fukasawa. Centrosome hyperamplification in human cancer: chromosome instability induced by p53 mutation and/or Mdm2 overexpression. *Oncogene* **18**, 1935-44. (1999).
106. N. H. Nielsen, C. Arnerlov, S. O. Emdin & G. Landberg. Cyclin E overexpression, a negative prognostic factor in breast cancer with strong correlation to oestrogen receptor status. *Br J Cancer* **74**, 874-80. (1996).
107. Y. Akama, W. Yasui, H. Yokozaki, H. Kuniyasu, K. Kitahara, T. Ishikawa & E. Tahara. Frequent amplification of the cyclin E gene in human gastric carcinomas. *Jpn J Cancer Res* **86**, 617-21. (1995).
108. Y. Dong, L. Sui, Y. Tai, K. Sugimoto, T. Hirao & M. Tokuda. Prognostic significance of cyclin E overexpression in laryngeal squamous cell carcinomas. *Clin Cancer Res* **6**, 4253-8. (2000).
109. M. Erlanson & G. Landberg. Prognostic implications of p27 and cyclin E protein contents in malignant lymphomas. *Leuk Lymphoma* **40**, 461-70. (2001).
110. M. Erlanson, C. Portin, B. Linderholm, J. Lindh, G. Roos & G. Landberg. Expression of cyclin E and the cyclin-dependent kinase inhibitor p27 in malignant lymphomas-prognostic implications. *Blood* **92**, 770-7. (1998).

111. Y. Hedberg, E. Davoodi, B. Ljungberg, G. Roos & G. Landberg. Cyclin E and p27 protein content in human renal cell carcinoma: clinical outcome and associations with cyclin D. *Int J Cancer* **102**, 601-7 (2002).
112. T. Noguchi, Y. Dobashi, H. Minehara, M. Itoman & T. Kameya. Involvement of cyclins in cell proliferation and their clinical implications in soft tissue smooth muscle tumors. *Am J Pathol* **156**, 2135-47 (2000).
113. H. Kuhling, P. Alm, H. Olsson, M. Ferno, B. Baldetorp, R. Parwaresch & P. Rudolph. Expression of cyclins E, A, and B, and prognosis in lymph node-negative breast cancer. *J Pathol* **199**, 424-31 (2003).
114. R. Ohashi, C. Gao, M. Miyazaki, K. Hamazaki, T. Tsuji, Y. Inoue, T. Uemura, R. Hirai, N. Shimizu & M. Namba. Enhanced expression of cyclin E and cyclin A in human hepatocellular carcinomas. *Anticancer Res* **21**, 657-62 (2001).
115. M. W. Datta, A. A. Renshaw, A. Dutta, M. A. Hoffman & K. R. Loughlin. Evaluation of cyclin expression in testicular germ cell tumors: cyclin E correlates with tumor type, advanced clinical stage, and pulmonary metastasis. *Mod Pathol* **13**, 667-72 (2000).
116. M. Sugimoto, N. Martin, D. P. Wilks, K. Tamai, T. J. Huot, C. Pantoja, K. Okumura, M. Serrano & E. Hara. Activation of cyclin D1-kinase in murine fibroblasts lacking both p21(Cip1) and p27(Kip1). *Oncogene* **21**, 8067-74 (2002).
117. C. Bouchard, K. Thieke, A. Maier, R. Saffrich, J. Hanley-Hyde, W. Ansorge, S. Reed, P. Sicinski, J. Bartek & M. Eilers. Direct induction of cyclin D2 by Myc contributes to cell cycle progression and sequestration of p27. *Embo J* **18**, 5321-33 (1999).
118. D. C. Guttridge, C. Albanese, J. Y. Reuther, R. G. Pestell & A. S. Baldwin, Jr. NF-kappaB controls cell growth and differentiation through transcriptional regulation of cyclin D1. *Mol Cell Biol* **19**, 5785-99 (1999).
119. C. J. Sherr & F. McCormick. The RB and p53 pathways in cancer. *Cancer Cell* **2**, 103-12 (2002).
120. E. L. DuPree, S. Mazumder & A. Almasan. Genotoxic Stress Induces Expression of E2F4, Leading to Its Association with p130 in Prostate Carcinoma Cells. *Cancer Res* **64**, 4390-3 (2004).
121. J. M. Trimarchi & J. A. Lees. Sibling rivalry in the E2F family. *Nat Rev Mol Cell Biol* **3**, 11-20 (2002).
122. K. Kozar, M. A. Ciemerych, V. I. Rebel, H. Shigematsu, A. Zagozdzon, E. Sicinska, Y. Geng, Q. Yu, S. Bhattacharya, R. T. Bronson, K. Akashi & P. Sicinski. Mouse development and cell proliferation in the absence of d-cyclins. *Cell* **118**, 477-91 (2004).
123. M. Malumbres, R. Sotillo, D. Santamaria, J. Galan, A. Cerezo, S. Ortega, P. Dubus & M. Barbacid. Mammalian Cells Cycle without the D-Type Cyclin-Dependent Kinases Cdk4 and Cdk6. *Cell* **118**, 493-504 (2004).
124. C. J. Sherr & J. M. Roberts. CDK inhibitors: positive and negative regulators of G1-phase progression. *Genes Dev* **13**, 1501-12. (1999).
125. Q. Yu, Y. Geng & P. Sicinski. Specific protection against breast cancers by cyclin D1 ablation. *Nature* **411**, 1017-21 (2001).
126. Y. Dobashi, A. Goto, M. Fukayama, A. Abe & A. Ooi. Overexpression of cdk4/cyclin D1, a possible mediator of apoptosis and an indicator of prognosis in human primary lung carcinoma. *Int J Cancer* **110**, 532-41 (2004).
127. M. Y. Niu, M. Menard, J. C. Reed, S. Krajewski & M. A. Pratt. Ectopic expression of cyclin D1 amplifies a retinoic acid-induced mitochondrial death pathway in breast cancer cells. *Oncogene* **20**, 3506-18 (2001).
128. T. Hideshima, P. L. Bergsagel, W. M. Kuehl & K. C. Anderson. Advances in biology of multiple myeloma: clinical applications. *Blood* **104**, 607-18 (2004).

129. J. C. Lung, J. S. Chu, J. C. Yu, C. T. Yue, Y. L. Lo, C. Y. Shen & C. W. Wu. Aberrant expression of cell-cycle regulator cyclin D1 in breast cancer is related to chromosomal genomic instability. *Genes Chromosomes Cancer* **34**, 276-84 (2002).
130. G. J. Boonen, B. A. van Oirschot, A. van Diepen, W. J. Mackus, L. F. Verdonck, G. Rijksen & R. H. Medema. Cyclin D3 regulates proliferation and apoptosis of leukemic T cell lines. *J Biol Chem* **274**, 34676-82 (1999).
131. M. Bai, E. Tsanou, N. J. Agnantis, A. Chaidos, D. Dimou, A. Skyrlas, S. Dimou, M. Vlychou, V. Galani & P. Kanavaros. Expression of cyclin D3 and cyclin E and identification of distinct clusters of proliferation and apoptosis in diffuse large B-cell lymphomas. *Histol Histopathol* **18**, 449-57. (2003).
132. J. K. Buolamwini. Cell cycle molecular targets in novel anticancer drug discovery. *Curr Pharm Des* **6**, 379-92 (2000).
133. D. X. Liu & L. A. Greene. Neuronal apoptosis at the G1/S cell cycle checkpoint. *Cell Tissue Res* **305**, 217-28 (2001).
134. O. Kranenburg, A. J. van der Eb & A. Zantema. Cyclin D1 is an essential mediator of apoptotic neuronal cell death. *Embo J* **15**, 46-54 (1996).
135. P. Sumrejkanchanakij, M. Tamamori-Adachi, Y. Matsunaga, K. Eto & M. A. Ikeda. Role of cyclin D1 cytoplasmic sequestration in the survival of postmitotic neurons. *Oncogene* **22**, 8723-30 (2003).
136. F. Xia & S. N. Powell. The molecular basis of radiosensitivity and chemosensitivity in the treatment of breast cancer. *Semin Radiat Oncol* **12**, 296-304 (2002).
137. G. Nishimura, M. Tsukuda, L. X. Zhou, S. Furukawa & Y. Baba. Cyclin D1 expression as a prognostic factor in advanced hypopharyngeal carcinoma. *J Laryngol Otol* **112**, 552-5 (1998).
138. M. Strasberg Rieber & M. Rieber. Suppression of cyclin D1 but not cdk4 or cyclin A with induction of melanoma terminal differentiation. *Biochem Biophys Res Commun* **216**, 422-7 (1995).
139. B. Driscoll, S. Buckley, L. Barsky, K. Weinberg, K. D. Anderson & D. Warburton. Abrogation of cyclin D1 expression predisposes lung cancer cells to serum deprivation-induced apoptosis. *Am J Physiol* **276**, L679-87 (1999).
140. M. Caraglia, P. Tagliaferri, A. Budillon & A. Abbruzzese. Post-translational modifications of eukaryotic initiation factor-5A (eIF-5A) as a new target for anti-cancer therapy. *Adv Exp Med Biol* **472**, 187-98 (1999).
141. S. S. Palakurthi, R. Fluckiger, H. Aktas, A. K. Changolkar, A. Shahsafaei, S. Harneit, E. Kilic & J. A. Halperin. Inhibition of translation initiation mediates the anticancer effect of the n-3 polyunsaturated fatty acid eicosapentaenoic acid. *Cancer Res* **60**, 2919-25 (2000).
142. S. Tsutsumi, T. Yanagawa, T. Shimura, T. Fukumori, V. Hogan, H. Kuwano & A. Raz. Regulation of cell proliferation by autocrine motility factor/phosphoglucose isomerase signaling. *J Biol Chem* **278**, 32165-72 (2003).
143. M. Malumbres & M. Barbacid. To cycle or not to cycle: a critical decision in cancer. *Nat Rev Cancer* **1**, 222-31 (2001).
144. G. Henneke, S. Koundrioukoff & U. Hubscher. Phosphorylation of human Fen1 by cyclin-dependent kinase modulates its role in replication fork regulation. *Oncogene* **22**, 4301-13 (2003).
145. G. Salazar, D. Liu, C. Liao, L. Batkiewicz, R. Arbing, S. S. Chung, K. Lele & D. J. Wolgemuth. Apoptosis in male germ cells in response to cyclin A1-deficiency and cell cycle arrest. *Biochem Pharmacol* **66**, 1571-9 (2003).
146. D. Liu, M. M. Matzuk, W. K. Sung, Q. Guo, P. Wang & D. J. Wolgemuth. Cyclin A1 is required for meiosis in the male mouse. *Nat Genet* **20**, 377-80 (1998).

147. R. Yang, T. Nakamaki, M. Lubbert, J. Said, A. Sakashita, B. S. Freyaldenhoven, S. Spira, V. Huynh, C. Muller & H. P. Koeffler. Cyclin A1 expression in leukemia and normal hematopoietic cells. *Blood* **93**, 2067-74 (1999).
148. M. Murphy, M. G. Stinnakre, C. Senamaud-Beaufort, N. J. Winston, C. Sweeney, M. Kubelka, M. Carrington, C. Brechot & J. Sobczak-Thepot. Delayed early embryonic lethality following disruption of the murine cyclin A2 gene. *Nat Genet* **15**, 83-6 (1997).
149. J. Faivre, M. Frank-Vaillant, R. Poulhe, H. Mouly, C. Jessus, C. Brechot & J. Sobczak-Thepot. Centrosome overduplication, increased ploidy and transformation in cells expressing endoplasmic reticulum-associated cyclin A2. *Oncogene* **21**, 1493-500 (2002).
150. K. Hiromura, J. W. Pippin, M. J. Blonski, J. M. Roberts & S. J. Shankland. The subcellular localization of cyclin dependent kinase 2 determines the fate of mesangial cells: role in apoptosis and proliferation. *Oncogene* **21**, 1750-8 (2002).
151. J. S. Knight & E. S. Robertson. Epstein-Barr virus nuclear antigen 3C regulates cyclin A/p27 complexes and enhances cyclin A-dependent kinase activity. *J Virol* **78**, 1981-91 (2004).
152. J. Wang, X. Chenivesse, B. Henglein & C. Brechot. Hepatitis B virus integration in a cyclin A gene in a hepatocellular carcinoma. *Nature* **343**, 555-7 (1990).
153. H. A. Rogoff, M. T. Pickering, F. M. Frame, M. E. Debatis, Y. Sanchez, S. Jones & T. F. Kowalik. Apoptosis associated with deregulated E2F activity is dependent on E2F1 and Atm/Nbs1/Chk2. *Mol Cell Biol* **24**, 2968-77 (2004).
154. J. K. Hsieh, D. Yap, D. J. O'Connor, V. Fogal, L. Fallis, F. Chan, S. Zhong & X. Lu. Novel function of the cyclin A binding site of E2F in regulating p53-induced apoptosis in response to DNA damage. *Mol Cell Biol* **22**, 78-93 (2002).
155. I. R. Bukholm, A. Husdal, J. M. Nesland, A. Langerod & G. Bukholm. Overexpression of cyclin A overrides the effect of p53 alterations in breast cancer patients with long follow-up time. *Breast Cancer Res Treat* **80**, 199-206 (2003).
156. Y. Ito, H. Yoshida, K. Nakano, Y. Takamura, K. Kobayashi, T. Yokozawa, F. Matsuzuka, N. Matsuura, K. Kuma & A. Miyauchi. Expression of G2-M modulators in thyroid neoplasms: correlation of cyclin A, B1 and cdc2 with differentiation. *Pathol Res Pract* **198**, 397-402 (2002).
157. Q. Zhou, Q. He & L. J. Liang. Expression of p27, cyclin E and cyclin A in hepatocellular carcinoma and its clinical significance. *World J Gastroenterol* **9**, 2450-4 (2003).
158. H. C. Shih, T. Shiozawa, K. Kato, T. Imai, T. Miyamoto, J. Uchikawa, T. Nikaido & I. Konishi. Immunohistochemical expression of cyclins, cyclin-dependent kinases, tumor-suppressor gene products, Ki-67, and sex steroid receptors in endometrial carcinoma: positive staining for cyclin A as a poor prognostic indicator. *Hum Pathol* **34**, 471-8 (2003).
159. S. Aaltomaa, P. Lipponen, M. Ala-Opas, M. Eskelinen, K. Syrjanen & V. M. Kosma. Expression of cyclins A and D and p21(waf1/cip1) proteins in renal cell cancer and their relation to clinicopathological variables and patient survival. *Br J Cancer* **80**, 2001-7 (1999).
160. K. Handa, M. Yamakawa, H. Takeda, S. Kimura & T. Takahashi. Expression of cell cycle markers in colorectal carcinoma: superiority of cyclin A as an indicator of poor prognosis. *Int J Cancer* **84**, 225-33 (1999).
161. M. Volm & R. Koomagi. Relevance of proliferative and pro-apoptotic factors in non-small-cell lung cancer for patient survival. *Br J Cancer* **82**, 1747-54 (2000).
162. M. Volm & W. Rittgen. Cellular predictive factors for the drug response of lung cancer. *Anticancer Res* **20**, 3449-58 (2000).

163. R. L. Huuhtanen, C. P. Blomqvist, T. O. Bohling, T. A. Wiklund, E. J. Tukiainen, M. Virolainen, B. Tribukait & L. C. Andersson. Expression of cyclin A in soft tissue sarcomas correlates with tumor aggressiveness. *Cancer Res* **59**, 2885-90 (1999).
164. B. Tian, T. Takasu & C. Henke. Functional role of cyclin A on induction of fibroblast apoptosis due to ligation of CD44 matrix receptor by anti-CD44 antibody. *Exp Cell Res* **257**, 135-44 (2000).
165. J. Q. Li, H. Miki, F. Wu, K. Saoo, M. Nishioka, M. Ohmori & K. Imaida. Cyclin A correlates with carcinogenesis and metastasis, and p27(kip1) correlates with lymphatic invasion, in colorectal neoplasms. *Hum Pathol* **33**, 1006-15 (2002).
166. I. R. Bukholm, G. Bukholm & J. M. Nesland. Over-expression of cyclin A is highly associated with early relapse and reduced survival in patients with primary breast carcinomas. *Int J Cancer* **93**, 283-7 (2001).
167. V. A. Florenes, R. S. Faye, G. M. Maelandsmo, J. M. Nesland & R. Holm. Levels of cyclin D1 and D3 in malignant melanoma: deregulated cyclin D3 expression is associated with poor clinical outcome in superficial melanoma. *Clin Cancer Res* **6**, 3614-20 (2000).
168. L. A. Porter, I. H. Cukier & J. M. Lee. Nuclear localization of cyclin B1 regulates DNA damage-induced apoptosis. *Blood* **101**, 1928-33 (2003).
169. M. Brandeis, I. Rosewell, M. Carrington, T. Crompton, M. A. Jacobs, J. Kirk, J. Gannon & T. Hunt. Cyclin B2-null mice develop normally and are fertile whereas cyclin B1-null mice die in utero. *Proc Natl Acad Sci U S A* **95**, 4344-9 (1998).
170. M. Shen, Y. Feng, C. Gao, D. Tao, J. Hu, E. Reed, Q. Q. Li & J. Gong. Detection of cyclin b1 expression in g(1)-phase cancer cell lines and cancer tissues by postsorting Western blot analysis. *Cancer Res* **64**, 1607-10 (2004).
171. P. J. Halloran & R. G. Fenton. Irreversible G2-M arrest and cytoskeletal reorganization induced by cytotoxic nucleoside analogues. *Cancer Res* **58**, 3855-65 (1998).
172. A. M. Ibrado, C. N. Kim & K. Bhalla. Temporal relationship of CDK1 activation and mitotic arrest to cytosolic accumulation of cytochrome C and caspase-3 activity during Taxol-induced apoptosis of human AML HL-60 cells. *Leukemia* **12**, 1930-6 (1998).
173. Y. H. Ling, U. Consoli, C. Tornos, M. Andreeff & R. Perez-Soler. Accumulation of cyclin B1, activation of cyclin B1-dependent kinase and induction of programmed cell death in human epidermoid carcinoma KB cells treated with taxol. *Int J Cancer* **75**, 925-32 (1998).
174. P. Chadebech, I. Truchet, L. Brichese & A. Valette. Up-regulation of cdc2 protein during paclitaxel-induced apoptosis. *Int J Cancer* **87**, 779-86 (2000).
175. A. Wang, N. Yoshimi, N. Ino, T. Tanaka & H. Mori. Overexpression of cyclin B1 in human colorectal cancers. *J Cancer Res Clin Oncol* **123**, 124-7 (1997).
176. K. Allan, R. C. Jordan, L. C. Ang, M. Taylor & B. Young. Overexpression of cyclin A and cyclin B1 proteins in astrocytomas. *Arch Pathol Lab Med* **124**, 216-20 (2000).
177. S. K. Banerjee, A. P. Weston, M. N. Zoubine, D. R. Campbell & R. Cherian. Expression of cdc2 and cyclin B1 in Helicobacter pylori-associated gastric MALT and MALT lymphoma : relationship to cell death, proliferation, and transformation. *Am J Pathol* **156**, 217-25 (2000).
178. J. C. Soria, S. J. Jang, F. R. Khuri, K. Hassan, D. Liu, W. K. Hong & L. Mao. Overexpression of cyclin B1 in early-stage non-small cell lung cancer and its clinical implication. *Cancer Res* **60**, 4000-4 (2000).
179. H. J. Park, J. C. Lyons, T. Ohtsubo & C. W. Song. Cell cycle progression and apoptosis after irradiation in an acidic environment. *Cell Death Differ* **7**, 729-38 (2000).
180. L. A. Porter, G. Singh & J. M. Lee. Abundance of cyclin B1 regulates gamma-radiation-induced apoptosis. *Blood* **95**, 2645-50 (2000).

181. F. B. Charrier-Savournin, M. T. Chateau, V. Gire, J. Sedivy, J. Piette & V. Dulic. p21-Mediated Nuclear Retention of Cyclin B1-Cdk1 in Response to Genotoxic Stress. *Mol Biol Cell* **15**, 3965-76 (2004).
182. M. Castedo, J. L. Perfettini, T. Roumier, K. Andreau, R. Medema & G. Kroemer. Cell death by mitotic catastrophe: a molecular definition. *Oncogene* **23**, 2825-37 (2004).
183. J. Boonstra. Progression through the G1-phase of the on-going cell cycle. *J Cell Biochem* **90**, 244-52. (2003).
184. M. H. Lee & H. Y. Yang. Regulators of G1 cyclin-dependent kinases and cancers. *Cancer Metastasis Rev* **22**, 435-49. (2003).
185. H. Gali-Muhtasib & N. Bakkar. Modulating cell cycle: current applications and prospects for future drug development. *Curr Cancer Drug Targets* **2**, 309-36. (2002).
186. A. M. Senderowicz. Small-molecule cyclin-dependent kinase modulators. *Oncogene* **22**, 6609-20 (2003).
187. A. Almasan, S. Linke, T. Paulson, L.-c. Huang & G. M. Wahl. Genetic instability as a consequence of inappropriate entry and progression through S-phase. *Cancer Metast. Rev.* **14**, 59-73. (1995).
188. P. A. Bunn, Jr. The potential role of proteasome inhibitors in the treatment of lung cancer. *Clin Cancer Res* **10**, 4263s-4265s. (2004).
189. M. M. Facchinetti, A. De Siervi, D. Toskos & A. M. Senderowicz. UCN-01-induced cell cycle arrest requires the transcriptional induction of p21(waf1/cip1) by activation of mitogen-activated protein/extracellular signal-regulated kinase kinase/extracellular signal-regulated kinase pathway. *Cancer Res* **64**, 3629-37. (2004).
190. Y. H. Jin, K. J. Yoo, Y. H. Lee & S. K. Lee. Caspase 3-mediated cleavage of p21WAF1/CIP1 associated with the cyclin A-cyclin-dependent kinase 2 complex is a prerequisite for apoptosis in SK-HEP-1 cells. *J Biol Chem* **275**, 30256-63 (2000).
191. R. Y. Poon & T. Hunter. Expression of a novel form of p21Cip1/Waf1 in UV-irradiated and transformed cells. *Oncogene* **16**, 1333-43. (1998).
192. O. Coqueret. New roles for p21 and p27 cell-cycle inhibitors: a function for each cell compartment? *Trends Cell Biol* **13**, 65-70. (2003).
193. T. G. Davies, D. J. Pratt, J. A. Endicott, L. N. Johnson & M. E. Noble. Structure-based design of cyclin-dependent kinase inhibitors. *Pharmacol Ther* **93**, 125-33 (2002).
194. T. G. Davies, J. Bentley, C. E. Arris, F. T. Boyle, N. J. Curtin, J. A. Endicott, A. E. Gibson, B. T. Golding, R. J. Griffin, I. R. Hardcastle, P. Jewsbury, L. N. Johnson, V. Mesguiche, D. R. Newell, M. E. Noble, J. A. Tucker, L. Wang & H. J. Whitfield. Structure-based design of a potent purine-based cyclin-dependent kinase inhibitor. *Nat Struct Biol* **9**, 745-9 (2002).
195. J. Litz, P. Carlson, G. S. Warshamana-Greene, S. Grant & G. W. Krystal. Flavopiridol potently induces small cell lung cancer apoptosis during S phase in a manner that involves early mitochondrial dysfunction. *Clin Cancer Res* **9**, 4586-94. (2003).
196. L. Cartee, R. Smith, Y. Dai, M. Rahmani, R. Rosato, J. Almenara, P. Dent & S. Grant. Synergistic induction of apoptosis in human myeloid leukemia cells by phorbol 12-myristate 13-acetate and flavopiridol proceeds via activation of both the intrinsic and tumor necrosis factor-mediated extrinsic cell death pathways. *Mol Pharmacol* **61**, 1313-21 (2002).
197. D. M. Kim, S. Y. Koo, K. Jeon, M. H. Kim, J. Lee, C. Y. Hong & S. Jeong. Rapid induction of apoptosis by combination of flavopiridol and tumor necrosis factor (TNF)-alpha or TNF-related apoptosis-inducing ligand in human cancer cell lines. *Cancer Res* **63**, 621-6 (2003).
198. A. Almasan & A. Ashkenazi. Apo2L/TRAIL: apoptosis signaling, biology, and potential for cancer therapy. *Cytokine Growth Factor Rev* **14**, 337-48 (2003).

199. J. E. Karp, D. D. Ross, W. Yang, M. L. Tidwell, Y. Wei, J. Greer, D. L. Mann, T. Nakanishi, J. J. Wright & A. D. Colevas. Timed sequential therapy of acute leukemia with flavopiridol: in vitro model for a phase I clinical trial. *Clin Cancer Res* **9**, 307-15 (2003).
200. K. Camphausen, K. J. Brady, W. E. Burgan, M. A. Cerra, J. S. Russell, E. E. Bull & P. J. Tofilon. Flavopiridol enhances human tumor cell radiosensitivity and prolongs expression of gammaH2AX foci. *Mol Cancer Ther* **3**, 409-16 (2004).
201. Z. N. Demidenko & M. V. Blagosklonny. Flavopiridol induces p53 via initial inhibition of Mdm2 and p21 and, independently of p53, sensitizes apoptosis-reluctant cells to tumor necrosis factor. *Cancer Res* **64**, 3653-60 (2004).
202. U. Raju, E. Nakata, K. A. Mason, K. K. Ang & L. Milas. Flavopiridol, a cyclin-dependent kinase inhibitor, enhances radiosensitivity of ovarian carcinoma cells. *Cancer Res* **63**, 3263-7 (2003).
203. S. Wittmann, P. Bali, S. Donapaty, R. Nimmanapalli, F. Guo, H. Yamaguchi, M. Huang, R. Jove, H. G. Wang & K. Bhalla. Flavopiridol down-regulates antiapoptotic proteins and sensitizes human breast cancer cells to epothilone B-induced apoptosis. *Cancer Res* **63**, 93-9 (2003).
204. N. R. Wall, D. S. O'Connor, J. Plescia, Y. Pommier & D. C. Altieri. Suppression of survivin phosphorylation on Thr34 by flavopiridol enhances tumor cell apoptosis. *Cancer Res* **63**, 230-5 (2003).
205. E. A. Monaco, 3rd, C. M. Beaman-Hall, A. Mathur & M. L. Vallano. Roscovitine, olomoucine, purvalanol: inducers of apoptosis in maturing cerebellar granule neurons. *Biochem Pharmacol* **67**, 1947-64. (2004).
206. C. Yu, M. Rahmani, Y. Dai, D. Conrad, G. Krystal, P. Dent & S. Grant. The lethal effects of pharmacological cyclin-dependent kinase inhibitors in human leukemia cells proceed through a phosphatidylinositol 3-kinase/Akt-dependent process. *Cancer Res* **63**, 1822-33. (2003).
207. M. E. Lane, B. Yu, A. Rice, K. E. Lipson, C. Liang, L. Sun, C. Tang, G. McMahon, R. G. Pestell & S. Wadler. A novel cdk2-selective inhibitor, SU9516, induces apoptosis in colon carcinoma cells. *Cancer Res* **61**, 6170-7. (2001).
208. S. R. Whittaker, M. I. Walton, M. D. Garrett & P. Workman. The Cyclin-dependent kinase inhibitor CYC202 (R-roscovitine) inhibits retinoblastoma protein phosphorylation, causes loss of Cyclin D1, and activates the mitogen-activated protein kinase pathway. *Cancer Res* **64**, 262-72 (2004).
209. W. Chen, J. Lee, S. Y. Cho & H. A. Fine. Proteasome-mediated destruction of the cyclin a/cyclin-dependent kinase 2 complex suppresses tumor cell growth in vitro and in vivo. *Cancer Res* **64**, 3949-57 (2004).
210. Y. N. Chen, S. K. Sharma, T. M. Ramsey, L. Jiang, M. S. Martin, K. Baker, P. D. Adams, K. W. Bair & W. G. Kaelin, Jr. Selective killing of transformed cells by cyclin/cyclin-dependent kinase 2 antagonists. *Proc Natl Acad Sci U S A* **96**, 4325-9 (1999).
211. W. Krek, G. Xu & D. M. Livingston. Cyclin A-kinase regulation of E2F-1 DNA binding function underlies suppression of an S phase checkpoint. *Cell* **83**, 1149-58 (1995).
212. M. Pruschy, C. Wirbelauer, C. Glanzmann, S. Bodis & W. Krek. E2F-1 has properties of a radiosensitizer and its regulation by cyclin A kinase is required for cell survival of fibrosarcoma cells lacking p53. *Cell Growth Differ* **10**, 141-6 (1999).
213. M. Pagano, G. Draetta & P. Jansen-Durr. Association of cdk2 kinase with the transcription factor E2F during S phase. *Science* **255**, 1144-7 (1992).

214. N. Mendoza, S. Fong, J. Marsters, H. Koeppen, R. Schwall & D. Wickramasinghe. Selective cyclin-dependent kinase 2/cyclin A antagonists that differ from ATP site inhibitors block tumor growth. *Cancer Res* **63**, 1020-4 (2003).
215. D. C. Nguyen, B. Parsa, A. Close, B. Magnusson, D. L. Crowe & U. K. Sinha. Overexpression of cell cycle regulatory proteins correlates with advanced tumor stage in head and neck squamous cell carcinomas. *Int J Oncol* **22**, 1285-90 (2003).
216. J. Jiang, C. B. Matranga, D. Cai, V. M. Latham, Jr., X. Zhang, A. M. Lowell, F. Martelli & G. I. Shapiro. Flavopiridol-induced apoptosis during S phase requires E2F-1 and inhibition of cyclin A-dependent kinase activity. *Cancer Res* **63**, 7410-22 (2003).

Chapter 7

# APOPTOSIS INDUCTION BY TUMOR-TARGETED TOXINS

Andrew Thorburn
*Wake Forest University*

Abstract: Tumor cell-targeted toxins are recombinant proteins that consist of a targeting domain that preferentially recognizes tumor cells and facilitates entry of the protein into the target cells with a bacterial or plant toxin that kills the tumor cells. A large number of toxins targeted against different kinds of tumor cell have been developed in recent years and the first such toxins are approved for use against specific cancers. In most cases, targeted toxins kill their target cells by inducing caspase-dependent apoptosis. However the mechanism by which the apoptotic machinery is activated may differ with different toxins and in different cell types. Moreover, recent work shows that the same toxin can kill different kinds of tumor cells through different molecular mechanisms.

Key words: Diphtheria toxin, tumor cell targeting, caspases, caspase-independent.

## 1. INTRODUCTION

Despite improvements in cancer treatment in recent years some tumors remain stubbornly resistant to available therapies. Even where useful treatments currently exist many patients respond poorly creating a continuing need to develop and understand the mode of action of new therapies that are targeted against particular tumor types. Tumor cell-targeted toxins represent one such kind of therapy and have been developed to treat a variety of tumors including various kinds of leukemia and lymphomas, indeed the first FDA approved targeted toxin, ONTAK, which consists of a diphtheria toxin fused to IL-2 has been used successfully to treat cutaneous T-cell lymphoma [1]. Promising results have also been



*M. Sluyser (ed.), Application of Apoptosis to Cancer Treatment,* 179-187.

obtained treating solid tumors especially brain tumors where the toxin is delivered directly to the tumor tissue. Targeted toxins that have been developed for brain cancers using targeting domains that react with receptors that recognize Interleukin13, transferrin, urokinase and IL4 [2-5]. Intratumoral administration of these agents can result in tumor regression and clinical remissions in glioblastoma multiforme patients that can last for years [2, 3]. This finding is encouraging because glioblastoma multiforme patients have a dismal prognosis even with aggressive radiation, chemotherapy and surgical treatments with median survival of only about 9 months and a five year survival rate of ~1% [6].

Targeted toxins are fusion proteins that combine a targeting molecule, which selectively binds to and enters tumor cells, with a toxin that kills the target cells. Tumor cells selectively take up the fusion protein through receptor-mediated endocytosis, the toxin portion is released from the endosome into the cytoplasm and the toxin kills the cell. Different kinds of targeting protein can be used including antibodies that recognize tumor cell-specific epitopes or growth factors that bind to cell surface receptors [7]. Toxins are derived from bacterial pathogens (e.g. diphtheria toxin, DT, or Pseudomonas exotoxin A, PE) or plants (e.g. ricin). These toxins block protein synthesis by different mechanisms. Ricin cleaves ribosomal RNA to disrupt the ribosome whereas DT and PE ADP-ribosylate the translation elongation factor 2 to prevent protein synthesis. The molecular mechanisms through which targeted toxins kill are incompletely understood [8]. In this chapter, we will review how these agents work and discuss how the differences in mechanism between tumor cell types, differences between different kinds of toxin and combinations of toxins and other agents may allow us to design tailored cancer treatment strategies.

## 2. TARGETED TOXINS CAN INDUCE CASPASE-DEPENDENT APOPTOSIS

Because the toxin molecules (DT, PE or ricin) that have been used to make targeted toxins inhibit protein synthesis and protein synthesis is an essential cellular activity, it seems obvious that the target cells should die in response to the toxin. However mechanistic studies on the way that tumor cells die when treated with these agents has provided several surprises. For example, toxins that inhibit protein synthesis by different mechanism can kill cells by distinct mechanisms suggesting that the way toxin-treated cancer cells die is determined not by the lack of protein synthesis but rather by the way that protein synthesis is inhibited. The best evidence for this arose from a non-biased screen to identify inhibitors of toxin-induced apoptosis.

Genetic selection for cDNAs that confer resistance to PE led to the isolation of specific cDNAs that could prevent PE-induced apoptosis [9]. These cDNAs also confer resistance to DT, but not ricin. One cDNA encoded an antisense fragment of cellular apoptosis susceptibility gene (CAS). Down regulation of endogenous CAS was responsible for resistance to toxin-induced apoptosis. However, CAS antisense had no effect on ADP-ribosylation or inhibition of protein synthesis [10]. Because CAS antisense blocks apoptosis by only some protein synthesis inhibitors and has no effect on protein synthesis inhibition itself, these data suggest that a toxin's ability to induce apoptosis depends on its mechanism of action rather than its ability to inhibit protein synthesis per se. As discussed below this theme, whereby different toxins seem to kill cells through different mechanisms even when all the drugs inhibit protein synthesis is becoming a common refrain in this field. These complexities provide scientific interest from a purely basic perspective– why would cells respond differently when two agents inhibit protein synthesis by different methods? Why do two different cells respond differently when the same agent inhibits protein synthesis? Moreover, they may also provide opportunities to maximize the anti-tumor effect when used clinically.

Targeted toxins can induce apoptosis as shown by caspase activation in the dying cells and classical apoptotic morphology [11]. This raises the question of which apoptosis signaling pathways are activated in the tumor cells. We possess a relatively sophisticated understanding of the apoptotic machinery [12] at least as regards caspase activation. Apoptotic caspases include "initiator caspases" (caspase-2, -8, -9 and -10) that start an apoptotic cascade and "effector caspases" (caspase-3, -6 and –7) that disassemble the cell. Caspases cleave specific substrates at a few sites [13] to alter the activity of the target protein resulting in the apoptotic phenotypes. Two main pathways leading to caspase activation have been characterized [14].

The extrinsic or death receptor pathway is activated by receptors of the Tumor Necrosis Factor Receptor superfamily [15]. These receptors contain an intracellular protein interaction domain called a death domain (DD) and induce apoptosis by forming a multiprotein complex called the Death-Inducing Signaling Complex (DISC). Upon ligand binding, activated death receptors recruit an adapter protein called Fas Associated Death Domain protein (FADD) [16]. FADD consists of two protein interaction domains: a DD and a death effector domain (DED). The DED interacts with a DED on the initiator procaspase-8. FADD binds to the Fas and TRAIL receptors (DR4 and DR5) receptor through interactions between the two death domains and activities that are regulated by the DED [17, 18]. This complex recruits the inactive pro-form of caspase-8. Aggregation of caspase-8 leads to dimerization, which activates protease activity [19-21]. For a recent review, see [22]. Initiator caspases activate effector caspases such as caspase-3 causing the cell to undergo apoptosis by cleaving specific substrates [13].

Diverse stress pathways cause release of mitochondrial proteins to activate the other well known apoptosis pathway– the "intrinsic" pathway [23]. A defining characteristic of this pathway is that anti-apoptotic members of the Bcl-2 family such as Bcl-2 and Bcl-xL inhibit caspase activation and apoptosis induced by stimuli that work through this pathway. Protein release occurs through mechanisms that are still unclear [24]. Released cytochrome c (cyt c) interacts with Apaf-1, pro-caspase 9 and dATP to form a complex called the apoptosome [25]. This complex dimerizes and activates caspase 9, which then activates effector caspases to induce apoptosis. Other released proapoptotic mitochondrial proteins include Apoptosis Inducing Factor (AIF) [26], Smac/Diablo [27, 28] and Endonuclease G [29] and Omi/HtrA2 [30-33]. Death receptors can activate the intrinsic pathway through cleavage of Bid, which translocates to mitochondria [34] providing a link between the extrinsic and intrinsic apoptosis pathways.

## 3. TARGETED TOXINS ACTIVATE DISTINCT APOPTOTIC PATHWAYS

Targeted toxins can activate caspase 3-like activities that cleave known caspase substrates such as Poly(ADP) ribose polymerase (PARP) [35, 36]. Furthermore, PE-immunotoxin-induced apoptosis was inhibited and cell viability increased by treatment of T cell leukemia cells with a caspase inhibitor, z-VAD.fmk [35]. Diphtheria toxin fusions with GMCSF can activate caspases in myeloid target cells [11, 37], even when these cells display multi-drug resistance to other chemotherapeutic agents [37]. These data suggest that caspase activation is an important aspect of targeted toxin-induced cell death however they do not determine which caspase activation pathway (intrinsic, extrinsic or another pathway) starts the process. A clue comes from experiments where Bcl-2 was overexpressed in MCF7 breast cancer cells and inhibited a PE-fused immunotoxin from inducing apoptosis [36]. However, while this result suggests that mitochondrial dysfunction is important in the response, these data do not discriminate between direct activation of the intrinsic pathway and activation after activation of the extrinsic death receptor pathway. Indeed, other data suggest that components of the extrinsic pathway may be important in toxin-induced tumor cell death. In myeloid leukemia cells, treated with a GMCSF-targeted diphtheria toxin protein both caspase-8 and caspase-9 were activated by the toxin. However, caspase-8, not caspase-9, was the apical caspase responsible for initiating the apoptosis pathway because while caspase-9 inhibition did not affect cell death, caspase-8 inhibition prevented subsequent caspase-9 activation, effector caspase activation and cell death [11]. The involvement of caspase-8 in toxin-induced death suggests that other components of the extrinsic pathway may be involved and this was confirmed by the finding that the adaptor protein FADD was required for GMCSF-DT-induced apoptosis. However, inhibition of death receptor signaling had no effect on toxin-induced death suggesting that this effect does not involve the death receptors themselves [11]. These data suggest that the targeted toxin may be able to activate the death receptor machinery but through a mechanism that is independent of receptor activation.

## 4. THE SAME TARGETED TOXIN CAN KILL DIFFERENT TUMOR CELLS BY DIFFERENT MECHANISMS

Different types of cancer sometimes overexpress the same receptor. For example the Epidermal Growth Factor (EGF) Receptor is expressed in many epithelial tumors such as breast cancer and is also frequently overexpressed in other tumor types such as glioblastoma. This allows one to study the mechanism of killing by the same targeted toxin in different tumor cell types. Recent unpublished work from our group shows that an EGF-diphtheria toxin protein kills epithelial tumor cells by activating caspases that lead to classical apoptosis with its associated hallmarks such as membrane blebbing and fragmentation into apoptotic bodies. The same toxin also kills glioma cells [38, 39]. However we do not detect caspase activation in glioma cells, which die without showing the hallmarks of caspase-dependent apoptosis. Moreover, unlike epithelial cells, caspase inhibitors do not affect EGF-diphtheria toxin-induced glioma cell death. These data indicate that different tumor cell types can activate different cell death pathways when treated with a targeted toxin. This result is interesting from a basic mechanistic perspective because it again implies that it is not the fact that protein synthesis is inhibited but rather the way that it is inhibited that regulates how the cells die. More importantly, these data also suggest that different tumor types may respond optimally to combinations of toxins with other agents and could develop different resistance mechanisms. For example, tumor cells that undergo classical apoptosis may develop resistance by inactivating caspases or other components of the extrinsic or intrinsic pathways while this would have no effect in glioma cells.

## 5. TARGETED TOXINS SYNERGIZE WITH OTHER ANTI-CANCER DRUGS

Targeted diphtheria toxins have been shown to synergize with standard chemotherapeutic agents such as AraC [40]. Other protein synthesis inhibitors such as ricin did not synergize with AraC in these studies providing yet more evidence that the way that the toxin works is more important than the fact that it inhibits protein synthesis in determining how it kills cells. Targeted toxins can also synergize with other targeted toxins. For example, EGF-targeted diphtheria toxin can synergize with IL13-targted pseudomonas exotoxin to kill glioma cells that possess both receptors [39].

Recent unpublished work from our group also demonstrates synergy with other "targeted" anti-cancer therapeutics. In this case we showed that combining EGF-targeted diphtheria toxin with an antibody that activates the TRAIL receptor DR5 led to synergistic cell killing. Interestingly, when the combination was used to treat glioma cells, which as mentioned above do not activate caspases in response to the toxin on its own, robust caspase activation occurred. Thus at least in the case of some tumor cell types, combining a targeted toxin with another anti-cancer agent can not only increase the amount of tumor cell death but can also change the way that the cells die. This could have practically important consequences for cancer therapy. One difference between classical, caspase-dependent apoptosis and other forms of cell death is that caspase-dependent apoptosis is associated with reduced inflammation. This might be good or bad depending on the circumstances. Tumor cell killing with increased inflammation might be more effective at reducing tumor burden because it stimulates host immune-mediated anti-tumor responses, which may work better when the tumor cells die by caspase-independent mechanisms [41]. In other situations, effects associated with increased inflammation such as tissue swelling might cause serious problems. This problem may be more important in specific tissues. For example, inflammation in subcutaneous tissues might be less problematic than inflammation and swelling in the brain.

## 6. SUMMARY

Targeted toxins have been shown to be effective anti-tumor agents in many preclinical models and are displaying efficacy in clinical trials and making their way into the clinic as approved drugs. These agents often but not always work by inducing caspase-dependent apoptosis, which at least in some cases appears to be achieved through activation of the death receptor signaling pathway but is also affected by Bcl-2 proteins. However, different toxins activate different cell death pathways and even the same toxin can activate different signaling pathways that lead to cell death in different cells. These complexities make it difficult for us to work out how any given toxin works and suggests that it may not be feasible to extrapolate based on one kind of toxin or one tumor type to determine how any particular toxin works against any particular tumor type. Added complications come from the finding that some toxins synergize with standard chemotherapy agents and with other targeted therapies such as TRAIL receptor agonists and that different targeted toxins can even synergize with each other in some but not all tumor cells. Further understanding of how these agents work should allow us to improve their use as anti-cancer treatments.

## REFERENCES

1. Frankel AE, Fleming DR, Powell BL, Gartenhaus R. DAB(389)IL2 (ONTAK((R))) fusion protein therapy of chronic lymphocytic leukaemia. *Expert Opin Biol Ther* 2003;3:179-86.
2. Laske DW, Youle RJ, Oldfield EH. Tumor regression with regional distribution of the targeted toxin TF-CRM107 in patients with malignant brain tumors. *Nat Med* 1997;3:1362-8.
3. Rand RW, Kreitman RJ, Patronas N, Varricchio F, Pastan I, Puri RK. Intratumoral administration of recombinant circularly permuted interleukin-4-Pseudomonas exotoxin in patients with high-grade glioma. *Clin Cancer Res* 2000;6:2157-65.
4. Vallera DA, Li C, Jin N, Panoskaltsis-Mortari A, Hall WA. Targeting urokinase-type plasminogen activator receptor on human glioblastoma tumors with diphtheria toxin fusion protein DTAT. *J Natl Cancer Inst* 2002;94:597-606.
5. Li C, Hall WA, Jin N, Todhunter DA, Panoskaltsis-Mortari A, Vallera DA. Targeting glioblastoma multiforme with an IL-13/diphtheria toxin fusion protein in vitro and in vivo in nude mice. *Protein Eng* 2002;15:419-27.
6. Davis FG, Freels S, Grutsch J, Barlas S, Brem S. Survival rates in patients with primary malignant brain tumors stratified by patient age and tumor histological type: an analysis based on Surveillance, Epidemiology, and End Results (SEER) data, 1973-1991. *J Neurosurg* 1998;88:1-10.
7. Frankel AE, Kreitman RJ, Sausville EA. Targeted toxins. *Clin Cancer Res* 2000;6:326-34.
8. Thorburn A, Thorburn J, Frankel AE. Induction of apoptosis by tumor cell-targeted toxins. *Apoptosis* 2004.
9. Brinkmann U, Brinkmann E, Pastan I. Expression cloning of cDNAs that render cancer cells resistant to Pseudomonas and diphtheria toxin and immunotoxins. *Mol Med* 1995;1:206-16.
10. Brinkmann U, Brinkmann E, Gallo M, Scherf U, Pastan I. Role of CAS, a human homologue to the yeast chromosome segregation gene CSE1, in toxin and tumor necrosis factor mediated apoptosis. *Biochemistry* 1996;35:6891-9.
11. Thorburn J, Frankel AE, Thorburn A. Apoptosis by leukemia cell-targeted diphtheria toxin occurs via receptor-independent activation of Fas-associated death domain protein. *Clin Cancer Res* 2003;9:861-5.

12. Hengartner MO. The biochemistry of apoptosis. *Nature* 2000;407:770-6.
13. Fischer U, Janicke RU, Schulze-Osthoff K. Many cuts to ruin: a comprehensive update of caspase substrates. *Cell Death Differ* 2003;10:76-100.
14. Budihardjo I, Oliver H, Lutter M, Luo X, Wang X. Biochemical pathways of caspase activation during apoptosis. *Annu Rev Cell Dev Biol* 1999;15:269-90.
15. Ashkenazi A, Dixit VM. Death receptors: signaling and modulation. *Science* 1998;281:1305-8.
16. Chinnaiyan AM, O'Rourke K, Tewari M, Dixit VM. FADD, a novel death domain-containing protein, interacts with the death domain of Fas and initiates apoptosis. *Cell* 1995;81:505-12.
17. Thomas LR, Henson A, Reed JC, Salsbury FR, Thorburn A. Direct binding of FADD to the TRAIL receptor DR5 is regulated by the death effector domain of FADD. *J Biol Chem* 2004;279:32780-32785.
18. Thomas L, Stillman D, Thorburn A. Regulation of Fas-associated death domain interactions by the death effector domain identified by a modified reverse two hybrid screen. *J. Biol. Chem.* 2002;277:34343-34348.
19. Boatright KM, Renatus M, Scott FL, Sperandio S, Shin H, Pedersen IM, Ricci JE, Edris WA, Sutherlin DP, Green DR, Salvesen GS. A unified model for apical caspase activation. *Mol Cell* 2003;11:529-41.
20. Donepudi M, Sweeney AM, Briand C, Grutter MG. Insights into the Regulatory Mechanism for Caspase-8 Activation. *Mol Cell* 2003;11:543-549.
21. Chang DW, Ditsworth D, Liu H, Srinivasula SM, Alnemri ES, Yang X. Oligomerization is a general mechanism for the activation of apoptosis initiator and inflammatory procaspases. *J Biol Chem* 2003;278:16466-9.
22. Thorburn A. Death Receptor-induced cell killing. *Cellular Signalling* 2004;16:139-144.
23. Wang X. The expanding role of mitochondria in apoptosis. *Genes Dev* 2001;15:2922-33.
24. Newmeyer DD, Ferguson-Miller S. Mitochondria: releasing power for life and unleashing the machineries of death. *Cell* 2003;112:481-90.
25. Li P, Nijhawan D, Budihardjo I, Srinivasula SM, Ahmad M, Alnemri ES, Wang X. Cytochrome c and dATP-dependent formation of Apaf-1/caspase-9 complex initiates an apoptotic protease cascade. *Cell* 1997;91:479-489.
26. Susin SA, Lorenzo HK, Zamzami N, Marzo I, Snow BE, Brothers GM, Mangion J, Jacotot E, Costantini P, Loeffler M, Larochette N, Goodlett DR, Aebersold R, Siderovski DP, Penninger JM, Kroemer G. Molecular characterization of mitochondrial apoptosis-inducing factor. *Nature* 1999;397:441-6.
27. Du C, Fang M, Li Y, Li L, Wang X. Smac, a mitochondrial protein that promotes cytochrome c-dependent caspase activation by eliminating IAP inhibition. *Cell* 2000;102:33-42.
28. Verhagen AM, Ekert PG, Pakusch M, Silke J, Connolly LM, Reid GE, Moritz RL, Simpson RJ, Vaux DL. Identification of DIABLO, a mammalian protein that promotes apoptosis by binding to and antagonizing IAP proteins. *Cell* 2000;102:43-53.
29. Li LY, Luo X, Wang X. Endonuclease G is an apoptotic DNase when released from mitochondria. *Nature* 2001;412:95-9.
30. Verhagen AM, Silke J, Ekert PG, Pakusch M, Kaufmann H, Connolly LM, Day CL, Tikoo A, Burke R, Wrobel C, Moritz RL, Simpson RJ, Vaux DL. HtrA2 promotes cell death through its serine protease activity and its ability to antagonise inhibitor of apoptosis proteins. *J Biol Chem* 2002;277:445-454.
31. Hegde R, Srinivasula SM, Zhang Z, Wassell R, Mukattash R, Cilenti L, DuBois G, Lazebnik Y, Zervos AS, Fernandes-Alnemri T, Alnemri ES. Identification of

Omi/HtrA2 as a mitochondrial apoptotic serine protease that disrupts IAP-caspase interaction. *J Biol Chem* 2002;277:432-438.

32. Martins LM, Iaccarino I, Tenev T, Gschmeissner S, Totty NF, Lemoine NR, Savopoulos J, Gray CW, Creasy CL, Dingwall C, Downward J. The serine protease Omi/HtrA2 regulates apoptosis by binding XIAP through a Reaper-like motif. *J Biol Chem* 2002;277:439-444.
33. Suzuki Y, Imai Y, Nakayama H, Takahashi K, Takio K, Takahashi R. A serine protease, htra2, is released from the mitochondria and interacts with xiap, inducing cell death. *Mol Cell* 2001;8:613-21.
34. Luo X, Budihardjo I, Zou H, Slaughter C, Wang X. Bid, a Bcl2 interacting protein, mediates cytochrome c release from mitochondria in response to activation of cell surface death receptors. *Cell* 1998;94:481-90.
35. Keppler-Hafkemeyer A, Kreitman RJ, Pastan I. Apoptosis induced by immunotoxins used in the treatment of hematologic malignancies. *Int J Cancer* 2000;87:86-94.
36. Keppler-Hafkemeyer A, Brinkmann U, Pastan I. Role of caspases in immunotoxin-induced apoptosis of cancer cells. *Biochemistry* 1998;37:16934-42.
37. Perentesis JP, Waddick KG, Bendel AE, Shao Y, Warman BE, Chandan-Langlie M, Uckun FM. Induction of apoptosis in multidrug-resistant and radiation-resistant acute myeloid leukemia cells by a recombinant fusion toxin directed against the human granulocyte macrophage colony-stimulating factor receptor. *Clin Cancer Res* 1997;3:347-55.
38. Liu TF, Cohen KA, Ramage JG, Willingham MC, Thorburn AM, Frankel AE. A diphtheria toxin-epidermal growth factor fusion protein is cytotoxic to human glioblastoma multiforme cells. *Cancer Res* 2003;63:1834-7.
39. Liu TF, Willingham MC, Tatter SB, Cohen KA, Lowe AC, Thorburn A, Frankel AE. Diphtheria Toxin-Epidermal Growth Factor Fusion Protein and Pseudomonas Exotoxin-Interleukin 13 Fusion Protein Exert Synergistic Toxicity against Human Glioblastoma Multiforme Cells. *Bioconjug Chem* 2003;14:1107-1114.
40. Kim CN, Bhalla K, Kreitman RJ, Willingham MC, Hall P, Tagge EP, Jia T, Frankel AE. Diphtheria toxin fused to granulocyte-macrophage colony-stimulating factor and Ara-C exert synergistic toxicity against human AML HL-60 cells. *Leuk Res* 1999;23:527-38.
41. Reiter I, Krammer B, Schwamberger G. Cutting edge: differential effect of apoptotic versus necrotic tumor cells on macrophage antitumor activities. *J Immunol* 1999;163:1730-2.

Chapter 8

# TARGETING CELL-DEATH PATHWAYS IN MULTIPLE MYELOMA: THERAPEUTIC IMPLICATIONS

Dharminder Chauhan, Ph.D and Kenneth Anderson, M.D.
*The Jerome Lipper Multiple Myeloma Center, Department of Medical Oncology, Dana Farber Cancer Institute, Harvard Medical School, Boston, MA 02115*

Abstract: Multiple myeloma (MM) remains fatal despite available therapies. The median survival of MM patients is 3-4 years with standard therapy and 4-5 years with high dose and transplantation therapy. Drug resistance occurs despite the use of novel anti-MM therapies. Mechanisms conferring chemoresistance include chromosomal abnormalities, the interaction of MM cells with their bone marrow (BM) microenvironment, and defects in apoptotic signaling pathways. Delineation of cell death and growth cascades using both genomics and proteomics provides the rationale to inhibit growth and/or amplify apoptotic signaling in MM cells, in order to enhance anti-MM activity of available therapies, prevent development of drug resistance, and improve patient outcome in MM.

Key words: Multiple Myeloma; apoptosis; growth; drug-resistance; novel therapies

## 1. INTRODUCTION

MM cells primarily localize in the BM, where various cytokines promote MM cell growth, survival, and prevent the cytotoxic effects of chemotherapy (1, 2). In particular, adhesion of MM cells to BM stromal cells (SCs) trigger transcription and secretion of cytokines, such as interleukin-6 (IL-6), and Insulin-like growth factor-I (IGF-I), and vascular endothelial growth factor (VEGF), which not only trigger proliferation of MM cells, but also inhibit drug-induced apoptosis of MM cells (3-9). The mechanisms whereby MM



*M. Sluyser (ed.), Application of Apoptosis to Cancer Treatment,* 189-208.

cells lose their ability to undergo apoptosis in response to chemotherapies involve constitutive activation of cellular proliferation pathways, as well as inhibition of cell death machinery. Delineation of both cytokine-induced growth cascades and chemotherapy-triggered apoptotic signaling pathways provides the rationale for combining anti-tumor agents to specifically target these signaling cascades, thereby inhibiting growth and amplifying apoptotic signaling.

Apoptosis or programmed cell death (PCD) is characterized by distinct morphological changes including membrane blebbing, nuclear condensation, appearance of apoptotic antibodies, and oligonucleosomal DNA (10-13). The induction of apoptosis involves activity of aspartate specific cysteine proteases or caspases (cysteinyl, aspartate-specific proteases), which can either inactivate or activate target substrates by proteolytic cleavage (14). Initiator caspases undergo autocatalytic processing, and then cleave and activate the downstream executioner caspases that orchestrate cell death (15). Other well known cellular targets of caspases includes poly (ADP ribose) polymerase (PARP) and DNA-dependent protein kinase. Direct evidence for involvement of caspases in apoptosis is derived from studies with baculovirus protein p35, which directly inhibits cysteine proteases and blocks activation of apoptosis (12, 13). Genetic and biochemical evidence indicates that apoptosis proceeds by two major cell death pathways: an intrinsic pathway that involves mitochondrial membrane permeabilization and release of several apoptogenic factors, such as cytochrome-c (cyto-c) (16, 17), Smac/DIABLO (18, 19), AIF (20), EndoG (21) and HtrA2/Omi (Suzuki 2001); and an extrinsic apoptotic signaling pathway that occurs primarily *via* caspase-8 activation.

Recent studies have shown that various anti-MM drugs induce apoptosis via both a mitochondria-dependent and mitochondria-independent manner; however, these signals converge to common downstream pathways (22, 23) (Figure 1). Anti-MM agent commonly trigger alterations in mitochondrial transmembrane potential (Deltapsim, $\Psi$m), activation of caspase-8, caspase-9, caspase-3 and PARP cleavage (22); however, these agents induce differential upstream signaling pathways. For example, Dex-induced apoptosis is independent of c-Jun NH2-terminal kinase (SAPK/JNK) pathways, whereas proteasome inhibitor or irradiation-induced apoptosis requires JNK activation (24). In addition, various anti-MM agents downregulate signaling pathways related to growth (Ras-Raf-MAPK; nuclear factor-kappa B (NF-κB), anti-apoptosis or drug-resistance (PI3K-Akt, BCL2, IAPs), and angiogenesis (PI3K-PKC) (23, 24). Together, these findings suggest that inhibition of apoptotic signaling is associated with downregulation of growth and survival signaling, and that therapeutic strategies combining two or more agents that simultaneously target these signaling pathways will have enhanced anti-MM activity.

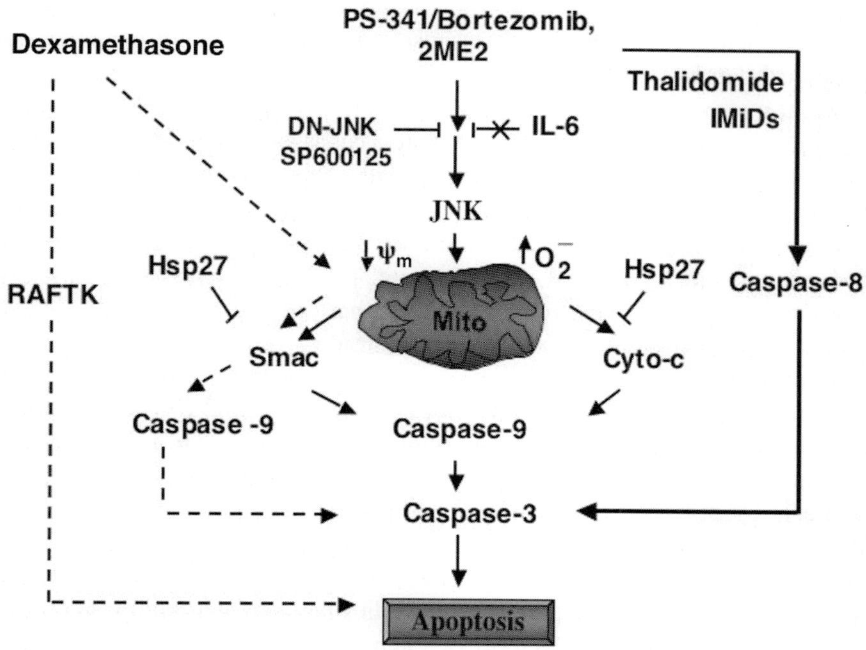


*Figure 1.* Diverse apoptosis-related signaling in MM cells. Bortezomib/PS-341-or 2ME2 induces activation of JNK, which translocates to mitochondria and facilitates the release of cyto-c and smac. Conversely, inhibition of JNK using dominant-negative (DN)-JNK or a biochemical inhibitor SP600125 abrogates cyto-c/smac release and apoptosis. In contrast to Bortezomib or 2ME2, Dexamethasone (Dex)-induced apoptosis is independent of JNK activation and accompanied by smac, but not cyto-c release (regular arrows: 2ME2 or Bortezomib-induced signal; dashed arrows: Dex-induced signal). Thalidomide and Revlimid (immunomodulatory agent) triggered-apoptosis is associated with activation of caspase-8.

## 2. MODULATION OF MITOCHONDRIAL APOPTOTIC SIGNALING

Mitochondria regulate generation of metabolic energy in the form of ATP, and have a well-established role during apoptosis (25). The mechanisms mediating cytochrome-c (cyto-c) release into the cytoplasm during apoptosis remains unclear; however, its been linked to the antagonistic activities of Bcl2 family members of proteins (26, 27). Once released into the cytoplasm from mitochondria, cyto-c binds to Apaf-1, which self-oligomerizes and recruits pro-caspase-9 to form apoptosome complexes, thereby allowing for autoprocessing of pro-caspase-9, with subsequent cleavage and activation of caspase-3 (28-30). Our studies in MM cells have shown that anti-MM agent-induced apoptosis is not always associated with activation of both caspase-8 and caspase-9. For example, our preclinical data showed that Thalidomide and Revlimid (immunomodulatory drug) trigger caspase-8 activation in MM cells (31, 32), whereas Dexamethasone (Dex) primarily activates caspase-9 (33, 34) (Figure 1). Combining Thalidomide or Revlimid with Dex therefore triggers dual apoptotic signaling and synergistic anti-MM activity (31, 35), providing the basis for a clinical trial of Thalidomide or Revlimid combined with Dex in MM patients (36). Overall response rate in these patients is remarkably improved compared to the treatment with either drug alone (36). Our recent studies have also shown that 2Methoxyestradiol (2ME2) and proteasome inhibitor PS-341/Bortezomib-induced apoptosis in MM cells is associated with activation caspase-9 *via* triggering the release of both cyto-c and Smac, whereas Dex activates caspase-9 through Smac, but not cyto-c, release (37, 38). Interleukin-6 (IL-6) inhibits Dex, but not 2ME2 or PS-341-induced, apoptosis by preventing the Smac release (37). These data further indicate diversity in drug-induced apoptotic signaling in MM cells and suggests that combining Dex with novel agents, which trigger cyto-c release will enhance its anti-MM activity (Figure 1). Other studies have shown that the inhibitor of apoptosis proteins (IAPs) suppresses apoptosis by directly binding to and inhibiting caspases (18). For example, XIAP, c-IAP-1, and c-IAP-2 bind to procaspase-9 and prevent its activation/processing, thereby blocking the downstream apoptosis-related events such as proteolytic cleavage of caspase-3, -6, and –7 (18). Our studies showed that treatment of MM cells with various anti-MM agents is associated with downregulation of IAPs (23); however, whether direct inhibition of IAPs in MM cells trigger apoptosis in these cells remains to be examined. Importantly, mitochondrial protein smac activates caspase-9 by overcoming inhibitory effects of inhibitors IAPs on caspase-9, suggesting the potential clinical utility of smac mimetics to compensate for defects in mitochondrial apoptotic signaling.

The extrinsic apoptotic pathway is activated through binding of ligand to cell surface receptors (39). Tumor necrosis factor receptor-1 (TNFR1) and the Fas receptor contains the death domains (DDs) and recruit DD-containing adaptor proteins TNFR1-associated DD (TRADD) and Fas-associated DD (FADD). Interaction between death domains of Fas and FADD trigger the autoactivation of pro-caspase-8 (40). During TNF-induced signaling, TRADD recruits FADD, followed by complex formation and release of TNFR1 to trigger pro-caspase-8 activation (41, 42). The receptor for TNF-related apoptosis-inducing ligand (TRAIL) activates pro-caspase-8 in a FADD-dependent fashion (43). Overall, caspase-8 and caspase-9 mediate major cellular death signaling, and the lack of activation of either or both caspases confers drug resistance. Conversely, using therapies such as smac mimetic or combining drugs that trigger both intrinsic as well as extrinsic pathways overcomes the defects in apoptotic signaling otherwise observed in response to a single agent.

## 3. ACTIVATION OF CLASSICAL STRESS RESPONSE PATHWAY *VIA* C-JUN NH2-TERMINAL KINASE (JNK)

Stress stimuli (TNF, Irradiation, sphingomyelinase and UV light) activate JNK and induce apoptosis; conversely, inhibition of JNK markedly abrogates stress-induced apoptosis (44-46). Importantly JNK facilitates the release of cyto-c during apoptosis (47). We and others have shown a role of JNK during chemotherapy-induced apoptosis in MM cells (37, 48). For example, Bortezomib- or 2ME2-induced apoptosis is associated with activation of JNK, translocation of JNK from cytosol to mitochondria, and release of both cyto-c and Smac solo from mitochondria to cytosol. Inhibition of JNK, either by using dominant-negative mutant (DN-JNK) or a specific biochemical inhibitor of JNK SP600125 abrogates both release of Smac and induction of apoptosis (37) (Figure 1). Mechanisms whereby JNK trigger the release of cyto-c involve modulation of anti-apoptotic protein Bcl2/Bclx in the outer membrane of mitochondria (49-51). BH3 members of the BCL2 family can also engage BAX and Bak to release cyto-c (52). The finding that Bax is required for JNK-dependent apoptosis (53), coupled with the requirement of JNK for cyto-c release (47), confirms the essential role of Bax in mediating JNK-dependent release of cyto-c. Our studies further shows that Bax cooperates with JNK in triggering the smac release; and conversely, Bortezomib-induced apoptosis is markedly abrogated in Bax knock-out cells. Together, these results suggest that JNK mediates the release of cyto-c and Smac during 2ME2 or Bortezomib-induced apoptosis in MM cells; however, not all apoptotic agents induce JNK activation and related mitochondrial pathways. For example, Dex-induced apoptosis in MM cells is JNK-independent (33), but is associated with Smac release (34). Our results are consistent with other studies demonstrating that anti-Fas-induced apoptosis is associated with the release of cyto-c in JNK-deficient MEF cells whereas UV-induced apoptosis in these same cells requires JNK activation for release of cyto-c (47). Thus JNK activation and JNK-dependent release of mitochondrial proteins may be stimulus specific (Figure 1). Nevertheless, these data implicate JNK as a potential therapeutic target to amplify apoptotic signaling in MM cells.

# 4. MODULATION OF MECHANISMS MEDIATING DRUG RESISTANCE

As noted above, chemotherapy achieves responses in many MM patients, but prolonged drug exposure may lead to the development of drug resistance. Dex is commonly used as therapy in MM; however, most MM patients become refractory to steroids after initial therapy (54). One of the major factors for the development of drug-resistance in MM is their binding to BMSCs, which confers cell-adhesion-mediated drug resistance (CAM-DR) (3, 55, 56). It is comprised of two components: First, adhesion of MM cells to fibronectin significantly decreases the anti-MM activities of DNA-damaging drugs (56); second, MM-bone marrow stromal cells (BMSCs) interaction increases cytokine secretion from BMSCs, which also prevents the cytotoxic effects of drugs (Figure 2). For example, IL-6 prevents Dex-induced apoptosis in MM cells (57, 58).

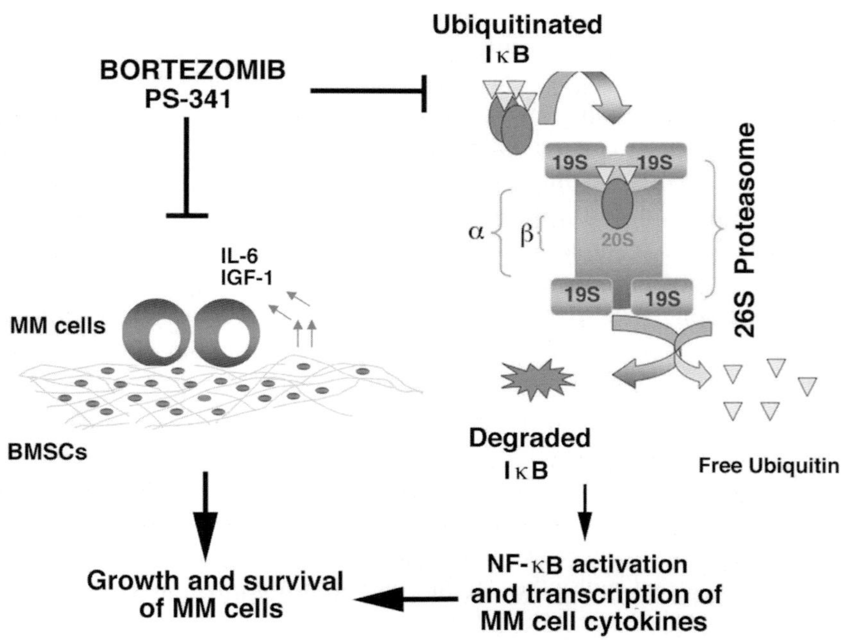


*Figure 2.* Bortezomib/PS-341 negatively regulates two major growth and survival pathways in MM cells. Bortezomib blocks the adhesion of MM cells to bone marrow stromal cells

(BMSCs), thereby inhibiting the adhesion-related transcription and secretion of cytokines. Bortezomib inhibits NF-κB activation by preventing proteasomal degradation of NF-κB inhibitory protein IκB.

Conversely, inhibition of the adhesion molecules either on MM or BMSCs enhances the anti-MM activity of chemotherapy (55). Alternatively, targeting the proteins mediating Dex-induced apoptosis or IL-6-induced protective signaling can also overcome the CAM-DR. Our prior study showed that Dex-induced apoptosis in MM cells require activation of related adhesion focal tyrosine kinase (RAFTK) (59) (Figure 1); and conversely, IL-6-mediated protection against Dex-induced apoptosis is mediated *via* induction of SHP2, a tyrosine phosphatase, which dephosphorylates RAFTK and thereby both inhibits Dex-induced apoptosis and promotes MM cell survival (58). Based on these findings, a small molecule inhibitor of SHP2 is under evaluation to enhance steroid sensitivity and overcome steroid-resistance in MM cells.

## 5. MODULATION OF NUCLEAR FACTOR-KAPPA B (NF-κB)

Our earlier studies showed that the gene promoters of various MM cell cytokines contain consensus-binding sites for the transcription factor NF-κB, which can be activated during cytokine and/or adhesion-induced proliferation and survival signaling in MM cells and BMSCs (6, 60). NF-κB remains inactivated in the cytoplasm due to its complex formation with inhibitory protein I-kappa B-alpha (IκB-α) (Figure 2). Upon growth or survival stimulation, IκB-α is phosphorylated, ubiquitinated, and degraded by proteasomes, leading to the disassociation of p50/65, and its translocation to the nucleus (61, 62). Activated NF-κB binds to the consensus sites present within the promoter region of various MM cytokines and thereby induces their transcription and secretion (6, 60, 62, 63). Conversely, several anti-MM agents including Dex, thalidomide, and Bortezomib inhibit NF-κB activity (Figure 2). Curcumin (diferuloylmethane) also decreases constitutive activation of IκB-α in MM cells, resulting in suppression of growth and induction of apoptosis in MM cells (64). To determine whether NF-κB directly affects MM cell growth, we utilized two specific inhibitors of NF-κB signaling pathway, SN-50 and I-kappa B Kinase (IKK) inhibitor (PS-1145). Our gene profiling studies showed high expression of telomerase reverse transcriptase (hTERT) catalytic subunit in MM cells compared to normals, and that hTERT protein interacts directly with NF-κB p65 (65). Importantly, treatment of MM cells with a cell-permeable specific inhibitor of NF-κB activity SN-50 blocked interaction of activated NF-κB with hTERT and its nuclear translocation, thereby inhibiting telomerase function (65). Another study demonstrated that SN-50 triggered apoptosis in MM cell lines and patient cells is also associated with: downregulation of Bcl-2, XIAP, cIAP-1 and -2, and survivin; upregulation of Bax; cytochrome c release; and activation of caspase-9, -3, but not caspase- 8 (66). Furthermore, SN50 sensitized MM cells to both TNF-alpha and TNF-related apoptosis-inducing ligand (TRAIL)/Apo2L. Finally, SN50 inhibited TNF-alpha-induced expression of another NF-κB target gene, intercellular adhesion molecule-1 (ICAM-1), which mediates homotypic adhesion of MM cells (67). In another study, we used an IKK inhibitor (PS-1145) and showed that either PS-1145 or Bortezomib blocked TNF-alpha-Induced NF-κB activation by inhibiting phosphorylation and degradation of IκB-α (35). Furthermore, Dex increased IκB protein and enhanced blockade of NF-κB activation by PS-1145. PS-1145 also blocks the protective effect of IL-6 against Dex-induced apoptosis. Moreover, PS-1145 inhibited both IL-6 secretion from BMSCs triggered by MM cell adhesion, and proliferation of MM cells adherent to BMSCs (35). However, in contrast to Bortezomib, PS-1145 only partially (20–50%) inhibited MM cell proliferation, suggesting that NF-κB blockade cannot account for all of the anti-MM activity of Bortezomib (35, 66). Together, these studies demonstrate that specific targeting of NF-κB overcomes the growth and survival advantage conferred by MM cell binding to BMSCs and cytokine secretion in the BM milieu. These preclinical findings have led to the clinical trials using such drugs in combination with conventional and novel therapies in MM.

## 6. BORTEZOMIB/PROTEASOME INHIBITOR PS-341 TARGETS NF-κB IN MM CELLS

One of the major mechanisms whereby proteasome inhibitors exert their growth inhibitory effects in cancer cells is by blocking NF-κB signaling (68). Moreover, patient MM-derived primary cells and BMSCs have upregulated NF-κB activity relative to normal cells (69). Furthermore, drug-sensitive MM cells show lower NF-κB activity than drug-resistant MM cells, suggesting that NF-κB confers chemoresistance (69). Elevated NF-κB levels have also been reported in MM cells derived from patients relapsing after chemotherapy (60). These findings indicate that NF-κB is a key regulator of growth and survival of MM cells in the BM milieu (Figure 2). Importantly, treatment of MM with proteasome inhibitor PS-341/Bortezomib prevents degradation of IκB, thereby blocking not only NF-κB activation, but also related cytokine production and the survival advantage for MM cells conferred by BMSCs (Figure 2).

Bortezomib downregulates NF-κB; however, NF-κB inhibition alone is unlikely to account for the total anti-MM activity of Bortezomib, as noted above (35, 66). Multiple genomics and proteomic studies have now established that Bortezomib affects various other signaling pathways. For example, Bortezomib-induced apoptosis is associated with these additional events (Figure 1): 1) activation of classical stress response proteins such as heat shock proteins, Hsp27, Hsp70 and Hsp90 (70, 71); 2) upregulation of c-Jun-NH2-terminal kinase (JNK) (37); 3) alteration of mitochondrial membrane potential and generation of reactive oxygen species (ROS) (72); 4) induction of intrinsic cell death pathway i.e, the release of mitochondrial proteins cytochrome-c/Smac into cytosol and activation of caspase-9 > caspase-3 cascade (22); 5) activation of extrinsic apoptotic signaling through Bid and caspase-8 cleavage (70); 6) impairment of DNA-dependent protein kinase (DNA-PK) (73), which is essential for the repair of DNA double-strand breaks; 7) inhibition of MM cell to BMSCs-host interaction, thereby block associated MM growth factor transcription and secretion from BMSCs (23); and 8) inhibition of MM cell growth factor-triggered MAPK and PI3-kinase/Akt signaling (74). Although many of these cellular events may appear correlative and common to other apoptotic agents, our studies have directly established the requirement of JNK activation during Bortezomib-induced MM cell death (37). Ongoing studies are evaluating the role of IκB using dominant-negative strategies and/or IκB knockout cells. Bortezomib may have additional substrates, which mediate normal cell growth and survival, and it is very likely that all the above signaling cascades mutually interact and contribute towards the overall response to Bortezomib in MM cells. Nevertheless, delineation of Bortezomib-triggered apoptotic signaling cascades allows us to combine Bortezomib with agents that utilize either similar or additional apoptotic pathways to enhance its cytotoxicity. For example, our proteomic analysis showed that Bortezomib impairs DNA repair mechanisms within MM cells (70), which has already translated into clinical trials using Bortezomib with DNA-damaging agents to treat relapsed/refractory MM.

## 7. THERAPEUTIC STRATEGIES TO OVERCOME BORTEZOMIB-RESISTANCE

Bortezomib induces apoptosis in MM cells; however, prolonged exposure is associated with toxicity and development of Bortezomib-resistance. Our recent studies have begun to unveil the mechanisms mediating Bortezomib-resistance. For example, our study showed that treatment with Bortezomib induces apoptosis in SUDHL6 (DHL6), but not

SUDHL4 (DHL4), lymphoma cells (71). Oligonucleotide array analysis demonstrated higher levels of transcripts of heat shock protein-27 (Hsp27) in DHL4 *versus* DHL6 cells, which was correlated with Hsp27 protein expression. Blocking Hsp27 using an anti-sense (AS) strategy restores the apoptotic response to Bortezomib in DHL4 cells; conversely, ectopic expression of wild type (WT) Hsp27 confers Bortezomib-resistance in Bortezomib-sensitive DHL6 cells. These findings provide the first direct evidence of Hsp27 protein conferring Bortezomib resistance. Moreover, MM cells obtained from patients refractory to Bortezomib treatment also show high levels of Hsp-27 expression. The mechanism(s) whereby Hsp-27 mediates Bortezomib-resistance are less known. We and others have shown that Hsp-27 negatively regulates the release of mitochondrial protein cytochrome-c and Smac, thereby blocking the intrinsic cell death-signaling pathway (75-77) (Figure 1). Further studies are required to determine whether inhibition of Hsp-27 using clinical grade specific inhibitors will enhance Bortezomib anti-MM activity and overcome drug-resistance. Besides Hsp-27, Bcl2 protein family members also confer drug-resistance in many cell types (78), and Bortezomib-triggered apoptosis in MM cells is partially abrogated by Bcl2 expression (70). Upregulated expression of inhibitors of apoptosis proteins (IAPs), such as XIAP, may also contribute to Bortezomib-resistance (70). It is likely that multiple mechanisms confer Bortezomib-resistance, suggesting that combinations of Bortezomib with conventional and/or novel agents will be required to overcome drug-resistance.

*In vitro* studies showed that combining Bortezomib with conventional agents such as Dex, Doxorubicin, Melphalan, or Mitoxantrone, triggers additive and/or synergistic anti-MM activity (69, 73, 79). Moreover, combined treatment of MM cells and of MM patient cells with Bortezomib and novel agents, such as Relvimid or triterpenoids CDDO-Imidazolide, also induces synergistic anti-MM activity (73, 80). For example, Bortezomib + CDDO-Im triggers synergistic apoptosis, even in Bortezomib-resistant MM cells from patients, thereby providing the preclinical basis for clinical protocols using this treatment regimen (80). Combination Bortezomib therapies have also bee effective in other cancers. For example, Bortezomib and irinotecan induces apoptosis in pancreatic tumor xenografts (81). Another study showed that Bortezomib prevents irinotecan-induced NF-κB activation, thereby increasing chemosensitivity and apoptosis in colorectal cancer cells in xenograft models (82). Finally, our recent oligonucleotide array analysis showed that mitochondrial peripheral benzodiazepine receptor is highly expressed in drug-resistant MM cells. PBRs, like Bcl2, are predominantly localized in mitocndria and function similar to Bcl2 i.e both block mitochondrial cell death signaling and confer drug-resistance.

Combining Bortezomib with an antagonist to PBR PK-11195 triggered synergistic anti-MM activity. These findings suggest that defects in mitochondrial proteins confer drug resistance, and that amplification of mitochondrial apoptotic signaling can increase cell death and overcome Bortezomib-resistance in MM cells. Understanding the mechanisms of drug sensitivity versus drug resistance provides the rationale for combination strategies designed to reduce attendant toxicity as well as overcome and/or prevent the development of drug-resistance.

## 8. ANTI-APOPTOTIC AND GROWTH SIGNALING

We and others have shown that various cytokines trigger MM cell growth signaling via Ras-Raf-MAPK pathways; anti-apoptotic/survival signaling via JAK-STAT, PI3K-Akt, Bcl2 and NF-κB pathways; and migration/angiogenesis via VEGFR-PI3K-PKC signaling pathways (24). Most drug-induced apoptosis is associated with simultaneous inhibition of growth promoting signaling, thereby enhancing its overall anti-MM activity of drug. Dex-induced apoptosis in MM cells is associated with downregulation of conventional Ras > Raf MAPK growth signaling, alongwith downregulation of p70rsk (33). Multiple studies have made similar observations using various conventional anti-MM drugs. We and others have shown that IL-6 and IGF induce growth of MM cells *via* Ras > Raf > MAPK and/or PI3-K/Akt signaling pathways in MM cells (48, 57, 83-86). Specifically, IL-6 induces activation of Akt/PKB activation and its downstream targets including Bad, GSK, and forkhead transcriptional factor (FKHR) whereas blockade of Akt activation by the PI3K inhibitor LY294002 abrogates both IL-6 triggered MEK/MAPK activation and proliferation; as well as resistance to Dex-induced apoptosis and associated caspase activation. We also showed that IL-6 triggered PI3-K/Akt signaling in MM.1S cells and downstream phosphorylation of FKHR, with related G1/S phase transition; conversely, LY294002 blocked this signaling, resulting in upregulation of p27 (KIP1) and G1 growth arrest. IGF-1 similarly stimulated sustained activation of NF-κB and Akt; induced phosphorylation of FKHRL-1; upregulated a series of intracellular anti-apoptotic proteins including FLIP, survivin, cIAP-2, A1/Bf1-1, and XIAP; and decreased Apo2L/ TRAIL-sensitivity of MM cells (9). A recent study showed that blockade of Ras-mediated signaling by Farnesy1 transferase inhibitors (FTIs) induces apoptosis in MM cells even in the presence of BMSCs. Blockade of cell cycle and growth signaling pathways by UCN- 01 and MEK/2 inhibitors induces apoptosis and overcome drug resistance in MM cells. Biochemical inhibitors of JAK-STAT pathways, piceatannol

(JAK1/STAT3 inhibitor) and tyrphostin AG490 (JAK2/STAT3 inhibitor), sensitize MM cells to apoptosis induced by various therapeutic agents including cisplatin, fludarabine, Adriamycin, and vinblastine. Importantly, delineation of these signaling cascades provides the basis for clinical trials combining conventional or novel agents to neutralize cell-surface receptor-triggered MM cell growth and survival. For example, our recent study showed that inhibition of IGF-1R using an IGF-1R tyrosine kinase inhibitor markedly blocks MM cell lines and patient cells growth *in vitro* and *in vivo* in mouse model (84). Furthermore, cotreatment of MM cells with IL-6 superantagonist Sant-7 and Dex triggers synergistic growth arrest and apoptosis (87). Humanized anti-interleukin-6 receptor monoclonal antibody has also been shown to induce apoptosis of both fresh and cloned human MM cells *in vitro* (88). Together, these studies suggest that specific targeting of intermediary molecules in growth and survival signaling provides an effective anti-MM therapy.

## 9. THERAPEUTIC APPLICATION OF DRUG-INDUCED GENE PROFILES

Our pilot studies using oligonucleotide arrays showed that Dex not only triggers transcription of genes involved in apoptotic signaling, but also simultaneously induced genes mediating anti-apoptotic signals (89). For example, treatment of MM cells with Dex induced an early transient increase in IL-6 receptor (IL-6R) transcripts and protein levels; this increased expression of IL-6R molecules on the MM cell surface allowed enhanced binding of soluble IL-6, thereby triggering growth/anti-apoptotic signaling in MM cells. Furthermore, an increase in the TGF-FR type II also occurred early after Dex treatment of MM cells, consistent with a defense mechanism whereby increased TGF-FRII expression on the MM cell surface facilitated enhanced TGF-F binding, thereby triggering transcription and secretion of IL-6 and related protection against Dex-induced apoptosis. These data suggest that combination of Dex with agents that block protective signaling may both enhance sensitivity and prevent development of Dex-resistance. For example, neutralizing antibodies against IL-6R or IGF-R disrupt Dex-induced protective signaling, thereby increasing the anti-MM activity of Dex (84). Alternatively, using Dex with specific biochemical inhibitors of IL-6 or IGF-induced growth signaling pathways triggers synergistic anti-MM effects. Similarly, our recent study showed that an inhibitor of VEGF receptor PTK-787, induces MM cell apoptosis (90). Another study identified CD40 also as a potential therapeutic target, since it induces secretion of pro-angiogenic factor VEGF in MM cell and trigger migration of these cells (91,

92). Collectively, these studies provide the preclinical rationale for clinical trials using agents that target cell surface molecules, together with other agents that directly activate nuclear and cytoplasmic apoptotic signaling, to enhance drug sensitivity and overcoming drug resistance.

## 10. MODULATION OF BCL2 PROTEIN

The Bcl2 family of proteins are functionally categorized into two distinct groups based on their pro- apoptotic (Bax, Bad, BCL-xs, Bik) or anti-apototic (Bcl2, Bcl-xl, Bcl-W, Mcl-1, A1) activities (78). Bcl2 contains four conserved Bcl2 homology (BH) domains with a hydrophobic sequence in the carboxyl-terminal region. The relative expression level of the pro- and anti-apoptotic proteins determines the degree of responsiveness to apoptotic inducers. Bcl2 is localized in the endoplasmic reticulum, nuclear membrane, and the outer mitochondrial membrane (93). Bcl2 is highly upregulated in many cancer cells, including MM, and confers drug resistance (78, 94). In the context of MM, peptides targeting the BH3 domain of Bcl2 enhance Bortezomib/PS-341-induced apoptosis whereas overexpression of BCL2 inhibits Bid cleavage and associated apoptosis (95). Another recent study demonstrated that an anti-sense oligodeoxynucleotide (ODN) complementary to the first six codons of the Bc12 mRNA, G3139 (oblimersen sodium; Genasense), downregulated expression of Bcl2 and enhanced sensitivity to Dex- and Doxorubicin-induced apoptosis in MM cells. These data support further clinical evaluation of G3139 therapy in MM (96). Anti-sense ODNs to Mcl1, but not BCL2 or Bclxs, induce apoptosis associated with a decreased mitochondrial membrane potential (97). Together, these findings suggest that inhibition of Bcl2 function increases ant-MM activity of conventional or novel drugs.

## 11. CONCLUSION

Characterization of growth and apoptotic signaling pathways allows identification of molecular targets to interrupt MM cell growth and trigger apoptosis. For example, anti-MM drugs induce apoptosis *via* release of mitochondrial proteins cyto-c or Smac, suggesting that active peptides against these molecules may sensitize MM cells to various anti-MM agents. Already our preliminary data suggest that smac mimetics trigger MM cell apoptosis and induce synergistic anti-MM activity when combined with Bortezomib. The finding that Dex-induced apoptosis occurs without cyto-c release or JNK activation provides the rationale for amplifying apoptotic signaling by combining Dex with novel agents that trigger both cyto-c release and JNK activation. Additionally, the combination of biochemical inhibitors of growth/survival signaling pathways with conventional or novel anti-MM agents enhances MM cell death. Gene profiling and proteomic studies have identified novel drugs that specifically target molecules on the MM cell surface (IL-6R, IGF-1R, VEGFR, CD40), cytoplasm (Hsp90, Hsp27, p38 MAPK, IKK/NF-κB), mitochondria (PK-11195, smac mimetics), ER (Bcl2) and nucleus (Telomestatin, HDAC inhibitors). Anti-sense strategy against anti-apoptotic signaling molecules (Mc11, Bc12, SHP2, or IAPs) will also enhance anti-MM activity of other agents. The successful development of Bortezomib therapy for MM has established proteasome inhibition as an effective therapeutic strategy for the treatment of relapsed/refractory MM. However, both intrinsic and acquired Bortezomib-resistance has already been observed, and ongoing studies are evaluating other novel proteasome inhibitors as well as combination strategies to overcome drug resistance and improve patient outcome in MM.

## ACKNOWLEDGEMENTS

Supported by: NIH grants 50947, CA 78373, SPORE P50 CA100707, P01 CA078378, Doris Duke Distinguished Clinical Research Scientist Award (KCA); Multiple Myeloma Research Foundation (MMRF) Senior Research Award (DC); The Myeloma Research Fund, and The Cure Myeloma Fund

## REFERENCES

1. R. T. Greenlee, T. Murray, S. Bolden, and P. A. Wingo, Cancer statistics, 2000, *CA Cancer J Clin* **50**(1), 7-33 (2000).
2. K. C. Anderson, Novel immunomodulatory therapies in the treatment of multiple myeloma, *Oncology (Huntingt)* **18**(8), 988-990 (2004).
3. K. C. Anderson, Moving disease biology from the lab to the clinic, *Cancer* **97**(3 Suppl), 796-801 (2003).
4. B. Klein, X. G. Zhang, M. Jourdan, J. Content, F. Houssiau, L. Aarden, M. Piechaczyk, and R. Bataille, Paracrine rather than autocrine regulation of myeloma-cell growth and differentiation by interleukin-6, *Blood* **73**(2), 517-526 (1989).
5. H. Uchiyama, B. A. Barut, A. F. Mohrbacher, D. Chauhan, and K. C. Anderson, Adhesion of human myeloma-derived cell lines to bone marrow stromal cells stimulates IL-6 secretion, *Blood* **82**3712-3720 (1993).
6. D. Chauhan, H. Uchiyama, Y. Akbarali, M. Urashima, K. Yamamoto, T. A. Libermann, and K. C. Anderson, Multiple myeloma cell adhesion-induced interleukin-6 expression in bone marrow stromal cells involves activation of NF-kappa B, *Blood* **87**(3), 1104-1112. (1996).

7. A. Lichtenstein, Y. Tu, C. Fady, R. Vescio, and J. Berenson, Interleukin-6 inhibits apoptosis of malignant plasma cells., *Cell Immunol* **162**248-255 (1995).
8. G. G. Freund, D. T. Kulas, and R. A. Mooney, Insulin and IGF-1 increase mitogenesis and glucose metabolism in the multiple myeloma cell line, RPMI 8226, *J Immunol* **151**1811 (1993).
9. C. S. Mitsiades, N. Mitsiades, V. Poulaki, R. Schlossman, M. Akiyama, D. Chauhan, T. Hideshima, S. P. Treon, N. C. Munshi, P. G. Richardson, and K. C. Anderson, Activation of NF-kappaB and upregulation of intracellular anti-apoptotic proteins via the IGF-1/Akt signaling in human multiple myeloma cells: therapeutic implications, *Oncogene* **21**(37), 5673-5683 (2002).
10. M. L. Coleman, E. A. Sahai, M. Yeo, M. Bosch, A. Dewar, and M. F. Olson, Membrane blebbing during apoptosis results from caspase-mediated activation of ROCK I, *Nat Cell Biol* **3**(4), 339-345 (2001).
11. S. Sahara, M. Aoto, Y. Eguchi, N. Imamoto, Y. Yoneda, and Y. Tsujimoto, Acinus is a caspase-3-activated protein required for apoptotic chromatin condensation, *Nature* **401**(6749), 168-173 (1999).
12. A. H. Willie, J. F. Kerr, and A. R. Currie, Cell death: the significance of apoptosis., *Int. Rev. Cytol.* **68**251-306 (1980).
13. M. Enari, H. Sakahira, H. Yokoyama, K. Okawa, A. Iwanatsu, and S. Natata, A caspase-activated DNase that degrades DNA during apoptosis, and its inhibitor ICAD, *Nature* **391**43-50 (1998).
14. B. B. Wolf, and D. R. Green, Suicidal tendencies: apoptotic cell death by caspase family proteinases, *J Biol Chem* **274**(29), 20049-20052 (1999).
15. N. A. Thornberry, and Y. Lazebnik, Caspases: enemies within., *Science* **281**1312-1316 (1998).
16. R. M. Kluck, and S. Kashibatta, A Lively Meeting of a Deathly Topic, *Apoptosis* **2**337-342 (1997).
17. J. Yang, X. Liu, K. Balla, C. N. Kim, N. M. Ibardo, J. Cai, T. i. Peng, D. P. Jones, and X. Wang, Prevention of apoptosis by Bcl-2-Release of Cytochrome-c from the mitochondria blocked., *Science* **275**1129-1136 (1997).
18. C. Du, M. Fang, Y. Li, L. Li, and X. Wang, Smac, a mitochondrial protein that promotes cytochrome c-dependent caspase activation by eliminating IAP inhibition, *Cell* **102**(1), 33-42. (2000).
19. A. M. Verhagen, P. G. Ekert, M. Pakusch, J. Silke, L. M. Connolly, G. E. Reid, R. L. Moritz, R. J. Simpson, and D. L. Vaux, Identification of DIABLO, a mammalian protein that promotes apoptosis by binding to and antagonizing IAP proteins, *Cell* **102**(1), 43-53 (2000).
20. S. A. Susin, H. K. Lorenzo, N. Zamzami, I. Marzo, B. E. Snow, G. M. Brothers, J. Mangion, E. Jacotot, P. Costantini, M. Loeffler, N. Larochette, D. R. Goodlett, R. Aebersold, D. P. Siderovski, J. M. Penninger, and G. Kroemer, Molecular characterization of mitochondrial apoptosis-inducing factor, *Nature* **397**(6718), 441-446 (1999).
21. L. Y. Li, X. Luo, and X. Wang, Endonuclease G is an apoptotic DNase when released from mitochondria, *Nature* **412**(6842), 95-99 (2001).
22. D. Chauhan, and K. C. Anderson, Mechanisms of cell death and survival in multiple myeloma (MM): Therapeutic implications, *Apoptosis* **8**(4), 337-343 (2003).
23. T. Hideshima, and K. C. Anderson, Molecular mechanisms of novel therapeutic approaches for multiple myeloma, *Nat Rev Cancer* **2**(12), 927-937 (2002).
24. D. Chauhan, T. Hideshima, and K. C. Anderson, Apoptotic signaling in multiple myeloma: therapeutic implications, *Int J Hematol* **78**(2), 114-120 (2003).

25. M. P. Mattson, and G. Kroemer, Mitochondria in cell death: novel targets for neuroprotection and cardioprotection, *Trends Mol Med* **9**(5), 196-205 (2003).
26. J. C. Reed, Bcl-2 family of proteins., *Oncogene* **17**3225-3236 (1998).
27. S. Willis, C. L. Day, M. G. Hinds, and D. C. Huang, The Bcl-2-regulated apoptotic pathway, *J Cell Sci* **116**(Pt 20), 4053-4056 (2003).
28. H. Zou, Y. Li, X. Liu, and X. Wang, An APAF-1.cytochrome c multimeric complex is a functional apoptosome that activates procaspase-9, *J Biol Chem* **274**(17), 11549-11556 (1999).
29. S. M. Srinivasula, M. Ahmad, T. Fernandes-Alnemri, and E. S. Alnemri, Autoactivation of procaspase-9 by Apaf-1-mediated oligomerization, *Mol Cell* **1**(7), 949-957 (1998).
30. K. Cain, S. B. Bratton, C. Langlais, G. Walker, D. G. Brown, X. M. Sun, and G. M. Cohen, Apaf-1 oligomerizes into biologically active approximately 700-kDa and inactive approximately 1.4-MDa apoptosome complexes, *J Biol Chem* **275**(9), 6067-6070 (2000).
31. T. Hideshima, D. Chauhan, Y. Shima, N. Raje, F. E. Davies, Y. T. Tai, S. P. Treon, B. Lin, R. L. Schlossman, P. Richardson, G. Muller, D. I. Stirling, and K. C. Anderson, Thalidomide and its analogs overcome drug resistance of human multiple myeloma cells to conventional therapy, *Blood* **96**(9), 2943-2950. (2000).
32. N. Mitsiades, C. S. Mitsiades, V. Poulaki, D. Chauhan, P. G. Richardson, T. Hideshima, N. C. Munshi, S. P. Treon, and K. C. Anderson, Apoptotic signaling induced by immunomodulatory thalidomide analogs in human multiple myeloma cells: therapeutic implications, *Blood* **99**(12), 4525-4530 (2002).
33. D. Chauhan, P. Pandey, A. Ogata, G. Teoh, S. Treon, M. Urashima, S. Kharbanda, and K. C. Anderson, Dexamethasone induces apoptosis of multiple myeloma cells in a JNK/SAP kinase independent mechanism, *Oncogene* **15**837-843 (1997).
34. D. Chauhan, T. Hideshima, S. Rosen, J. C. Reed, S. Kharbanda, and K. C. Anderson, Apaf-1/cytochrome c-independent and Smac-dependent induction of apoptosis in multiple myeloma (MM) cells, *J Biol Chem* **276**(27), 24453-24456. (2001).
35. T. Hideshima, D. Chauhan, P. Richardson, C. Mitsiades, N. Mitsiades, T. Hayashi, N. Munshi, L. Dong, A. Castro, V. Palombella, J. Adams, and K. C. Anderson, NF-kappa B as a therapeutic target in multiple myeloma, *J Biol Chem* **28**28 (2002).
36. A. Palumbo, A. Bertola, P. Falco, R. Rosato, F. Cavallo, L. Giaccone, S. Bringhen, P. Musto, P. Pregno, T. Caravita, G. Ciccone, and M. Boccadoro, Efficacy of low-dose thalidomide and dexamethasone as first salvage regimen in multiple myeloma, *Hematol J* **5**(4), 318-324 (2004).
37. D. Chauhan, G. Li, T. Hideshima, K. Podar, C. Mitsiades, N. Mitsiades, N. Munshi, S. Kharbanda, and K. C. Anderson, JNK-dependent Release of Mitochondrial Protein, Smac, during Apoptosis in Multiple Myeloma (MM) Cells, *J Biol Chem* **278**(20), 17593-17596 (2003).
38. D. Chauhan, G. Li, D. Auclair, T. Hideshima, K. Podar, N. Mitsiades, C. Mitsiades, L. B. Chen, N. Munshi, S. Saxena, and K. C. Anderson, 2-Methoxyestardiol and bortezomib/proteasome-inhibitor overcome dexamethasone-resistance in multiple myeloma cells by modulating Heat Shock Protein-27, *Apoptosis* **9**(2), 149-155 (2004).
39. G. Screaton, and X. N. Xu, T cell life and death signalling via TNF-receptor family members, *Curr Opin Immunol* **12**(3), 316-322 (2000).
40. A. M. Chinnaiyan, K. O'Rourke, M. Tewari, and V. M. Dixit, FADD, a novel death domain-containing protein, interacts with the death domain of Fas and initiates apoptosis, *Cell* **81**(4), 505-512 (1995).
41. A. M. Chinnaiyan, and V. M. Dixit, The cell-death machine, *Curr. Biol.* **6**555-562 (1996).

42. H. Hsu, H.-B. Shu, M.-G. Pan, and D. V. Goeddel, TRADD-TRAF2 and TRADD-FADD interactions define two distinct TNF receptor 1 signal transduction pathways, *Cell* **84**299-308 (1996).
43. P. Schneider, M. Thome, K. Burns, J. L. Bodmer, K. Hofmann, T. Kataoka, N. Holler, and J. Tschopp, TRAIL receptors 1 (DR4) and 2 (DR5) signal FADD-dependent apoptosis and activate NF-kappaB, *Immunity* **7**(6), 831-836 (1997).
44. J. M. Kyriakis, and J. Avruch, Sounding the alarm: Protein kinase cascades activated by stress and inflammation, *J. Biol. Chem.* **271**24313-24316 (1996).
45. B. Derijard, M. Hibi, I. H. Wu, T. Barrett, B. Su, T. Deng, M. Karin, and R. J. Davis, JNK1: A protein kinase stimulated by UV light and Ha-Ras that binds and phosphorylates the c-Jun activation domain, *Cell* **76**1025-1037 (1994).
46. M. Verheij, R. Bose, X. H. Lin, B. Yhao, W. D. Jarvis, S. Grant, M. J. Birrer, E. Szabo, L. I. Zon, J. M. Kyriakis, A. Haimovitz-Friedman, Z. Fuks, and R. N. Kolesnick, Requirement for ceramide-initiated SAPK/JNK signalling in stress-induced apoptosis, *Nature* **380**75-79 (1996).
47. C. Tournier, P. Hess, D. D. Yang, J. Xu, T. K. Turner, A. Nimnual, D. Bar-Sagi, S. N. Jones, R. A. Flavell, and R. J. Davis, Requirement of JNK for stress-induced activation of the cytochrome c-mediated death pathway, *Science* **288**(5467), 870-874 (2000).
48. D. Chauhan, S. Kharbanda, A. Ogata, M. Urashima, G. Teoh, M. Robertson, D. W. Kufe, and K. C. Anderson, Interleukin-6 inhibits Fas-induced apoptosis and stress-activated protein kinase activation in multiple myeloma cells, *Blood* **89**227-234 (1997).
49. G. Kroemer, The proto-oncogene Bcl-2 and its role in regulating apoptosis, *Nature Medicine* **3**614-620 (1997).
50. G. Kroemer, and J. C. Reed, Mitochondrial control of cell death, *Nat Med* **6**(5), 513-519 (2000).
51. S. Kharbanda, S. Saxena, K. Yoshida, P. Pandey, M. Kaneki, Q. Wang, K. Cheng, Y. N. Chen, A. Campbell, T. Sudha, Z. M. Yuan, J. Narula, R. Weichselbaum, C. Nalin, and D. Kufe, Translocation of SAPK/JNK to mitochondria and interaction with Bcl-x(L) in response to DNA damage, *J Biol Chem* **275**(1), 322-327 (2000).
52. M. C. Wei, W. X. Zong, E. H. Cheng, T. Lindsten, V. Panoutsakopoulou, A. J. Ross, K. A. Roth, G. R. MacGregor, C. B. Thompson, and S. J. Korsmeyer, Proapoptotic BAX and BAK: a requisite gateway to mitochondrial dysfunction and death, *Science* **292**(5517), 727-730 (2001).
53. K. Lei, A. Nimnual, W. X. Zong, N. J. Kennedy, R. A. Flavell, C. B. Thompson, D. Bar-Sagi, and R. J. Davis, The Bax subfamily of Bcl2-related proteins is essential for apoptotic signal transduction by c-Jun NH(2)-terminal kinase, *Mol Cell Biol* **22**(13), 4929-4942 (2002).
54. K. C. Anderson. Plasma Cell Tumors. In: Holland JF, Frei III E, Bast RC, Kufe DW, Norton DL, Weichselbaum RR, eds. *Cancer Medicine*. 4th ed. Baltimore: Williams and Wilkins; in press.
55. L. A. Hazlehurst, S. A. Enkemann, C. A. Beam, R. F. Argilagos, J. Painter, K. H. Shain, S. Saporta, D. Boulware, L. Moscinski, M. Alsina, and W. S. Dalton, Genotypic and phenotypic comparisons of de novo and acquired melphalan resistance in an isogenic multiple myeloma cell line model, *Cancer Res* **63**(22), 7900-7906 (2003).
56. T. H. Landowski, N. E. Olashaw, D. Agrawal, and W. S. Dalton, Cell adhesion-mediated drug resistance (CAM-DR) is associated with activation of NF-kappa B (RelB/p50) in myeloma cells, *Oncogene* **22**(16), 2417-2421 (2003).
57. D. Chauhan, H. Uchiyama, M. Urashima, K. Yamamoto, and K. C. Anderson, Regulation of interleukin-6 in multiple myeloma and bone marrow stromal cells, *Stem Cells* **15**35-39 (1995).

58. D. Chauhan, P. Pandey, T. Hideshima, S. Treon, N. Raje, F. E. Davies, Y. Shima, Y. T. Tai, S. Rosen, S. Avraham, S. Kharbanda, and K. C. Anderson, SHP2 mediates the protective effect of interleukin-6 against dexamethasone-induced apoptosis in multiple myeloma cells, *J Biol Chem* **275**(36), 27845-27850. (2000).
59. D. Chauhan, T. Hideshima, P. Pandey, S. Treon, G. Teoh, N. Raje, S. Rosen, N. Krett, H. Husson, S. Avraham, S. Kharbanda, and K. C. Anderson, RAFTK/PYK2-dependent and -independent apoptosis in multiple myeloma cells, *Oncogene* **18**(48), 6733-6740. (1999).
60. R. Feinman, J. Koury, M. Thames, B. Barlogie, and J. Epstein, Role of NF-kB in the rescue of multiple myeloma cells from glucocorticoids-induced apoptosis by Bcl-2., *Blood* **93**3044-3052 (1999).
61. I. Alkalay, A. Yaron, A. Hatzubai, S. Jung, A. Avraham, O. Gerlitz, I. Pashut-Lavon, and Y. Ben-Neriah, In vivo stimulation of I kappa B phosphorylation is not sufficient to activate NF-kappa B, *Mol Cell Biol* **15**(3), 1294-1301 (1995).
62. M. Karin, Y. Yamamoto, and Q. M. Wang, The IKK NF-kappa B system: a treasure trove for drug development, *Nat Rev Drug Discov* **3**(1), 17-26 (2004).
63. M. Ogawa, T. Nishiura, K. Oritani, H. Yoshida, M. Yoshimura, Y. Okajima, J. Ishikawa, K. Hashimoto, I. Matsumura, Y. Tomiyama, and Y. Matsuzawa, Cytokines prevent dexamethasone-induced apoptosis via the activation of mitogen-activated protein kinase and phosphatidylinositol 3-kinase pathways in a new multiple myeloma cell line, *Cancer Res* **60**(15), 4262-4269 (2000).
64. A. C. Bharti, N. Donato, S. Singh, and B. B. Aggarwal, Curcumin (diferuloylmethane) down-regulates the constitutive activation of nuclear factor-kappa B and Ikappa Balpha kinase in human multiple myeloma cells, leading to suppression of proliferation and induction of apoptosis, *Blood* **101**(3), 1053-1062 (2003).
65. M. Akiyama, T. Hideshima, T. Hayashi, Y. T. Tai, C. S. Mitsiades, N. Mitsiades, D. Chauhan, P. Richardson, N. C. Munshi, and K. C. Anderson, Nuclear factor-kappaB p65 mediates tumor necrosis factor alpha-induced nuclear translocation of telomerase reverse transcriptase protein, *Cancer Res* **63**(1), 18-21 (2003).
66. N. Mitsiades, C. S. Mitsiades, V. Poulaki, D. Chauhan, P. G. Richardson, T. Hideshima, N. Munshi, S. P. Treon, and K. C. Anderson, Biologic sequelae of nuclear factor-kappaB blockade in multiple myeloma: therapeutic applications, *Blood* **99**(11), 4079-4086 (2002).
67. M. M. Kawano, N. Huang, H. Tanaka, H. Ishikawa, A. Sakai, O. Tanabe, M. Nobuyoshi, and A. Kuramoto, Homotypic cell aggregations of human myeloma cells with ICAM-1 and LFA-1 molecules, *Br J Haematol* **79**(4), 583-588 (1991).
68. J. Adams, The proteasome: a suitable antineoplastic target, *Nat Rev Cancer* **4**(5), 349-360 (2004).
69. M. H. Ma, H. H. Yang, K. Parker, S. Manyak, J. M. Friedman, C. Altamirano, Z. Q. Wu, M. J. Borad, M. Frantzen, E. Roussos, J. Neeser, A. Mikail, J. Adams, N. Sjak-Shie, R. A. Vescio, and J. R. Berenson, The Proteasome Inhibitor PS-341 Markedly Enhances Sensitivity of Multiple Myeloma Tumor Cells to Chemotherapeutic Agents, *Clin Cancer Res* **9**(3), 1136-1144 (2003).
70. N. Mitsiades, C. S. Mitsiades, V. Poulaki, D. Chauhan, G. Fanourakis, X. Gu, C. Bailey, M. Joseph, T. A. Libermann, S. P. Treon, N. C. Munshi, P. G. Richardson, T. Hideshima, and K. C. Anderson, Molecular sequelae of proteasome inhibition in human multiple myeloma cells, *Proc Natl Acad Sci U S A* **99**(22), 14374-14379 (2002).
71. D. Chauhan, G. Li, R. Shringarpure, K. Podar, Y. Ohtake, T. Hideshima, and K. C. Anderson, Blockade of Hsp27 overcomes Bortezomib/proteasome inhibitor PS-341 resistance in lymphoma cells, *Cancer Res* **63**(19), 6174-6177 (2003).

72. D. Chauhan, L. Guilan, M. Sattler, T. Hideshima, K. Podar, C. Mitsiades, N. Mitsiades, N. Munshi, and K. C. Anderson, Superoxide-Dependent and Independent Mitochondrial Signaling During Apoptosis in Multiple Myeloma (MM) cells, *Oncogene (In Press)* (2003).
73. N. Mitsiades, C. S. Mitsiades, P. G. Richardson, V. Poulaki, Y. T. Tai, D. Chauhan, G. Fanourakis, X. Gu, C. Bailey, M. Joseph, T. A. Libermann, R. Schlossman, N. C. Munshi, T. Hideshima, and K. C. Anderson, The proteasome inhibitor PS-341 potentiates sensitivity of multiple myeloma cells to conventional chemotherapeutic agents: therapeutic applications, *Blood* **101**(6), 2377-2380 (2003).
74. T. Hideshima, C. Mitsiades, M. Akiyama, T. Hayashi, D. Chauhan, P. Richardson, R. Schlossman, K. Podar, N. C. Munshi, N. Mitsiades, and K. C. Anderson, Molecular mechanisms mediating antimyeloma activity of proteasome inhibitor PS-341, *Blood* **101**(4), 1530-1534 (2003).
75. D. Chauhan, G. Li, T. Hideshima, K. Podar, C. Mitsiades, N. Mitsiades, L. Catley, Y. T. Tai, T. Hayashi, R. Shringarpure, R. Burger, N. Munshi, Y. Ohtake, S. Saxena, and K. C. Anderson, Hsp27 inhibits release of mitochondrial protein Smac in multiple myeloma cells and confers dexamethasone resistance, *Blood* **102**(9), 3379-3386 (2003).
76. C. G. Concannon, A. M. Gorman, and A. Samali, On the role of Hsp27 in regulating apoptosis, *Apoptosis* **8**(1), 61-70 (2003).
77. J. M. Bruey, C. Ducasse, P. Bonniaud, L. Ravagnan, S. A. Susin, C. Diaz-Latoud, S. Gurbuxani, A. P. Arrigo, G. Kroemer, E. Solary, and C. Garrido, Hsp27 negatively regulates cell death by interacting with cytochrome c, *Nat Cell Biol* **2**(9), 645-652 (2000).
78. S. Cory, and J. M. Adams, The Bcl2 family: regulators of the cellular life-or-death switch, *Nat Rev Cancer* **2**(9), 647-656 (2002).
79. T. Hideshima, P. Richardson, D. Chauhan, V. J. Palombella, P. J. Elliott, J. Adams, and K. C. Anderson, The proteasome inhibitor PS-341 inhibits growth, induces apoptosis, and overcomes drug resistance in human multiple myeloma cells, *Cancer Res* **61**(7), 3071-3076. (2001).
80. D. Chauhan, G. Li, K. Podar, T. Hideshima, R. Shringarpure, L. Catley, C. Mitsiades, N. Munshi, Y. T. Tai, N. Suh, G. W. Gribble, T. Honda, R. Schlossman, P. Richardson, M. B. Sporn, and K. C. Anderson, The bortezomib/proteasome inhibitor PS-341 and triterpenoid CDDO-Im induce synergistic anti-multiple myeloma (MM) activity and overcome bortezomib resistance, *Blood* **103**(8), 3158-3166 (2004).
81. S. A. Shah, M. W. Potter, T. P. McDade, R. Ricciardi, R. A. Perugini, P. J. Elliott, J. Adams, and M. P. Callery, 26S proteasome inhibition induces apoptosis and limits growth of human pancreatic cancer, *J Cell Biochem* **82**(1), 110-122 (2001).
82. J. C. Cusack, Jr., R. Liu, M. Houston, K. Abendroth, P. J. Elliott, J. Adams, and A. S. Baldwin, Jr., Enhanced chemosensitivity to CPT-11 with proteasome inhibitor PS-341: implications for systemic nuclear factor-kappaB inhibition, *Cancer Res* **61**(9), 3535-3540 (2001).
83. T. Hideshima, N. Nakamura, D. Chauhan, and K. C. Anderson, Biologic sequelae of interleukin-6 induced PI3-K/Akt signaling in multiple myeloma, *Oncogene* **20**(42), 5991-6000. (2001).
84. C. Mitsiades, N. Mitsiades, C. McMullan, V. Poulaki, R. Shringarpure, T. Hideshima, D. Chauhan, M. Joseph, T. A. Libermann, C. Garcia-Echeverria, M. Person, F. Hoffman, K. Anderson, and A. Kung, Inhibition of the insulin-like growth factor-1 tyrosine kinase activity as a therapeutic strategy for multiple myeloma, other hematologic malignancies, and solid tumors, *Cancer cell* Immediate early publication February 26 (2004).

85. Y. W. Qiang, E. Kopantzev, and S. Rudikoff, Insulinlike growth factor-I signaling in multiple myeloma: downstream elements, functional correlates, and pathway cross-talk, *Blood* **99**(11), 4138-4146 (2002).
86. Y. W. Qiang, L. Yao, G. Tosato, and S. Rudikoff, Insulin-like growth factor I induces migration and invasion of human multiple myeloma cells, *Blood* **103**(1), 301-308 (2004).
87. P. Tassone, S. Forciniti, E. Galea, R. Savino, M. C. Turco, P. Iacopino, P. Tagliaferri, G. Morrone, G. Ciliberto, and S. Venuta, Synergistic induction of growth arrest and apoptosis of human myeloma cells by the IL-6 super-antagonist Sant7 and Dexamethasone, *Cell Death Differ* **7**(3), 327-328. (2000).
88. T. Hirata, C. Shimazaki, T. Sumikuma, E. Ashihara, H. Goto, T. Inaba, Y. Koishihara, and M. Nakagawa, Humanized anti-interleukin-6 receptor monoclonal antibody induced apoptosis of fresh and cloned human myeloma cells in vitro, *Leuk Res* **27**(4), 343-349 (2003).
89. D. Chauhan, D. Auclair, E. K. Robinson, T. Hideshima, G. Li, K. Podar, D. Gupta, P. Richardson, R. L. Schlossman, N. Krett, L. B. Chen, N. C. Munshi, and K. C. Anderson, Identification of genes regulated by dexamethasone in multiple myeloma cells using oligonucleotide arrays, *Oncogene* **21**(9), 1346-1358 (2002).
90. B. Lin, K. Podar, D. Gupta, Y. T. Tai, S. Li, E. Weller, T. Hideshima, S. Lentzsch, F. Davies, C. Li, E. Weisberg, R. L. Schlossman, P. G. Richardson, J. D. Griffin, J. Wood, N. C. Munshi, and K. C. Anderson, The vascular endothelial growth factor receptor tyrosine kinase inhibitor PTK787/ZK222584 inhibits growth and migration of multiple myeloma cells in the bone marrow microenvironment, *Cancer Res* **62**(17), 5019-5026 (2002).
91. Y. T. Tai, K. Podar, D. Gupta, B. Lin, G. Young, M. Akiyama, and K. C. Anderson, CD40 activation induces p53-dependent vascular endothelial growth factor secretion in human multiple myeloma cells, *Blood* **99**(4), 1419-1427. (2002).
92. Y. T. Tai, K. Podar, N. Mitsiades, B. Lin, C. Mitsiades, D. Gupta, M. Akiyama, L. Catley, T. Hideshima, N. C. Munshi, S. P. Treon, and K. C. Anderson, CD40 induces human multiple myeloma cell migration via phosphatidylinositol 3-kinase/AKT/NF-kappa B signaling, *Blood* **101**(7), 2762-2769 (2003).
93. M. D. Jacobson, J. F. Burne, M. P. King, T. Miyashita, J. C. Reed, and M. C. Raff, Bcl-2 blocks apoptosis in cells lacking mitochondrial DNA, *Nature* **361**365-369 (1993).
94. P. Liston, W. G. Fong, and R. G. Korneluk, The inhibitors of apoptosis: there is more to life than Bcl2, *Oncogene* **22**(53), 8568-8580 (2003).
95. N. Mitsiades, C. S. Mitsiades, P. G. Richardson, C. McMullan, V. Poulaki, G. Fanourakis, R. Schlossman, D. Chauhan, N. C. Munshi, T. Hideshima, V. M. Richon, P. A. Marks, and K. C. Anderson, Molecular sequelae of histone deacetylase inhibition in human malignant B cells, *Blood* **101**(10), 4055-4062 (2003).
96. N. W. van de Donk, M. M. Kamphuis, M. van Dijk, H. P. Borst, A. C. Bloem, and H. M. Lokhorst, Chemosensitization of myeloma plasma cells by an antisense-mediated downregulation of Bcl-2 protein, *Leukemia* **17**(1), 211-219 (2003).
97. S. Derenne, B. Monia, N. M. Dean, J. K. Taylor, M. J. Rapp, J. L. Harousseau, R. Bataille, and M. Amiot, Antisense strategy shows that Mcl-1 rather than Bcl-2 or Bcl-x(L) is an essential survival protein of human myeloma cells, *Blood* **100**(1), 194-199 (2002).

Chapter 9

# CEPHALOSTATIN 1-INDUCED APOPTOSIS IN TUMOR CELLS

*Selective Induction of Smac/DIABLO Release*

Verena M. Dirsch and Angelika M. Vollmar
*Department of Pharmacy, University of Munich, Butenandtstr. 5-13, 81377 Munich, Germany*

Abstract: Cephalostatin 1 is a marine bis-steroidal compound with strong (mean panel $GI_{50}$: 2.2 (± 1.21) x $10^{-9}$ M) and particular cytotoxic activity as shown in the *in vitro* 60-cell line screen by the National cancer Institute (USA). Cephalostatin 1 was shown to trigger cell death in human Jurkat T cells with classical morphological (cell shrinkage, membrane blebbing, DNA fragmentation) and biochemical (phosphatidylserine translocation, activation of caspases) signs of apoptosis. Cephalostatin 1-induced apoptosis, however, differs significantly from apoptosis induced by classical chemotherapeutic drugs in that it selectively leads to Smac/DIABLO (second mitochondria-derived activator of caspases/direct inhibitor of apoptosis-binding protein with a low isoelectric point) release from mitochondria. Cytochrome c and apoptosis inducing factor (AIF) is retained in the mitochondrial intermembrane space. Selectively released Smac/DIABLO translocates into the cytosol where it binds to XIAP displacing previously bound caspase-9. Caspase-9 does not associate with Apaf-1 and thus appears to be processed without requirement for apoptosome formation. This unique mitochondrial signaling by cephalostatin 1 is accompanied by mitochondrial matrix condensation instead of mitochondrial swelling. Swelling is assumed to precede outer mitochondrial membrane permeabilization in response to induction of the mitochondrial (intrinsic) apoptosis pathway. These results characterize cephalostatin 1 as a remarkable experimental chemotherapeutic agent with a unique mechanism of action. Further studies will show whether cephalostatin 1 will be a good candidate to be developed against chemoresistant types of cancer.

Key words: cephalostatin, apoptosis, Smac/DIABLO, mitochondria, caspase 9, XIAP



*M. Sluyser (ed.), Application of Apoptosis to Cancer Treatment,* 209-221.

# 1. INTRODUCTION

Natural products play a significant role in the discovery and development of drugs for the treatment of human disease. This is particularly evident in the area of cancer, where more than 60% of all drugs are shown to be of natural origin[1]. In this context especially compounds of marine origin gained more and more attention over the past decades.

Why are compounds of marine origin promising candidates for anti-cancer drug development? Many marine organisms are soft-bodied and live sedentary. Thus, they developed the ability to produce toxic compounds or to obtain them from marine microorganisms as a chemical defense. The toxicity of these compounds needs to be extremely potent since they are released and diluted into water[2].

Although today the potency of marine compounds especially for the development of new anti-cancer drugs is recognized still difficulties, such as the availability in quantities high enough for preclinical and clinical studies, have to be overcome.

## 1.1 Cephalostatins

The cephalostatins are a group of structurally strongly related bis-streoidal compounds isolated from the marine organism *Cephalodiscus gilchristi* Ridewood (Cephalodiscidae). The structure of cephalostatin 1 is depicted in Figure 1. Up to now, 19 derivatives could be characterized. The cephalostatins belong to the most cytotoxic marine natural products ever tested by the National Cancer Institute (NCI)/USA. All cephalostatins show the same unique cytotoxicity profile in the NCI-60 panel (see Figure 2), which displays the growth inhibitory potency of substances against 60 cancer cell lines of diverse origin[3,4]. Although the cytotoxicity profiles of the cephalostatins do not differ, their potency varies depending on the chemical composition. Cephalostatin 1 proved to be the most potent form in the NCI-60 panel. The two-day NCI-60 screen yielded a mean panel $GI_{50}$ concentration of 2.2 (±1.21) x $10^{-9}$ M for cephalostatin 1[5,6]. Interestingly, the obtained cephalostatin cytotoxicity fingerprint did not show comparable correlations to any members of the standard agent database suggesting that the differential cytotoxicity of cephalostatins derives from a unprecedented, but as yet undefined, mechanism of action[7]. Beyond *in vitro* testings, cephalostatin 1 was proven to be effective in several xenografts, such as melanoma, sarcoma, in leukemia and even in a human mammary carcinoma model[8].

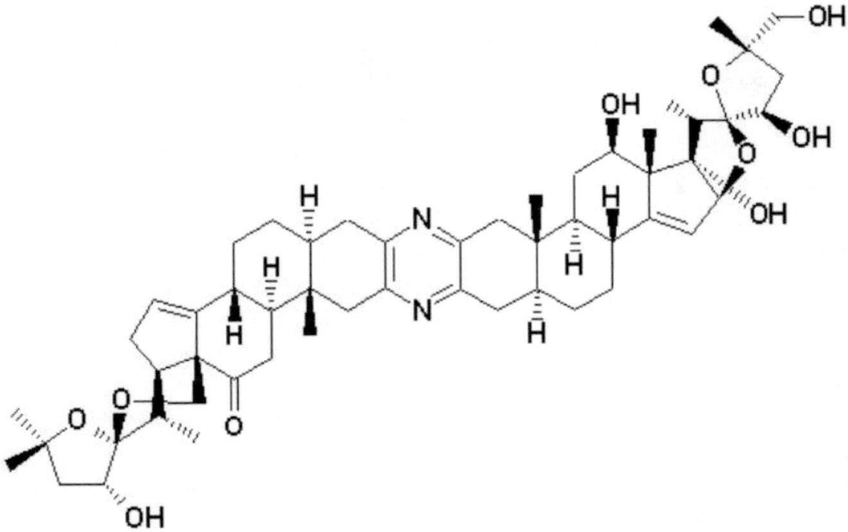

*Figure 1.* Chemical structure of cephalostatin 1

## 1.2 Apoptosis signaling

Cancer is characterized by an imbalance between cell proliferation and cell death[9]. In order to maintain proper tissue homeostasis cells possess intrinsic mechanisms of self destruction accounting for the so called programmed cell death or apoptosis. Defects in apoptosis inducing pathways may lead to neoplastic transformation. Moreover, since chemotherapeutic drugs induce death in malignant cells by triggering apoptosis, defects in apoptosis signaling pathways may lead to resistance in chemotherapy[10,11].

A current view of stress- or chemotherapy-induced apoptosis places mitochondria into the center of the apoptosis signaling network since in most forms of apoptosis the induction of mitochondrial membrane permeabilization (MMP) defines the point of no return. However, other organelles or cellular compartments (e.g. the endoplasmatic reticulum (ER), the cytoskeleton, lysosomes, and the nucleus) may contribute to apoptosis signaling by sensing damage or integrating pro-apoptotic signals[12].

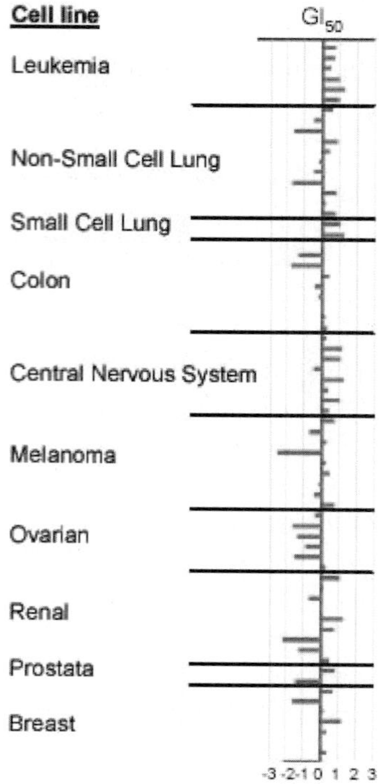


*Figure 2.* Fingerprint of the cytotoxic profile of cephalostatin 1 as evaluated in the NCI-60 screen. $GI_{50}$, 50% growth inhibition. The zero value represents the mean $GI_{50}$ obtained from all cell lines tested. Negative bars, cells are less sensitive, positive bars, cells are more sensitive towards cephalostatin 1 compared to the mean value. Adapted from[13]

MMP as a central event in chemotherapy-induced apoptosis is tightly controlled by members of the Bcl-2 family[14]. Pro-apoptotic members of the Bcl-2 family comprise the multidomain Bax family (Bax, Bak, Bok) and the BH3-only family (Bid, Bim, Bad, Bmf, Noxa, Puma etc.). Pro-survival members are represented by Bcl-2, its close homologous Bcl-xL and Bcl-w, A1, Mcl-1. The pro-apoptotic BH3-only members appear to be involved in death initiation upstream of the Bax family proteins since they are unable to kill in the absence of Bax and Bak. Either Bax or Bak seem to be necessary for apoptosis in many cell types. Both proteins are considered to function mainly at the mitochondrion by the formation of pores triggering MMP and the release of death-inducing molecules from the mitochondrial intermembrane space[14-16].

Pro-apoptotic molecules released from mitochondria as a consequence of MMP are cytochrome c, Smac/DIABLO (second mitochondria-derived activator of caspases/direct inhibitor of apoptosis-binding protein with a low isoelectric point); Omi/Htra2 (high temperature requirement protein A2), and AIF (apoptosis-inducing factor). The classical apoptosis pathway is initiated by cytochrome c. It prompts the ATP-dependent assembly of cytochrome c, Apaf-1 and procaspase-9 into a holoenzyme complex called 'apoptosome', which leads to efficient processing and activation of the initiator caspase-9[17]. Initially activated caspase-9 activates downstream caspases, such as caspase-3, which results in DNA fragmentation and apoptosis. The proteins Smac/DIABLO or Omi/Htra2 act as additional mechanism to promote caspase activation by binding to and antagonizing IAPs (inhibitor of apoptosis proteins). Members of the IAP protein family can counteract both the activation and the activity of caspases. XIAP (X-linked inhibitor of apoptosis), cIAP1, cIAP2 and NAIP (neuronal apoptosis inhibitor protein) antagonize cell death by inhibiting caspase-3 and caspase-7 activity. XIAP, in addition, blocks both caspase-9 activity and activation. From all IAP proteins XIAP is the most efficient caspase inhibitor. Smac/DIABLO promotes caspase activation by displacing caspase-9 complexed with XIAP[18]. AIF once released translocates to the nucleus where it induces caspase-independent chromatin condensation.

## 1.3 Mechanisms of chemoresistance

The central role of mitochondria in drug-induced apoptosis implicate that alterations in mitochondria-associated signaling may play an important role in drug resistance. Mutations or modifications in Bcl-2 family protein expression can alter the response towards drugs in experimental systems. Moreover, clinical correlative studies suggest that high level expression of anti-apoptotic Bcl-2 proteins confers a chemoresistant phenotype on cancer cells, including AML, ALL, CLL, multiple myeloma, prostate carcinoma, malignant brain tumors, and neuroblastoma[19]. Several experimental studies provided support that IAPs inhibit apoptosis in response to chemotherapy. *In vitro*, XIAP, cIAP1, or cIAP2 suppressed apoptosis induced by cisplatin, cytarabine, TRAIL, staurosporine, or $\gamma$-irradiation. Increased IAP expression correlated with a poor therapy response in myeloid leukemia cells. Overexpression of an other IAP family protein, survivin, is associated with adverse prognosis in several tumors, such as neuroblastoma, AML, colon, lung and esophagus carcinoma[19,20]. Strategies to use Smac/DIABLO to overcome IAP-mediated inhibition of chemoresistance were already tested: a Smac-penetratin fusion peptide was shown to bind to XIAP and cIAP1 enhancing drug-induced caspase activation suggesting that it might be indeed able to overcome antiapoptotic effects of IAPs[21]. Jia et al.[22] demonstrated that transfection of Smac increased the sensitivity of leukemic cells to TRAIL- and UV-induced apoptosis and activation of caspases.

# 2. CEPHALOSTATIN 1-MEDIATED APOPTOSIS

New drugs that induce apoptosis by mechanisms differing from those of classical chemotherapeutic drugs might provide a chance to overcome chemoresistance. The unique differential cytotoxicity profile of cephalostatins obtained from the 60 cell-line *in vitro* screen of the NCI provided a good chance that cephalostatins may utilize new molecular mechanism to trigger cell death. Thus, cephalostatin 1 (1 μM)-induced mechanism of cell death induction was characterized using the well established Jurkat leukemia T cells (clone J16) as a model[23,24].

## 2.1 Morphology

The overall morphology of cephalostatin 1-induced cell death in leukemia Jurkat T cells corresponded very well to classical signs of apoptosis: cell shrinkage and increased cellular granularity as determined by flow cytometry due to the formation of apoptotic bodies. Hoechst staining of the nucleus and fluorescence microscopy revealed classical DNA fragmentation and chromatin condensation in response to cephalostatin 1 treatment. Analysis of death morphology by electron microscopy, however, indicated some differences to classical apoptotic organelle morphology (Figure 3): nuclei appeared condensed and fragmented and the nuclear membranes dilated (D, blue arrows). Surprisingly, the mitochondria in cephalostatin 1-treated cells appeared smaller with increased electron density compared to untreated cells suggesting a condensation of the mitochondrial matrix (C, arrows). The cristae appeared blurred and swollen (C, insert). The ER membranes are enlarged and vesicles which are presumably ER-derived accumulate in the cytoplasm (E, asterisks)[25]. An increased density of the mitochondrial matrix is at odds with the concept that mitochondrial swelling finally leads to rupture of the outer mitochondrial membrane and release of cytochrome c[26]. In addition to nuclei and mitochondria, the ER seemed to be affected by cephalostatin treatment.

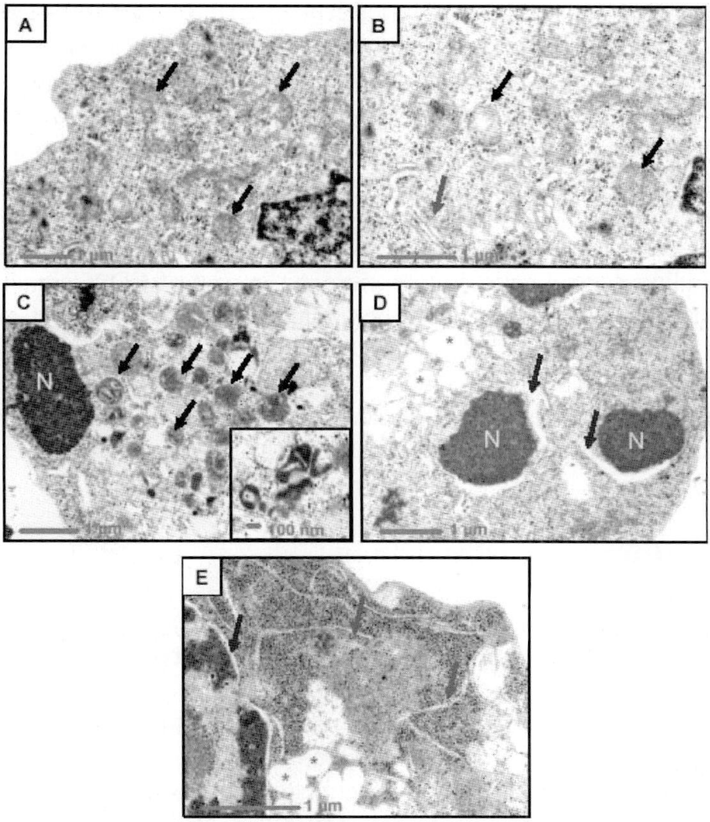


*Figure 3.* Electron microscopic analysis of cephalostatin 1-mediated mitochondrial changes. Cells were left untreated (A, B) or stimulated with cephalostatin 1 (1 μM, 16 h) (C, D, E). A, C, 1 : 5,000 fold magnification; B, D, E 1 : 10,000 fold magnification. Black arrows indicate mitochondria, blue arrows point to enlarged nuclear membranes and red arrows denote ER membranes. Asterisks demonstrate the vesicles that accumulate in the cytoplasm upon cephalostatin 1 treatment. N, nucleus[25].

## 2.2 Biochemistry

Cephalostatin 1 triggered phosphatidylserine translocation to the outside of the plasma membrane which is an early sign of apoptosis. Phosphatidylserine is an aminophospholipid normally located in the cytosolic layer of the plasma membrane and is translocated to the outside of the membrane by scramblases under apoptotic conditions. Once outside, it serves as "eat me" signal for phagocytes and other neighbouring cells thus triggering removal of the apoptotic cells.

Cephalostatin 1 treatment of Jurkat leukemia T cells led to processing of caspase-8, caspase-9 and caspase-3. Activation of caspases was essential in cephalostatin 1-induced apoptosis since the pan-caspase inhibitor zVAD.fmk (25 μM) was able to completely block DNA fragmentation. Cephalostatin 1 activated caspase-8 did not act as initiator caspase upstream of mitochondria based on the following observations: a) caspase-8 processing occurred significantly later (16 h) than caspase-3 activation (8 h); b) Jurkat caspase-8 knockout cells responded equally well towards cephalostatin 1 than the corresponding control cells. Furthermore, an impact of CD95 signaling that is mediated via caspase-8[23] could be excluded in cephalostatin 1-mediated apoptosis: Jurkat cells deficient in CD95 (Jurkat R)[27] showed no difference in their reactivity towards cephalostatin 1 compared to control cells. Finally, any contribution of caspases upstream of mitochondria could be excluded since the pan-caspase inhibitor zVAD.fmk (25 μM) had no impact on the release of the mitochondrial protein Smac/DIABLO.

## 2.3 Mitochondrial signaling

The specific morphology of Jurkat T cell mitochondria in response to cephalostain 1 implicated a particular mitochondrial signaling of cephalostatin. Indeed, although cephalostatin 1 led to dissipation of the mitochondrial membrane potential (increasing from 8 to 24 h) as visualized by staining with the mitochondria-targeted dye JC-1[28], cytochrome c was not released from the mitochondrial intermembrane space. Thus, obviously cephalostatin 1 induced inner but not outer mitochondrial membrane permeabilization (MMP). Moreover, also AIF was not released from mitochondria into the cytosol. Surprisingly, Smac/DIABLO was heavily unleashed already after 2 h of cephalostatin treatment. This result obtained by western blot analysis of mitochondrial as well as cytosolic fractions was confirmed by confocal microscopy. This particular selectivity for the release of Smac/DIABLO from mitochondria was not restricted to cephalostatin 1. The structurally strongly related cephalostatin 2 showed the same behavior[29].

In order to clarify whether Smac/DIABLO is necessary for cephalostatin 1-induced apoptosis, Jurkat T cells overexpressing the anti-apoptotic Bcl-xL protein were employed[30]. Overexpression of Bcl-xL strongly delayed Smac/DIABLO release from mitochondria by about 14 h. This delay was also seen in cephalostatin 1-induced DNA fragmentation suggesting that, indeed, Smac/DIABLO is necessary for cephalostatin 1-triggered apoptosis. Inner MMP was not affected by overexpression of Bcl-xL probably indicating that inner MMP is just a secondary event within the cephalostatin 1-induced signaling cascade.

Interestingly, overexpression of Bcl-2 was by far not as effective in delaying cephalostatin 1-mediated DNA fragmentation as Bcl-xL. This was found to be due to inactivation of Bcl-2 by cephalostatin-induced Bcl-2 phosphorylation[25].

Smac/DIABLO is reported to act by sequestering inhibitors of apoptosis (see 1.2). Indeed we found, that Smac/DIABLO released in response to cephalostatin 1 binds to XIAP as shown by immunoprecipitation and western blot analysis. Consequently, caspase-9 was found to be released from XIAP. This suggests that Smac/DIABLO released by cephalostatin 1 may aid in the activation of caspase-9.

Nevertheless, this leaves the question of how caspase-9 is activated in the absence of cytochrome c. Caspase-9 is activated by its recruitment into the apoptosome. Apoptosome formation usually requires mitochondria-derived cytochrome c, the cytosolic factor Apaf-1, ATP and procaspase-9 (see 1.2). Therefore, we tested, whether caspase-9 activation by cephalostatin 1 treatment involves an association of caspase-9 with Apaf-1. As demonstrated by immunoprecipitation and western blot, cephalostatin 1 does not induce a binding between caspase-9 and Apaf-1 suggesting that caspase-9 activation in response to cephalostatin occurs independently of the formation of an apoptosome.

This raises the question as to whether caspase-9 acts as crucial caspase initiating cephalostatin 1-mediated apoptosis. Employment of the caspase-9 inhibitor zLEHDfmk (50 μM) followed by cephalostatin 1 treatment showed that caspase-3 activity was abrogated indicating that caspase-3 is indeed activated by caspase-9. However, pretreatment with the inhibitor led to an inhibition of etoposide-induced apoptosis but only to a slight decrease in DNA fragmentation mediated by cephalostatin 1. This indicates that caspase-9 contributes to but is not essential for cephalostatin 1-triggered apoptosis.

These results provide the current view we have of cephalostatin 1-induced apoptosis signaling, depicted in Figure 4.

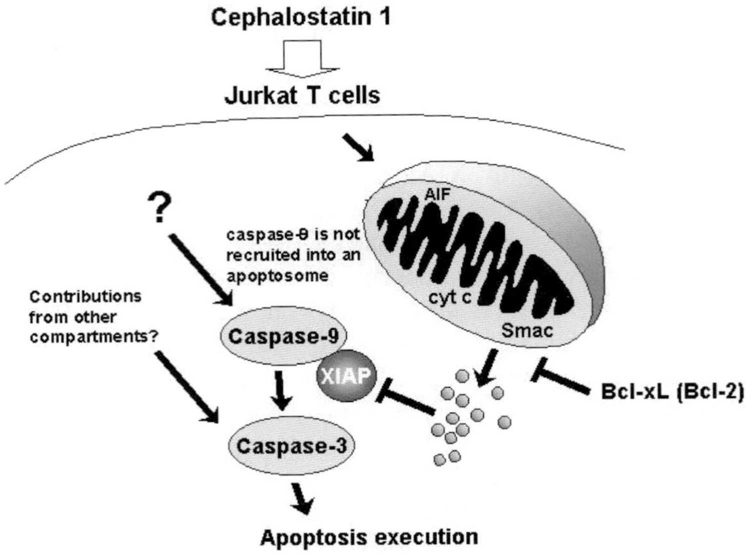


*Figure 4.* Scheme of cephalostatin-mediated apoptosis signaling in Jurkat leukemia T cells

# 3. DISCUSSION AND OUTLOOK

The most prominent feature of cephalostatin 1-mediated apoptosis signaling is the triggering of selective Smac/DIABLO release from mitochondria. The release was found to occur independently of caspase activation. The molecular players, however, contributing to Smac/DIABLO liberation are not yet identified. A conceivable contribution may come from JNK activation which is activated early in cephalostatin 1-induced apoptosis[25]. JNK was shown to produce a distinct Bid cleavage product in TNFα-induced apoptosis (jBid) which is able to induce selective Smac release without concomitant cytochrome c leakage[31]. Truncated Bid (tBid) cross-linking caspase-8 activation to mitochondrial activation has been shown to induce a conformational change of Bax and Bak whereupon these trigger the release of proteins from the mitochondrial intermembrane space. Obviously, jBid effects Smac release in a different manner since concomitant cytochrome c release was not detected. It is conceivable that different cleavage products of Bid lead to another form of conformational change in Bax and Bak which might result in different pore formation.

A further important observation was that cephalostatin 1 activates caspase-9 without the formation of an apoptosome. Recent publications propose that under certain conditions, Smac/DIABLO may be sufficient to initiate a caspase-9-dependent apoptotic pathway in the absence of cytochrome c and an apoptosome formation[32,33]. However, caspase-9 inhibitor studies indicated that caspase-9 may not be the essential initiator of the cephalostatin 1-induced caspase cascade. Further experiments using caspase-9 knockout cells will show whether cephalostatin 1 is able to trigger apoptosis in the absence of caspase-9. If this is the case, cephalostatin may be most useful in cancers where the formation of an apoptosome and subsequent caspase-9 activation is impaired. Right now studies are under way using carcinoma cells with defined defects in mitochondrial signaling.

Although overexpression of Bcl-xL was found to delay both apoptosis and Smac/DIABLO release, at present, it can not be excluded that besides Smac/DIABLO other signaling molecules from cellular compartments sequestering Bcl-xL, such as the ER contribute to caspase activation. This view is fuelled by the fact that caspase-9 seems not to be the initiator caspase in cephalostatin 1-triggered apoptosis. Thus, the initiator signaling may come from a compartment other than the mitochondria and Smac/DIABLO will allow full caspase activation by replacing XIAP from caspase-3 and/or caspase-7[18].

Overall, these first results characterizing the apoptosis-inducing mechanism of cephalostatin 1 clearly indicate a unique apoptosis signaling triggered by this marine bis-steroidal compound. Further studies will show whether cephalostatin will prove to be valuable in either treating chemoresistant tumors or in sensitizing tumors against chemotherapy. In either case, we have identified a tool that will allow us further insight into apoptosis signaling.

## REFERENCES


1. D.J. Newman, G.M. Cragg, and K.M. Snader, Natural products as sources of new drugs over the period 1981-2002, *J. Nat. Prod.* **66**, 1022-1037 (2003).
2. B. Haefner, Drugs from the deep: marine natural products as drug candidates, *Drug Discov. Today* **8**, 536-544 (2003).
3. M.R. Grever, S.A. Schepartz, and B.A: Chabner, The National Cancer Institute: cancer drug discovery and development program, *Semin. Oncol.* **19**, 622-638 (1992).
4. A. Monks, D.A. Scudiero, G.S. Johnson, K.D. Paull, and E.A: Sausville, The NCI anti-cancer drug screen: a smart screen to identify effectors of novel targets. *Anticancer Drug Des.* **12**, 553-541 (1997).
5. G.R. Pettit, R. Tan, J.-P. Xu, Y. Ichihara, M.D. Williams, and M.R. Boyd, Antineoplastic agents. 398. Isolation and structure elucidation of cephalostatins 18 and 19, *J. Nat. Prod.* **61**, 955-958 (1998).

6. T.G. LaCour, C. Guo, S. Ma, J.U. Jeong, M.R. Boyd, S. Matsunaga, N. Fusetani, and P.L. Fuchs, On topography and functionality in the B-D rings of cephalsostatin cytotoxins. *Bioorg. Med. Chem. Lett.* **9**, 2587-2592 (1999).
7. M.R. Boyd, The NCI in vitro anticancer drug discovery screen. Concept, implementation, and operation, 1985-1995, in: *Anticancer Drug Development Guide – Preclinical screening, clinical trials, and approval*, B. A. Teicher, ed., Humana Press, Totowa, pp.23-42 (1997).
8. G.R. Pettit, Antineoplastic agents. 317. Marine animal and terrestrial plant anticancer constituents, *Pure & Appl. Chem.* **66**, 2271-2281 (1994).
9. D. Hanahan, and R.A. Weinberg, The hallmarks of cancer, *Cell* **100**, 57-70 (2000).
10. S.H. Kaufmann, and W.C: Earnshaw, Induction of apoptosis by cancer chemotherapy, *Exp. Cell. Res.* **256**, 42-49 (2000).
11. F.H. Igney, and P.H. Krammer, Death and anti-death: tumour resistance to apoptosis,. *Nat. Rev. Cancer* **2**: 277-288 (2002).
12. K.F. Ferri, and G. Kroemer, G., Organelle-specific initiation of cell death pathways. *Nat. Cell Biol.* **3**, E255-E263 (2001).
13. http://dtp.nci.nih.gov/dtpstandard/dwindex/index.jsp
14. S. Cory, and J.M-Adams, The Bcl-2 family: regulators of the cellular life-or-death switch,. *Nat Rev. Cancer* **2**, 647-656 (2002).
15. L. Coultas, and A. Strasser, The role of the Bcl-2 protein family in cancer, *Semin. Cancer Biol.* **13**, 115-123 (2003).
16. M. Jäättelä, Multiple cell death pathways as regulators of tumour initiation and progression, *Oncogene* **23**, 2746-2756 (2004).
17. I. Herr, and K.-M. Debatin, Cellular stress response and apoptosis in cancer therapy. *Blood* **98**, 2603-2614 (2001).
18. X. Saelens, N. Festjens, L. Vande Walle, M. van Gurp, G. van Loo, and P. Vandenabeele, Toxic proteins released from mitochondria in cell death, *Oncogene* **23**, 2861-2874 (2004).
19. K.-M. Debatin, Apoptosis pathways in cancer and cancer therapy, *Cancer Immunol Immunother.* **53**, 153-159 (2004).
20. G.S. Salvesen, and C.S: Duckett, IAP proteins: blocking the road to death's door, *Nat. Rev. Mol. Cell Biol.* **3**, 401-410 (2002).
21. C.R. Arnt, M.V. Chiorean, M.P. Heldebrant, G.J. Gores, and S.H. Kaufmann, S.H., Synthetic Smac/DIABLO peptides enhance the effects of chemotherapeutic agents by binding to XIAP and cIAP1 in situ, *J. Biol. Chem.* **277**, 44236-44243 (2002).
22. L. Jia, Y. Patwari, S.M Kelsey, S.M. Srinivasula, S.G. Agrawal, E.S. Alnemri, and A.C. Newland, Role of Smac in human leukaemic cell apoptosis and proliferation, *Oncogene* **22**, 1589-1599 (2003).
23. Scaffidi, C., Kirchhoff, S., Krammer, P.H., and Peter, M.E., 1999, Apoptosis signaling in lymphocytes, *Curr. Opin. Immunol.* **11**: 277-85.
24. V.M. Dirsch, I.M. Müller, S.T. Eichhorst, G.R. Pettit, Y. Kamano, M. Inoue, J.-P. Xu, Y. Ichihara, G. Wanner, and A.M. Vollmar, Cephalostatin 1 selectively triggers the releaseof Smac/DAIBLO and subsequent apoptosis that is characterized by an increased density of the mitochondrial matrix, *Cancer Res.* **63**, 8869-8876 (2003).
25. I.M. Müller, Characterisation of apoptosis signal transduction induced by the marine compound cephalostatin 1 in leukemic cells, Thesis, University of Munich (2004) (http://edoc.ub.uni-muenchen.de/archive/00002061/).
26. M.O. Hengartner, The biochemistry of apoptosis, *Nature* **407**, 770-776 (2000).

27. M.E. Peter, J. Dhein, A. Ehret, S. Hellbardt, H. Walczak, G. Moldenhauer, and P.H. Krammer, APO-1 (CD95)-dependent and -independent antigen receptor-induced apoptosis in human T and B cell lines. *Int. Immunol.*, **7**, 1873-1877 (1995).
28. P. Bernardi, L. Scorrano, R. Colonna, V. Petronilli, and F. Di Lisa, F. Mitochondria and cell death. Mechanistic aspects and methodological issues. *Eur. J. Biochem.*, **264**, 687-701 (1999).
29. A. Rudy, V.M. Dirsch, and A.M. Vollmar, unpublished results (2004).
30. H. Walczak, A. Bouchon, H. Stahl, and P.H. Krammer, Tumor necrosis factor-related apoptosis-inducing ligand retains its apoptosis-inducing capacity on Bcl-2- or Bcl-$x_L$-overexpressing chemotherapy-resistant tumor cells, *Cancer Res.*, **60**, 3051-3057 (2000).
31. Y. Deng, X. Ren, L. Yang, Y. Lin, and X. Wu , A JNK-dependent pathway is required for TNFα-induced apoptosis, *Cell* **115**, 61-70 (2003).
32. D. Chauhan, T. Hideshima, S. Rosen, J.C. Reed, S. Kharbanda, and K.C. Anderson, Apaf-1/Cytochrome c-independent induction of apoptosis in muliple myeloma (MM) cells, *J. Biol. Chem.* **276**: 24453-24456 (2001).
33. I.A. McNeish, S. Bell, T. McKay, T. Tenev, M. Marani, and N.R. Lemoine, Expression of Smac/DIABLO in ovarian carcinoma cells induces apoptosis via a caspase-9-mediated pathway, *Exp. Cell Res.* **286**, 186-198 (2003).

Chapter 10

# HAMLET; A NOVEL TOOL TO IDENTIFY APOPTOTIC PATHWAYS IN TUMOR CELLS

Caroline Düringer, Oskar Hallgren, Lotta Gustafsson, Jenny Petterson, Ann-Kristin Mossberg, Taras Manilov and Catharina Svanborg
*Department of Microbiology, Immunology and Glycobiology (MIG), Institute of Laboratory Medicine, Lund University, Sölvegatan 23, SE-223 62 Lund, Sweden*

Abstract: Tumor cells often carry mutations in genes that control cell survival, and become resistant to signals that trigger cell death. Yet, some cell death pathways remain intact in tumor cells. If identified, these pathways might be exploited to selectively remove tumor cells. HAMLET (human α-lactalbumin made lethal to tumor cells) is a protein-lipid complex derived from human milk that activates cell death programs in tumor cells but not in healthy differentiated cells. We use HAMLET as a tool to identify apoptosis and apoptosis-like cell death mechanisms in tumor cells and to understand if these mechanisms differ between tumor and healthy cells. HAMLET interacts with the cell surface, translocates into the cytoplasm and accumulates in cell nuclei, where it disrupts the chromatin. Recent *in vivo* studies have shown that HAMLET maintains the tumoricidal activity in glioblastoma, papilloma and bladder cancer models, with no significant side effects. The results suggest that HAMLET should be explored as a new therapeutic agent with selectivity for the tumor and with little toxicity for adjacent healthy tissue. Such therapies are a much-needed complement to conventional treatments, to reduce the side effects and improve the selectivity.

Key words: HAMLET; lactalbumin; cancer; apoptosis; apoptosis-like; programmed cell death; alpha- histone; p53; bcl-2; caspase; glioblastoma; HPV

## 1. INTRODUCTION

Programmed cell death (PCD) is the mirror image of proliferation and is crucial to remove ageing cells during tissue development. In addition, cells



*M. Sluyser (ed.), Application of Apoptosis to Cancer Treatment,* 223-246.

that are redundant, seriously damaged or dangerous may be removed by PCD without disrupting tissue function. During the last decade, classical apoptosis has been defined as a process driven largely by mitochondrial perturbation and involving the activation of an effector protease (caspase) cascade. More recently, several alternative pathways of programmed cell death have been studied and as a result, the apoptosis concept has become more diversified. HAMLET-induced cell death shows features of classical apoptosis but many aspects of the cellular response to HAMLET are unconventional, suggesting that HAMLET activates PCD in tumor cells through other death pathways. This review focuses on mechanisms of HAMLET induced cell death and the use of HAMLET to remove tumor cells *in vivo*.

Naturally occurring cell death in developing tissues was first reported in 1842 by Vogt and in 1885 Flemming described the morphology of cells undergoing ''chromatolysis'' (1). In 1972, Kerr, Wyllie and Currie(2) coined the term apoptosis based on morphological criteria (see refs 3 and 4 for details on the history of cell death). The apoptotic cells showed shrinkage, membrane blebbing, nuclear condensation, chromatin aggregation, detachment and the formation of membrane-bound vesicles called apoptotic bodies. The apoptotic cells or cell residues were rapidly recognized and digested by neighbouring cells and tissue macrophages and no inflammatory response occurred(2, 3). Later, specific biochemical responses were associated with apoptosis such as degradation of DNA into fragments of discrete sizes(3, 4) and exposure of phagocytic markers(5). As the molecular mechanisms have been elucidated, the term apoptosis is used to describe a defined molecular system for cell death activation, regulation and execution, involving specific groups of proteins. Horvitz and colleagues showed that apoptosis is a highly regulated and controlled genetic suicide program in the nematode *C. elegans* (6) and homologues of *C. elegans* apoptotic proteins have been identified in humans even though the mammalian cell death program is more complex. PCD is defined as an active process that depends on signals or activities within the dying cell (7, 8) but all cells undergoing PCD do not fulfil the criteria for classical apoptosis. Cell death is now seen as a continuum from classical apoptosis to necrosis-like death. Jäättelä and colleagues(7, 9) classify variants of PCD based on the morphology and degree of chromatin condensation (Figure 1) with classical apoptosis resulting in fragmented, compact chromatin of discrete and simple geometrical shapes (Figure 1A). The nuclear changes are accompanied by cell shrinkage, cytoplasmic blebbing, apoptotic body formation and exposure of phagocytic markers. Apoptosis-like PCD is characterised by fragmented and condensed but less compact chromatin of more irregular shapes (Figure 1B). Any combination of other apoptotic features may be present, but the cells express phagocytic markers prior to lysis. Necrosis-like PCD occurs with very little chromatin condensation and the cell may lyse before phagocytic markers are displayed (Figure 1C). Most dying cells activate multiple signalling pathways and the resulting mode of cell death is the sum of these pathways. Initiation signals originate from the outside of the cell or are created by intracellular disturbances. These signals propagate through the cell and converge on the mitochondria which control the response through the balance between pro- and anti-death molecules in the mitochondrial outer membrane(10). If death signals predominate, execution of the cell death program is initiated. If not, the cell is rescued. The cancer cell may thus achieve resistance to cell death stimuli at many different stages in the process.

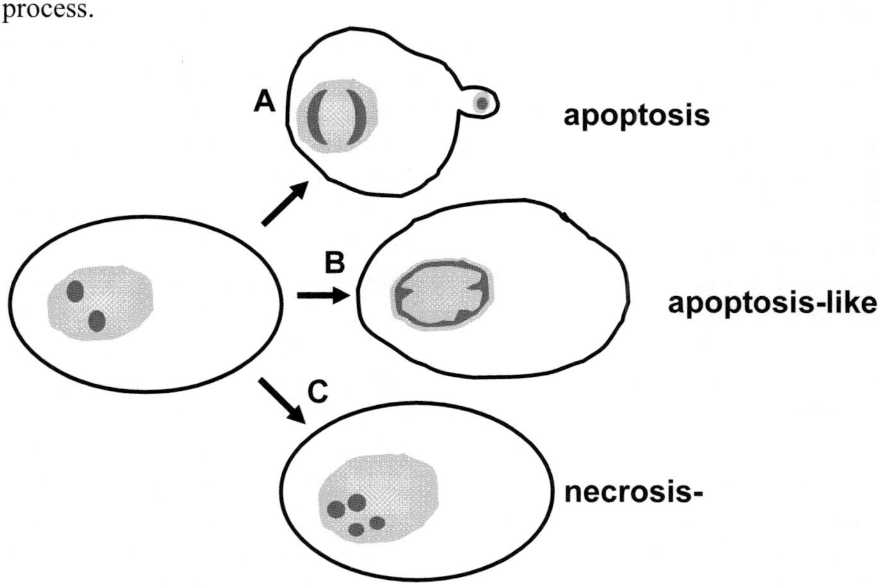


***Figure** 1*. Classification of PCD based on morphology and degree of condensation of chromatin. A. Apoptosis results in compact chromatin condensed to discrete simple geometrical shapes. **B.** Apoptosis-like PCD is accompanied by less compact and more complicated structures. **C.** Necrosis-like PCD occurs with no or very little chromatin condensation(7, 9). HAMLET causes both A and B.

## 2. HAMLET

Complex formed by human α-lactalbumin and oleic acid (Figure 2) that induces tumor cell death but leaves healthy, differentiated cells unaffected(11, 12). The activity of HAMLET was discovered by serendipity while using breast milk fractions to inhibit bacterial adherence to a lung carcinoma line. In addition to blocking adherence, one milk fraction killed the cells(11). The active molecular complex was shown to contain α-lactalbumin(12). This globular 14.2 kDa protein has four α-helices and a triple-stranded anti-parallel β-sheet and a high affinity calcium binding site(13). Calcium binding is required for the protein to fold into the native conformation(14). α-Lactalbumin is the most abundant protein in human milk and functions as a coenzyme in lactose synthesis(15). The native form of α-lactalbumin had no apoptotic effect, however but the milk fraction had been exposed to low pH, which causes a change in the tertiary conformation of α-lactalbumin(12). HAMLET was shown to consist of α-lactalbumin in a partially unfolded state characterised by native-like secondary structure and fluctuating tertiary packing(13, 16). As partially unfolded states of α-lactalbumin are not normally stable at physiological conditions, we assumed that the active complex must contain a cofactor that stabilised the altered conformation. .The cofactor was identified as oleic acid (C18:1 9cis). Other lipids, with exception for C18:1 11 c, did not form such complexes with α-lactalbumin(17).

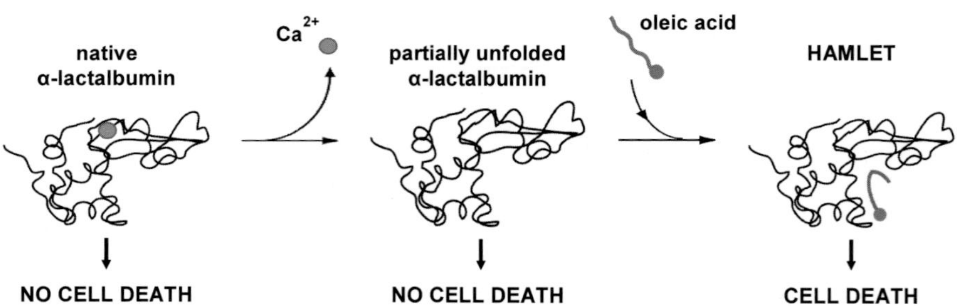


*Figure 2.* The fatty acid component is important for the biological effect of HAMLET. Both the conformational change in alpha-lactalbumin induced by the removal of calcium and the specific association of oleic acid is required for the cell death-inducing effect of HAMLET.

Conversion to the cell death-inducing form was achieved when α-lactalbumin in its calcium-free state was combined with oleic acid and HAMLET was defined as the conversion product of human apo α-lactalbumin and oleic acid(18).

## 2.1 Cellular effects of HAMLET

The cellular spectrum of HAMLET is very broad and to date, more than 40 tumor cell lines and tissues have been found to undergo PCD (19). Based on the sensitivity to HAMLET, cells can be placed in three main categories. Lymphoid cells are the most sensitive carcinomas are intermediately sensitive and healthy cells are resistant.

The sub-cellular distribution of HAMLET was investigated to form hypotheses about the mechanism of tumor cell death. Initially, biotinylated protein was used in confocal microscopy and radio-labelled protein was used in sub-cellular fractionation(20). More recently, the sub-cellular distribution has been reassessed using Alexa-flour labelled HAMLET in real-time confocal microscopy (Figure 3). HAMLET binds with similar efficiency to the surface of both healthy cells and tumor cells, but enters the cytoplasm of the tumor cells and travels to the nucleus where it accumulates. Healthy cells, in contrast, do not internalise the protein and only small amounts of Alexa-HAMLET were detected in the cytoplasm. Finally, HAMLET does not accumulate in the nuclei in healthy cells. We conclude that tumor cells take up larger amounts of HAMLET than healthy cells, and that the protein complex travels through the cytoplasm to cell nuclei(20). We have used the difference in sub-cellular localisation between tumor cells and healthy cells as a basis to explore the cellular targets for HAMLET in tumor cells.

## 2.2 Initiation pathways

Cell death can be initiated by a variety of external stimuli, and the effect of HAMLET has been compared to these known PCD stimuli. HAMLET binds in a patchy, granular manner to both tumor cells and healthy cells, suggesting that the ability to bind *per se* is not crucial for the difference in sensitivity. Cell surface receptors for HAMLET have not been identified but it is likely that HAMLET recognises receptors formed by conserved cell surface molecules(21). A rapid and powerful pathway to PCD is induced by death receptor ligands, which activate death receptors (DRs) at the plasma membrane. The DRs are transmembrane proteins of the tumor necrosis factor (TNF) receptor super-

Family with an intracellular death domain (DD)(22, 23), which can couple the receptor either to a cell death cascade or to a kinase pathway that turns on gene expression. Fas (also called CD95 or Apo1) and TNFR1 (TNF receptor 1)(24) are stimulated by binding of Fas ligand (FasL or CD95L) to the extracellular DR domain(25) recruits adaptor proteins in the signal transduction cascade. Stimulation of the DRs may induce apoptotic or necrotic cell death.

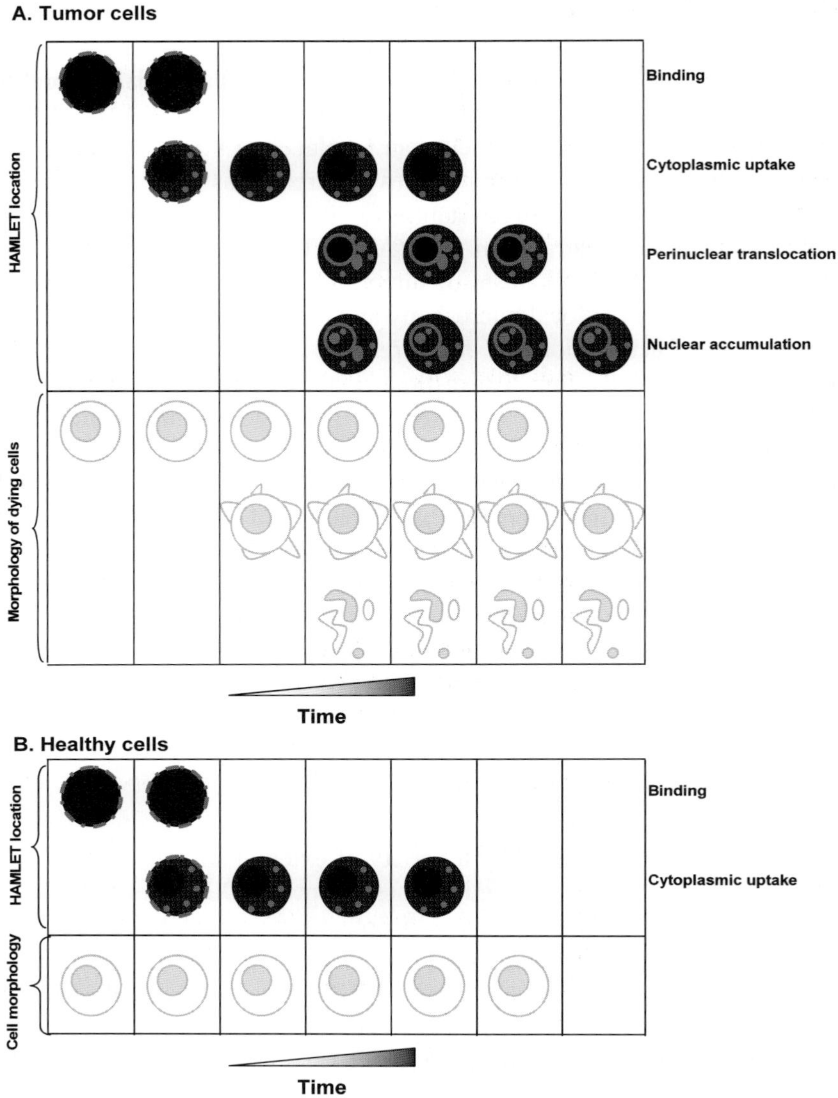


***Figure 3.*** Subcellular localisation of HAMLET. A. Tumor cells and B. Healthy cells

Apoptosis signalling from Fas is mediated by the caspase family of proteases through one of two major pathways depending on the cell type(26). Mitochondria and induces release of inter-membrane space proteins(27, 28). Alternatively, apoptotic cell death can be mediated through the recruitment of the protein Daxx to the DR and activation of the apoptosis signal-regulating kinase 1 (ASK1) and Jun N-terminal kinase (JNK) pathways(29, 30). In addition, stimulation of Fas and TNFR1 can lead to necrotic cell death through production of ROS(31).

The Fas DR pathway is not activated by HAMLET, as shown by the broad cellular spectrum, the caspase independence (see below) and by inhibition experiments using specific anti-CD95 antibodies(32). Type I cells directly activate the effector caspase-3 through caspase-8 and cell death proceeds independently of the Bcl-2 family members and the mitochondria. In type II cells, the signal is amplified through the mitochondria through proteolytic activation of Bid, which translocates to

## 2.3 HAMLET, mitochondria and the Bcl-2 family

The mitochondrial inter-membrane space contains several proteins of the cell death machinery. If released into the cytosol, these proteins participate in the execution of the cell death program and this marks the irreversible phase of death and must be carefully regulated. Mitochondrial outer membrane integrity is controlled by the Bcl-2 family of proteins, which share one or more of four Bcl-2 homology domains (33-35) (BH1 to BH4) (reviewed in (36)). Bcl-2 and Bcl-$X_L$ are death inhibitory proteins whereas the multidomain (BH1-3) proteins Bax and Bak and the "BH3-only" proteins and Bid, Bad, Bim, Noxa and Puma all promote death. The Bcl-2 family proteins are present both on intracellular membranes such as the ER and mitochondrial outer membrane and in the cytosol. The proposed function of Bcl-2 and Bcl-$X_L$ is to stabilise the membranes whereas Bax and its relatives act by perturbing the mitochondrial outer membrane to release inter-membrane space proteins. The BH3-only proteins sense cellular damage and death stimuli and act upstream by inhibiting Bcl-2 or activating Bax and Bak. The pro- and anti-death members regulate the activity of each other by forming inactive heterodimers and eventually, the ratio of pro- and anti-death members of the Bcl-2 family present at the mitochondrial membrane determines how the cell will respond to a death or survival stimulus (reviewed in (9, 36, 37)).

Several models for mitochondrial membrane permeabilisation have been proposed (reviewed in (37-39)). Based on the structural similarity between Bcl-$X_L$ and the pore-forming subunit of diphtheria toxin, it is suggested that Bax and Bak form specific pores in cooperation with Bid and mitochondrial lipids. Pore formation does not irreversibly perturb mitochondrial function, but triggers death pathways by releasing proteins and ions. A second model proposes permeability transition (PT) with loss of mitochondrial membrane potential by opening of the PT pore. PT can be induced by mitochondrial toxins, reactive oxygen species (ROS) and $Ca^{2+}$ released from the ER. PT pore opening results in osmotic swelling and selective rupture of the outer membrane and Bcl-2 proteins have been shown to interact with and regulate the activity of the PT pore. A third model describes destabilisation of the outer membrane by mitochondrial fission. Mitochondria are present in interphase as a tubular network, but in preparation for mitosis the network is fragmented to facilitate distribution to the daughter cells. Evidence supporting this model includes the presence of mitochondrial fission during PCD, co-localisation between Bax and fission proteins and resistance to PCD of fission protein knock out cells.

Some cells appear to undergo classical apoptosis in response to HAMLET with cell shrinkage, nuclear condensation and DNA fragmentation(11). Based on these observations, we investigated the role of mitochondria and ability of the Bcl-2 family to regulate the response to HAMLET. HAMLET was shown to co-localise with mitochondria in the cytoplasm and to bind isolated mitochondria, inducing membrane depolarisation, swelling and cytochrome c release in a calcium dependent fashion (32, 40). We have exposed cell lines differing in Bcl-2 expression to HAMLET, and monitored cytochrome c release, caspase activity and survival. Surprisingly, cell death was unrelated to the level of Bcl-2 expression, determined by Western blots, and over-expression of Bcl-2 or Bcl-$X_L$ in transfected cells did not save them from death following exposure to HAMLET, nor did it reduce the level of cytosolic cytochrome c or activated caspases(19).

## 2.4 HAMLET and caspases

Proteases involved in the execution of the cell death program are shown in Figure 4. The effector caspases are activated by cleavage by initiator caspases or other proteases and in most systems, the activation of a potent effector caspase cascade requires release of molecules from the mitochondrial intermembrane space. When released to the cytosol, cytochrome c associates with Apaf-1 and dATP to form the apoptosome (41, 42) which recruits and activates procaspase-9(43), which proceeds to activate caspase-3(44). Caspase-3 (supported by caspase-6 and -7) is responsible for producing several features of the apoptotic cell death morphology(45). The cytosolic inhibitors of apoptosis (IAPs) protein family (46) prevent unwanted caspase activation in the cytosol during normal cellular conditions and their inhibitory effect is neutralised by two other proteins released from the inter-membrane space, Smac/DIABLO (Second mitochondrial activator of caspases/direct inhibitor of apoptosis (IAP)-binding protein with low pI)(47, 48) and Omi/HtrA2(49, 50).

Caspase-3 executes the cell death program by proteolytically activating or inactivating important structural and functional cellular proteins. Nuclear fragmentation and condensation are the results of cleavage of DNA, lamin, topoisomerases, linker histones and other structural proteins in the nucleus(51-55). During apoptosis, chromatin is degraded to discrete fragments of 50-200 kbp (high molecular weight fragments) and internucleosomal fragments of approximately 200 bp(56), mediated by the caspase activated DNase (CAD) (57). In non-apoptotic cells, CAD exists in an inactive complex with its inhibitor ICAD. During apoptosis, caspase-3-mediated cleavage of ICAD releases CAD, allowing it to cleave chromatin. Active blebbing and detachment is mediated by the activation of p21-activated kinase(58, 59) and gelsolin(60), leading to disruption of the actin filament network. Efficient removal of the dying cell without leakage of cellular contents to the surrounding tissue is ensured by the exposure of phagocytic markers, of which the best described is the externalisation of the inner plasma membrane phospholipid phosphatidylserine(61).

The caspases play an important role also in PCD signalling and execution (reviewed in (45)). Initiator (caspase-2, -8, -9, -10 and -12) and effector (caspase-3, -6 and -7) caspases are present in the cell as inactive pro-enzymes and are activated in response to cell death stimuli.

Caspase 2, 3 and 9 activities were detected in HAMLET treated cells but the response to HAMLET was significantly lower than to other cell death stimuli. Cleavage of caspase substrates such as PARP, lamin B and α-fodrin has also been detected(32). Tumor cell death in response to HAMLET was shown to proceed even in the absence of caspase activity, however. The pan- caspase-inhibitor zVAD-fmk was shown to block inter-nucleosomal DNA fragmentation, but failed to rescue the cells from death (19). This suggested to us that tumor cells exposed to HAMLET activate caspase-independent PCD mechanisms. Subcellular fractionation was first used to identify cellular targets for HAMLET. HAMLET was shown to bind three low molecular weight proteins in nuclear extracts. These proteins were identified as histones H2B, H3 and H4 and the specificity of HAMLET for histones was confirmed using purified bovine histones in a far western assay, where HAMLET bound to H2B, H3 and H4, weakly to H2A and not at all to H1. In addition, HAMLET recognised several proteins in a membrane fraction, but there was no binding to the mitochondria-containing and cytosolic fractions. The lack of binding of HAMLET to proteins in mitochondrial extracts was interesting, as co-localisation studies had suggested such an association in cellular systems. The specificity for histones suggested that HAMLET might interact with chromatin in tumor cell nuclei. The eukaryotic genome consists of almost two meters of double stranded DNA and to fit in a cell nucleus of approximately 10 μm, DNA is packed into chromatin. Chromatin is a dynamic structure and its properties can be modulated by a variety of mechanisms, such as DNA methylation, histone modification or replacement with histone variants and binding of non-histone chromosomal proteins. The dynamic structure of chromatin allows both for compaction of DNA and local accessibility to proteins involved in essential processes such as DNA replication, transcription and repair (62). Adjacent nucleosomes are connected by linker DNA and the linker histone H1. The linker histones and the core histone tails cooperate in packaging of the nucleosome array into a 30 nm thick fibre(63, 64). Additional higher-order levels of packaging are then required to form functional states of chromatin through the phases of the cell cycle (reviewed in (65)). HAMLET was shown to interact with natively folded and biologically functional histones.

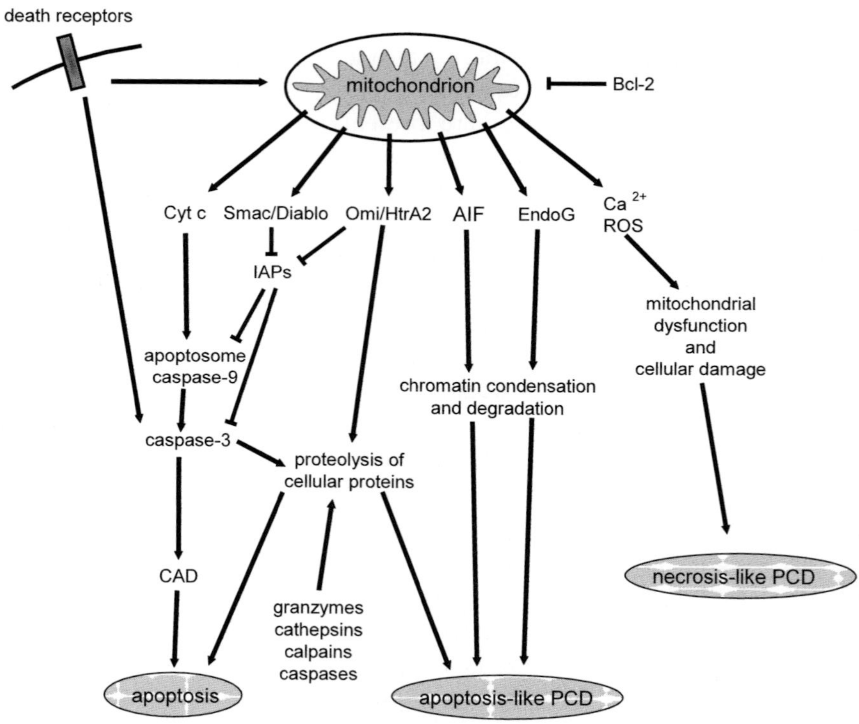


***Figure 4.*** Execution pathways. Most cell death stimuli converge on the mitochondrion where they initiate release of death-promoting molecules from the intermembrane space. These molecules can be divided into two groups, those that activate caspases directly or indirectly (cytochrome c and Smac/DIABLO) and those that function independent of caspases (AIF and Endonuclease G). Omi/HtrA2 is involved in both caspase-mediated death as an inhibitor of IAPs and in caspase-independent mechanisms through its serine protease function. There are

three general execution routes leading to the main PCD variants. Activation of the caspase cascade, either by the DISC or the apoptosome, leads to apoptotic PCD. Secondly, AIF and endonuclease G cooperate with non-caspase proteases to produce an apoptosis-like morphology which is caspase independent. Third, a functional defect of mitochondrion leads to ROS production and Ca2+ fluctuation and necrosis-like PCD. HAMLET activates a weak caspase response. Bcl-2 does not regulate this response (Hallgren et al. in preparation).

In affinity chromatography, HAMLET bound H2A, H2B, H3 and H4 and BIAcore assays showed rapid binding with very slow dissociation. Mixing of histones with HAMLET in solution resulted in precipitation of the proteins, suggesting a high affinity of the binding. Both denatured and native histones were precipitated by HAMLET, with a preference for H3 and H4.

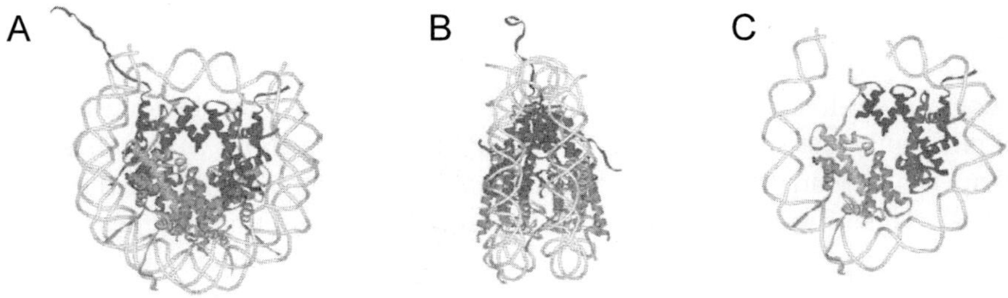


***Figure 5.*** Structure of the nucleosome core particel. H3 is shown in blue, H4 in green, H2A in yellow and H2B in pink. The complete core particle with 146 bp of DNA wrapped around a histone octamer. A. Front and B. side view. C. A half nucleosome core with only one copy of each histone and C-terminal part of H3 from the other half is shown to illustrate the histone-histone interactions in the octamer. HAMLET binds histones, nucleosomes and chromatin.

The *in vivo* relevance of these interactions was demonstrated in HeLa tumor cells expressing GFP-tagged histones. Alexa-HAMLET co-localised with the GFP-histones in cell nuclei and induced changes in the global chromatin structure. The chromatin was condensed to the nuclear periphery or to large, spherical structures. HAMLET was present in both of these chromatin patterns. The core histones are small, basic proteins that are highly conserved during evolution. They consist of two main structural regions, the histone tail(s) and the conserved histone-fold region, a tertiary structure of three α-helices joined by two loops shared by all the core histones(66, 67). The fold mediates both histone-histone and histone-DNA interactions in the nucleosome core. With the exception of H2A which has tails on both sides, the tails are amino terminal to the histone-fold region. They contain sites for post-translational modifications, which are important for the assembly of nucleosomes, transcriptional regulation and altering of chromatin structure and activation state during the cell cycle. Histone modifications have been proposed to constitute a ''histone code'' that is read by structural chromatin proteins and transcription factors(68). Several histone-binding proteins have been shown to act as chaperones in chromatin assembly and remodelling, histone storage and sperm decondensation(69, 70). In addition to inhibiting unspecific interactions between histones and DNA, the chaperones shuttle histones from their site of synthesis to sites of chromatin assembly (reviewed in (69)). The chaperone Asf1 interacts with other proteins involved in remodelling, transcription, cell cycle regulation and histone modification, suggesting that histone chaperones may regulate gene expression and cell growth (reviewed in (71)).

The histone binding properties of HAMLET were compared to that of a known histone chaperone, Nucleosome Assembly Protein 1 (NAP-1). We allowed NAP-1 or HAMLET to interact with histones and added short DNA fragments to the reaction. NAP-1 induced a concentration-dependent assembly of nucleosomes, whereas HAMLET completely prevented nucleosome formation. This shows that the affinity for histones differs significantly between HAMLET and physiological histone chaperones. The chaperones bind histones with affinities low enough to allow delivery to DNA. The affinity of HAMLET for histones apparently exceeds that of DNA and as a result, if HAMLET comes in contact with chromatin, it can disrupt the association between histones and DNA and thus perturb the structure and function of chromatin.

We conclude that HAMLET interacts with all structural and functional conformations of histones, from denatured proteins to natively folded soluble histones and histones in nucleosomes and this allows HAMLET to interfere with histone biosynthesis and function at several steps. In the nucleus, HAMLET could remove histones from DNA or directly bind nucleosomes and impair their function. Alternatively, the binding of HAMLET to chromatin could induce DNA damage. It has been observed that defects in chromatin assembly can lead to double strand DNA breaks and activation of the S-phase checkpoint. The mechanism behind this is not clear but it can be due to lack of stabilisation behind the replication fork(72). HAMLET may also interact with newly synthesised histones at the ribosome or with cytoplasmic histones associated with chaperones. As the affinity of HAMLET for histones appears to be higher than that of the chaperones, HAMLET could compete with chaperones for the histones and prevent their correct folding and transport to the nucleus. This would void the chromatin assembly machinery of histones and induce chromatin damage.

## 2.5 HAMLET-induced cell death is independent of p53

In other models, damage to the cell nucleus has been shown to trigger cell death programs (Figure 6). The major mediator of PCD signalling in the nucleus is the p53 tumor suppressor protein, a transcription factor situated at the centre of a highly connected network that controls cell growth and genome stability (reviewed in (73, 74)) and more than 50% of all human tumors lack functional p53(75). Inactivation of p53 occurs by mutations in the p53 gene itself or in other genes whose products interact with and control p53 activity. The level of p53 in the cell is mainly controlled by its degradation by the ubiquitine-proteasome system and by MDM2, that targets p53 for degradation(76). MDM2 is a transcriptional target for p53, resulting in a feedback loop in which p53 and MDM2 levels regulate each other. Double strand DNA breaks activate the checkpoint kinases ATM and Chk2, which phosphorylate the N-terminal tail of p53 and reduce its affinity for MDM2(77). Oncogene expression (Ras and Myc) increases the expression and stability of ARF, which binds MDM2 and sequesters it in the nucleolus(78). Chemotherapeutic drugs, UV radiation and protein kinase inhibitors activate ATR kinase and casein kinase II that also phosphorylate p53 and induce its dissociation from MDM2(79). In addition to protein stabilisation, posttranslational modifications are required for p53 activity. Phosphorylation, acetylation and ribosylation of the C-terminal domain activates the DNA binding site of p53, which is masked in the inactive protein, allowing for the transcriptional activities of p53(74). This activation can also be induced by antibodies or drugs(80), presenting new ways to restore p53 function in tumor cells. p53 protects cells and organisms against uncontrolled growth and genomic instability by inducing cell cycle arrest, DNA repair or cell death through both transcription-dependent and independent mechanisms(81). p53 promotes PCD at several stages by stimulating expression of Bax, Fas, ROS-producing enzymes and apoptosis protease activating factor 1 (Apaf-1) and reducing the expression of Bcl-2 (reviewed in (82)). In addition, p53 localises to mitochondria in response to cell death stimuli and promotes apoptosis(83). This effect is independent of protein expression and is suggested to depend on a direct interaction with Bcl-2 family proteins(84, 85). Other tumor suppressive effects of p53 include inhibition of angiogenesis and induction of cell differentiation(73).

As HAMLET targets the chromatin, we expected p53 to influence the death program. Surprisingly, however, HAMLET was shown to kill tumor cells regardless of their p53 status. Tumor cells with different p53 status did not differ in HAMLET sensitivity, as shown with cells with no, mutated or wild type p53. Furthermore, we found no relationship of HAMLET sensitivity to *p53* genotype in different tumor cells. Furthermore, *p53* deletions or gain of function mutations did not change the HAMLET sensitivity of tumor cell lines.

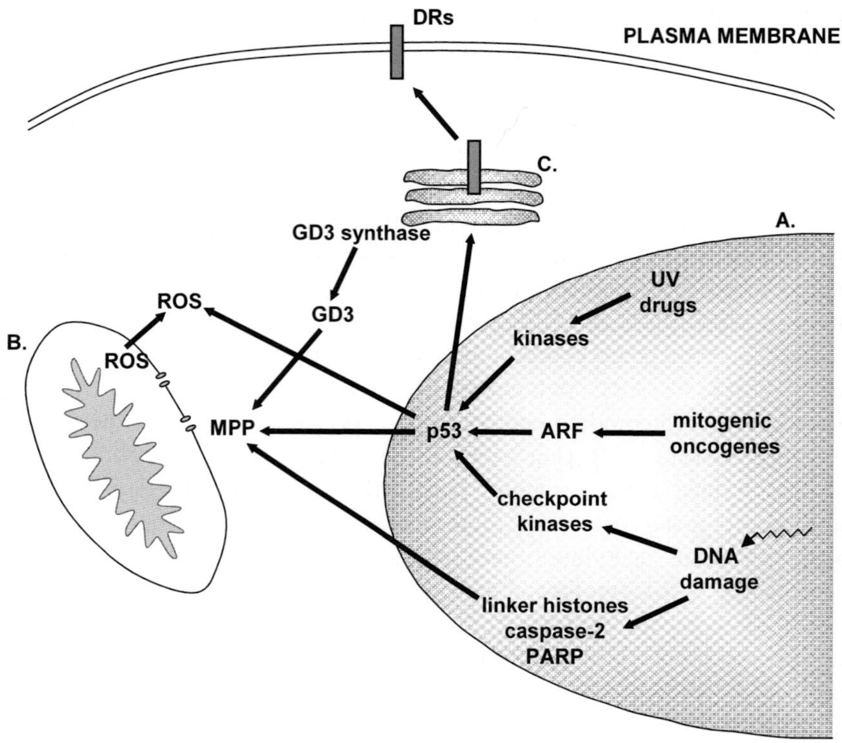


***Figure 6.*** Response to DNA damage. A. In the nucleus. B. Mitochondrial activation. C. Upregulation of death receptors. HAMLET disrupts the chromatin but cell death is independent of p53.

Other links between nuclear stress and PCD induction include activation of nuclear caspase-2 by low doses of the topoisomerase inhibitor etoposide. The activated caspase induces release of death promoting molecules from mitochondria(86). It is not known how caspase-2 is activated in the nucleus but it has been proposed that there is an adaptor complex in the nucleus similar to the DISC. DNA double strand breaks caused by X-ray irradiation induces release and mitochondrial translocation of the linker histone H1.2(87). H1.2 is released from the nucleus by a p53 dependent mechanism and induces release of mitochondrial intermembrane space molecules. Caspase-independent cell death in response to DNA damage can be mediated by the poly-(ADP-ribose) polymerase (PARP). PARP functions in a DNA damage surveillance network and facilitates DNA repair. Its functions in cell death are unclear, but it induces release of the apoptosis-inducing factor (AIF) from mitochondria in a caspase-independent manner(88). HAMLET induced cell death is independent of caspase 2, however. We conclude that the nuclear response to HAMLET does not involve p53, that caspase-2 may be activated but that HAMLET-induced cell death is caspase-2 independent.

## 2.6 Is there a nuclear pathway to HAMLET-induced cell death?

The difference in nuclear localisation between sensitive and resistant cells and the very high affinity of HAMLET for histones suggest that this interaction may play a role in tumor cell death. Disturbances of chromatin structure and function could activate cell death programs by several mechanisms. First, the interaction of HAMLET with chromatin could damage the DNA itself, but this is less likely, as mechanisms activated by DNA damage such as p53 and caspase activation are not important in HAMLET-induced cell death. Histone acetylases have been implicated in signalling to the cell death machinery in response to DNA damage(89) and should be explored._Second, HAMLET may trigger cell death directly by disturbing chromatin structure and function. Chromatin assembly is not only required for replication of the DNA, but is essential for DNA repair and transcription. By disturbing chromatin assembly, HAMLET may interfere with the metabolic machinery of the cell and induce a general stress response resulting in cell death. Alternatively, a specific surveillance system for chromatin integrity could transmit signals to cell death effectors. In yeast, an Asf1 deletion (histone chaperone) was shown to arrest cells in G2/M phase, followed by cell death with morphological features of both apoptosis and necrosis (90) but the human Asf-1 homologue has additional functions and it is difficult to extrapolate from yeast.

Regardless of mechanism, the HAMLET model suggests that molecules which disturb chromatin structure and function may trigger cell death. The broad effect presumes that HAMLET attacks fundamental functions of the tumor cell such as the chromatin assembly machinery. This would ensure the death of the targeted cell, independently of the status of the classical apoptotic machinery, whose key players are often inactivated in tumor cells. On the other hand, the nuclear pathway might function as a back up to ensure the death of all tumor cells, in concert with signals from the cell surface or in the cytoplasm.

# 3. HAMLET AND TUMOR CELL APOPTOSIS *IN VIVO*

## 3.1 Human skin papillomas

Human skin papillomas (HPVs) were chosen to perform the first HAMLET treatment in humans(91). Papillomas are pre-malignant lesions of the skin and mucosal surfaces(92), and treatments are often ineffective or destructive(93). The HPV vaccine development shows great promise as a prevention of HPV infection, but it is not therapeutic. In our study, HAMLET was applied topically once a day for three weeks in a double blind, placebo-controlled protocol (Figure 7). The lesions were measured and photographed once a week and a reduction in papilloma volume by >75% was considered significant. The results showed that HAMLET treatment reduced the papilloma volume in 100% (20/20) patients compared to 15% in the placebo group ($p<0.001$). No adverse reactions were reported, and there was no significant difference in efficacy between immuno-competent and immuno-suppressed patients. Results were confirmed by treatment of the placebo group.

We conclude that HAMLET retains its tumoricidal activity *in vivo,* and propose that topical HAMLET application should be explored as a new treatment against papillomas.

## 3.2 Local HAMLET treatment of brain tumors

Glioblastomas (GBMs) account for more than 60% of all primary brain tumors, and have the most unfavorable prognosis with a mean survival time of less than 1 year(94). The tumors grow invasively and are inaccessible to surgical removal due to their and diffuse infiltrating growth. GBM cells undergo apoptosis in response to HAMLET *in vitro,* but are insensitive to α-lactalbumin, and differentiated brain cells are resistant to this effect.

New experimental treatments of GBM are best investigated by xeno-transplantation of human glioma biopsies into the nude rat brain The xenografts show the infiltrative growth characteristic of human tumors(95). Prior to intracranial injection of the tumor cells, human tumor biopsies are allowed to form spheroids *in vitro* as an intermediate step to obtain standardized inocula of tumor cells. After xeno-transplantation, the rats develop pressure symptoms and large tumor masses can be detected by MRI scans.

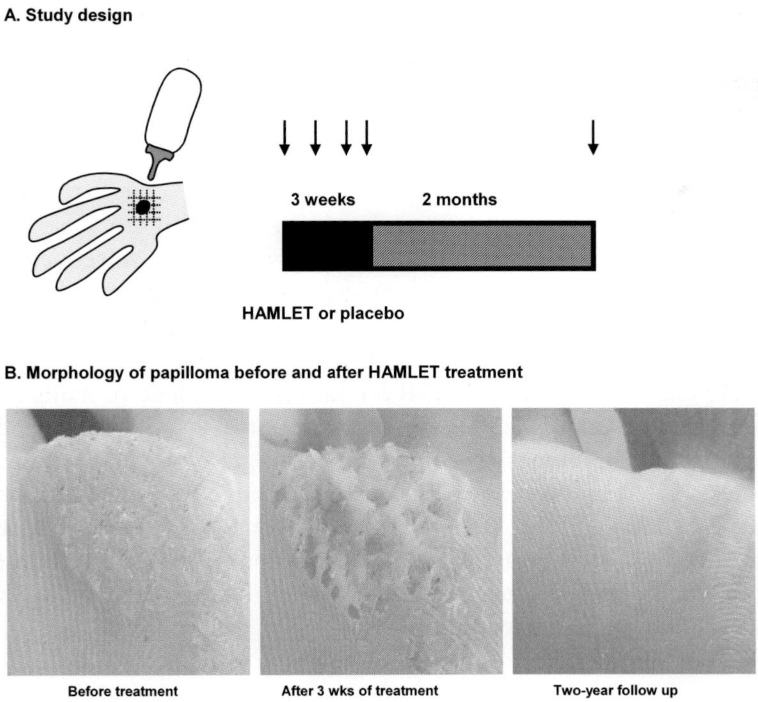


C. Treatment effect of HAMLET contra placebo (Number of patients)

| | Effect | No effect | Total |
|---|---|---|---|
| HAMLET | 20 | 0 | 20 |
| Placebo | 3 | 17 | 20 |

Effect = a mean volume decrease of ≥75%, $p < 0.001$, after 3 weeks of treatment

***Figure 7.*** HAMLET treatment of human skin papillomas. A. Placebo controlled study for 3 weeks with 2 months and 2 years follow up. B. Skin papillomas before and after treatment. C. Outcome of HAMLET versus placebo. (data summerised from New England Journal of medicine(91) )

The effect of HAMLET on tumor tissue was investigated in this model. HAMLET was then administered by convection-enhanced delivery with α-lactalbumin as a control. HAMLET was shown to inhibit tumor development as shown by a delayed onset of pressure symptoms. Rats receiving α-lactalbumin developed symptoms significantly earlier than the HAMLET treated animals ($p < 0.001$). The TUNEL assay showed extensive apoptosis in the tumor but the surrounding tissue did not show TUNEL-labeling.

Furthermore, the infusion of HAMLET did not harm the normal brain and did not produce any neurological symptoms. We conclude that HAMLET has the potential to act as a selective inducer of apoptosis in patients with malignant gliomas.

## 4. CONCLUSIONS

New cancer treatments should ideally trigger apoptosis selectively in tumor cells. HAMLET offers a new model in which to approach this question. HAMLET induces cell death in tumor cells, but leaves fully differentiated cells unaffected. The cells die with a mixture of classical and apoptosis-like characteristics and HAMLET kills a wide range of malignant cells *in vitro.* Interestingly, HAMLET maintains this activity *in vivo*, in patients with skin papillomas and in a rat xeno-graft model of human glioblastomas suggesting that HAMLET can be selective for tumor cells and that HAMLET may control tumor progression without apparent tissue toxicity. HAMLET thus shows great promise as a tool to activate cell death pathways in tumor cells and as a new therapeutic, with the advantage of selectivity for tumor cells and low toxicity.

## ACKNOWLEDGMENTS

This study was supported by the American Cancer Society, The Swedish Cancer Society, the Swedish Pediatric Cancer Foundation, The Medical Faculty (Lund University), The Segerfalk Foundation, The Royal Physiographic Society and The Anna-Lisa, Sven-Eric Lundgren Foundation (Lund, Sweden) and The Norwegian Cancer Society and Centre de Reserche Public de la Santé Luxembourg.

## REFERENCES

1. Flemming, W. 1885. Über die bildung von richtungsfiguren in säugethiereiern beim untergang graaf'scher follikel. *Arch. Anat. Physiol:221.*
2. Kerr, J. F. R., A. H. Wyllie, and A. R. Currie. 1972. Apoptosis: A basic biological phenomenon with wide-ranging implications in tissue kinetics. *Br. J. Cancer 26:239.*
3. Wyllie, A. H., J. F. Kerr, and A. R. Currie. 1980. Cell death: the significance of apoptosis. *Int Rev Cytol 68:251.*
4. Arends, M. J., R. G. Morris, and A. H. Wyllie. 1990. Apoptosis: The role of endonuclease. *Am J Pathol 136:593.*

5. Fadok, V. A., D. R. Voelker, P. A. Campbell, J. J. Cohen, D. L. Bratton, and P. M. Henson. 1992. Exposure of phosphatidylserine on the surface of apoptotic lymphocytes triggers specific recognition and removal by macrophages. *J Immunol 148:2207.*
6. Ellis, H. M., and H. R. Horvitz. 1986. Genetic control of programmed cell death in the nematode C. elegans. *Cell 44:817.*
7. Leist, M., and M. Jaattela. 2001. Four deaths and a funeral: from caspases to alternative mechanisms. *Nat Rev Mol Cell Biol 2:589.*
8. Lockshin, R. A., and Z. Zakeri. 2001. Programmed cell death and apoptosis: origins of the theory. *Nat Rev Mol Cell Biol 2:545.*
9. Jaattela, M. 2002. Programmed cell death: many ways for cells to die decently. *Ann Med 34:480.*
10. Susin, S. A., N. Zamzami, and G. Kroemer. 1998. Mitochondria as regulators of apoptosis: doubt no more. *Biochim Biophys Acta 1366:151.*
11. Håkansson, A., B. Zhivotovsky, S. Orrenius, H. Sabharwal, and C. Svanborg. 1995. Apoptosis induced by a human milk protein. *Proc Natl Acad Sci U S A 92:8064.*
12. Svensson, M., H. Sabharwal, A. Hakansson, A. K. Mossberg, P. Lipniunas, H. Leffler, C. Svanborg, and S. Linse. 1999. Molecular characterization of alpha-lactalbumin folding variants that induce apoptosis in tumor cells. *J Biol Chem 274:6388.*
13. Permyakov, E. A., and L. J. Berliner. 2000. alpha-Lactalbumin: structure and function. *FEBS Lett 473:269.*
14. Rao, K. R., and K. Brew. 1989. Calcium regulates folding and disulfide-bond formation in alpha- lactalbumin. *Biochem Biophys Res Commun 163:1390.*
15. Brew, K., T. C. Vanaman, and R. L. Hill. 1968. The role of alpha-lactalbumin and the A protein in lactose synthetase: a unique mechanism for the control of a biological reaction. *Proc Natl Acad Sci U S A 59:491.*
16. Kuwajima, K. 1996. The molten globule state of alpha-lactalbumin. *Faseb J 10:102.*
17. Svensson, M., A. K. Mossberg, J. Pettersson, S. Linse, and C. Svanborg. 2003. Lipids as cofactors in protein folding: stereo-specific lipid-protein interactions are required to form HAMLET (human alpha-lactalbumin made lethal to tumor cells). *Protein Sci 12:2805.*
18. Svensson, M., A. Hakansson, A. K. Mossberg, S. Linse, and C. Svanborg. 2000. Conversion of alpha-lactalbumin to a protein inducing apoptosis. *Proc Natl Acad Sci U S A 97:4221.*
19. Svanborg, C., H. Agerstam, A. Aronson, R. Bjerkvig, C. Duringer, W. Fischer, L. Gustafsson, O. Hallgren, I. Leijonhuvud, S. Linse, A. K. Mossberg, H. Nilsson, J. Pettersson, and M. Svensson. 2003. HAMLET kills tumor cells by an apoptosis-like mechanism--cellular, molecular, and therapeutic aspects. *Adv Cancer Res 88:1.*
20. Håkansson, A., J. Andreasson, B. Zhivotovsky, D. Karpman, S. Orrenius, and C. Svanborg. 1999. Multimeric alpha-lactalbumin from human milk induces apoptosis through a direct effect on cell nuclei. *Exp Cell Res 246:451.*
21. Håkansson, A. 1999. Apoptosis induced by a human milk protein complex. Cellular and structural studies in tumour cells and bacteria. In *Microbiology, Immunology and Glycobiology*. Lund University, Lund, Sweden, p. 93.
22. Tartaglia, L. A., T. M. Ayres, G. H. Wong, and D. V. Goeddel. 1993. A novel domain within the 55 kd TNF receptor signals cell death. *Cell 74:845.*
23. Nagata, S. 1997. Apoptosis by death factor. *Cell 88:355.*
24. Smith, C. A., T. Farrah, and R. G. Goodwin. 1994. The TNF receptor superfamily of cellular and viral proteins: activation, costimulation, and death. *Cell 76:959.*

25. Huang, B., M. Eberstadt, E. T. Olejniczak, R. P. Meadows, and S. W. Fesik. 1996. NMR structure and mutagenesis of the Fas (APO-1/CD95) death domain. *Nature 384:638.*
26. Scaffidi, C., S. Fulda, A. Srinivasan, C. Friesen, F. Li, K. J. Tomaselli, K. M. Debatin, P. H. Krammer, and M. E. Peter. 1998. Two CD95 (APO-1/Fas) signaling pathways. *Embo J 17:1675.*
27. Luo, X., I. Budihardjo, H. Zou, C. Slaughter, and X. Wang. 1998. Bid, a Bcl2 interacting protein, mediates cytochrome c release from mitochondria in response to activation of cell surface death receptors. *Cell 94:481.*
28. Li, H., H. Zhu, C. J. Xu, and J. Yuan. 1998. Cleavage of BID by caspase 8 mediates the mitochondrial damage in the Fas pathway of apoptosis. *Cell 94:491.*
29. Yang, X., R. Khosravi-Far, H. Y. Chang, and D. Baltimore. 1997. Daxx, a novel Fas-binding protein that activates JNK and apoptosis. *Cell 89:1067.*
30. Chang, H. Y., H. Nishitoh, X. Yang, H. Ichijo, and D. Baltimore. 1998. Activation of apoptosis signal-regulating kinase 1 (ASK1) by the adapter protein Daxx. *Science 281:1860.*
31. Holler, N., R. Zaru, O. Micheau, M. Thome, A. Attinger, S. Valitutti, J. L. Bodmer, P. Schneider, B. Seed, and J. Tschopp. 2000. Fas triggers an alternative, caspase-8-independent cell death pathway using the kinase RIP as effector molecule. *Nat Immunol 1:489.*
32. Kohler, C., A. Håkansson, C. Svanborg, S. Orrenius, and B. Zhivotovsky. 1999. Protease activation in apoptosis induced by MAL. *Exp. Cell Res. 249:260.*
33. Tsujimoto, Y., L. R. Finger, J. Yunis, P. C. Nowell, and C. M. Croce. 1984. Cloning of the chromosome breakpoint of neoplastic B cells with the t(14;18) chromosome translocation. *Science 226:1097.*
34. Vaux, D. L., S. Cory, and J. M. Adams. 1988. Bcl-2 gene promotes haemopoietic cell survival and cooperates with c-myc to immortalize pre-B cells. *Nature 335:440.*
35. Hockenbery, D., G. Nunez, C. Milliman, R. D. Schreiber, and S. J. Korsmeyer. 1990. Bcl-2 is an inner mitochondrial membrane protein that blocks programmed cell death. *Nature 348:334.*
36. Cory, S., D. C. Huang, and J. M. Adams. 2003. The Bcl-2 family: roles in cell survival and oncogenesis. *Oncogene 22:8590.*
37. Hengartner, M. O. 2000. The biochemistry of apoptosis. *Nature 407:770.*
38. Newmeyer, D. D., and S. Ferguson-Miller. 2003. Mitochondria: releasing power for life and unleashing the machineries of death. *Cell 112:481.*
39. van Gurp, M., N. Festjens, G. van Loo, X. Saelens, and P. Vandenabeele. 2003. Mitochondrial intermembrane proteins in cell death. *Biochem Biophys Res Commun 304:487.*
40. Kohler, C., V. Gogvadze, A. Hakansson, C. Svanborg, S. Orrenius, and B. Zhivotovsky. 2001. A folding variant of human alpha-lactalbumin induces mitochondrial permeability transition in isolated mitochondria. *Eur J Biochem 268:186.*
41. Zou, H., W. J. Henzel, X. Liu, A. Lutschg, and X. Wang. 1997. Apaf-1, a human protein homologous to C. elegans CED-4, participates in cytochrome c-dependent activation of caspase-3 [see comments]. *Cell 90:405.*
42. Zou, H., Y. Li, X. Liu, and X. Wang. 1999. An APAF-1.cytochrome c multimeric complex is a functional apoptosome that activates procaspase-9. *J Biol Chem 274:11549.*
43. Li, P., D. Nijhawan, I. Budihardjo, S. M. Srinivasula, M. Ahmad, E. S. Alnemri, and X. Wang. 1997. Cytochrome c and dATP-dependent formation of Apaf-1/caspase-9 complex initiates an apoptotic protease cascade. *Cell 91:479.*

44. Acehan, D., X. Jiang, D. G. Morgan, J. E. Heuser, X. Wang, and C. W. Akey. 2002. Three-dimensional structure of the apoptosome: implications for assembly, procaspase-9 binding, and activation. *Mol Cell 9:423.*
45. Degterev, A., M. Boyce, and J. Yuan. 2003. A decade of caspases. *Oncogene 22:8543.*
46. Deveraux, Q. L., N. Roy, H. R. Stennicke, T. Van Arsdale, Q. Zhou, S. M. Srinivasula, E. S. Alnemri, G. S. Salvesen, and J. C. Reed. 1998. IAPs block apoptotic events induced by caspase-8 and cytochrome c by direct inhibition of distinct caspases. *Embo J 17:2215.*
47. Verhagen, A. M., P. G. Ekert, M. Pakusch, J. Silke, L. M. Connolly, G. E. Reid, R. L. Moritz, R. J. Simpson, and D. L. Vaux. 2000. Identification of DIABLO, a mammalian protein that promotes apoptosis by binding to and antagonizing IAP proteins. *Cell 102:43.*
48. Du, C., M. Fang, Y. Li, L. Li, and X. Wang. 2000. Smac, a mitochondrial protein that promotes cytochrome c-dependent caspase activation by eliminating IAP inhibition. *Cell 102:33.*
49. Suzuki, Y., Y. Imai, H. Nakayama, K. Takahashi, K. Takio, and R. Takahashi. 2001. A serine protease, HtrA2, is released from the mitochondria and interacts with XIAP, inducing cell death. *Mol Cell 8:613.*
50. Verhagen, A. M., J. Silke, P. G. Ekert, M. Pakusch, H. Kaufmann, L. M. Connolly, C. L. Day, A. Tikoo, R. Burke, C. Wrobel, R. L. Moritz, R. J. Simpson, and D. L. Vaux. 2002. HtrA2 promotes cell death through its serine protease activity and its ability to antagonize inhibitor of apoptosis proteins. *J Biol Chem 277:445.*
51. Robertson, J. D., S. Orrenius, and B. Zhivotovsky. 2000. Review: nuclear events in apoptosis [In Process Citation]. *J Struct Biol 129:346.*
52. Lazebnik, Y. A., A. Takahashi, R. D. Moir, R. D. Goldman, G. G. Poirier, S. H. Kaufmann, and W. C. Earnshaw. 1995. Studies of the lamin proteinase reveal multiple parallel biochemical pathways during apoptotic execution. *Proc. Natl. Acad. Sci. U S A 92:9042.*
53. Rao, L., D. Perez, and E. White. 1996. Lamin proteolysis facilitates nuclear events during apoptosis. *J Cell Biol 135:1441.*
54. Casiano, C. A., S. J. Martin, D. R. Green, and E. M. Tan. 1996. Selective cleavage of nuclear autoantigens during CD95 (Fas/APO-1)-mediated T cell apoptosis. *J Exp Med 184:765.*
55. Greidinger, E. L., D. K. Miller, T. T. Yamin, L. Casciola-Rosen, and A. Rosen. 1996. Sequential activation of three distinct ICE-like activities in Fas-ligated Jurkat cells. *FEBS Lett 390:299.*
56. Nagata, S. 2000. Apoptotic DNA fragmentation. *Exp Cell Res 256:12.*
57. Enari, M., H. Sakahira, H. Yokoyama, K. Okawa, A. Iwamatsu, and S. Nagata. 1998. A caspase-activated DNase that degrades DNA during apoptosis, and its inhibitor ICAD [see comments] [published erratum appears in Nature 1998 May 28;393(6683):396]. *Nature 391:43.*
58. Lee, N., H. MacDonald, C. Reinhard, R. Halenbeck, A. Roulston, T. Shi, and L. T. Williams. 1997. Activation of hPAK65 by caspase cleavage induces some of the morphological and biochemical changes of apoptosis. *Proc Natl Acad Sci U S A 94:13642.*
59. Rudel, T., and G. M. Bokoch. 1997. Membrane and morphological changes in apoptotic cells regulated by caspase-mediated activation of PAK2. *Science 276:1571.*
60. Kothakota, S., T. Azuma, C. Reinhard, A. Klippel, J. Tang, K. Chu, T. J. McGarry, M. W. Kirschner, K. Koths, D. J. Kwiatkowski, and L. T. Williams. 1997. Caspase-3-

generated fragment of gelsolin: effector of morphological change in apoptosis. *Science 278:294.*
61. Maderna, P., and C. Godson. 2003. Phagocytosis of apoptotic cells and the resolution of inflammation. *Biochim Biophys Acta 1639:141.*
62. Wolffe, A. P. 1998. *Chromatin structure and function.* Academic Press, San Diego.
63. Allan, J., N. Harborne, D. C. Rau, and H. Gould. 1982. Participation of core histone "tails" in the stabilization of the chromatin solenoid. *J Cell Biol 93:285.*
64. Carruthers, L. M., and J. C. Hansen. 2000. The core histone N termini function independently of linker histones during chromatin condensation. *J Biol Chem 275:37285.*
65. Horn, P. J., and C. L. Peterson. 2002. Molecular biology. Chromatin higher order folding--wrapping up transcription. *Science 297:1824.*
66. Arents, G., R. W. Burlingame, B. C. Wang, W. E. Love, and E. N. Moudrianakis. 1991. The nucleosomal core histone octamer at 3.1 A resolution: a tripartite protein assembly and a left-handed superhelix. *Proc Natl Acad Sci U S A 88:10148.*
67. Arents, G., and E. N. Moudrianakis. 1995. The histone fold: a ubiquitous architectural motif utilized in DNA compaction and protein dimerization. *Proc Natl Acad Sci U S A 92:11170.*
68. Jenuwein, T., and C. D. Allis. 2001. Translating the histone code. *Science 293:1074.*
69. Philpott, A., T. Krude, and R. A. Laskey. 2000. Nuclear chaperones. *Semin Cell Dev Biol 11:7.*
70. Verreault, A. 2000. De novo nucleosome assembly: new pieces in an old puzzle. *Genes Dev 14:1430.*
71. Haushalter, K. A., and J. T. Kadonaga. 2003. Chromatin assembly by DNA-translocating motors. *Nat Rev Mol Cell Biol 4:613.*
72. Ye, X., A. A. Franco, H. Santos, D. M. Nelson, P. D. Kaufman, and P. D. Adams. 2003. Defective S phase chromatin assembly causes DNA damage, activation of the S phase checkpoint, and S phase arrest. *Mol Cell 11:341.*
73. Vogelstein, B., D. Lane, and A. J. Levine. 2000. Surfing the p53 network. *Nature 408:307.*
74. Vousden, K. H. 2002. Activation of the p53 tumor suppressor protein. *Biochim Biophys Acta 1602:47.*
75. Hollstein, M., D. Sidransky, B. Vogelstein, and C. C. Harris. 1991. p53 mutations in human cancers. *Science 253:49.*
76. Kubbutat, M. H., S. N. Jones, and K. H. Vousden. 1997. Regulation of p53 stability by Mdm2. *Nature 387:299.*
77. Carr, A. M. 2000. Cell cycle. Piecing together the p53 puzzle. *Science 287:1765.*
78. Sherr, C. J., and J. D. Weber. 2000. The ARF/p53 pathway. *Curr Opin Genet Dev 10:94.*
79. Meek, D. W. 1999. Mechanisms of switching on p53: a role for covalent modification? *Oncogene 18:7666.*
80. Bykov, V. J., N. Issaeva, A. Shilov, M. Hultcrantz, E. Pugacheva, P. Chumakov, J. Bergman, K. G. Wiman, and G. Selivanova. 2002. Restoration of the tumor suppressor function to mutant p53 by a low-molecular-weight compound. *Nat Med 8:282.*
81. Vousden, K. H. 2000. p53: death star. *Cell 103:691.*
82. Sax, J. K., and W. S. El-Deiry. 2003. p53 downstream targets and chemosensitivity. *Cell Death Differ 10:413.*
83. Marchenko, N. D., A. Zaika, and U. M. Moll. 2000. Death signal-induced localization of p53 protein to mitochondria. A potential role in apoptotic signaling. *J Biol Chem 275:16202.*

84. Chipuk, J., K. Kuwana, L. Bouchier-Hayes, D. Newmeyer, M. Schuler, and D. R. Green. 2003. p53 directly activates bax. In *Programmed cell death*, Cold Spring Harbor, New York, p. 160.
85. Chipuk, J. E., U. Maurer, D. R. Green, and M. Schuler. 2003. Pharmacologic activation of p53 elicits Bax-dependent apoptosis in the absence of transcription. *Cancer Cell 4:371.*
86. Robertson, J. D., M. Enoksson, M. Suomela, B. Zhivotovsky, and S. Orrenius. 2002. Caspase-2 acts upstream of mitochondria to promote cytochrome c release during etoposide-induced apoptosis. *J Biol Chem 277:29803.*
87. Konishi, A., S. Shimizu, J. Hirota, T. Takao, Y. Fan, Y. Matsuoka, L. Zhang, Y. Yoneda, Y. Fujii, A. I. Skoultchi, and Y. Tsujimoto. 2003. Involvement of histone H1.2 in apoptosis induced by DNA double-strand breaks. *Cell 114:673.*
88. Yu, S. W., H. Wang, M. F. Poitras, C. Coombs, W. J. Bowers, H. J. Federoff, G. G. Poirier, T. M. Dawson, and V. L. Dawson. 2002. Mediation of poly(ADP-ribose) polymerase-1-dependent cell death by apoptosis-inducing factor. *Science 297:259.*
89. Ikura, T., V. V. Ogryzko, M. Grigoriev, R. Groisman, J. Wang, M. Horikoshi, R. Scully, J. Qin, and Y. Nakatani. 2000. Involvement of the TIP60 histone acetylase complex in DNA repair and apoptosis. *Cell 102:463.*
90. Yamaki, M., T. Umehara, T. Chimura, and M. Horikoshi. 2001. Cell death with predominant apoptotic features in Saccharomyces cerevisiae mediated by deletion of the histone chaperone ASF1/CIA1. *Genes Cells 6:1043.*
91. Gustafsson, L., I. Leijonhufvud, A. Aronsson, A. K. Mossberg, and C. Svanborg. 2004. Treatment of skin papillomas with topical alpha-lactalbumin-oleic acid. *N Engl J Med 350:2663.*
92. zur Hausen, H. 2002. Papillomaviruses and cancer: from basic studies to clinical application. *Nat Rev Cancer 2:342.*
93. Gibbs, S., I. Harvey, J. Sterling, and R. Stark. 2002. Local treatments for cutaneous warts: systematic review. *Bmj 325:461.*
94. Gundersen, S., K. Lote, and E. Hannisdal. 1996. Prognostic factors for glioblastoma multiforme--development of a prognostic index. *Acta Oncol 35 Suppl 8:123.*
95. Engebraaten, O., G. O. Hjortland, H. Hirschberg, and O. Fodstad. 1999. Growth of precultured human glioma specimens in nude rat brain. *J Neurosurg 90:125.*

Chapter 11

# CANCER IMMUNOTHERAPY: ON THE TRAIL OF A CURE?

Thomas S. Griffith[1,2,3], Troy J. Kemp[1,2], Aaron T. Ludwig[1], Jill M. Moore[1], and Rebecca L. VanOosten[1,2]
*Department of Urology1, Interdisciplinary Graduate Program in Immunology2, and The Prostate Cancer Research Program of the Holden Comprehensive Cancer Center3, University of Iowa, 375 Newton Road, Iowa City, IA 52242*

Abstract: Since its discovery in 1995, numerous studies have investigated the potential of using TRAIL (TNF-related apoptosis-inducing ligand) as an alternative cancer therapeutic, since it is a potent inducer of apoptosis in tumor cells but not in normal cells and tissues. As a consequence, a great deal is known about TRAIL/TRAIL receptor expression, the molecular mechanism of TRAIL receptor signaling, and methods of altering tumor cell sensitivity to TRAIL-induced apoptosis. Translating the preclinical TRAIL studies into the clinic is beginning, with the hope that TRAIL will retain all of its tumoricidal activity against human primary tumors *in situ* with no toxic side effects.

Key words: TRAIL; apoptosis; tumor; adenovirus; immunotherapy

## 1. INTRODUCTION

Cell death can generally be classified into two forms based on morphological and biochemical criteria: necrosis and apoptosis[1-3]. Necrosis is the nonphysiological or passive type of cell death that is usually caused by extreme trauma or injury to the cell[4]. It generally affects cells in groups rather than single cells, and evokes inflammation when it develops *in vivo*. Cells undergoing necrosis normally have problems maintaining proper plasma membrane function, so they can no longer regulate osmotic pressure. The cells swell and rupture, spilling their cellular contents into the



*M. Sluyser (ed.), Application of Apoptosis to Cancer Treatment,* 247-270.

surrounding tissue, resulting in the nonspecific cellular destruction that leads to an inflammatory response necessary to remove the debris and begin tissue repair.

The other morphological pattern of cell death, apoptosis, is a more subtle process characterized by numerous cellular changes[3]. This type of cell death usually affects scattered single cells, and is characterized histologically by the formation of small, spherical cytoplasmic fragments, some of which contain pyknotic remnants of nuclei. These changes begin with cellular condensation that leads to the cytoplasm becoming overcrowded with organelles. The nucleus quickly develops condensed chromatin that associates with the nuclear envelope. These nuclear changes are accompanied by fragmentation of the cellular DNA into a ladder of regular subunits, the result of random double-stranded breaks in the linker regions between nucleosomes[5]. While these intracellular changes are taking place, the cell membrane becomes ruffled and blebbed in a process called zeiosis[6]. During this process, the blebs break off into apoptotic bodies. Finally, the fate of the apoptotic cell is to be phagocytosed by surrounding cells before it can rupture and release its potentially inflammatory contents. Thus, apoptotic cell death is well suited for a role in tissue homeostasis, since it can result in the deletion of cells with little tissue disruption.

The rapid phagocytosis of apoptotic cells suggests that alterations in the plasma membrane occur early in the death process, which target these cells for removal, and the phagocytes involved in the clearance of apoptotic cells have been demonstrated to use several mechanisms to recognize target cells[7]. For example, the $\alpha_v\beta_3$ vitronectin receptor, recognition of thrombospondin, membrane carbohydrate changes, and changes in cell membrane net surface charge have all been described as recognition elements for phagocytic cells[7]. Another possible marker for apoptotic cell identification is through changes in membrane composition involving the loss of cell membrane phospholipid asymmetry[8, 9]. Normal plasma membranes exhibit a marked phospholipid asymmetry, with phosphatidylcholine and sphingomyelin predominantly on the outer membrane leaflet and most of the phosphatidylethanolamine and phosphatidylserine on the inner membrane leaflet[10]. This asymmetry is maintained by an ATP- and $Mg^{2+}$ -dependent translocase mediating the inward transport of negatively charged phospholipids and/or association of phosphatidylserine with membrane skeletal proteins[11]. During apoptosis, cell membrane leaflet composition changes, with a redistribution of negatively charged phosphatidylserine to the outer leaflet. Surface exposure of phosphatidylserine occurs with chromatin condensation, precedes the increase in membrane permeability, and constitutes one of the principal targets of phagocyte recognition.

Upon the identification of these basic characteristics, apoptotic cell death was identified to be a normal physiological process associated with a variety of basic systems, such as embryonic development, tissue remodeling, immune regulation, and tumor regression, and is regulated *in situ* by many extracellular and intracellular signals[12, 13]. Survival signals from the environment and intracellular sensor molecules that monitor cellular integrity normally regulate a cell's apoptotic machinery. For example, in the event a cell loses contact with its surroundings or sustains irreparable internal damage, the apoptotic death process is commonly initiated. Additionally, apoptosis helps to balance the number of cells within a multicellular organism that arise through normal cellular proliferation and, thus, maintain the normal homeostatic nature of the organism. Just as with apoptosis, cell proliferation is a highly regulated process with numerous regulatory proteins. A variety of growth factors and proto-oncogenes serve as positive regulators of cell cycle progression. In contrast, there are multiple tumor suppressor genes that function to oppose uncontrolled cell proliferation by inhibiting the activity of the proto-oncogenes. Thus, a delicate balance in the rates of cellular proliferation and death must be maintained for homeostasis, such that malfunctions in proliferation or death results in a plethora of disease states.

Cancer is probably the most common and studied disease associated with a defect or inhibition of the apoptotic process. Tumor cells from a wide array of human malignancies have a decreased ability to undergo apoptosis, and the development of therapeutic agents targeting the apoptosis pathway within tumor cells have consequently become heavily investigated. Of those areas being studied, the induction of apoptotic tumor cell death through an active, instructive process mediated by death receptors, cell surface receptors that transmit apoptotic signals when bound by their cognate death ligands, is of great therapeutic interest. The receptors mediating the death signal belong to the tumor necrosis factor (TNF) receptor superfamily, which is characterized by similar, cysteine-rich extracellular domains[14, 15]. Additionally, all death receptors contain a homologous cytoplasmic sequence called the death domain (DD), which serves as a recognition point for the apoptotic machinery[15-17]. The ligands for the death receptors belong to the TNF family of cytokines, a group of molecules that influence a variety of immunological functions. For example, two of the most studied death ligands are TNF and Fas ligand (FasL), which induce apoptosis in many physiological events, such as autoimmunity, activation-induced cell death (AICD), immune privilege, and evasion of tumors from the immune system[18-22].

## 2. TRAIL IS A POTENT INDUCER OF TUMOR CELL APOPTOSIS

TRAIL (TNF-related apoptosis inducing ligand) is another TNF family member that is capable of inducing apoptosis and has recently received great attention because of its therapeutic potential. While searching an expressed sequence tag (EST) database using a conserved sequence contained in many TNF family members, Wiley *et al.* were the first to identify an EST that was then used to clone the full length cDNA for TRAIL[23] (a.k.a. Apo-2 ligand[24]). The extracellular domain of TRAIL was found to be most homologous to Fas ligand (28% a.a. identity), but also had significant identity to TNF (23%), lymphotoxin (LT)-α (23%), and LT-β (22%). Whereas the homology of TRAIL to other TNF family members may be considered low, examination of the crystal structure of monomeric TRAIL found it to be similar to that of TNF, TNF-β, and CD40 ligand[25]. TRAIL monomers contain two antiparallel β-pleated sheets that form a β sandwich core framework, and the monomers interact with other TRAIL monomers in a head-to-tail fashion, to form a bell-shaped trimer[25]. This oligomerization enhances TRAIL activity as studies with recombinant soluble TRAIL found that the most biologically active form was multimeric, or crosslinked, rather than monomeric[23].

Early studies identified two unique characteristics of TRAIL. First, TRAIL-induced apoptosis only in tumorigenic or transformed cells and not normal cells[23]. And as with the other death-inducing members of the TNF family (i.e. FasL and TNF), cells undergoing TRAIL-induced death exhibited many of the hallmarks of apoptosis, including DNA fragmentation, expression of pro-phagocytic signals (i.e. phosphatidylserine) on the cell membrane, and cleavage of multiple intracellular proteins by caspases[23, 24, 26, 27]. Soluble TRAIL is tumoricidal for over 75% of the more than sixty hematopoietic and non-hematopoietic tumor cell lines tested *in vitro*, suggesting that TRAIL could be used as a broad-spectrum, anti-tumor molecule *in vivo*[23, 24, 28, 29]. TRAIL may be important in the AICD of T cells during HIV infection[30]. Peripheral blood human T cells express TRAIL after CD3 crosslinking and type I IFN stimulation, perhaps also contributing to the AICD of T cells in the natural setting[31]. In addition, human NK, monocytes, and dendritic cells express TRAIL following cytokine stimulation, transforming them into potent killers of tumor cells[32-34]. Recent work from our laboratory has revealed that CpG-containing oligonucleotides are also potent inducers of TRAIL on human PBMC (especially monocytes and B cells) via an IFN-α-dependent mechanism (Figure 1A and [35]). Second, in contrast to other TNF family members whose expression is tightly regulated and often transiently expressed, mRNA for TRAIL is detected in a wide range of tissues, including peripheral blood lymphocytes, spleen, thymus, prostate, ovary, small intestine, colon and placenta[23].

**(A)**

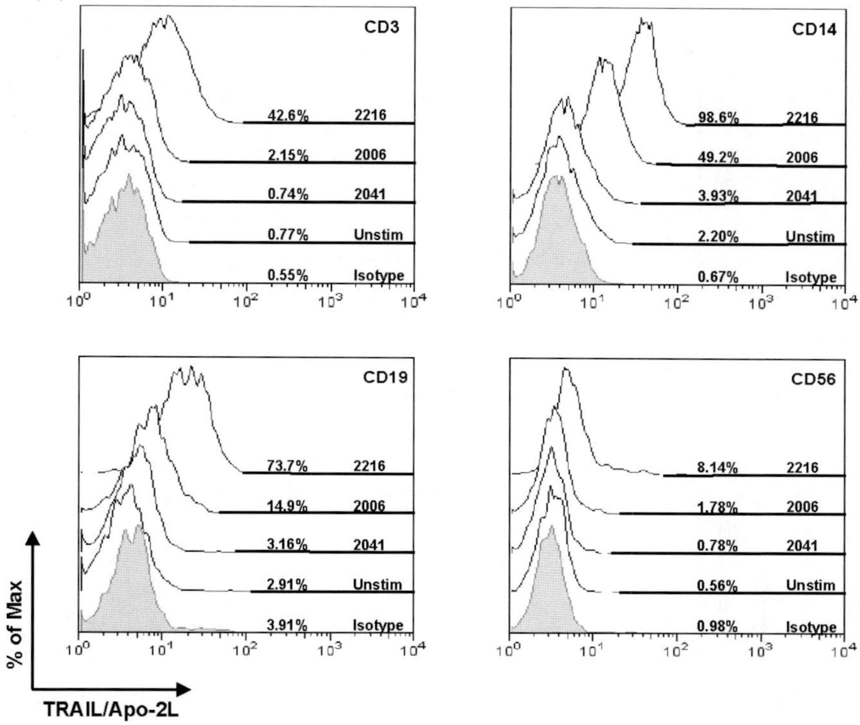


*Figure 1.* **(A)** TRAIL expression on human PBMC (CD3+, CD14+, CD19+, and CD56+ cells) after incubation for 24 h in the absence or presence of CpG-B ODN 2006, CpG-B ODN control 2041, or CpG-A ODN 2216 (1 μg/ml). Histograms represent 104 gated cells in all conditions, and viability was greater than 95% as assessed by propidium iodide exclusion.

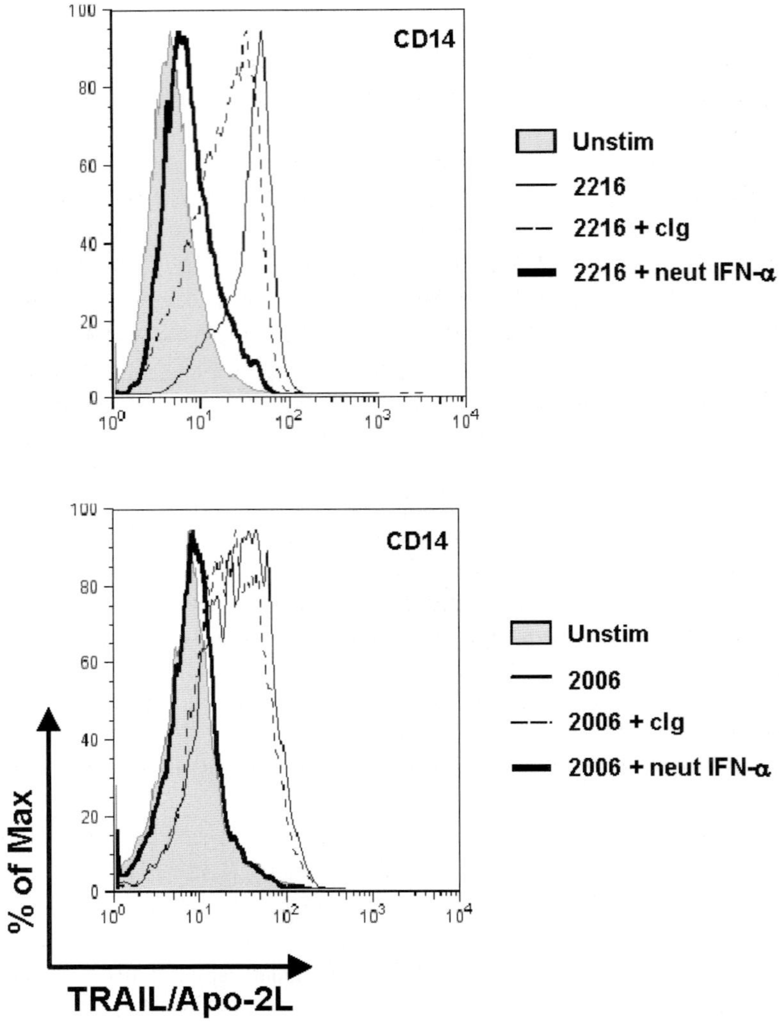


*Figure 1.* **(B)** IFN-α neutralization abrogates CpG ODN-induced TRAIL expression on CD14+ monocytes. PBMC were incubated for 24 h in the absence or presence of either 1 μg/ml CpG-A 2216 or CpG-B 2006, IFN-α neutralizing antiserum (10,000 NU/ml), or a nonspecific control Ig. Cells were then analyzed by flow cytometry for TRAIL surface expression on CD14+ cells. Histograms represent 104 gated cells in all conditions, and viability was greater than 95% as assessed by propidium iodide exclusion.

## 3. TRAIL RECEPTOR SIGNAL TRANSDUCTION

As mentioned, apoptotic cell death can be triggered by a variety of agents acting on different cellular receptors, most notably by the death receptors of the TNF receptor superfamily. Unlike Fas ligand and TNF, which interact with a single or pair of receptors, respectively, TRAIL specifically binds to five distinct receptors: DR4[36], DR5/TRAIL-R2[37-39], TRID/DcR1/TRAIL-R3[37, 38, 40], TRAIL-R4/DcR2[28, 41] (hereafter referred to as TRAIL-R1, -R2, -R3, and -R4, respectively), and osteoprotegerin (OPG)[42]. Both TRAIL-R1 and TRAIL-R2 contain a cytoplasmic death domain, and crosslinking by TRAIL or receptor-specific mAb activates the apoptosis signaling pathway in sensitive cells[27, 36-39]. In contrast, neither TRAIL-R3 (which is GPI linked) nor TRAIL-R4 (which is a type I membrane protein) contains a complete cytoplasmic death domain, and neither can mediate apoptosis upon ligation[28, 37, 38, 40, 41]. OPG is a soluble receptor capable of binding to TRAIL *in vitro* and blocking TRAIL-induced apoptosis[42]. Because TRAIL-R3, -R4, and OPG bind to TRAIL without directly signaling for cell death, it was initially proposed that these receptors inhibit TRAIL-induced apoptosis by acting either as membrane-bound or soluble antagonistic receptors[37, 38, 40, 42] or via transduction of an anti-apoptotic signal [28]. Therefore, the presence or absence of TRAIL-R3, TRAIL-R4, and/or OPG was thought to determine whether a cell is resistant or sensitive, respectively, to TRAIL-induced apoptosis[37, 38, 41, 42]. Further investigation of many tumor cell lines, however, disproved this theory as the sole mechanism regulating TRAIL-sensitivity and resistance[26, 27].

While the proximal signaling events have been extensively studied for Fas and TNFR1, initial evaluation of the TRAIL receptor signaling pathway proved difficult, yielding contradictory results from different laboratories. It was not until recently that the molecular signal transduction events leading to apoptosis induced via TRAIL/TRAIL receptor interaction were elucidated. Death receptor, be it TRAIL-R1/R2, Fas, or TNF-R1, crosslinking leads to the formation of a multiprotein structure called the death-inducing signaling complex (DISC[43]) that includes the death receptor, the death adapter protein Fas-associated death domain protein (FADD[44]), and the proteolytic cysteine protease procaspase-8. All death receptors share a homophilic protein/protein intracellular domain, called the DD[16, 17], which is required to attract specialized apoptosis signaling molecules that often contain a DD themselves. FADD interacts directly with the DD of TRAIL-R1, TRAIL-R2, Fas, and indirectly with the DD of TNF-R1 through TRADD (TNF Receptor-associated death domain protein). In a homotypic interaction, the DD of FADD binds to the DD of TRAIL-R1 or –R2. The death effector domain (DED) of FADD, in turn, interacts with the DED of procaspase-8[45]. Procaspase-8 is proteolytically cleaved and activated at the DISC, which then initiates the apoptosis executing caspase cascade. The downstream executioner caspases, caspase-3, -6, and -7, are then activated to cleave the numerous structural and regulatory proteins that maintain cellular integrity. A schematic representation of the TRAIL-R1/-R2 signal transduction pathway is depicted in Figure 2.

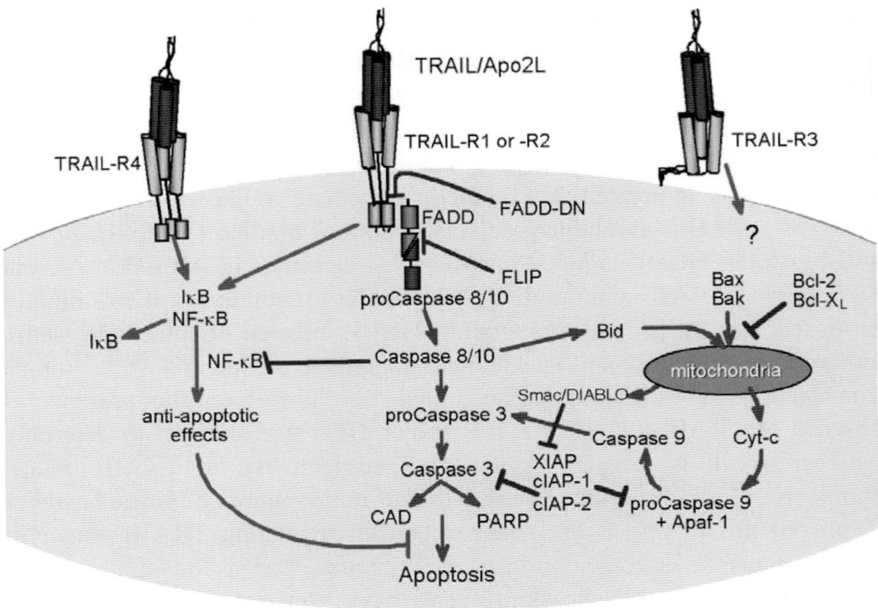


*Figure 2.* Schematic representation of TRAIL receptor signaling within a tumor cell.

Caspases are a family of cysteine proteases with cleavage specificity for aspartic acid residues[46]. All caspases share a similar structure and are synthesized as inactive precursors that consist of a prodomain at its amino terminus, a large subunit the middle of the molecule, and a carboxy-terminal small subunit. Caspase activation proceeds by proteolytic cleavage of the constitutively expressed proform, where one cleavage event removes the prodomain and the other separates the two subunits. The intriguing nature of the cleavage sites indicates that these molecules can undergo autoactivation, as well as activating each other in an enzymatic cascade similar to the coagulation and complement cascades. Although the catalytic domain is located within the large subunit, it is only active when associated with the small subunit. In fact, crystallographic studies demonstrate active caspases are tetramers made by the association of two heterodimers and, consequently, contain two independent catalytic sites[47, 48]. The conclusion that caspases are the central executioners in the apoptotic process was made from the following observations: 1) caspase activation is a very early event that occurs in all forms of apoptotic cell death; 2) viral or oncogenic protein-mediated inhibition of caspase activation, or inhibition with small peptides, blocks all the morphological features seen in apoptosis; and 3) addition of recombinant caspases in cell-free systems results in the same apoptotic cytoplasmic and nuclear changes. However, not all caspases serve as effector molecules within this signaling pathway. Whereas caspases-3, -6, and –7 have been directly implicated with the execution of apoptotic cells, caspases-8, -9, -10, and –2 are initiator or regulatory caspases, such that their activities do not directly account for the morphological features of apoptosis. Instead, the primary role of the initiator/regulator caspases is to function as signaling molecules, transducing stress signals that activate the effector caspases. The fact that the proteolytic activity of active caspases within a cell is responsible for its demise led to the conclusion that the main function of the caspases was to destroy intracellular proteins in such a way that the cell would not be able to function any longer. While this is generally true, some caspase substrates are activated upon cleavage, for example cleavage of the Bcl-2 family member Bid facilitates its targeting to the mitochondria, leading to cytochrome c release[49].

## 4. PHYSIOLOGICAL ROLES FOR TRAIL

It quickly became clear from the numerous *in vitro* studies that TRAIL was a potent inducer to tumor cell apoptosis, but there was nothing known regarding the normal physiological activities of TRAIL *in vivo*. In an attempt to address this question, Sedger *et al.*[50] generated TRAIL-gene targeted (-/-) mice. These mice develop normally and display no defects in lymphoid or myeloid cell homeostasis or function. Despite the fact that TRAIL binds OPG and OPG-overexpressing mice have increased bone density, TRAIL$^{-/-}$ mice showed no evidence of aberrant bone growth. The one abnormality observed in these mice, however, was that they were more susceptible to tumor burden. Interestingly, while the TRAIL deficiency resulted in a significant biological disadvantage for controlling the growth of TRAIL-sensitive tumor *in vivo,* the mice did not have an increased tendency to spontaneously develop tumors compared to normal mice. Shortly thereafter, a series of studies were published clearly and elegantly demonstrating the importance of TRAIL in natural tumor immunosurveillance[51-55]. These studies also highlighted the importance of TRAIL-expressing NK cells in the elimination of tumors *in vivo.* These observations are critical in the development of future immune-based therapies for cancer.

Subsequent reports have provided a glimpse into additional physiological roles for TRAIL. First, TRAIL appears to be important in controlling susceptibility to certain autoimmune diseases. Chronic blockade of TRAIL, by using $TRAIL^{-/-}$ mice or soluble TRAIL-R2:Fc in mice, exacerbated the development and severity of symptoms present in models of experimental autoimmune encephalomyelitis, collagen-induced arthritis, and diabetes compared to wild-type mice[56-59]. Based on the observations in the collagen-induced arthritis model, it was also proposed that TRAIL was also important in proper thymocyte development[57]. This conclusion has been met with some controversy, though, as contradictory results have been reported[60].

The importance for FasL, another potent apoptosis-inducing TNF family member, in maintaining immune privilege was highlighted by studies in the eye, testis, and placenta[21, 61-63]. Activated lymphocytes entering these areas are promptly killed via Fas-Fas ligand interactions. Additionally, FasL expression within the eye is an important contributor to the success of corneal transplants. Because of its ability to also kill activated lymphocytes and having highest homology with Fas ligand, investigation into the potential of the TRAIL-TRAIL receptor system to protect these same sites from immune attack was explored. Indeed, TRAIL performs many of the same functions within the eye and placenta. Within the eye, TRAIL is constitutively expressed on numerous ocular structures, including the cornea and retina[64], suggesting a role for TRAIL in tumor surveillance within the eye. Ocular expression of TRAIL may also explain the paucity of clinical cases of ocular tumors; however, it is unknown whether TRAIL plays any role in the success of corneal transplants. Studies investigating the expression of TRAIL in first trimester placentas found prominent expression in syncytiotrophoblasts, where it was localized primarily to the apical brush border[65]. In addition, TRAIL was present on villous stroma and stromal cells, particularly the Hofbauer cells (placental macrophages), amnion epithelial cells, and maternal decidual cells. Expression was low to absent in fibroblastic mesenchymal cells and endothelial cells. In addition to the high level of TRAIL expression, trophoblasts also expressed significant levels of the non-signaling TRAIL receptor, TRAIL-R3. Based on the expression of these two molecules, it was concluded that TRAIL may be an important contributor to immune tolerance during pregnancy.

As previously mentioned, TRAIL expression can be induced on T cells, NK cells, DC, and Mϕ following IFN-α and IFN-γ stimulation[31-34]. Oligodeoxynuclotides containing CpG motifs (CpG ODN) have profound immunostimulatory effects including anti-tumor immunity[66], and recognition of CpG DNA by the immune system has evolved as a defense mechanism against bacterial (intracellular) pathogens[67]. CpG ODN activate numerous immune cell populations and drive the immune response toward a Th1 phenotype[66]. Recently, we showed that human PBMC stimulated by CpG ODN induce an anti-tumor response by upregulating TRAIL expression on multiple PBMC populations ($CD4^+$ and $CD8^+$ T cells, $CD19^+$ B cells, $CD56^+$ NK cells, and $CD14^+$ monocytes) via IFN-α produced by plasmacytoid dendritic cells (pDC)[35]. Of similar interest, *Mycobacterial* species contain among the highest CpG nucleotide content of all prokaryotes, and much of the pioneering work on immunostimulatory DNA was performed on BCG-mediated tumor resistance studies[66]. Because of our studies with CpG ODN and the CpG nucleotide content within *Mycobacterial* species, we proposed that the mechanism for the anti-tumor activity of *Mycobacterium bovis* bacillus Calmette-Guérin (BCG) therapy for bladder cancer[68] was analogous to that of CpG ODN, where the cytokine response, specifically IFN, to BCG leads to the upregulation of TRAIL. This indeed proved to be the case, since we detected the presence of soluble, functional TRAIL protein in the urine of patients following BCG immunotherapy, and observed increased levels of TRAIL in those patients that responded favorably to BCG therapy[69]. When we analyzed urine of patients undergoing BCG therapy to determine which of the voided cells were expressing TRAIL, there was high expression of TRAIL on voided neutrophils (PMN) during the time period examined (3-5 h post instillation). Given these observations, we hypothesized that *in vitro* stimulation of PMN with BCG induced the expression of TRAIL. To test the hypothesys, PMN were stimulated with BCG, IFN-α, or a combination of BCG and IFN-α for 5 h, after which the PMN were assessed for TRAIL expression. As illustrated in Figure 3, PMN stimulated *in vitro* with BCG were the only cells to express TRAIL. Interestingly, there was no additive or synergistic response when the combination of BCG and IFN-α was used for stimulation. While still preliminary, these results may signal a new and significant role for the massive granulocyte influx following BCG instillation.

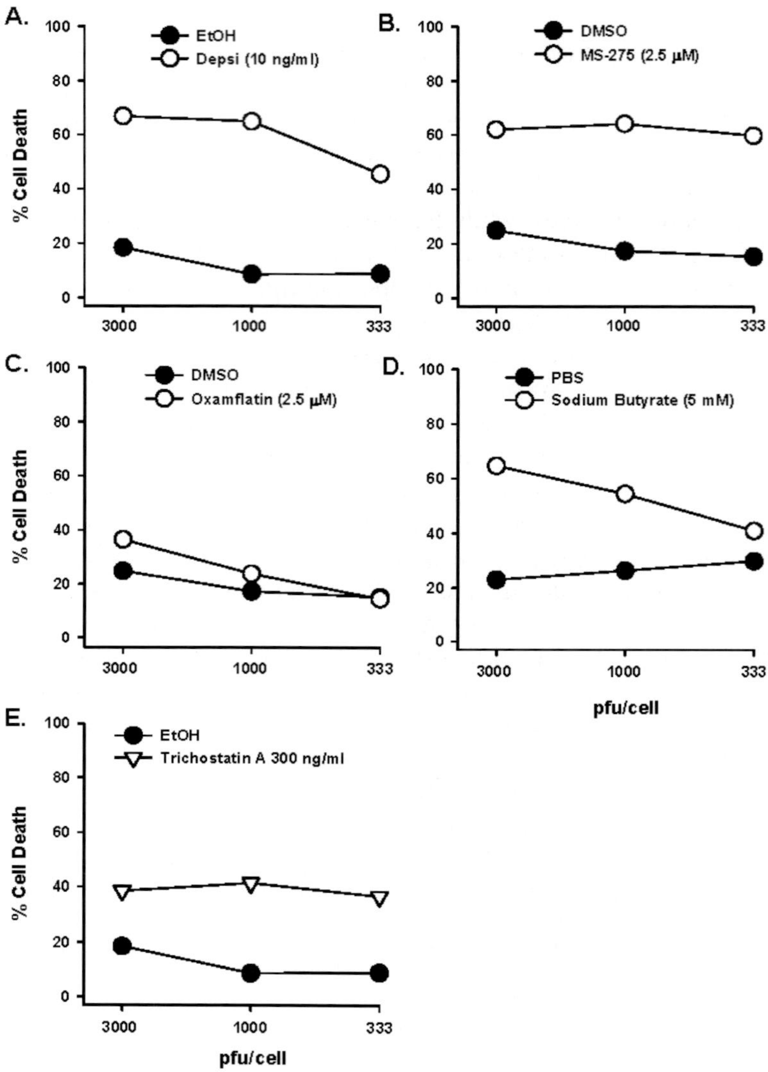


*Figure 3.* Augmentation of Ad5-TRAIL-induced death of DU-145 human prostate tumor cells by histone deacetylase inhibitors. 96-well flat-bottom microtiter plates were seeded with DU-145 cells (2 X 104 cells/well) and allowed to adhere for at least 6 h. Cells were incubated in (A) depsipeptide, (B) MS-275, (C) oxamflatin, (D) sodium butyrate, or (E) trichostatin A at the indicated concentrations for 16 h before adding Ad5-TRAIL. DU-145 cells treated with Ad5-TRAIL alone are indicated as the "medium" group in each panel. Cell viability was then determined after 24 h by crystal violet staining. Each value represents the mean of 3 wells.

# 5. TRAIL AS AN ANTITUMOR THERAPEUTIC

Since the first report describing TRAIL[23], nearly all of the published studies have focused on its anti-tumor activities *in vitro* and *in vivo*, with the goal of developing this molecule into an antitumor therapeutic agent. As described above, early *in vitro* analyses demonstrated that recombinant, soluble TRAIL possessed the unique ability of inducing apoptosis in a broad range of tumor cell lines, while having little to no activity against normal cells[23, 24, 26]. Based on these *in vitro* results, it was suggested that TRAIL could be used as an anti-tumor therapy without the side effects seen with other TNF family members, namely FasL and TNF. The first reports examining the antitumor activity of TRAIL *in vivo* also evaluated its toxicity against normal tissues. Walczak *et al.*[70] compared the effects of intravenous administration of soluble TRAIL or FasL into immunocompetent mice. Injection of TRAIL (either human or murine, 200 μg) did not result in any observable toxic effects, whereas mice given human FasL (50 μg) were dead within 1 h due to massive hepatocellular degeneration, necrosis, and hemorrhage. Histologic examination of tissues obtained from mice treated for a longer period (500 μg/d for 14 days) also demonstrated no toxicity. Further investigation into the potential side effects of TRAIL was performed in nonhuman primates by Ashkenazi *et al.*[71]. The exposure of cynomolgus monkeys to TRAIL at 0.1-10 mg/kg/d over 7 days did not induce detectable toxicity. In contrast, TNF induced severe toxicity at 0.003 mg/kg/d. Interestingly, certain recombinant forms of soluble TRAIL have been shown to induce apoptosis in human hepatocytes *in vitro*[72, 73], raising some concerns about the therapeutic potential of TRAIL. The explanation for the observed hepatocyte killing may actually lie in the form of TRAIL used. A TRAIL monomer of native sequence contains a single cysteine, Cys-230, and when forming trimeric TRAIL the cysteines from three monomers are close to each other and chelate $Zn^{2+}$ [25, 74]. In contrast, the poly-His tagged recombinant TRAIL version that demonstrated toxicity toward hepatocytes *in vitro*[72, 73] had a low $Zn^{2+}$ content and displayed an aberrant structure compared to native TRAIL[75]. These results suggest that the question of liver toxicity was nothing more than an *in vitro* observation entirely dependent upon the form of TRAIL used, and the use of untagged TRAIL (a.k.a. Apo2L/TRAIL.0[75]) in a therapeutic setting in humans will not display any toxicity.

These results clearly demonstrated the safety of large, systemic doses of TRAIL, leading to further studies that evaluated the antitumor activity *in vivo*. For these studies, immunocompromised CB.17 SCID mice were subcutaneously injected with human tumor cells, followed by intraperitoneal or intravenous injections of soluble TRAIL starting at various days after tumor implantation[70, 71]. Multiple doses of TRAIL beginning the day after tumor implantation suppressed tumor outgrowth, with many animals becoming tumor-free. Further studies have demonstrated that the *in vivo* antitumor activity of TRAIL is enhanced when combined with chemotherapeutics, such as camptothecin and 5-fluoruracil, and ionizing radiation[71, 76, 77]. The antitumor activity of TRAIL following systemic administration in these experiments demonstrated that TRAIL could interact with the primary tumor and, potentially, any metastases that would normally be difficult to detect and/or treat. One major drawback to these findings, however, was that large amounts of TRAIL (up to 500 μg/d) were required to inhibit tumor formation, since most of the protein was cleared within 5 hours[70]. Furthermore, the systemic administration of TRAIL appeared to be most successful when administered shortly after tumor implantation[70, 71, 76, 77]. Although these previous results show promise, successful treatment relied on repeated, systemic administration of large amounts of soluble TRAIL to have an effect. Yet, the administration of equivalent doses of recombinant TRAIL protein into humans may be problematic. Thus, the development of an alternative means of delivery may increase the relative activity of TRAIL such that larger, more established tumors may be eradicated as efficiently as smaller tumors.

An alternative approach, which was first reported from our laboratory, is the delivery of the TRAIL gene using a nonreplicative adenoviral vector[78]. Localized therapy of solid tumors has been successful in a number of settings. For example, the treatment of prostate cancer with local (intraprostatic) regimens is also common practice, with cryotherapy, brachytherapy, and several experimental viral-based studies serving as current treatment options[79-81]. Of the viral-based therapies, published data indicate minimal toxicity for adenovirus injection into the prostate up to doses of $10^{11}$ pfu[82]. As demonstrated in the studies performed to date, transfer of the TRAIL gene by Ad5-TRAIL into human prostate tumor cells *in vitro* and *in vivo* led to the rapid transcription and translation of the transferred TRAIL gene into functional TRAIL protein that, when expressed on the cell surface, induced apoptotic death in TRAIL-sensitive tumor cell targets but not normal cells[78, 83].

Though our results showed Ad5-TRAIL was effective in suppressing tumor outgrowth *in vivo*, additional debulking of the tumor by increasing the tumoricidal activity of Ad5-TRAIL should only make the task of removing any residual tumor cells by the immune system easier. Methods to enhance the distribution and expression of recombinant genes are currently being investigated. We have performed preliminary studies with a number of histone deacetylase (HDAC) inhibitors, MS-275, depsipeptide, oxamflatin, sodium butyrate, and trichostatin A. These HDAC inhibitors display differential abilities to increase the susceptibility of prostate tumor cells to adenoviral infection, augment the level of adenoviral transgene expression *in vitro* and *in vivo,* and increase the tumoricidal activity of Ad5-TRAIL (Figure 3). Reports describing the efficacy of TRAIL-based therapies combined with histone deacetylase inhibitors are beginning to appear in the literature[84-88], which will open new avenues of therapy for cancer.

## 6. MECHANISMS REGULATING SENSITIVITY/ RESISTANCE TO TRAIL-INDUCED APOPTOSIS

Despite TRAIL's ability to induce apoptosis in a wide variety and number of human tumor cell lines, there are many reports describing the existence of TRAIL-resistant tumor cells. Consequently, investigation into the mechanism for tumor cell resistance to TRAIL-induced apoptosis has become an intense area of research. It is also believed that studies examining TRAIL-resistant tumor cells may reveal clues to explain the profound level of resistance shown by normal cells and tissues to TRAIL-induced apoptosis. TRAIL mRNA is constitutively expressed in a wide variety of tissue and cell types[23], suggesting that the restricted expression of the different TRAIL receptors is what regulates the induction of TRAIL-mediated apoptosis. However, as previously discussed, mRNA for the four TRAIL receptors has been detected in a wide range of normal cells and tissues, and each of the four TRAIL receptors is capable of binding TRAIL with comparable affinity (less than 1 nanomolar)[28, 40]. Therefore, the initial hypothesis to explain a cell's ability to respond to TRAIL was that the non-death-inducing TRAIL receptors (TRAIL-R3 and –R4) were acting as "decoys", and were the chief molecules determining whether a cell was resistant or sensitive to TRAIL-induced death[37, 38, 41]. This hypothesis seemed logical and drew support from experiments utilizing TRAIL-sensitive cells overexpressing either TRAIL-R3 or -R4, resulting in an inhibition of TRAIL-induced apoptotic cell death[37, 38, 41]. Rather surprisingly, though, immunohistochemical analysis of several human tissues (brain, colon, heart, liver, lung, kidney, and testis) found TRAIL-R3 expression only within the brain, heart, liver, and testis[89]. Additionally, TRAIL-R4 may be more effective than TRAIL-R3 in protecting target cells to TRAIL-induced death[28], perhaps because TRAIL-R4 ligation activates NF-κB, a known inhibitor of death-ligand-induced apoptosis[90, 91]. NF-kB activation can prevent cells from undergoing TNF-$\alpha$-induced cell death, probably by up-regulating expression of a gene or group of genes whose products are anti-apoptotic[90, 91]. However, ligation of TRAIL-R1 and -R2 also results in activation of NF-kB[92, 93], though still resulting in apoptotic cell death. Thus, such explanations cannot fully account for resistance to TRAIL-induced apoptosis. While TRAIL-R3 and/or -R4 expression may indeed be a means of regulating TRAIL-mediated apoptosis, analysis of TRAIL receptor mRNA and protein expression in a panel of human tumor cell lines by RT-PCR, immunoblotting, and flow cytometry have indicated no correlation between TRAIL resistance and TRAIL-R3/-R4 mRNA expression.

A more defendable hypothesis involves the differential expression of any of a number of pro- and anti-apoptotic proteins within the tumor cell that help to regulate the signals generated from trimerized TRAIL-R1 and/or –R2. Initial experiments examining the apoptotic signaling events originating from TRAIL-R1 and –R2 were based on their similarity to the well-characterized death receptors Fas and TNFR-1. As discussed earlier, induction of the cell death upon Fas or TNFR-1 trimerization begins with the binding of FADD directly (as for Fas) or indirectly (as for in TNFR-1 by TRADD binding of FADD) to the DD of these receptors[94]. Caspase-8 is recruited to the receptor complex, which is activated by autocatalysis that, then, activates other caspases leading to cell death[94, 95]. The same events occur when TRAIIL-R1 or –R2 are trimerized. Thus, the issue of the intracellular levels of pro- and anti-apoptotic proteins and their correlation with TRAIL sensitivity has received a great deal of attention in attempting to explain differences in tumor cell sensitivity to TRAIL. Molecules such as FLIP, Bcl-2 family members, inhibitors of apoptosis (IAP) proteins, Akt, and Toso, have all been implicated in regulating the TRAIL receptor signal transduction pathway[26, 96-102]. Furthermore, identification of numerous chemotherapeutic drugs that specifically regulate the levels of these proteins, as well as others important in the TRAIL receptor signaling pathway, has been shown to alter tumor cell sensitivity to TRAIL. It is likely that therapy that combines agents that target apoptosis-regulatory proteins with TRAIL will be necessary to make TRAIL a feasible alternative approach for treating cancer.

A third option for explaining the differences in tumor cell sensitivity to TRAIL-induced apoptosis focuses on cell cycle progression. Apoptosis of dividing cells occurs at various stages of the cell cycle depending on the cell type and/or death-inducing stimulus. For example, TNF- and Fas ligand-induced apoptosis is influenced by cell cycle stage. Moreover, inappropriate regulation of the cell cycle machinery can also result in induction of apoptosis. A recent study by Jin *et al.*[103] using SW480 colon cancer and H460 lung cancer cell lines determined that arresting these cells at the $G_0/G_1$ phase resulted in a higher level of sensitivity to TRAIL-induced apoptosis compared to the same cells arrested at other cell cycle phases. The mechanism by which $G_1$-arrested cells display increased sensitivity to

TRAIL remains to be determined. One possible explanation may lie in the levels of anti-apoptotic proteins during the different cell cycle phases. Though using a different system, Algeciras-Schimnich *et al.*[104] found that activated T cells arrested in $G_1$ phase contained high levels of FLIP protein, which correlated with an increase in TCR-induced apoptosis observed in other cell cycle phases. It is possible that FLIP and/or other pro- or anti-apoptotic proteins fluctuate in a similar fashion in tumor cells.

Tumors cells are commonly found to contain genetic abnormalities that account for the production of mutated proteins. Thus, it is possible that some tumors possess mutations that inhibit TRAIL-induced apoptosis and the blockade of TRAIL-induced apoptosis facilitates the survival of the tumor cells. The *TRAIL-R2* gene has been mapped to human chromosome 8p21[105], a region that is frequently the site of losses of heterozygosity in many types of cancer[106]. Mutations of the *TRAIL-R2* gene have been identified in head and neck cancer, non-small cell lung cancer, breast cancer, non-Hodgkin's lymphoma, colorectal cancer, gastric cancer, and hepatocellular carcinoma[107-113]. In each case, the mutations have been located within the DD, which is important for connecting the receptor to the apoptotic signal transduction pathway. The genes for the four TRAIL receptors are tightly clustered on human chromosome 8p21-22[28, 39, 40], suggesting they evolved relatively recently via gene duplication. It remains to be determined if the other TRAIL receptor genes are also mutated in some cancers that alter tumor cell sensitivity to TRAIL-induced apoptosis.

## 7. CONCLUDING THOUGHTS

The development of TRAIL as a potential anti-cancer therapeutic arose from its ability to potently induce apoptosis in tumor cells, while having little or no detectable cytotoxic effects on normal cells and tissues. The identification of five distinct receptors that can bind TRAIL has significantly increased the potential complexity of this receptor/ligand system in terms of understanding normal physiological functions and as a therapeutic molecule. Not only will it be essential to optimize the form of TRAIL used in therapy, as well as the mode of delivery, it may be even more important to understand the mechanisms by which tumor cells regulate their sensitivity to TRAIL-induced apoptosis. At the same time, elucidating the molecular basis that protects normal cells and tissues from TRAIL will also be essential in formulating TRAIL-based therapies for cancer that will not be toxic to normal cells within the body. Since its identification nearly 10 years ago, there has been a great deal learned how TRAIL functions within the body in tumor and non-tumor scenarios. Future studies will continue to develop TRAIL into the cancer therapeutic many believe it has.

## ACKNOWLEDGEMENTS

The authors wish to thank Drs. Ben Elzey and Timothy Ratliff for careful reading of the manuscript. This work was supported by a Department of Defense Prostate Cancer Research Program New Investigator Award (PC010599), an Alliance for Cancer Gene Therapy Young Investigator Award, and a University of Iowa Carver College of Medicine Medical Research Initiative Grant.

## REFERENCES

1. Cohen JJ. Apoptosis. Immunol Today 1993;14(3):126-130.
2. Wyllie AH, Kerr JF, Currie AR. Cell death: the significance of apoptosis. Int Rev Cytol 1980;68:251-306.
3. Vaux DL, Strasser A. The molecular biology of apoptosis. Proc Natl Acad Sci U S A 1996;93(6):2239-2244.
4. Kerr JF, Wyllie AH, Currie AR. Apoptosis: a basic biological phenomenon with wide-ranging implications in tissue kinetics. Br J Cancer 1972;26(4):239-257.
5. McConkey DJ, Nicotera P, Orrenius S. Signalling and chromatin fragmentation in thymocyte apoptosis. Immunol Rev 1994;142:343-363.
6. Godman GC, Miranda AF, Deitch AD, Tanenbaum SW. Action of cytochalasin D on cells of established lines. III. Zeiosis and movements at the cell surface. J Cell Biol 1975;64(3):644-667.
7. Savill J, Fadok V, Henson P, Haslett C. Phagocyte recognition of cells undergoing apoptosis. Immunol Today 1993;14(3):131-136.
8. Fadok VA, Voelker DR, Campbell PA, Cohen JJ, Bratton DL, Henson PM. Exposure of phosphatidylserine on the surface of apoptotic lymphocytes triggers specific recognition and removal by macrophages. J Immunol 1992;148(7):2207-2216.
9. Martin SJ, Reutelingsperger CP, McGahon AJ, et al. Early redistribution of plasma membrane phosphatidylserine is a general feature of apoptosis regardless of the initiating stimulus: inhibition by overexpression of Bcl-2 and Abl. J Exp Med 1995;182(5):1545-1556.
10. van Meer G. Transport and sorting of membrane lipids. Curr Opin Cell Biol 1993;5(4):661-673.
11. Bevers EM, Tilly RH, Senden JM, Comfurius P, Zwaal RF. Exposure of endogenous phosphatidylserine at the outer surface of stimulated platelets is reversed by restoration of aminophospholipid translocase activity. Biochemistry 1989;28(6):2382-2387.
12. Steller H. Mechanisms and genes of cellular suicide. Science 1995;267(5203):1445-1449.
13. Thompson CB. Apoptosis in the pathogenesis and treatment of disease. Science 1995;267(5203):1456-1462.

14. Armitage RJ. Tumor necrosis factor receptor superfamily members and their ligands. Curr Opin Immunol 1994;6(3):407-413.
15. Smith CA, Farrah T, Goodwin RG. The TNF receptor superfamily of cellular and viral proteins: activation, costimulation, and death. Cell 1994;76(6):959-962.
16. Tartaglia LA, Ayres TM, Wong GH, Goeddel DV. A novel domain within the 55 kd TNF receptor signals cell death. Cell 1993;74(5):845-853.
17. Itoh N, Nagata S. A novel protein domain required for apoptosis. Mutational analysis of human Fas antigen. J Biol Chem 1993;268(15):10932-10937.
18. Cerami A, Beutler B. The role of cachectin/TNF in endotoxic shock and cachexia. Immunol Today 1988;9(1):28-31.
19. Alderson MR, Tough TW, Davis-Smith T, et al. Fas ligand mediates activation-induced cell death in human T lymphocytes. J Exp Med 1995;181(1):71-77.
20. Zheng L, Fisher G, Miller RE, Peschon J, Lynch DH, Lenardo MJ. Induction of apoptosis in mature T cells by tumour necrosis factor. Nature 1995;377(6547):348-351.
21. Griffith TS, Brunner T, Fletcher SM, Green DR, Ferguson TA. Fas ligand-induced apoptosis as a mechanism of immune privilege. Science 1995;270(5239):1189-1192.
22. Hahne M, Rimoldi D, Schroter M, et al. Melanoma cell expression of Fas(Apo-1/CD95) ligand: implications for tumor immune escape. Science 1996;274(5291):1363-1366.
23. Wiley SR, Schooley K, Smolak PJ, et al. Identification and characterization of a new member of the TNF family that induces apoptosis. Immunity 1995;3(6):673-682.
24. Pitti RM, Marsters SA, Ruppert S, Donahue CJ, Moore A, Ashkenazi A. Induction of apoptosis by Apo-2 ligand, a new member of the tumor necrosis factor cytokine family. J. Biol. Chem. 1996;271(22):12687-12690.
25. Cha SS, Kim MS, Choi YH, et al. 2.8 A resolution crystal structure of human TRAIL, a cytokine with selective antitumor activity. Immunity 1999;11(2):253-261.
26. Griffith TS, Chin WA, Jackson GC, Lynch DH, Kubin MZ. Intracellular regulation of TRAIL-induced apoptosis in human melanoma cells. J. Immunol. 1998;161(6):2833-2840.
27. Griffith TS, Rauch CT, Smolak PJ, et al. Functional analysis of TRAIL receptors using monoclonal antibodies. J. Immunol. 1999;162(5):2597-2605.
28. Degli-Esposti MA, Dougall WC, Smolak PJ, Waugh JY, Smith CA, Goodwin RG. The novel receptor TRAIL-R4 induces NF-kappaB and protects against TRAIL-mediated apoptosis, yet retains an incomplete death domain. Immunity 1997;7(6):813-820.
29. Griffith TS, Lynch DH. TRAIL: a molecule with multiple receptors and control mechanisms. Curr. Opin. Immunol. 1998;10(5):559-563.
30. Katsikis PD, Garcia-Ojeda ME, Torres-Roca JF, et al. Interleukin-1b converting enzyme-like protease involvement in Fas-induced and activation-induced peripheral blood T cell apoptosis in HIV infection. TNF-related apoptosis-inducing ligand can mediate activation-induced T cell death in HIV infection. J. Exp. Med. 1997;186:1365-1372.
31. Kayagaki N, Yamaguchi N, Nakayama M, Eto H, Okumura K, Yagita H. Type I interferons (IFNs) regulate tumor necrosis factor-related apoptosis-inducing ligand (TRAIL) expression on human T cells: A novel mechanism for the antitumor effects of type I IFNs. J. Exp. Med. 1999;189(9):1451-1460.
32. Zamai L, Ahmad M, Bennett IM, Azzoni L, Alnemri ES, Perussia B. Natural killer (NK) cell-mediated cytotoxicity: differential use of TRAIL and Fas ligand by immature and mature primary human NK cells. J. Exp. Med. 1998;88:2375-2380.
33. Griffith TS, Wiley SR, Kubin MZ, Sedger LM, Maliszewski CR, Fanger NA. Monocyte-mediated tumoricidal activity via the tumor necrosis factor-related cytokine, TRAIL. J. Exp. Med. 1999;189(8):1343-1354.

34. Fanger NA, Maliszewski CR, Schooley K, Griffith TS. Human dendritic cells mediate cellular apoptosis via tumor necrosis factor-related apoptosis-inducing ligand (TRAIL). J. Exp. Med. 1999;190(8):1155-1164.
35. Kemp TJ, Elzey BD, Griffith TS. Plasmacytoid dendritic cell-derived IFN-alpha induces TNF-related apoptosis-inducing ligand/Apo-2L-mediated antitumor activity by human monocytes following CpG oligodeoxynucleotide stimulation. J Immunol 2003;171(1):212-218.
36. Pan G, O'Rourke K, Chinnaiyan AM, et al. The receptor for the cytotoxic ligand TRAIL. Science 1997;276(5309):111-113.
37. Pan G, Ni J, Wei YF, Yu G, Gentz R, Dixit VM. An antagonist decoy receptor and a death domain-containing receptor for TRAIL. Science 1997;277(5327):815-818.
38. Sheridan JP, Marsters SA, Pitti RM, et al. Control of TRAIL-induced apoptosis by a family of signaling and decoy receptors. Science 1997;277(5327):818-821.
39. Walczak H, Degli-Esposti MA, Johnson RS, et al. TRAIL-R2: a novel apoptosis-mediating receptor for TRAIL. EMBO J. 1997;16(17):5386-5397.
40. Degli-Esposti MA, Smolak PJ, Walczak H, et al. Cloning and characterization of TRAIL-R3, a novel member of the emerging TRAIL receptor family. J. Exp. Med. 1997;186:1165-1170.
41. Marsters SA, Sheridan JP, Pitti RM, et al. A novel receptor for Apo2L/TRAIL contains a truncated death domain. Curr. Biol. 1997;7(12):1003-1006.
42. Emery JG, McDonnell P, Burke MB, et al. Osteoprotegerin is a receptor for the cytotoxic ligand TRAIL. J Biol Chem 1998;273(23):14363-14367.
43. Kischkel FC, Hellbardt S, Behrmann I, et al. Cytotoxicity-dependent APO-1 (Fas/CD95)-associated proteins form a death-inducing signaling complex (DISC) with the receptor. Embo J 1995;14(22):5579-5588.
44. Chinnaiyan AM, O'Rourke K, Tewari M, Dixit VM. FADD, a novel death domain-containing protein, interacts with the death domain of Fas and initiates apoptosis. Cell 1995;81(4):505-512.
45. Medema JP, Scaffidi C, Kischkel FC, et al. FLICE is activated by association with the CD95 death-inducing signaling complex (DISC). Embo J 1997;16(10):2794-2804.
46. Thornberry NA. Caspases: key mediators of apoptosis. Chem Biol 1998;5(5):R97-103.
47. Rotonda J, Nicholson DW, Fazil KM, et al. The three-dimensional structure of apopain/CPP32, a key mediator of apoptosis. Nat Struct Biol 1996;3(7):619-625.
48. Walker NP, Talanian RV, Brady KD, et al. Crystal structure of the cysteine protease interleukin-1 beta-converting enzyme: a (p20/p10)2 homodimer. Cell 1994;78(2):343-352.
49. Li H, Zhu H, Xu CJ, Yuan J. Cleavage of BID by caspase 8 mediates the mitochondrial damage in the Fas pathway of apoptosis. Cell 1998;94(4):491-501.
50. Sedger LM, Glaccum MB, Schuh JC, et al. Characterization of the in vivo function of TNF-alpha-related apoptosis-inducing ligand, TRAIL/Apo2L, using TRAIL/Apo2L gene-deficient mice. Eur. J. Immunol. 2002;32(8):2246-2254.
51. Takeda K, Hayakawa Y, Smyth MJ, et al. Involvement of tumor necrosis factor-related apoptosis-inducing ligand in surveillance of tumor metastasis by liver natural killer cells. Nat. Med. 2001;7(1):94-100.
52. Smyth MJ, Cretney E, Takeda K, et al. Tumor necrosis factor-related apoptosis-inducing ligand (TRAIL) contributes to interferon gamma-dependent natural killer cell protection from tumor metastasis. J. Exp. Med. 2001;193(6):661-670.
53. Takeda K, Smyth MJ, Cretney E, et al. Critical role for tumor necrosis factor-related apoptosis-inducing ligand in immune surveillance against tumor development. J. Exp. Med. 2002;195(2):161-169.

54. Cretney E, Takeda K, Yagita H, Glaccum M, Peschon JJ, Smyth MJ. Increased susceptibility to tumor initiation and metastasis in TNF-related apoptosis-inducing ligand-deficient mice. J. Immunol. 2002;168(3):1356-1361.
55. Seki N, Hayakawa Y, Brooks AD, et al. Tumor necrosis factor-related apoptosis-inducing ligand-mediated apoptosis is an important endogenous mechanism for resistance to liver metastases in murine renal cancer. Cancer Res. 2003;63(1):207-213.
56. Hilliard B, Wilmen A, Seidel C, Liu TS, Goke R, Chen Y. Roles of TNF-related apoptosis-inducing ligand in experimental autoimmune encephalomyelitis. J Immunol 2001;166(2):1314-1319.
57. Lamhamedi-Cherradi SE, Zheng SJ, Maguschak KA, Peschon J, Chen YH. Defective thymocyte apoptosis and accelerated autoimmune diseases in TRAIL-/- mice. Nat Immunol 2003;4(3):255-260.
58. Mi QS, Ly D, Lamhamedi-Cherradi SE, et al. Blockade of tumor necrosis factor-related apoptosis-inducing ligand exacerbates type 1 diabetes in NOD mice. Diabetes 2003;52(8):1967-1975.
59. Lamhamedi-Cherradi SE, Zheng S, Tisch RM, Chen YH. Critical roles of tumor necrosis factor-related apoptosis-inducing ligand in type 1 diabetes. Diabetes 2003;52(9):2274-2278.
60. Cretney E, Uldrich AP, Berzins SP, Strasser A, Godfrey DI, Smyth MJ. Normal thymocyte negative selection in TRAIL-deficient mice. J Exp Med 2003;198(3):491-496.
61. Griffith TS, Yu X, Herndon JM, Green DR, Ferguson TA. CD95-induced apoptosis of lymphocytes in an immune privileged site induces immunological tolerance. Immunity 1996;5(1):7-16.
62. Bellgrau D, Gold D, Selawry H, Moore J, Franzusoff A, Duke RC. A role for CD95 ligand in preventing graft rejection. Nature 1995;377(6550):630-632.
63. Hunt JS, Vassmer D, Ferguson TA, Miller L. Fas ligand is positioned in mouse uterus and placenta to prevent trafficking of activated leukocytes between the mother and the conceptus. J Immunol 1997;158(9):4122-4128.
64. Lee HO, Herndon JM, Barreiro R, Griffith TS, Ferguson TA. TRAIL: a mechanism of tumor surveillance in an immune privileged site. J. Immunol. 2002;169(9):4739-4744.
65. Phillips TA, Ni J, Pan G, et al. TRAIL (Apo-2L) and TRAIL receptors in human placentas: implications for immune privilege. J Immunol 1999;162(10):6053-6059.
66. Krieg AM. CpG motifs in bacterial DNA and their immune effects. Annu. Rev. Immunol. 2002;20:709-760.
67. Elkins KL, Rhinehart-Jones TR, Stibitz S, Conover JS, Klinman DM. Bacterial DNA containing CpG motifs stimulates lymphocyte-dependent protection of mice against lethal infection with intracellular bacteria. J Immunol 1999;162(4):2291-2298.
68. Morales A, Eidinger D, Bruce AW. Intracavitary Bacillus Calmette-Guerin in the treatment of superficial bladder tumors. J Urol 1976;116(2):180-183.
69. Ludwig AT, Moore JM, Luo Y, et al. Tumor necrosis factor-related apoptosis-inducing ligand: a novel mechanism for Bacillus Calmette-Guerin-induced antitumor activity. Cancer Res 2004;64(10):3386-3390.
70. Walczak H, Miller RE, Ariail K, et al. Tumoricidal activity of tumor necrosis factor-related apoptosis-inducing ligand in vivo. Nat. Med. 1999;5(2):157-163.
71. Ashkenazi A, Pai RC, Fong S, et al. Safety and antitumor activity of recombinant soluble Apo2 ligand. J. Clin. Invest. 1999;104(2):155-162.
72. Jo M, Kim TH, Seol DW, et al. Apoptosis induced in normal human hepatocytes by tumor necrosis factor-related apoptosis-inducing ligand. Nat Med 2000;6(5):564-567.

73. Ozoren N, Kim K, Burns TF, Dicker DT, Moscioni AD, El-Deiry WS. The caspase 9 inhibitor Z-LEHD-FMK protects human liver cells while permitting death of cancer cells exposed to tumor necrosis factor-related apoptosis-inducing ligand. Cancer Res 2000;60(22):6259-6265.
74. Hymowitz SG, O'Connell MP, Ultsch MH, et al. A unique zinc-binding site revealed by a high-resolution X-ray structure of homotrimeric Apo2L/TRAIL. Biochemistry 2000;39(4):633-640.
75. Lawrence D, Shahrokh Z, Marsters S, et al. Differential hepatocyte toxicity of recombinant Apo2L/TRAIL versions. Nat Med 2001;7(4):383-385.
76. Gliniak B, Le T. Tumor necrosis factor-related apoptosis-inducing ligand's antitumor activity in vivo is enhanced by the chemotherapeutic agent CPT-11. Cancer Res. 1999;59(24):6153-6158.
77. Chinnaiyan AM, Prasad U, Shankar S, et al. Combined effect of tumor necrosis factor-related apoptosis-inducing ligand and ionizing radiation in breast cancer therapy. Proc. Natl. Acad. Sci. U. S. A. 2000;97(4):1754-1759.
78. Griffith TS, Anderson RD, Davidson BL, Williams RD, Ratliff TL. Adenoviral-mediated transfer of the TNF-related apoptosis-inducing ligand/Apo-2 ligand gene induces tumor cell apoptosis. J. Immunol. 2000;165(5):2886-2894.
79. Denis LJ. The role of active treatment in early prostate cancer. Radiother Oncol 2000;57(3):251-258.
80. Stone NN, Stock RG. Prostate brachytherapy in patients with prostate volumes >/= 50 cm(3): dosimetic analysis of implant quality. Int J Radiat Oncol Biol Phys 2000;46(5):1199-1204.
81. Steiner MS, Gingrich JR. Gene therapy for prostate cancer: where are we now? J Urol 2000;164(4):1121-1136.
82. Herman JR, Adler HL, Aguilar-Cordova E, et al. In situ gene therapy for adenocarcinoma of the prostate: a phase I clinical trial. Hum Gene Ther 1999;10(7):1239-1249.
83. Griffith TS, Broghammer EL. Suppression of tumor growth following intralesional therapy with TRAIL recombinant adenovirus. Mol Ther 2001;4(3):257-266.
84. Inoue H, Shiraki K, Ohmori S, et al. Histone deacetylase inhibitors sensitize human colonic adenocarcinoma cell lines to TNF-related apoptosis inducing ligand-mediated apoptosis. Int J Mol Med 2002;9(5):521-525.
85. Zhang XD, Gillespie SK, Borrow JM, Hersey P. The histone deacetylase inhibitor suberic bishydroxamate: a potential sensitizer of melanoma to TNF-related apoptosis-inducing ligand (TRAIL) induced apoptosis. Biochem Pharmacol 2003;66(8):1537-1545.
86. Rosato RR, Almenara JA, Dai Y, Grant S. Simultaneous activation of the intrinsic and extrinsic pathways by histone deacetylase (HDAC) inhibitors and tumor necrosis factor-related apoptosis-inducing ligand (TRAIL) synergistically induces mitochondrial damage and apoptosis in human leukemia cells. Mol Cancer Ther 2003;2(12):1273-1284.
87. Guo F, Sigua C, Tao J, et al. Cotreatment with histone deacetylase inhibitor LAQ824 enhances Apo-2L/tumor necrosis factor-related apoptosis inducing ligand-induced death inducing signaling complex activity and apoptosis of human acute leukemia cells. Cancer Res 2004;64(7):2580-2589.
88. Kim YH, Park JW, Lee JY, Kwon TK. Sodium butyrate sensitizes TRAIL-mediated apoptosis by induction of transcription from the DR5 gene promoter through Sp1 sites in colon cancer cells. Carcinogenesis 2004.

89. Spierings DC, de Vries EG, Vellenga E, et al. Tissue distribution of the death ligand TRAIL and its receptors. J Histochem Cytochem 2004;52(6):821-831.
90. Beg AA, Baltimore D. An essential role for NF-kappaB in preventing TNF-alpha-induced cell death. Science 1996;274(5288):782-784.
91. Van Antwerp DJ, Martin SJ, Kafri T, Green DR, Verma IM. Suppression of TNF-alpha-induced apoptosis by NF-kappaB. Science 1996;274(5288):787-789.
92. Chaudhary PM, Eby M, Jasmin A, Bookwalter A, Murray J, Hood L. Death receptor 5, a new member of the TNFR family, and DR4 induce FADD-dependent apoptosis and activate the NF-kappaB pathway. Immunity 1997;7(6):821-830.
93. Schneider P, Thome M, Burns K, et al. TRAIL receptors 1 (DR4) and 2 (DR5) signal FADD-dependent apoptosis and activate NF-kappaB. Immunity 1997;7(6):831-836.
94. Wallach D, Boldin M, Goncharov T, et al. Exploring cell death mechanisms by analyzing signaling cascades of the TNF/NGF receptor family. Behring Inst Mitt 1996(97):144-155.
95. Chinnaiyan AM, Dixit VM. The cell-death machine. Curr Biol 1996;6(5):555-562.
96. Fulda S, Meyer E, Debatin KM. Inhibition of TRAIL-induced apoptosis by Bcl-2 overexpression. Oncogene 2002;21(15):2283-2294.
97. Deng Y, Lin Y, Wu X. TRAIL-induced apoptosis requires Bax-dependent mitochondrial release of Smac/DIABLO. Genes Dev 2002;16(1):33-45.
98. Griffith TS, Fialkov JM, Scott DL, et al. Induction and regulation of tumor necrosis factor-related apoptosis-inducing ligand/Apo-2 ligand-mediated apoptosis in renal cell carcinoma. Cancer Res. 2002;62(11):3093-3099.
99. Ng CP, Bonavida B. X-linked inhibitor of apoptosis (XIAP) blocks Apo2 ligand/tumor necrosis factor-related apoptosis-inducing ligand-mediated apoptosis of prostate cancer cells in the presence of mitochondrial activation: sensitization by overexpression of second mitochondria-derived activator of caspase/direct IAP-binding protein with low pl (Smac/DIABLO). Mol Cancer Ther 2002;1(12):1051-1058.
100. Shiiki K, Yoshikawa H, Kinoshita H, et al. Potential mechanisms of resistance to TRAIL/Apo2L-induced apoptosis in human promyelocytic leukemia HL-60 cells during granulocytic differentiation. Cell Death Differ 2000;7(10):939-946.
101. Chen X, Thakkar H, Tyan F, et al. Constitutively active Akt is an important regulator of TRAIL sensitivity in prostate cancer. Oncogene 2001;20(42):6073-6083.
102. Nesterov A, Lu X, Johnson M, Miller GJ, Ivashchenko Y, Kraft AS. Elevated AKT activity protects the prostate cancer cell line LNCaP from TRAIL-induced apoptosis. J Biol Chem 2001;276(14):10767-10774.
103. Jin Z, Dicker DT, El-Deiry WS. Enhanced sensitivity of G1 arrested human cancer cells suggests a novel therapeutic strategy using a combination of simvastatin and TRAIL. Cell Cycle 2002;1(1):82-89.
104. Algeciras-Schimnich A, Griffith TS, Lynch DH, Paya CV. Cell cycle-dependent regulation of FLIP levels and susceptibility to Fas-mediated apoptosis. J. Immunol. 1999;162(9):5205-5211.
105. MacFarlane M, Ahmad M, Srinivasula SM, Fernandes-Alnemri T, Cohen GM, Alnemri ES. Identification and molecular cloning of two novel receptors for the cytotoxic ligand TRAIL. J Biol Chem 1997;272(41):25417-25420.
106. Pineau P, Nagai H, Prigent S, et al. Identification of three distinct regions of allelic deletions on the short arm of chromosome 8 in hepatocellular carcinoma. Oncogene 1999;18(20):3127-3134.
107. Pai SI, Wu GS, Ozoren N, et al. Rare loss-of-function mutation of a death receptor gene in head and neck cancer. Cancer Res 1998;58(16):3513-3518.

108. Lee SH, Shin MS, Kim HS, et al. Alterations of the DR5/TRAIL receptor 2 gene in non-small cell lung cancers. Cancer Res 1999;59(22):5683-5686.
109. Shin MS, Kim HS, Lee SH, et al. Mutations of tumor necrosis factor-related apoptosis-inducing ligand receptor 1 (TRAIL-R1) and receptor 2 (TRAIL-R2) genes in metastatic breast cancers. Cancer Res 2001;61(13):4942-4946.
110. Lee SH, Shin MS, Kim HS, et al. Somatic mutations of TRAIL-receptor 1 and TRAIL-receptor 2 genes in non-Hodgkin's lymphoma. Oncogene 2001;20(3):399-403.
111. Arai T, Akiyama Y, Okabe S, Saito K, Iwai T, Yuasa Y. Genomic organization and mutation analyses of the DR5/TRAIL receptor 2 gene in colorectal carcinomas. Cancer Lett 1998;133(2):197-204.
112. Park WS, Lee JH, Shin MS, et al. Inactivating mutations of KILLER/DR5 gene in gastric cancers. Gastroenterology 2001;121(5):1219-1225.
113. Jeng YM, Hsu HC. Mutation of the DR5/TRAIL receptor 2 gene is infrequent in hepatocellular carcinoma. Cancer Lett 2002;181(2):205-208.

Chapter 12

# HISTONE DEACETYLASE INHIBITORS AS A TREATMENT OF TRAIL-RESISTANT CANCERS

Jiri Neuzil[1,2],*, Ladislav Andera[2], and Brian Gabrielli[3]

*[1]Apoptosis Research Group, School of Health Science, Griffith University, Southport, Qld, Australia, [2]Laboratory of Cell Signalling and Apoptosis, Institute of Molecular Genetics, Czech Academy of Sciences, Prague, Czech Republic, and [3]Centre for Immunology and Cancer Research, Princess Alexandra Hospital, University of Queensland, Brisbane, Qld, Australia*

Abstract: Histone deacetylas inhibitors (HDIs) are a novel prospective group of potential anti-cancer agents, some of them being tested in a clinical setting. The major mode of action of these compounds is inhibition of the cell cycle transition and, consequently, differential regulation of a number of genes necessary for cell proliferation, while favouring expression of genes rather associated with anti-proliferative pathways and with cell death signalling. This implies a possible role of HDIs in adjuvant treatment of tumours resistant to other agents, operating via a different molecular mechanism. The TNF-related apoptosis-inducing ligant (TRAIL) is a promising immunological inducer of apoptosis efficient against a variety of tumours with a remarkably selective mode of action. However, some malignancies are resistant to TRAIL treatment. In this paper, we review our current knowledge on HDI-mediated sensitisation of TRAIL-non-responsive tumours to this apoptogen and suggest a future clinical potential of HDIs and TRAIL in cancer management.

Key words: Apoptosis, cell cycle arrest, death receptor, histone deacetylase inhibitor

## 1. INTRODUCTION

Recent decade has witnessed considerable progress in our understanding of molecular biology of cancer, resulting in novel strategies of neoplastic disease management. Of special interest is immunotherapy of cancer since



*M. Sluyser (ed.), Application of Apoptosis to Cancer Treatment,* 271-291.

'classical' approaches, including radiotherapy and pharmacological agents, are often resulting in secondary deleterious effects and their targets, the cancer cells, constantly mutate to become refractory to these treatments.

Several novel anti-neoplastic strategies emerged in recent years, including the use of ligands interacting with their receptors, often preferentially expressed on malignant cells. This may be best exemplified by the tumour necrosis factor-related apoptosis-inducing ligand (TRAIL) that binds to its cognate death receptors or decoy receptors. Since the latter are frequently over-expressed on normal rather than malignant tissues, TRAIL is highly selective for cancer cells, although other mechanisms are likely to be involved that determine the remarkable specificity of the immunological apoptogen. However, not all cancers are susceptible to TRAIL for a variety of reasons, including a shift in the ratio of the anti-apoptotic and the pro-apoptotic mediators in favour of the former. Strategies are being developed to 'rectify' this considerable complication of potential treatment, which would open the 'normal' TRAIL pro-apoptotic signalling routes. These include combinatorial approach such that cancers resistant to TRAIL are sensitized by small molecules, an approach that may prove to have significant clinical relevance.

## 2. MOLECULAR MECHANISM OF TRAIL SIGNALLING

TRAIL (Apo2L) is a member of the tumour necrosis factor-α (TNF-α) superfamily that was originally discovered and cloned based on its sequence homology to the Fas ligand and TNF-α[1,2]. TRAIL is predominantly expressed by immature NK cells[3], mature NK cells stimulated with IL-2 and interferons[3-5] and type I IFN-activated human peripheral blood T cells[4], dendritic cells and monocytes[6,7]. In contrast to the expression of TRAIL, which is restricted to immune cells, its receptors are found on virtually all cell types. Five TRAIL receptors have been identified in human cells, with two of them (TRAIL-R1/DR4, TRAIL-R2/DR5) inducing apoptosis, the other three (decoy receptor-1, DcR1, also known as TRAIL-R3, DcR2/TRAIL-R4, osteoprotegerin) serving as inhibitory decoy receptors (reviewed in Ref. 8). The fundamental function of TRAIL appears to be anti-tumour surveillance, *i.e.* induction of tumour cell apoptosis. Mice with targeted disruption of TRAIL develop normally but have increased susceptibility to tumour allografts and show rapid metastatic spread[9,10]. The suggested role of TRAIL in the negative selection of T cells is still controversial and has not been confirmed in recent reports[11,12].

Upon binding to its pro-apoptotic, death domain-containing receptors (DR4, DR5), TRAIL induces rapid formation of the death-inducing signalling complex (DISC) around the intracellular domains of these receptors. The TRAIL receptor DISC contains the adaptor FADD (Fas-associated death domain) protein and the apoptosis initiator caspases-8 and/or -10 [13-15]. FLICE inhibitory protein (FLIP), an inactive analogue and an inhibitor of caspase-8/-10, is also a component of the death receptors DISC complexes[16]. Over-expression of its short or long variant blocks DISC-induced activation of initiator caspases and, consequently, inhibits TRAIL-induced apoptosis (reviewed in Ref. 17). It has been suggested that physiological levels of FLIP long ($FLIP_L$) promote pro-caspase-8 self-processing at Fas and presumably at the TRAIL receptor DISC[18]. Activated initiator caspases then cleave the Bcl-2 homology-3 (BH3)-only protein Bid, and the truncated Bid (tBid) then translocates to mitochondria. Here tBid, either alone or in association with pro-apoptotic proteins Bax and/or Bak, triggers mitochondrial events including a drop in the mitochondrial inner trans-membrane potential, release of cytochrome c, Smac/Diablo, Omi/Hrt2a, the apoptosis-inducing factor (AIF) and other apoptosis-promoting factors[19-22]. As an alternative to apoptosis, TRAIL can also induce necrotic cell death, most probably through receptor-interacting protein (RIP)-dependent signalling[23,24].

In addition to induction of apoptosis or necrosis, activated TRAIL receptors can unleash signalling leading to activation of the nuclear factor-κB (NF-κB), mitogen-activated protein (MAP) and stress kinases (Figure 1). TRAIL-induced activation of NF-κB is apparently dependent on the adapter protein RIP and counteracts TRAIL-induced apoptosis[25-27]. In agreement with this, suppression of NF-κB activity enhances TRAIL-induced apoptosis in tumour cells of various origins[28-30]. TRAIL-mediated activation of extracellular-regulated kinase-1/2 (ERK1/2) also belongs to the anti-apoptotic arm of TRAIL signalling, and suppression of ERK1/2 activation enhances the pro-apoptotic action of TRAIL, similar to inhibition of NF-κB[31,32]. Interestingly, TRAIL-induced activation of ERK1/2 may play a role in the survival and proliferation of primary human vascular endothelial cells or in maturation of normal erythroid progenitors[33]. ERK1/2 activation by TRAIL occurs in two waves. The early activation of ERK1/2 is caspase-independent and peaks at 10-15 min after addition of TRAIL, the second wave starts at 40-50 min and is caspase-dependent (L.A., unpublished). However, ERK1/2 activation does not occur in all TRAIL-responsive cells. In contrast to ERK1/2 activation, TRAIL-induced phoshorylation of c-Jun N-terminal kinases (JNK) is caspase-dependent and occurs at 50-70 min after administration of TRAIL[26,34,35]. JNK activation is not essential for

TRAIL-induced apoptosis, although it contributes to it[36]. RIP kinase apparently plays an important role in TRAIL-induced activation of JNKs[35].

Thus, the complex network of signalling pathways of TRAIL apoptosis is regulated at multiple levels, including promoters and inhibitors of the overall process. In cancers, the anti-apoptotic factors are often over-expressed, causing the neoplastic disease to be refractory to TRAIL treatment. Therefore, there has been a search for strategies that would overcome this resistance. Histone deacetylase (HDAC) inhibitors (HDI) have emerged recently as clinically interesting compounds that can sensitise resistant cells to the TRAIL complex mechanisms.

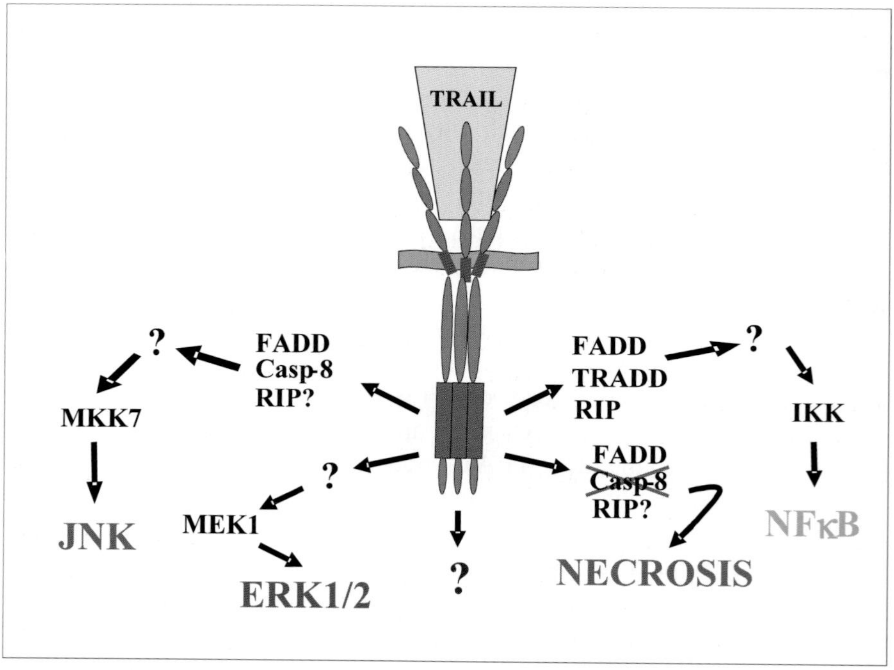


*Figure 1.* TRAIL-induced non-apoptotic signalling. In addition to DISC-dependent apoptosis TRAIL engagement of its pro-apoptotic receptors triggers in condition- and cell-specific manner other signalling pathways. Most of them (activation of JNK, NF-κB, and necrosis) depend on FADD-RIP connection. Activation of ERK1/2 requires MEK1 but there are no data as yet about its upstream, TRAIL-dependent activators.

## 3. HISTONE DEACETYLASE INHIBITORS AS ANTI-CANCER AGENTS

The potential of HDIs as anti-cancer drugs has been evaluated in recent reviews[37,38]. In this section, we will provide a very brief overview of their mechanisms of action, as these agents may promote TRAIL-induced apoptosis in tumour cells. In fact, reports are emerging now (see below), that this is, indeed, the case.

There are several structural classes of HDIs, including hydroxamic acids, cyclic peptides, short chain fatty acids and benzamides, the most widely studied being the hydroxamic acids, exemplified by suberoylanilide hydroxamic acid (SAHA), suberic bishydroxamic acid (SBHA) and NVP-LAQ824. All HDIs inhibit class I (HDAC1-3 and 8) and class II histone deacetylases (HDAC4-7, 9 and 10), but are inactive towards the $NAD^+$-dependent class III HDACs (SIRT 1-7). The class I and II HDACs are relatively widely expressed, have distinctive localisations and binding partners, and specific functions are emerging for the individual isoforms[39,40]. Most HDIs display little specificity towards particular class I and II isoforms, although a number of these drugs are incapable of inhibiting the tubulin deacetylase HDAC6[41,42], demonstrating the potential to produce isoform-specific HDIs.[43-45] The term histone deacetylase is patently inaccurate, as an increasing number of non-histone proteins are being identified as targets for acetylation, including many transcriptional regulators, *eg* p53, E2F, NF-κB[46,47].

Treatment of cells with HDIs can result in differentiation, cell cycle arrest and apoptosis, depending on the HDI and dose used, and the cell type. The cell cycle and apoptotic effects have engaged most interest in terms of the anti-cancer properties of these drugs. A valuable characteristic of HDIs as potential anti-cancer agents is their tumour selectivity. These drugs induce cell death in a wide variety of immortalised, virally transformed and tumour cell lines, but have little toxicity towards normal cell lines and tissues[48]. The molecular basis of this selectivity appears to be the functional status of a G2 phase cell cycle checkpoint, which appears to be defective in a high proportion of tumour cell lines but intact in normal cells[49]. This selectivity is observed not only in actively cycling cells, but also in non-proliferating cells[50] and is likely to underlie the relative lack of dose limiting toxicity in both pre-clinical and clinical trials[37,48,51].

Histone acetylation is generally associated with increased transcriptional activity, although transient histone acetylation is also linked to DNA replication and repair processes[52]. Treatment of cells with HDIs produces a rapid increase in histone acetylation, a consequence of the dynamic histone acetylation and deacetylation in the transcriptionally active euchromatin. The hypo-acetylated, transcriptionally silenced heterochromatin is only transiently acetylated during S phase, a process associated with deposition of chromatin onto the newly replicated heterochromatin DNA[53]. Acetylation of other non-histone proteins is also increased with HDI treatment[42,54], but the major focus of studies on the effects of HDIs has been on changes in transcription.

An almost universal effect of HDI treatment is the up-regulated expression of the cyclin dependent kinase (CDK) inhibitor p21*Waf1/Cip1*, a direct consequence of hyper-acetylation of its promotor region and increased transcription of the gene[55,56]. The transcriptional up-regulation is independent of p53 but does appear to require ataxia telangiectasia mutated (ATM) protein[57].P21*Waf1/Cip1* up-regulation is associated with the G1 phase cell cycle arrest observed in most cell lines treated with HDIs[49,58], but is not solely responsible for the G1 arrest observed. Up-regulation of other G1 phase CDK inhibitors[59-61] and down-regulation of G1/S phase cyclins[49,54,62-64] has also been reported and is likely to contribute to the G1 arrest. Whereas the increased p21*Waf1/Cip1* levels contribute to the anti-proliferative activity of HDIs, they paradoxically reduce the cytotoxicity of HDI treatment. This is due to the ability of p21*Waf1/Cip1* to block the activation of caspase-3, thereby blocking a major executioner caspase function.[65] Deletion or knock-down HDI-induced p21*Waf1/Cip1* expression increase the sensitivity of cells to the cytotoxic effects of HDI treatment and favour apoptosis[58,66-69].

As yet no single mechanism for HDI-induced cell death has been identified. HDI treatment causes cells to undergo aberrant mitosis and results in impairment of normal partitioning of the replicated chromosomes, the catastrophic mitotic failure triggering apoptosis[42]. However, HDI-induced apoptosis is also observed in non-cycling cells, which in asynchronously cycling cell populations that up-regulate p21*Waf1/Cip1* and undergo G1 arrest can account for a majority of cells, indicating that there are other stresses imposed on cells by HDI treatment that trigger an apoptotic response[50].

There also appears to be no single apoptotic mechanism favoured, and both intrinsic and extrinsic pathways, as well as caspase-independent pathways, have been reported. The mechanism utilised may again be cell type-, HDI- and dose-dependent. One common theme is that the drugs disrupt the integrity of the mitochondrial membrane. This is accompanied by loss of mitochondrial membrane potential and release into the cytoplasm of cytochrome c and Smac/DIABLO which potentiate caspase activation[50,68,71-74]. Over-expression of the anti-apoptotic Bcl-2 reduces HDI-induced cell death in all cases reported, the extent of protection possibly reflecting the degree of over-expression, although other factors may be involved[50,70,71,72-75]. Another common feature is the cleavage and activation of the pro-apoptotic BH3-only protein Bid[50,73,74], although in some cases this is a late event in apoptosis[68,72],while it is not detected at all in other cases[75]. HDI-induced apoptosis is also variably dependent on caspase activity. Activation of initiator caspases-9, -8, and -2 and the executioner caspase-3 have been widely observed[50,66,76-78]. Caspase-independent apoptosis has also been reported using the pan-caspase inhibitor z-VAD.fmk[58,73,74]. This may be a consequence of the relative insensitivity of caspase-2 to this inhibitor and its activity being responsible for the cleavage of Bid and the mitochondrial membrane disruption[78,79], or alternatively a calpain-dependent mechanism may be involved[74]. Extrinsic apoptotic pathways involving Fas, the Fas ligand and the death receptor pathways have also been demonstrated in some cases, associated with increased expression of Fas and/or Fas ligand[80,81]. Others have failed to detect either changes in the levels of these proteins or contributions of the extrinsic pathway to HDI-induced apoptosis[68,71,73].

The apoptotic pathways may be determined by the up- and down-regulated expression of pro- and anti-apoptotic proteins after HDI treatment. As with all of the mechanisms discussed above, the effects are variable, although where changes were detected a common theme was down-regulation of anti-apoptotic factors such as Bcl-2, Bcl-$x_L$, Mcl-1 and the FADD inhibitor FLIP, and up regulation of pro-apoptotic factors, including Bax and Bad[76,82]. In many cases these changes in protein expression levels are related to changes in transcription of the genes, which may in turn be a direct consequence of HDI-induced histone hyper-acetylation. The heterogeneity of transcriptional changes following treatment of different cell lines with different HDIs is well demonstrated in a recent report[83]. This study compared cell treatment with equipotent doses of three HDIs and revealed that only 13 genes were consistently up- or down-regulated across all cell lines by the HDIs when the expression of 6,800 genes was examined by micro-array analysis. In any single cell line treated with any single HDI, there are clearly a number of pro- and anti-apoptotic genes up- and down-regulated in response to HDIs[59]. The contribution of any of these expression changes to single agent HDI-induced apoptosis is unknown, although a number of groups have demonstrated that *de novo* protein synthesis is required for HDI-induced apoptosis as evidence that HDI-induced changes in protein expression and transcriptional changes are required[73,75,78]. All of these experimental systems used cycling cell populations, and inhibiting transcription or translation will also block cell cycle progression. Therefore, it is possible that the block in apoptosis is a consequence of cell cycle arrest, and cells not undergoing the HDI-induced mitotic failure do not signal apoptosis[42,49]. Cell death after mitotic failure was independent of the HDI used and observed in a number of cell lines[42] and there is good evidence that the selective cytotoxicity of HDIs may be, at least in some cases, independent of transcriptional changes.[84]

# 4. INHIBITION OF HISTONE DEACETYLASE AS A MODE OF SENSITISATION OF CANCER CELLS TO TRAIL

The major effect of HDIs is deregulation of cell cycle transition, thereby leading to differential expression of a number of genes directly or indirectly linked to apoptosis induction/progression. Resistance of cancer cells to TRAIL stems from regulation of expression of genes crucial for initiation and execution of TRAIL-dependent apoptosis. Assuming two major modes of TRAIL signalling, *i.e.* the strictly receptor-mediated/extrinsic pathway and the route in which mitochondrial mediators are involved (the intrinsic pathway), several types of genes can compromise TRAIL apoptotic killing, and their down-/up-regulation by HDIs may sensitise the cells to the immunological apoptogen[85]. The first paper reporting this phenomenon for colon adenocarcinomas was published a few years ago, but the mechanism was not elucidated at that stage[86].

The basic features of TRAIL signalling and the possible points where resistance may develop and that may serve as targets for intervention are depicted in Figure 2. Resistance to TRAIL involves both the intrinsic and extrinsic pathway, depending on the cell type, *i.e.* whether the cell preferentially responds to TRAIL by signalling directly from caspase-8 to the executioner caspases, or whether, due to low DISC level, the low activity of caspase-8 results in cleavage of Bid and activation of the ensuing mitochondrial mediators.

Obviously, the expression of TRAIL death and decoy receptors is critical for the down-stream activity of the apoptogen and, sometimes, low expression of the former and high expression of the latter compromises cell response. A very recent report provides evidence that HDIs can overcome this complication by changing the ratio of the death receptors to the decoy receptors in favour of the latter by means of up-regulation of DR5 in several malignant cell lines[87]. Since the effect has been reported for trichostatin A, sodium butyrate and SAHA, differential regulation of TRAIL receptors, exemplified by up-regulation of DR5, may be a feature inherent to multiple HDIs. It has been published recently that both DR4 and DR5 are up-regulated by the HDI NVP-LAQ824 in leukemic cells[88]. The same report also showed down-regulation of the caspase-8 inhibitor $FLIP_L$. These findings, together with the data on SAHA –dependent up-regulation of pro-caspase-8 in melanoma cells[82], indicate that multiple HDIs efficiently unmask the early stages of TRAIL pro-apoptotic signalling. Since activation of TRAIL death receptors leads to activation of the pro-survival NF-κB, a process that may be transient in some cases[29], and thereby up-regulate expression of the inhibitor of apoptosis protein (IAP) family members, this paradoxically, can counteract apoptotic signalling from TRAIL death receptors by inhibiting activity of executioner caspases. Several IAPs, including Survivin and XIAP have been shown to by down-regulated by SAHA and LAQ824 in melanoma[82] and leukemic cells[88], respectively, suggesting yet another point of sensitisation of the TRAIL pathway. Moreover, there are emerging data pointing to induction of cytosolic re-localisation of the IAP antagonist Smac/Diablo by butyrate, SAHA and SBHA in leukemic[89] and mesothelioma cells (J.N., unpublished), adding yet another levels of alternative/antagonistic events pertinent to TRAIL signalling that are affected by HDIs.

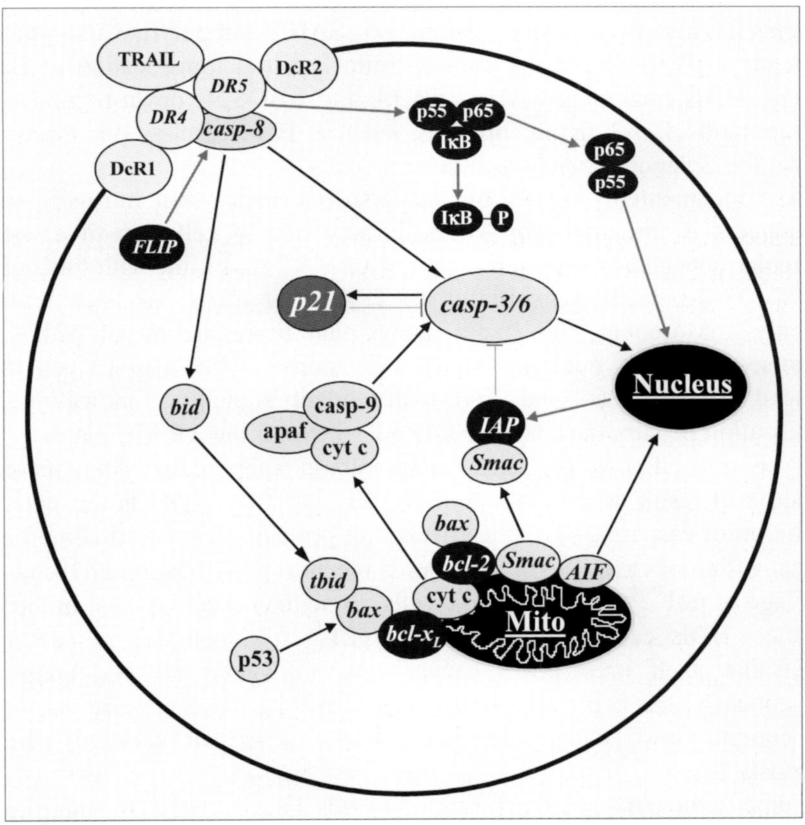


*Figure 2.* Apoptotic pathways in TRAIL signalling and development of TRAIL resistance. Following interaction with their cognate receptors, DR4 and DR5, pro-caspase-8 is recruited to FADD, which leads to activation of the caspase. At low level of DISC, caspase-8 cleaves Bid and tBid then activates apoptosis via the intrinsic, mitochondrial mode. At high level of DISC, caspase-8 directly activates terminal caspases that lead to the entry of the cells into the execution phase. The possible points of resistance to TRAIL include high level of expression of decoy receptors, mutations in the caspase-8 gene, activation of NF-κB that leads to expression of the caspase inhibitors IAPs, as well as the preferentially expression of the

mitochondrial anti-apoptotic proteins, including Bcl-xL and Bcl-2. A number of differentially expressed genes, sensitising cells to TRAIL, are a p21Wafl/Cip1-dependent, p53-independent cell cycle arrest due to the activity of HDIs. The names in italics indicate genes/proteins affected by HDIs and sensitising cells to TRAIL apoptosis.

Smac/Diablo re-localisation is a result of the effect of HDIs on the mitochondrial stability, which is another prominent target of HDI-mediated sensitisation of cancer cells to TRAIL. There has been some effort to understand the effect of HDIs on the mitochondrial targeted pro- and anti-apoptotic proteins. Up-regulation of the pro-apoptotic proteins Bak, Bid, Bim and Bax by SAHA has been reported for melanoma cells, and sensitisation of the cells to TRAIL due to the differential expression of these proteins has also been shown[82]. Moreover, SAHA and butyrate also sensitise leukemic cells to TRIAL by causing mitochondrial translocation of Bax.[89] Several HDIs, including SAHA, SBHA and LAQ824, down-regulated the anti-apoptotic Bcl-2 family proteins, such as Bcl-2, Bcl-$x_L$ and Mcl-1, as shown for melanoma[82] and mesothelioma cells[90].

As documented, expression of both extrinsic and intrinsic signal transducers is modulated by HDIs, largely due to cell growth arrest, in particular dependent on up-regulation of $p21^{Waf1/Cip1}$. In line with this notion, $p21^{Waf1/Cip1}$-dependent sensitisation to TRAIL by HDIs has been reported recently[91]. Moreover, this paper shows that up-regulation of $p21^{Waf1/Cip1}$ sensitised the cells not only to TRAIL apoptosis but also to cell death induced via the Fas- and TNF-α-dependent signalling, indicating that sensitisation of cancer cells by HDIs goes beyond the TRAIL pathway and may be utilised also for other immunologic apoptogens. An apparently paradoxical result of up-regulation of $p21^{Waf1/Cip1}$ by HDIs is the potential inhibition of caspases, since the checkpoint protein is known to compromise the activation of caspase-3[65]. Based on a recent report showing HDI-induced cleavage of $p21^{Waf1/Cip1}$ in leukemic cells[89], one may speculate that in the first instance, HDIs cause up-regulation of $p21^{Waf1/Cip1}$, resulting in cell cycle arrest that itself predisposes cancer cells to TRAIL-induced apoptosis. Subsequently, the cell cycle inhibitor is cleaved so that caspase activity is not compromised; thus, once arrested, the cells can undergo efficient apoptosis.

Some of the points of sensitisation towards TRAIL by HDIs, suggested in Figure 2, are the basis for the clinical utilisation of the intriguing partnership of these agents as discussed below. The current knowledge of genes regulated by HDIs resulting in sensitisation of cancer cells to TRAIL is summarised in Table I. Undoubtedly, novel targets of HDIs and their role in sensitisation of cancer cells to TRAIL are being discovered even during publication of this review, so that the genes listed in Table I will be incomplete in near future.

*Table 1.* Genes regulated by HDIs that sensitise cells to TRAIL apoptosis

| Gene | Type of regulation | Role in cell physiology | HDI type | Reference |
|---|---|---|---|---|
| AIF | Re-localisation | Mitochondrial destabilisation | Butyrate, SAHA | 89 |
| Bak | Up-regulation | Mitochondrial destabilisation | SAHA | 82 |
| Bim | Up-regulation | Mitochondrial destabilisation | SAHA | 82 |
| Bax | Re-localisation | Mitochondrial destabilisation | Butyrate, SAHA | 89 |
| Bcl-$x_L$ | Down-regulation | Mitochondrial destabilisation | SAHA; LAQ824; SBHA | 82; 88; 90 |
| Bcl-2 | Down-regulation | Mitochondrial destabilisation | SAHA; LAQ824 | 82; 88 |
| Bid | Up-regulation | Mitochondrial destabilisation | SAHA | 82 |
| | Cleavage | Mitochondrial destabilisation | Butyrate, SAHA | 89 |
| DR4 | Up-regulation | Facilitates TRAIL binding | LAQ824 | 97 |
| DR5 | Up-regulation | Facilitates TRAIL binding | Trichostatin, butyrate, SAHA; LAQ824 | 87, 88 |
| Mcl-1 | Down-regulation | Mitochondrial destabilisation | SAHA | 82 |
| pro-caspase-3 | Up-regulation | Apoptosis sensitisation | SAHA | 82 |
| pro-caspase-8 | Up-regulation | Apoptosis sensitisation | SAHA | 82 |
| $p21^{Waf1/Cip1}$ | Up-regulation | Cell cycle arrest | LAQ824; butyrate | 88; 91 |
| | Cleavage | Suppression of caspase-inhibitory activity | Butyrate, SAHA | 89 |
| $p27^{Kip1}$ | Up-regulation | Cell cycle arrest | LAQ824 | 88 |
| Smac/Diablo | Re-localisation | Suppression of IAP activity | Butyrate, SAHA; SBHA | 95; J.N. unpubl. |
| Survivin | Down-regulation | Inhibition of caspases | LAQ824 | 88 |
| XIAP | Down-regulation | Inhibition of caspases | SAHA; LAQ824 | 82; 88 |

# 5. CLINICAL ASPECTS OF SENSITISATION OF CANCER CELLS BY HDIS TO TRAIL

HDIs are undergoing clinical trials as single agent therapy for multiple indications. So far, they appear well tolerated with few dose limiting toxicities and have demonstrated *in vivo* effects on histone acetylation[52,92-94]. TRAIL has been demonstrated to have no toxicity towards normal tissue, although some forms of recombinant soluble TRAIL had toxicity towards primary human hepatocytes and brain cells[95-97]. These two classes of potential anti-cancer agents have the desirable feature of tumour selective toxicity, with minimal toxicity towards normal tissue, thus reducing the potential for dose limiting side effects during treatment. The ability of these two modalities to synergise *in vitro* is somewhat unexpected as they target different pathways to apoptosis, especially when the synergy is at doses below the levels required for either agent separately to induce any of the markers of apoptosis. However, the involvement of the mitochondrial death pathway in the combined HDI/TRAIL treatment has been demonstrated by the ability of anti-apoptotic proteins Bcl-2 and Bcl-$x_L$ to reduce if not completely block apoptosis by the combined treatments[88-90]. The mechanisms by which the synergy occurs appear to be varied, although a common theme appears to be the up regulation of death receptor pathway components and down-regulation of anti-apoptotic proteins[82,88-90].However, the most appealing factor of the synergy is that the effects of the combination of drugs are observed at lower doses than are required for either of the individual agents to produced significant levels of apoptosis. These lower doses are far more achievable *in vivo*, and also have a reduced potential for toxic side effects.

TRAIL has shown synergy with a number of common chemotherapeutic drugs in cell lines studies *in vitro* and in xenograft models. This appears to be most prominent in tumours with wild type p53 as DR5, one of the TRAIL receptors is a transcriptional target of p53[98]. HDIs can synergise with a number of drugs, although in most cases this appears to be by blocking HDI-induced $p21^{Waf1/Cip1}$ expression, which blocks caspase-3 activation and apoptosis[58,99,100]. Interestingly, TRAIL does not block HDI-induced $p21^{Waf1/Cip1}$ expression[89]. Indeed, the HDI-induced $p21^{Waf1/Cip1}$, which causes a G1 arrest, may be part of the synergistic mechanism as tumour cells appear more sensitive to TRAIL when arrested in G1 than other cell cycle phases[101].

No spectrum of tumour types has as yet been identified as more or less sensitive to HDI monotherapy either *in vitro* or in clinical trials. However, a range of defects in death receptor signalling pathway including DR4 deletions and DR5 mutations have been identified in tumours including lung, head and neck, gastric cancers and melanomas making them refractory to

TRAIL[102,103]. But, at least in the case of melanoma, there is good evidence that HDI treatment can sensitise these cells to TRAIL-induced apoptosis.

## 6. FUTURE PERSPECTIVES

There is a great hope that novel immunological apoptogens, epitomised by TRAIL, will overcome problems encountered by conventional cancer therapy. One of the most promising features of agents like TRAIL is their high toxicity for cancer cells, while being largely non-toxic to normal cells and tissues. However, multiple cancer cells are relatively resistant to TRAIL due to mutations in crucial genes, often resulting in lack of their expression or expression of dysfunctional proteins. Also, there have been reports showing differential expression of genes that compromises TRAIL's apoptogenic signalling. A frequently used regimen of cancer therapy is combinatorial treatment using synergistic chemotherapeutics with different modes of action so that lower concentrations of the individual agents can be used, lowering the potential of their secondary deleterious activity, and that one agent sensitises the cells to the other. Sometimes, chemotherapy is combined with radiation treatment. Nevertheless, these approaches have proved unsatisfactory in a variety of cancers due to additional mutations that can overcome even this type of treatment. TRAIL has the potential of becoming an anti-cancer agent of choice for reasons discussed in this review. The more widespread use of TRAIL in therapy for a range of types of cancer requires a combinatorial approach, so that the less responsive or resistant cancers will be efficiently eliminated. HDIs have proven efficient inducers of apoptosis as single agents and have clearly demonstrated potential to synergise with TRAIL to target the destruction of tumour cells.

Recent data provide evidence that these agents act as sensitisers of malignant cells to TRAIL-induced apoptosis. Inevitably, these intriguing compounds and their synergism will be a focus of a number of studies, both using cell culture and pre-clinical models. Several clinical trials, currently underway, may result in development of novel, specific anti-cancer strategies efficient across the spectrum of tumour types due to combination of the unique apoptogenic features of both TRAIL and histone deacetylase inhibitors.

## ACKNOWLEDGEMENTS

Work on this manuscript was partially supported by the Queensland Cancer Fund grants to J.N. and by grants K5020115 (Czech Academy of Sciences) and LN00A26 (Czech Ministry of Education) to L.A.

## REFERENCES

1. Wiley, S.R., Schooley, K., Smolak, P.J., Din, W.S., Huang, C.P., Nicholl, J.K., Sutherland, G.R., Smith, T.D., Rauch, C., Smith, C.A., et al. Identification and characterization of a new member of the TNF family that induces apoptosis. *Immunity* 3, 673–682 (1995).
2. Pitti, R.M., Marsters, S.A., Ruppert, S., Donahue, C.J., Moore, A., and Ashkenazi, A. Induction of apoptosis by Apo-2 ligand, a new member of the tumor necrosis factor cytokine family. *J. Biol. Chem.* 271, 12687–12690 (1996).
3. Zamai, L., Ahmad, M., Bennett, I.M., Azzoni, L., Alnemri, E.S., and Perussia, B. Natural killer (NK) cell-mediated cytotoxicity: differential use of TRAIL and Fas ligand by immature and mature primary human NK cells. J. Exp. Med. *188*, 2375–2380 (1998).
4. Kayagaki, N., Yamaguchi, N., Nakayama, M., Eto, H., Okumura, K., and Yagita, H. Type I interferons (IFNs) regulate tumor necrosis factor-related apoptosis-inducing ligand (TRAIL) expression on human T cells: a novel mechanism for the antitumor effects of type I IFNs. *J. Exp. Med.* 189, 1451–1460 (1999).
5. Kayagaki, N., Yamaguchi, N., Nakayama, M., Kawasaki, A., Akiba, H., Okumura, K., and Yagita, H. Involvement of TNF-related apoptosis-inducing ligand in human CD4 T cell-mediated cytotoxicity. J. Immunol. 162, 2639–2647 (1999).
6. Fanger, N.A., Maliszewski, C.R., Schooley, K., and Griffith, T.S. Human dendritic cells mediate cellular apoptosis via tumor necrosis factor-related apoptosis-inducing ligand (TRAIL). *J. Exp. Med.* 190, 1155–1164 (1999).
7. Griffith, T.S., Wiley, S.R., Kubin, M.Z., Sedger, L.M., Maliszewski, C.R., and Fanger, N.A. Monocyte-mediated tumoricidal activity via the tumor necrosis factor-related cytokine TRAIL. *J. Exp. Med.* 189, 1343–1354 (1999).
8. LeBlanc, H.N., and Ashkenazi, A. Apo2L/TRAIL and its death and decoy receptors. *Cell Death Diff.* 10, 66-75 (2003).
9. Cretney, E., Takeda, K., Yagita, H., Glaccum, M., Peschon, J.J., and Smyth, M.J. Increased susceptibility to tumor initiation and metastasis in TNF-related apoptosis-inducing ligand-deficient mice. *J. Immunol.*168, 1356–1361 (2002).
10. Seki, N., Hayakawa, Y., Brooks, A.D., Wine, J., Wiltrout, R.H., Yagita, H., Tanner, J.E., Smyth, M.J., and Sayers, T.J. Tumor necrosis factor-related apoptosis-inducing ligand-mediated apoptosis is an important endogenous mechanism for resistance to liver metastases in murine renal cancer. *Cancer Res.* 63, 207-213 (2003).
11. Lamhamedi-Cherradi, S.E., Zheng, S.J., Maguschak, K.A., Peschon, J., and Chen, Y.H. Defective thymocyte apoptosis and accelerated autoimmune diseases in TRAIL$^{-/-}$ mice. *Nat. Immunol.* 4, 255-260 (2003).
12. Cretney, E., Uldrich, A.P., Berzins, S.P., Strasser, A., Godfrey, D.I., and Smyth, M.J. Normal thymocyte negative selection in TRAIL-deficient mice. *J. Exp. Med.* 198, 491-496 (2003).

13. Bodmer, J.L., Holler, N., Reynard, S., Vinciguerra, P., Schneider, P., Juo, P., Blenis, J., and Tschopp, J. TRAIL receptor-2 signals apoptosis through FADD and caspase-8. *Nat. Cell Biol.* 2, 241–243 (2000).
14. Sprick, M.R., Weigand, M.A., Rieser, E., Rausch, C.T., Juo, P., Blenis, J., Krammer, P.H., and Walczak, H. FADD/MORT1 and caspase-8 are recruited to TRAIL receptors 1 and 2 and are essential for apoptosis mediated by TRAIL receptor 2. *Immunity* 12, 599–609 (2000).
15. Kischkel, F.C., Lawrence, D.A., Tinel, A., LeBlanc, H., Virmani, A., Schow, P., Gazdar, A., Blenis, J., Arnott, D., and Ashkenazi, A. Death receptor recruitment of endogenous caspase-10 and apoptosis initiation in the absence of caspase-8. *J. Biol. Chem.* 276, 46639-46646 (2001).
16. Thome, M., Schneider, P., Hofmann, K., Fickenscher, H., Meinl, E., Neipel, F., Mattmann, C., Burns, K., Bodmer, J.L., Schroter, M., Scaffidi, C., Krammer, P.H., Peter, M.E., and Tschopp, J. Viral FLICE-inhibitory proteins (FLIPs) prevent apoptosis induced by death receptors. *Nature* 386, 517-521 (1997).
17. Roth, W. and Reed, J.C. FLIP protein and TRAIL-induced apoptosis. *Vitam. Horm.* 67, 189-206 (2004).
18. Micheau, O., Thome, M., Schneider, P., Holler, N., Tschopp, J., Nicholson, D.W., Briand, C., and Grutter, M.G. The long form of FLIP is an activator of caspase-8 at the Fas death-inducing signaling complex. *J. Biol. Chem.* 277, 45162-45171 (2002).
19. Yamada, H., Tada-Oikawa, S., Uchida, A., and Kawanishi, S. TRAIL causes cleavage of bid by caspase-8 and loss of mitochondrial membrane potential resulting in apoptosis in BJAB cells. *Biochem. Biophys. Res. Commun.* 265, 130-133 (1999).
20. Zhang, X.D., Zhang, X.Y,, Gray, C.P., Nguyen, T., and Hersey, P. Tumor necrosis factor-related apoptosis-inducing ligand-induced apoptosis of human melanoma is regulated by Smac/DIABLO release from mitochondria. *Cancer Res.* 61, 7339-7348 (2001).
21. Deng, Y., Lin, Y., and Wu, X. TRAIL-induced apoptosis requires Bax-dependent mitochondrial release of Smac/DIABLO. *Genes Dev.* 16, 33-45 (2002).
22. Kandasamy, K., Srinivasula, S.M., Alnemri, E.S., Thompson, C.B., Korsmeyer, S.J., Bryant, J.L, and Srivastava, R.K. Involvement of pro-apoptotic molecules Bax and Bak in tumor necrosis factor-related apoptosis-inducing ligand (TRAIL)-induced mitochondrial disruption and apoptosis: differential regulation of cytochrome c and Smac/DIABLO release. *Cancer Res.* 63, 1712-1721 (2003).
23. Holler, N., Zaru, R., Micheau, O., Thome, M., Attinger, A., Valitutti, S., Bodmer, J.L., Schneider, P., Seed, B., and Tschopp, J. Fas triggers an alternative, caspase-8-independent cell death pathway using the kinase RIP as effector molecule. *Nat. Immunol.* 1, 489-495 (2002).
24. Kemp, T.J., Kim, J.S., Crist, S.A., and Griffith, T.S. Induction of necrotic tumor cell death by TRAIL/Apo-2L. *Apoptosis* 8, 587-599 (2003).
25. Schneider, P., Thome, M., Burns, K., Bodmer, J.L., Hofmann, K., Kataoka, T., Holler, N., and Tschopp, J. TRAIL receptors 1 (DR4) and 2 (DR5) signal FADD-dependent apoptosis and activate NF-κB. *Immunity* 7, 831-836 (1997).
26. Hu, W.H., Johnson, H., and Shu, H.B. Tumor necrosis factor-related apoptosis-inducing ligand receptors signal NF-κB and JNK activation and apoptosis through distinct pathways. *J. Biol. Chem.* 274, 30603-30610 (1999).
27. Harper, N., Farrow, S.N., Kaptein, A., Cohen, G.M., and MacFarlane, M. Modulation of tumor necrosis factor apoptosis-inducing ligand- induced NF-κB activation by inhibition of apical caspases. *J. Biol. Chem.* 276, 34743-34752 (2001).

28. Ghosh, S.K., Wood, C., Boise, L.H., Mian, A.M., Deyev, V.V., Feuer, G., Toomey, N.L., Shank, N.C., Cabral, L., Barber, G.N., and Harrington, W.J. Potentiation of TRAIL-induced apoptosis in primary effusion lymphoma through azidothymidine-mediated inhibition of NF-κB. *Blood* 101, 2321-2327 (2003).
29. Dalen, H., and Neuzil, J. α-Tocopheryl succinate sensitises a T lymphoma cell line to TRAIL-induced apoptosis by suppressing NF-κB activation. *Br. J. Cancer* 88, 153-158 (2003).
30. Chawla-Sarkar, M., Bauer, J.A., Lupica, J.A., Morrison, B.H., Tang, Z., Oates, R.K., Almasan, A., DiDonato, J.A., Borden, E.C., and Lindner, D.J. Suppression of NF-κB survival signaling by nitrosylcobalamin sensitizes neoplasms to the anti-tumor effects of Apo2L/TRAIL. *J. Biol. Chem.* 278, 39461-39469 (2003).
31. Tran, S.E., Holmstrom, T.H., Ahonen, M., Kahari, V.M., and Eriksson, J.E. MAPK/ERK overrides the apoptotic signaling from Fas, TNF, and TRAIL receptors. *J. Biol. Chem.* 276, 16484-16490 (2001).
32. Zhang, X.D., Borrow, J.M., Zhang, X.Y., Nguyen, T., and Hersey, P. Activation of ERK1/2 protects melanoma cells from TRAIL-induced apoptosis by inhibiting Smac/DIABLO release from mitochondria. *Oncogene* 22, 2869-2881 (2003).
33. Secchiero, P., Gonelli, A., Carnevale, E., Milani, D., Pandolfi, A., Zella, D., and Zauli, G. TRAIL promotes the survival and proliferation of primary human vascular endothelial cells by activating the Akt and ERK pathways. *Circulation* 107, 2250-2256 (2003).
34. Muhlenbeck, F., Schneider, P., Bodmer, J.L., Schwenzer, R., Hauser, A., Schubert, G., Scheurich, P., Moosmayer, D., Tschopp, J., and Wajant, H. The tumor necrosis factor-related apoptosis-inducing ligand receptors TRAIL-R1 and TRAIL-R2 have distinct cross-linking requirements for initiation of apoptosis and are non-redundant in JNK activation. *J. Biol. Chem.* 275, 32208-32213 (2000).
35. Lin, Y., Devin, A., Cook, A., Keane, M.M., Kelliher, M., Lipkowitz, S., and Liu, Z.G. The death domain kinase RIP is essential for TRAIL (Apo2L)-induced activation of IκB kinase and c-Jun N-terminal kinase. *Mol. Cell. Biol.* 20, 6638-6645 (2000).
36. Herr, I., Wilhelm, D., Meyer, E., Jeremias, I., Angel, P., and Debatin, K.M. JNK/SAPK activity contributes to TRAIL-induced apoptosis. *Cell Death Diff.* 6, 130-135 (2000).
37. Gabrielli, B.G., Johnstone, R.W., and Saunders, N.A. Identifying molecular targets mediating the anticancer activity of histone deacetylase inhibitors: a work in progress. *Curr. Cancer Drug Targets* 2, 337-353 (2002).
38. Johnstone, R.W. Histone-deacetylase inhibitors: novel drugs for the treatment of cancer. *Nat. Rev. Drug Discov.* 1, 287-99 (2002).
39. de Ruijter, A.J., van Gennip, A.H., Caron, H.N., Kemp, S., and van Kuilenburg, A.B. Histone deacetylases (HDACs): characterization of the classical HDAC family. *Biochem. J.* 370, 737-749 (2003).
40. Yang, X.J., and Seto, E. Collaborative spirit of histone deacetylases in regulating chromatin structure and gene expression. *Curr. Opin. Genet. Dev.* 13, 143-153 (2003).
41. Hubbert, C., Guardiola, A., Shao, R., Kawaguchi, Y., Ito, A., Nixon, A., Yoshida, M., Wang, X.F., and Yao, T.P. HDAC6 is a microtubule-associated deacetylase. *Nature* 417, 455-458 (2002).
42. Warrener, R., Beamish, H., Burgess, A., Waterhouse, N.J., Giles, N., Fairlie, D., and Gabrielli, B. Tumor cell-selective cytotoxicity by targeting cell cycle checkpoints. *FASEB J.* 17, 1550-1552 (2003).
43. Furumai, R., Komatsu, Y., Nishino, N., Khochbin, S., Yoshida, M., and Horinouchi, S. Potent histone deacetylase inhibitors built from trichostatin A and cyclic tetrapeptide antibiotics including trapoxin. *Proc. Natl. Acad. Sci. USA* 98, 87-92 (2001).

44. Haggarty, S.J., Koeller, K.M., Wong, J.C., Grozinger, C.M., and Schreiber, S. L. Domain-selective small-molecule inhibitor of histone deacetylase 6 (HDAC6)-mediated tubulin deacetylation. *Proc. Natl. Acad. Sci. USA* 100, 4389-4394 (2003).
45. Hu, E., Dul, E., Sung, C.M., Chen, Z., Kirkpatrick, R., Zhang, G.F., Johanson, K., Liu, R., Lago, A., Hofmann, G. et al. Identification of novel isoform-selective inhibitors within class I histone deacetylases. *J. Pharmacol. Exp. Ther.* 307, 720-728 (2003).
46. Kouzarides, T. Acetylation: a regulatory modification to rival phosphorylation? *EMBO J.* 19, 1176-1179 (2000).
47. Polevoda, B., and Sherman, F. The diversity of acetylated proteins. *Genome Biol.* 3, reviews0006. Epub 2002 Apr 30 (2002).
48. Qiu, L., Kelso, M.J., Hansen, C., West, M.L., Fairlie, D.P., and Parsons, P.G. Anti-tumour activity in vitro and in vivo of selective differentiating agents containing hydroxamate. *Br. J. Cancer* 80, 1252-1258 (1999).
49. Qiu, L., Burgess, A., Fairlie, D.P., Leonard, H., Parsons, P.G., and Gabrielli, B.G. Histone deacetylase inhibitors trigger a G2 checkpoint in normal cells that is defective in tumor cells. *Mol. Biol. Cell* 11, 2069-2083 (2000).
50. Burgess, A., Ruefli, A., Beamish, H., Warrener, R., Saunders, N.A., Johnstone, R., and Gabrielli, B.G Histone deacetylase inhibitors target proliferating and non-proliferating tumour cells. *Oncogene*, in press (2004).
51. Kelly, W.K., Richon, W.M., O'Connor, O., Curley, T., MacGregor-Curtelli, B., Yong, W., Klang, M., Schwartz, L., Richardson, S., Rosa, E., Drobnjak, M., Cordon-Cordo, C., Chiao, J.H., Rifkind, R., Marks, P.A., and Scher, H. Phase I clinical trial of histone deacetylase inhibitor: suberoylanilide hydroxamic acid administered intravenously. *Clin. Cancer Res.* 9, 3578-3588 (2003).
52. Mello, J.A., and Almouzni, G. The ins and outs of nucleosome assembly. *Curr. Opin. Genet. Dev.* 11, 136-141 (2001).
53. Taddei, A., Roche, D., Sibarita, J.B., Turner, B.M., and Almouzni, G. Duplication and maintenance of heterochromatin domains. *J. Cell Biol.* 147, 1153-1166 (1999).
54. Brinkmann, H., Dahler, A. L., Popa, C., Serewko, M.M., Parsons, P.G., Gabrielli, B.G., Burgess, A.J., and Saunders, N.A. (2001) Histone hyperacetylation induced by histone deacetylase inhibitors is not sufficient to cause growth inhibition in human dermal fibroblasts. J. Biol. Chem. 276, 22491-22499.
55. Richon, V.M., Sandhoff, T.W., Rifkind, R.A., and Marks, P.A. Histone deacetylase inhibitor selectively induces $p21^{Waf1}$ expression and gene-associated histone acetylation. *Proc. Natl. Acad. Sci. USA* 97, 10014-10019 (2000).
56. Sambucetti, L.C., Fischer, D.D., Zabludoff, S., Kwon, P.O., Chamberlin, H., Trogani, N., Xu, H., and Cohen, D. Histone deacetylase inhibition selectively alters the activity and expression of cell cycle proteins leading to specific chromatin acetylation and anti-proliferative effects. *J. Biol. Chem.* 274, 34940-34947 (1999).
57. Ju, R., and Muller, M.T. Histone deacetylase inhibitors activate $p21^{Waf1}$ expression via ATM. *Cancer Res.* 63, 2891-2897 (2003).
58. Burgess, A.J., Pavey, S., Warrener, R., Hunter, L.J., Piva, T.J., Musgrove, E.A., Saunders, N.A., Parsons, P. G., and Gabrielli, B.G. Up-regulation of $p21^{Waf1/Cip1}$ by histone deacetylase inhibitors reduces their cytotoxicity. *Mol. Pharmacol.* 60, 828-837 (2001).
59. Mitsiades, C.S., Mitsiades, N.S., McMullan, C.J., Poulaki, V., Shringarpure, R., Hideshima, T., Akiyama, M., Chauhan, D., Munshi, N., Gu, X., et al. Transcriptional signature of histone deacetylase inhibition in multiple myeloma: biological and clinical implications. *Proc. Natl. Acad. Sci. USA* 101, 540-545 (2004).

60. Sato, N., Fukushima, N., Maitra, A., Matsubayashi, H., Yeo, C. J., Cameron, J.L., Hruban, R.H., Goggins, M., Chambers, A.E., Banerjee, S. et al. Discovery of novel targets for aberrant methylation in pancreatic carcinoma using high-throughput microarrays. Histone acetylation-mediated regulation of genes in leukaemic cells. Gene expression profiling of multiple histone deacetylase (HDAC) inhibitors: defining a common gene set produced by HDAC inhibition in T24 and MDA carcinoma cell lines. *Cancer Res*. 63, 3735-3742 (2003).
61. Yokota, T., Matsuzaki, Y., Miyazawa, K., Zindy, F., Roussel, M.F., and Sakai, T. (2004) Histone deacetylase inhibitors activate INK4d gene through Sp1 site in its promoter. Oncogene 26, 26.
62. Kim, Y.B., Lee, K.H., Sugita, K., Yoshida, M., and Horinouchi, S. Oxamflatin is a novel antitumor compound that inhibits mammalian histone deacetylase. *Oncogene* 18, 2461-2470 (1999).
63. Sandor, V., Senderowicz, A., Mertins, S., Sackett, D., Sausville, E., Blagosklonny, M.V., and Bates, S.E. P21-dependent g(1) arrest with downregulation of cyclin D1 and upregulation of cyclin E by the histone deacetylase inhibitor FR901228. *Br. J. Cancer* 83, 817-825 (2000).
64. Wharton, W., Savell, J., Cress, W.D., Seto, E., and Pledger, W.J. Inhibition of mitogenesis in Balb/c-3T3 cells by Trichostatin A. Multiple alterations in the induction and activation of cyclin-cyclin-dependent kinase complexes. *J. Biol. Chem*. 275, 33981-33987 (2000).
65. Suzuki, A., Tsutomi, Y., Miura, M., and Akahane, K. Caspase 3 inactivation to suppress Fas-mediated apoptosis: identification of binding domain with p21 and ILP and inactivation machinery by p21. *Oncogene* 18, 1239-1244 (1999).
66. Rahmani, M., Yu, C., Dai, Y., Reese, E., Ahmed, W., Dent, P., and Grant, S. Co-administration of the heat shock protein 90 antagonist 17-allylamino-17-demethoxygeldanamycin with suberoylanilide hydroxamic acid or sodium butyrate synergistically induces apoptosis in human leukemia cells. *Cancer Res*. 63, 8420-8427 (2003).
67. Rahmani, M., Yu, C., Reese, E., Ahmed, W., Hirsch, K., Dent, P., and Grant, S. Inhibition of PI-3 kinase sensitizes human leukemic cells to histone deacetylase inhibitor-mediated apoptosis through p44/42 MAP kinase inactivation and abrogation of $p21^{Cip1/Waf1}$ induction rather than AKT inhibition. *Oncogene* 22, 6231-6242 (2003).
68. Rosato, R.R., Almenara, J.A., and Grant, S. The histone deacetylase inhibitor MS-275 promotes differentiation or apoptosis in human leukemia cells through a process regulated by generation of reactive oxygen species and induction of $p21^{Cip1/Waf1}$. *Cancer Res.* 63, 3637-3645 (2003).
69. Rosato, R.R., Almenara, J.A., Yu, C., and Grant, S. Evidence of a functional role for $p21^{Waf1/Cip1}$ down-regulation in synergistic antileukemic interactions between the histone deacetylase inhibitor sodium butyrate and flavopiridol. *Mol. Pharmacol*. 65, 571-581 (2004).
70. Vrana, J.A., Decker, R.H., Johnson, C.R., Wang, Z., Jarvis, W.D., Richon, V.M., Ehinger, M., Fisher, P.B., and Grant, S. Induction of apoptosis in U937 human leukemia cells by suberoylanilide hydroxamic acid (SAHA) proceeds through pathways that are regulated by Bcl-2/Bcl-$x_L$, c-Jun, and $p21^{Cip1}$, but independent of p53. *Oncogene* 18, 7016-7025 (1999).
71. Bernhard, D., Ausserlechner, M.J., Tonko, M., Loffler, M., Hartmann, B.L., Csordas, A., and Kofler, R. Apoptosis induced by the histone deacetylase inhibitor sodium butyrate in human leukemic lymphoblasts. *FASEB J.* 13, 1991-2001 (1999).

72. Henderson, C., Mizzau, M., Paroni, G., Maestro, R., Schneider, C., and Brancolini, C. Role of caspases, Bid, and p53 in the apoptotic response triggered by histone deacetylase inhibitors trichostatin-A (TSA) and suberoylanilide hydroxamic acid (SAHA). *J. Biol. Chem.* 278, 12579-12589 (2003).
73. Ruefli, A.A., Ausserlechner, M.J., Bernhard, D., Sutton, V.R., Tainton, K.M., Kofler, R., Smyth, M.J., and Johnstone, R.W. The histone deacetylase inhibitor and chemotherapeutic agent suberoylanilide hydroxamic acid (SAHA) induces a cell-death pathway characterized by cleavage of Bid and production of reactive oxygen species. *Proc. Natl. Acad. Sci. USA* 98, 10833-10838 (2001).
74. Mitsiades, N., Mitsiades, C.S., Richardson, P.G., McMullan, C., Poulaki, V., Fanourakis, G., Schlossman, R., Chauhan, D., Munshi, N.C., Hideshima, T. et al. Molecular sequelae of histone deacetylase inhibition in human malignant B cells. *Blood* 101, 4055-4062 (2003).
75. Zhang, X.D., Gillespie, S.K., Borrow, J.M., and Hersey, P. The histone deacetylase inhibitor suberic bishydroxamate regulates the expression of multiple apoptotic mediators and induces mitochondria-dependent apoptosis of melanoma cells. *Mol. Cancer Ther.* 3, 425-435 (2004).
76. Henderson, C., and Brancolini, C. Apoptotic pathways activated by histone deacetylase inhibitors: implications for the drug-resistant phenotype. *Drug Resist. Update* 6, 247-256 (2003).
77. Medina, V., Edmonds, B., Young, G.P., James, R., Appleton, S., and Zalewski, P.D. Induction of caspase-3 protease activity and apoptosis by butyrate and trichostatin A (inhibitors of histone deacetylase): dependence on protein synthesis and synergy with a mitochondrial/cytochrome c-dependent pathway. *Cancer Res.* 57, 3697-3707 (1997).
78. Peart, M.J., Tainton, K.M., Ruefli, A.A., Dear, A.E., Sedelies, K.A., O'Reilly, L.A., Waterhouse, N.J., Trapani, J.A., and Johnstone, R.W. Novel mechanisms of apoptosis induced by histone deacetylase inhibitors. *Cancer Res.* 63, 4460-4471 (2003).
79. Guo, Y., Srinivasula, S.M., Druilhe, A., Fernandes-Alnemri, T., and Alnemri, E.S. Caspase-2 induces apoptosis by releasing pro-apoptotic proteins from mitochondria. *J. Biol. Chem.* 277, 13430-13437 (2002).
80. Glick, R.D., Swendeman, S.L., Coffey, D.C., Rifkind, R.A., Marks, P.A., Richon, V.M., and La Quaglia, M.P. Hybrid polar histone deacetylase inhibitor induces apoptosis and CD95/CD95 ligand expression in human neuroblastoma. *Cancer Res.* 59, 4392-4399 (1999).
81. Kwon, S.H., Ahn, S.H., Kim, Y.K., Bae, G.U., Yoon, J.W., Hong, S., Lee, H.Y., Lee, Y.W., Lee, H.W., and Han, J.W. Apicidin, a histone deacetylase inhibitor, induces apoptosis and Fas/Fas ligand expression in human acute promyelocytic leukemia cells. *J. Biol. Chem.* 277, 2073-2080 (2002).
82. Zhang, X.D., Gillespie, S.K., Borrow, J.M., and Hersey, P. The histone deacetylase inhibitor suberic bishydroxamate: a potential sensitizer of melanoma to TNF-related apoptosis-inducing ligand (TRAIL) induced apoptosis. *Biochem. Pharmacol.* 66, 1537-1545 (2003).
83. Glaser, K.B., Staver, M.J., Waring, J.F., Stender, J., Ulrich, R.G., and Davidsen, S.K. Gene expression profiling of multiple histone deacetylase (HDAC) inhibitors: defining a common gene set produced by HDAC inhibition in T24 and MDA carcinoma cell lines. *Mol. Cancer Ther.* 2, 151-163 (2003).
84. Lindemann, R.K., Gabrielli, B.G., and Johnstone, R.W. Histone-deacetylase inhibitors for the treatment of cancer. *Cell Cycle*, in press (2004).

85. Marks, P., Rifkind, R. A., Richon, V. M., Breslow, R., Miller, T., and Kelly, W. K. Histone deacetylases and cancer: causes and therapies. *Nat. Rev Cancer* 1, 194-202 (2001).
86. Inoue, H., Shiraki, K., Ohmori, S., Sakai, T., Deguchi, M., Yamanaka, T., and Nakano, T. Histone deacetylase inhibitors sensitize human colonic adenocarcinoma cell lines to TNF-related apoptosis inducing ligand-mediated apoptosis. *Int. J. Mol. Med.* 9, 521-525 (2002).
87. Nakata, S., Yoshida, T., Horinaka, M., Shiraishi, T., Wakada, M., and Sakai, T. Histone deacetylase inhibitors upregulate death receptor 5/TRAIL-R2 and sensitize apoptosis by TRAIL/APO2-L in human malignant tumor cells. *Oncogene*, in press (2004).
88. Chopin, V., Slomianny, C., Hondermarck, H., and Le Beurhis, S. Synergistic induction in breast cancer cells by co-treatment with butyrate and TNF-α, TRAIL, or anti-Fas agonist antibody involves enhancement of death receptors' signaling and requires $p21^{Waf1}$. *Exp. Cell Res.* 298, 460-573 (2004).
89. Piekarz, R.L., Robey, R.W., Zhang, Z., Kayastha, G., Sayah, A., Abdeldaim, A.H., Torrico, S., and Bates, S.E. T-cell lymphoma as a model for the use of histone deacetylase inhibitors in cancer therapy: impact of depsipeptide on molecular markers, therapeutic targets, and mechanistic resistance. *Blood* 103, 4636-3643 (2003).
90. Sandor, V., Bakke, S., Robey, R.W., Kang, M.H., Blagosklonny, M.V., Bender, J., Brooks, R., Piekarz, R.L., Tucker, E., Figg, W.D., Chan, K.K., Goldspiel, B., Fojo, A.T., Balcerzak, S.P., and Bates, S.E. Phase I trial of the histone deacetylase inhibitor, depsipeptide (FR901228, NSC 630176), in patients with refractory neoplasms. *Clin. Cancer Res.* 8, 718-728 (2002).
91. Marks, P.A., Miller, T., and Richon, V.M. Histone deacetylase. *Curr. Opin. Pharmacol.* 3, 344-351 (2003).
92. Jo, M., Kim, T.H., Seol, D.W., Esplen, J.E., Dorko, K., Billiar, T.R., and Strom, S.C. Apoptosis induced in normal human hepatocytes by tumor necrosis factor-related apoptosis-inducing ligand. *Nat. Med.* 6, 564-567 (2000).
93. Nitsch, R., Bechmann, I., Deisz, R.A., Haas, D., Lehmann, T.N., Wendling, U., and Zipp, F. Human brain-cell death induced by tumour-necrosis-factor-related apoptosis-inducing ligand (TRAIL). *Lancet* 356, 827-828 (2000).
94. Lawrence, D., Shahrokh, Z., Marsters, S., Achilles, K., Shih, D., Mounho, B., Hillan, K., Totpal, K., DeForge, L., Schow, P., Hooley, J., Sherwood, S., Pai, R., Leung, S., Khan, L., Gliniak, B., Bussiere, J., Smith, C.A., Strom, S.S., Kelley, S., Fox, J.A., Thomas, D., and Ashkenazi, A. Differential hepatocyte toxicity of recombinant ApoL/TRAIL versions. *Nat. Med.* 7, 383-385 (2001).
95. Rosato, R.R., Almenara, J. A., Dai, Y., and Grant, S. Simultaneous activation of the intrinsic and extrinsic pathways by histone deacetylase inhibitors and tumor necrosis factor-related apoptosis-inducing ligand synergistically induces mitochondrial damage and apoptosis in human leukemic cells. *Mol. Cancer Ther.* 2, 1273-1284 (2003).
96. Neuzil, J., Gellert, N., and Sweattenham, E. Sensitisation of malignant mesothelioma to TRAIL apoptosis by inhibition of histone deacetylase: an important role of Bcl-$x_L$ down-regulation. *Biochem. Biophys. Res. Commun.* 314, 186-191 (2004).
97. Guo, F., Sigua, C., Tao, J., Bali, P., George, P., Li, Y., Wittmann, S., Moscinski, L., Atadja, P., and Bhalla, K. Co-treatment with histone deacetylase inhibitor LAQ824 enhances Apo-2L/tumor necrosis factor-related apoptosis-inducing ligand-induced death inducing signaling complex activity and apoptosis of human acute leukemia cells. *Cancer Res.* 64, 2580-2589 (2004).

98. Wang, S., and El-Deiry, W.S. Requirement of p53 targets in chemosensitization of colonic carcinoma to death ligand therapy. *Proc. Natl. Acad. Sci. USA.* 100, 15095-15100 (2003).
99. Rosato, R.R., Wang, Z., Gopalkrishnan, R.V., Fisher, P.B., and Grant, S. Evidence of a functional role for the cyclin-dependent kinase-inhibitor $p21^{Waf1/Cip1/Mda6}$ in promoting differentiation and preventing mitochondrial dysfunction and apoptosis induced by sodium butyrate in human myelomonocytic leukemia cells (U937). *Int. J. Oncol.* 19, 181-191 (2001).
100. Rosato, R.R., Almenara, J.A., Cartee, L., Betts, V., Chellappan, S.P., and Grant, S. The cyclin-dependent kinase inhibitor flavopiridol disrupts sodium butyrate-induced $p21^{Waf1/Cip1}$ expression and maturation while reciprocally potentiating apoptosis in human leukemia cells. *Mol. Cancer Ther.* 1, 253-266 (2002).
101. Jin, Z., Dicker, D.T., and El-Deiry, W.S. Enhanced sensitivity of G1 arrested human cells suggest a novel therapeutic strategy using a combination of simvastatin and TRAIL. *Cell Cycle* 1, 82-89 (2002).
102. Ozoren N., and El-Deiry, W.S. Cell surface death receptor signaling in normal and cancer cells. *Semin. Cancer Biol.* 13, 135-147 (2003).
103. Hersey, P., and Zhang, X.D. How melanoma cells evade TRAIL-induced apoptosis. *Nat. Rev. Cancer* 1, 142-150 (2001)

Chapter 13

# BASIS OF CELL KILL FOLLOWING CLINICAL RADIOTHERAPY

Oliver Faulhaber and Robert G Bristow*
*Departments of Medical Biophysics and Radiation Oncology, University of Toronto and Ontario Cancer Institute/Princess Margaret Hospital (University Health Network), 610 University Avenue, Toronto, Ontario CANADA M5G2M9*

Abstract: Over one half of cancer patients are treated radiotherapy. Technological advances in radiation targeting and concurrent chemotherapy continue to improve clinical radiotherapy outcome. Modern radiotherapy clinical trials are ongoing with novel molecular-targeted agents, requiring an increased understanding of cell death signals in a tissue-specific manner. Herein, we critically appraise the relative roles of apoptosis, mitotic catastrophe and terminal growth arrest in relation to final clonogenic cell kill following radiotherapy. Mitotic catastrophe and terminal growth arrest form the basis of the majority of cell kill during radiotherapy for common epithelial tumors (e.g. prostate, breast, lung, etc.) whereas more sensitive tumors (e.g. lymphomas or germ cell tumors) undergo apoptosis. Targeting of apoptotic, cell cycle checkpoint and DNA repair pathways may further augment cell kill from all three death pathways. Using intra-treatment biopsies or non-invasive imaging may soon allow for prediction of individual patient response and judicial selection of molecular targeting based on specific tumor cell signaling.

Key words: radiobiology, apoptosis, mitotic catastrophe, cell cycle arrest, radiotherapy, clonogenic survival, survivin, p53, ceramide, senescence

## 7. INTRODUCTION

Since the discoveries by Roentgen and Curie more than a century ago, the biological effects of ionizing radiation have played a major role in diagnostic and therapeutic medicine. Radical radiotherapy, alone or in combination with chemotherapy, can be curative for a number of tumor sites



*M. Sluyser (ed.), Application of Apoptosis to Cancer Treatment,* 293-320.

including tumors of the cervix, head and neck, prostate, bladder, lung, rectum, testes and lymphoma. Radiotherapy is delivered by interstitial brachytherapy implants or by external beam radiotherapy using linear accelerators (i.e., high energy x-rays and electron beams), $^{60}Co$ sources (γ–rays produced by radioactive decay due to unstable nuclei) or charged particle accelerators. Improvements in radiotherapy outcome have been driven by sophisticated treatment planning methods allowing for improved physical targeting methods (e.g. conformal or intensity-modulated radiotherapy (IMRT)) and through increasing use of combined-modality therapy (i.e., radiotherapy concurrent with chemotherapy or hormone therapy). Future improvements in radiotherapy outcome will be the result of the judicial use of molecular biomarkers and genomic "fingerprinting") to drive the selection of molecular-targeted drugs and improve local tumor control [1] and requires an increased understanding of cell death responses in a dose-, time- and tissue-specific manner.

The understanding of curative fractionated radiotherapy versus palliative radiotherapy is an important concept. Radiation dose is measured in grays (Gy) as the amount of energy absorbed per unit mass. Typical radiocurative regimens require a series of daily radiation dose fractions of 1.8 to 2Gy over 6 to 8 weeks to achieve total doses in the order of 60 to 80Gy. Higher dose per fractions can be used in palliative settings (e.g. 8Gy single dose or 5 to 10 fractions of 3 to 5Gy) to relieve symptoms resulting from compression of surrounding tissues. Curative fractionated regimens attempt to maximize the *therapeutic ratio* in which a maximum dose is delivered to the tumor in an effort to sterilize all stem cells (clonogens) capable of tumor regrowth, but is a dose that is tolerable dose to the normal tissues within the irradiation volume. The final level of cell killing during a period of fractionated radiotherapy is related to multiple factors termed the 5 **R**'s of radiotherapy: the intrinsic **r**adiosensitivity of the normal and tumor cells; the **r**edistribution of cells within the cell cycle between treatments; the **r**eoxygenation of hypoxic cells during the course of radiotherapy; the **r**epopulation of normal and tumor cells during radiotherapy; the repair of normal tissues between each radiotherapy fraction [2].

Many studies have solely used large single radiation doses (e.g. 10 to 20 Gy) to assess cell death responses following irradiation. This may activate different signaling and cell death pathways from other experiments using more clinically-relevant doses of 2 Gy which are used during radiocurative fractionated protocols. The sole use of high doses of radiation to study cell death mechanisms may explain some of the difficulties in extrapolating cell biological data *in vitro* to the clinical outcome of patients *in vivo*.

This review will first provide a general background regarding the cellular response to ionizing radiation in relation to intracellular damage sensing and signaling, irradiation-induced cell cycle arrests and DNA repair. We will then use these background concepts to critically appraise the role of three main cell death mechanisms in the normal and tumor tissue response to radiotherapy: (i) apoptosis, (ii) mitotic catastrophe and (iii) terminal growth arrest (tumor senescence). Where possible, we will intersperse basic biological data with clinically-relevant examples in a tumor type-specific manner and highlight potential molecular-targeted therapies which might augment cell kill and improve clinical outcome.

# 8. MOLECULAR RESPONSES TO RADIOTHERAPY

Ionizing radiation rapidly generates chemically-reactive free radicals (i.e. molecules with unpaired electrons) that randomly interact with DNA, RNA and proteins. Both DNA and the plasma cell membrane are targeted. Free radical attack on DNA occurs as clusters of multiply-damaged sites containing DNA single- or double-strand breaks (DNA-ssb or DNA-dsbs), altered or lost DNA bases and DNA-DNA or DNA-protein cross-links. It has been estimated that $10^5$ ionizations occur within a diploid cell per 1Gy and leads to approximately 1000-3000 DNA-DNA or DNA-protein cross-links, 1000 damaged DNA bases, 500-1000 single-strand and 25 to 50 double-strand DNA breaks [3]. This damage activates sensing and transduction pathways involved in cell cycle checkpoint control and DNA repair (e.g. ATM, MRE11, RAD50, DNA-PKcs, 53BP1, p53, CHK2). Damage to the plasma membrane drives signal transduction through the ceramide-sphingomyelin pathway (discussed below) [4].

Immediately following cellular damage, key gene expression patterns emerge driving signal transduction at the plasma membrane or DNA in an attempt to repair sub-lethal damage or activate cell death in response to lethal (i.e. non-repairable) damage [5]. This includes the activation of cell stress-related proteins including early growth response factor (EGR-1), MAP-kinases, SAP-kinases and RAS-RAF signaling proteins. In addition to these stress responses, increased expression of growth factors and cytokines required for normal tissue and organ survival under potentially genotoxic conditions is observed (e.g. platelet-derived growth factor (PDGF), transforming growth factor beta (TGF-$\beta$), basic fibroblast growth factor (bFGF), tumor necrosis factor (TNF-$\alpha$)). Prolonged expression of cytokines may be the basis for late tissue damage following radiotherapy. Recent cDNA microarray analyses have shown that radiation-induced gene expression is cell-type specific, dose-dependent, and varies under *in vitro* versus *in vivo* conditions for the same cells [6]. This observation supports the concept that the "molecular profiling" of tumor and normal tissues derived directly from patients may predict relevant radiation-stress responses amenable to tumor radiosensitization or normal tissue radioprotection.

Oncogene activation or aberrant (mutant) tumor suppressor gene expression prior to irradiation may predict for tumor cell radioresistance. For example, increased radiation survival in vitro is observed following transfection of activated *Raf*, *Ras*, *c-Myc* or mutated *p53* alleles, either alone or in combination [7]. Using specific pathway inhibitors, it has become clear that radioresistance can arise at multiple points within the *Ras* oncogene signaling pathway. This includes aberrant signaling by proteins upstream of RAS (e.g. epidermal growth factor receptor (EGFR)) or by aberrant signaling within pathways downstream to RAS (e.g. RAF-MEK-ERK and PI3-K/AKT/PKB). All have all been clinked to tumor radioresistance in both pre-clinical and clinical studies [8, 9]. As a consequence, the radiosensitization of tumor cells can be achieved by the inhibition of activated oncogene function using drug inhibitors or by genetic approaches using anti-sense or small interfering RNA (siRNA). For example, pharmacologic inhibitors of RAS protein prenylation (e.g. farnesyl transferase inhibitors or FTIs) initially radiosensitized human breast, lung, colon and bladder cancer cell lines or xenografts activated *H-* or *K-ras* genes. Subsequently, a recent Phase I trial using the FTI L-778,123 has shown promising tumor responses with minimal toxicity in advanced lung and head and neck cancer [10]. Similarly, the inhibition of EGFR using cetuximab-C225 (a monoclonal anti-EGFR antibody), gefinitib-ZD1839 (a tyrosine kinase inhibitor of EGFR signaling), alone or in combination, are being actively investigated in pre-clinical and clinical studies based on their ability to radiosensitize tumors *in vivo* [9].

As part of the DNA damage sensing cascade, ionizing radiation initiates the rapid phosphorylation of the histone, H2AX (denoted γH2AX when phosphorylated) at the sites of DNA damage(see Figure 1)[11]. Nuclear foci at the site of damage, each containing thousands of γH2AX molecules, can be detected using immunofluorescent microscopy. Initially, the MRE11/RAD50/NBS1 complex senses DNA damage activating the kinase activity of ATM and DNA-PK within the first hour of irradiation [12]. Two main pathways of DNA-dsb exist in human cells. Homologous recombination (HR; including the concerted actions of RAD51 protein and RAD52, RAD54, RAD51B/C/D, XRCC3, RPA and BRCA2) is the preferential repair pathway during the S and G2 phases of the cell cycle. In contrast, non-homologous end-joining (NHEJ; includes the DNA-$PK_{CS}$, Ku70/Ku80, Ligase IV and XRCC4 proteins) is the preferred pathway for cells during the G1 phase of the cell cycle [13, 14]. Focal DNA damage and bidning by DNA-dsb repair proteins therefore acts as a beacon for the subsequent phosphorylation and signal transduction of other proteins involved in cell cycle control (e.g. 53BP1, BRCA1, MDC1, SMC1, p53, CHK2). Residual γH2AX foci at times greater than 12 hours following irradiation represent non-repairable DNA damage leading to cell death. [15].

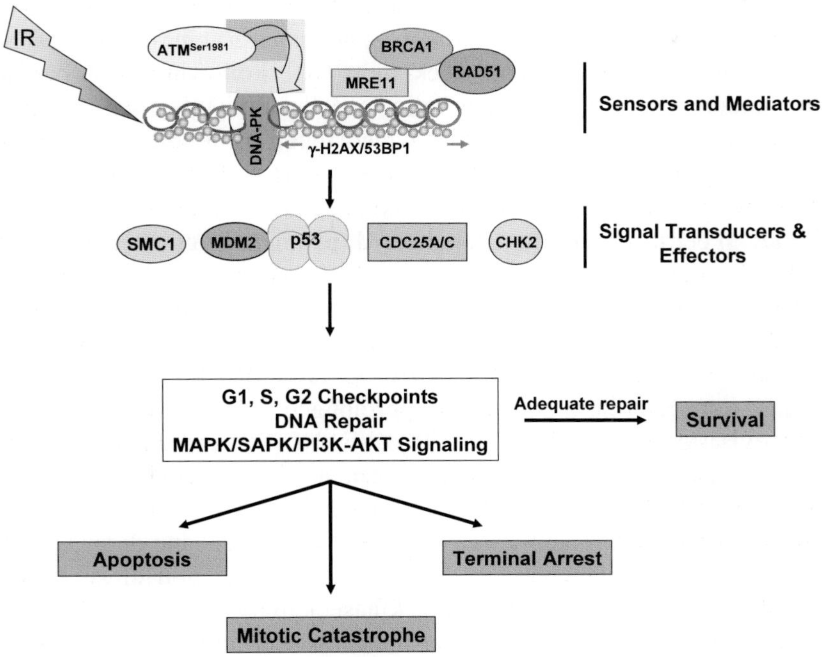


*Figure 1.* Sensing and signaling of DNA damage within the nucleus following radiotherapy. Ionizing radiation [IR] causes direct damage to DNA, plasma membrane and cell organelles (discussed in detail in the text). Shown here is a summary of nuclear events following cellular exposure to ionizing radiation. DNA damage in the form of DNA-breaks or DNA base modification can activate the ATM and DNA-PK PI3-kinase complexes to phosphorylate histone H2AX (i.e. γH2AX). This serves as a biomarker of the number of DNA breaks at a given time post-irradiation. DNA breaks recruit a number of DNA sensing and repair factors including ATM, RAD51/52, DNA-PK (Ku70/80 hetero-dimer along with the DNA-PKcs catalytic subunit) and BRCA1, all within one hour of irradiation. Subsequent transduction of damage signals via MAPK, SAPK, PI3-K/AKT signaling in addition to the CHK2, p53, SMC1 and CDC25A/C cell cycle checkpoint proteins, co-ordinates DNA repair during the G1, S and G2 cell cycle arrests. For cells that can appropriately repair sub-lethal DNA damage, cells can re-enter the cell cycle and cycle to division. For cells that sustain lethal amounts of DNA damage, or can not adequately repair sub-lethal DNA damage due to a repair- or checkpoint-deficiency, cell death occurs. This can occur by activating specific pathways, dependent on the level of DNA damage and cell or tissue type: apoptosis, mitotic catastrophe and/or terminal growth arrest. Tumor cell populations may respond to fractionated radiotherapy with one pathway, or a combination of cell death pathways, depending on tumor pathology and the tumor microenvironment (i.e. hypoxia, pH gradients, growth factor availability). The characteristics of each radiation death pathway are summarized in Table 1.

In response to irradiation, human tumor and normal cells delay their progression through the cell cycle and can arrest within G1/S, intra-S and G2/M checkpoints. Checkpoint control presumably allows for the repair of DNA damage prior to DNA replication or mitosis, although it is controversial whether altered checkpoint control is a direct determinant of radiosensitivity [2]. Checkpoints may prevent genetic instability in daughter cell progeny. The ATM (*ataxia telangiectasia mutated*) kinase plays a major role in the initiation of all three checkpoints [12]. Following DNA damage, ATM undergoes auto-phosphorylation, followed by its subsequent phosphorylation of p53 and CHK2, which together inhibit pRB-E2F and CDC25A-Cyclin E complexes required for the G1 to S transition and DNA replication during S-phase [16]. The radiation-induced G1 arrest is therefore abrogated in cells that lack functional ATM, p53, CHK2 or pRB proteins. In fibroblasts, irradiation may cause induction of a permanent terminal growth arrest (i.e. radiation-induced senescence) associated with increased expression of the cycle-dependent kinase inhibitors $p21^{WAF}$, $p16^{INK4a}$ and senescence-associated β-galactosidase (SA-β-gal) [17-20].

The intra-S checkpoint is prolonged by radiation-induced phosphorylation of the BRCA1, NBS1, FANC-D2 and SMC1 proteins. Phosphorylation of SMC1 (a cohesin protein) is activated by the MRN complex in irradiated areas of damage and may act to silence replication origins until the DNA breaks are repaired [12]. The onset and duration of the G2 delay during radiotherapy is secondary to the function of ATM, p53, CHK1, 14-3-3-σ, 53BP1 and CDC25C kinase activity. These are required to activate the Cyclin B-CDC2 complex for mitosis. Irradiation also decreases the expression, stability or nuclear accumulation of Cyclin B, thereby preventing the formation of nuclear Cyclin B-CDC2 complexes [2, 21]. DNA repair activity has been detected to occur during the radiation-induced G2 delay and the relative success of repair can directly correlated to the extent of radiation survival [22]. Therefore, strategies which shorten the G2 arrest may inhibit the repair of sub-lethal damage and activate cell death pathways of apoptosis or mitotic catastrophe (discussed below) [23].

The molecular responses outlined in this section interact to control the extent of cell death within normal or tumor tissues. Additionally, irradiated cells can liberate growth factors or proteins which are toxic to surrounding non-irradiated cells (i.e. the bystander effect). Inhibition of the reproductive ability of tumor cells is an important endpoint of these molecular responses, especially those that integrate to cause cell death at clinically-relevant doses (~2 Gy). A major aim of clinical radiotherapy is to sterilize all tumor stem cells (i.e. tumor clonogens) which are capable of re-growing the tumor [2]. In the next section, we review the evidence for apoptosis, mitotic catastrophe and terminal growth arrest as components of clonogenic radiation cell kill.

# 9. CELL DEATH DURING RADIOTHERAPY

Many types of cells do not show morphological evidence of radiation damage until they attempt to divide. A cell that has lost its ability to generate a "clone" is regarded as having been killed, even though it may undergo a few divisions or remain intact for a substantial period post-irradiation. The morphology of the cell at the time of cell lysis following irradiation can be either apoptotic or necrotic. Lethally damaged cells may: (1) undergo cell lysis secondary to apoptosis, (2) undergo up to four abortive mitotic cycles and then finally undergo cell lysis as a result of mitotic catastrophe or (3), undergo a permanent (terminal) growth arrest such as that observed for irradiated fibroblasts. The phenotypic and genotypic characteristics for each pathway are reviewed in Table 1.

Radiotherapeutic success is dependent on clonogenic cell killing. For that reason, strategies that target the above pathways and result in increased tumor clonogenic kill may lead to new therapies based on tumor cell radiosensitization. The clonogenic cell survival assay quantifies the survival of stem cells (clonogens) following irradiation. Following irradiation, the cells are incubated for a pre-determined number of days based on their proliferative rate (usually 7-14 days). Those cells that retain unlimited proliferative capacity will divide to form discrete colonies of cells containing more than 50 cells (i.e., capable of at least 5 divisions following irradiation).The ratio of the number of colonies formed relative to the number of cells initially plated in the untreated versus irradiated dishes, can be used to calculate the fraction of cells surviving the treatment (i.e. the cell surviving fraction). Measured or calculated radiobiologic parameters that describe the survival of cells following doses of ~ 2Gy (the SF2 value) reflect clinically-relevant cell death. Importantly, SF2 values derived from patient biopsies have been correlated to the clinical local control following radiotherapy supporting the concept that intrinsic radiosensitivity (and therefore cell death pathways) are an important determinant of clinical response [2]. Typically SF2 ranges between 0.2 and 0.8. After 30 fractionated treatments, this would translate into a range of final tumor cell survival between $1\text{x } 10^{-30}$ to $1.2 \text{ x } 10^{-3}$, respectively ! Therefore, any radiosensitization strategy that can decrease the SF2 value may have large effects on clinical response, due to the exponential nature of cell kill by radiotherapy.

Non-clonogenic assays can estimate the relative radiosensitivity of cells, although short-term and long-term assays do not always correlate within the same cell line [2]. Radiation-induced apoptosis can be quantified by: morphologic analysis, annexin-V or TUNEL staining within cultures or tissue sections; the extent of cells within the sub-G1 fraction using flow cytometry and/or relative DNA fragmentation patterns (see Table 1).

*Table 1.* Cell Death Mechanisms In Tumor Cells Following Radiotherapy

| | **Apoptosis** | **Necrosis** | **Mitotic Catastrophe** | **Terminal growth arrest** |
|---|---|---|---|---|
| Definition | Programmed cell death | Pathologic cell death | Cell death due to mitotic failure | Radiation-induced senescence |
| 1. Morphological changes | 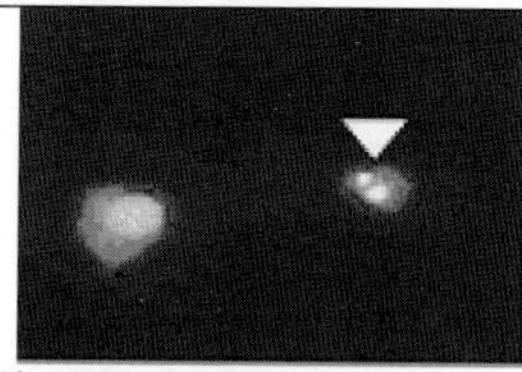 | 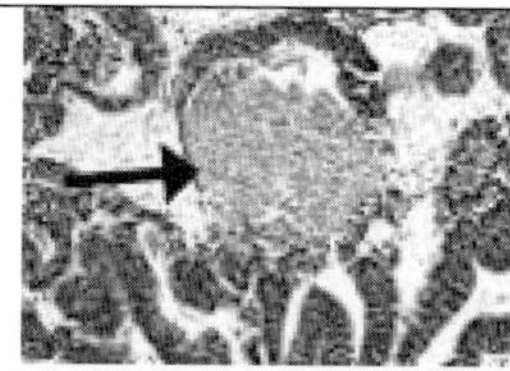 | 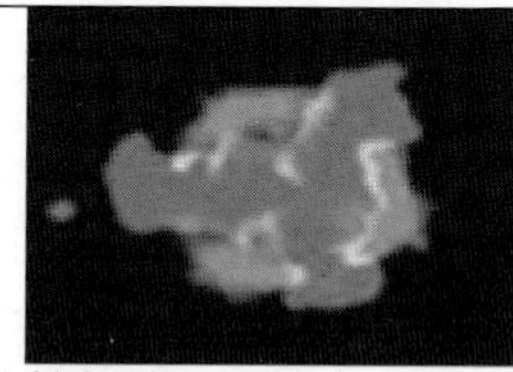 | 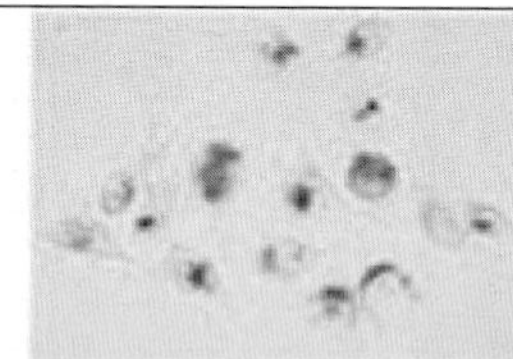 |
| Nucleus | Chromatin condensation, fragmented nuclei (apoptotic bodies contained in nuclear envelope), specific DNA fragmentation | Non-specific degradation of DNA and nuclear envelope breakdown | Multiple micronuclei, abnormal mitoses with mis-aggregated chromosomes | Senescence-associated heterochromatic foci |
| Cytoplasm | Shrunken and fragmented cytoplasm | Mitochondrial and cellular swelling, organelle degradation, inceased vacuolation | Increased cell size | Increased granularity and flattening of cells |
| Cell membrane | Blebbing | Swelling, rupture | Intact | Intact |
| Biochemical changes | Caspase-dependent | Passive | Caspase-independent, abnormal CDK1/cyclin B activation | Senescence-associated β-galactosidase activity (SA-β-gal+) |
| Tissue Inflammation ? | No | Yes | Probable | Yes |
| 2. Gene expression following irradiation | Radiation-induced apoptosis stimulated by overexpression of myc, bax, p53, SAPK and TRAIL. Inhibited by BCL-2, MTp53, PTEN/Akt. | Unknown | Stimulated by deficiencies of G1, G2 and prophase checkpoint proteins (e.g. p53, ATM, ATR, Chk1/2, Cdc25A/B/C, Plk1, Prk, BUB1/2, MAD and survivin) | Stimulated by telomere shortening, hTERT expression, p53, $p21^{WAF1}$ and $p16^{INK4a}$ |
| 3. Correlation with clinical radioresponse | Probable in germ cell tumors, ovarian cancer and lymphomas; unlikely in epithelial tumors | Not a reliable endpoint | Probable correlation in most epithelial tumors | Needs further study, but probable in normal epithelial tissue responses and tumors |
| 4. Detection methods | TUNEL or Annexin-V staining, caspase activation, specific DNA fragmentation, sub-G1 peak on flow cytometry, nuclear or H&E morphology | Electron microscopy and H&E morphology, detection of inflammation in surrounding tissue | Nuclear morphology, assays for mitotic markers | Staining for SA-β-gal activity, flow cytometric proliferation assays, increased p53, $p16^{INK4A}$ levels and morphology |

Another non-clonogenic assay is the MTT assay, which measures relative cell growth in control versus irradiated populations over a 3 to 5 period following treatment. Cell viability is determined using colorimetric assessment of reduction of a tetrazolium compound and is useful for cell lines that die by primarily by apoptosis or rapid mitotic catastrophe. The clonogenic assay remains a favored assay as it represents cumulative cell survival following radiation as a result of all types of cell death [24, 25].

## 9.1 Radiation-induced apoptosis: the p53 and ceramide signaling pathways

In highly-radiosensitive cells (such as lymphocytes, spermatocytes, thymocytes, and salivary gland epithelium and the cancers that arise from them), irradiation causes the cells to undergo classical apoptosis associated with cell membrane blebbing, the formation of nuclear apoptotic bodies, and DNA "laddering". Depending on the type of cell, the intracellular target(s) for the induction of the apoptotic response can be either the plasma membrane or DNA, or both (see Figure 2) [26]. The reason(s) why some cell types undergo extensive radiation-induced apoptosis within a few hours after irradiation, whilst others do not, may relate to the relative expression and function of proteins activated within the *extrinsic* or *intrinsic* pathways of apoptosis. The reader is referred to other chapters within this book for details regarding these molecular pathways. The *extrinsic* pathway is initiated by TRAIL, a member of the TNF superfamily, which induces apoptosis in response to external stimuli through clustering of receptors, DR4 and DR5 in the plasma membrane thereby forming the death-inducing signaling complex (DISC). Subsequent recruitment of the adaptor molecule FADD, and activation of Caspases 8 and 10, leads to Caspase 3 cleavage of death-effector proteins (reviewed in [27]). In lung and prostate models, ionizing radiation enhances the therapeutic potential of recombinant TRAIL *in vitro* and *in vivo.* This can be abrogated by the BCL-2 protein demonstrating cross-talk between intrinsic and extrinsic signals[27, 28]. Recombinant TRAIL induces apoptosis in a variety of cancer cell lines regardless of p53 status and may therefore be an effective radiosensitizer if associated with increased tumor clonogen killing [29].

The *intrinsic* mitochondrial-based pathway can initiate apoptosis in response to radiation-associated DNA breaks, checkpoint defects, hypoxia or membrane damage. Both nuclear signals (e.g. ATM-p53) and membrane signals (e.g. ceramide) damage can activate this pathway Following pro-apoptotic activation of BAX (a member of the *Bcl-2* gene superfamily), the mitochondrial membrane is engaged release cytochrome *c* into the cytosol. Cytochrome c then binds the adaptor APAF-1, forming an 'apoptosome' that activates the apoptosis-initiating protease Caspase 9 which activates Caspases 3, 6, 7 and 8 to effect the final morphology of apoptotic cell death.

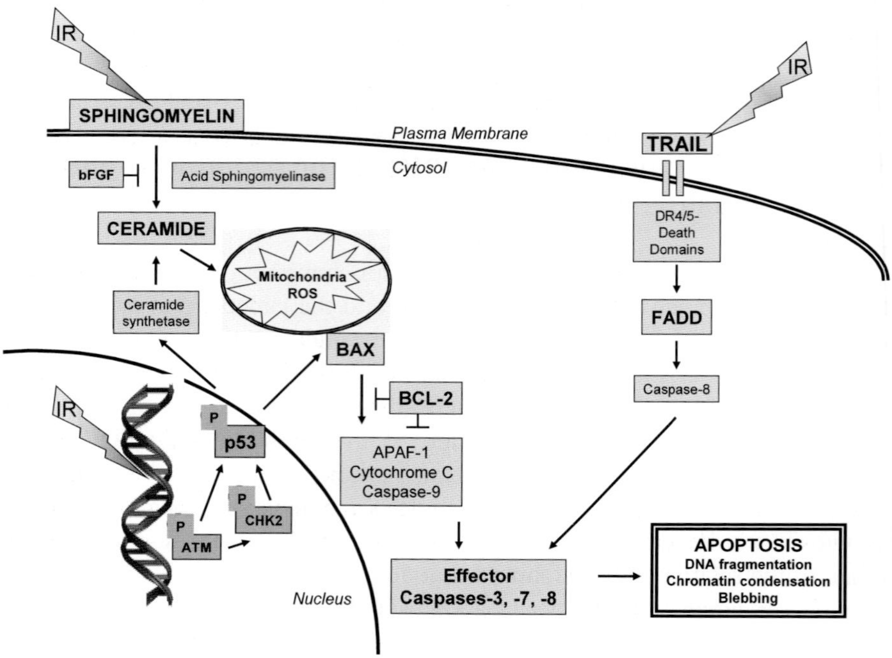


*Figure 2.* Activation of intrinsic and extrinsic pathways of apoptosis by radiotherapy. In exquisitely radiosensitive cells (thymocytes, lymphocytes, germ cells and the tumors arising from these cells), radiation can induce apoptosis using the intrinsic apoptosis pathway as a result of DNA damage by phosphorylation and activation of the ATM-CHK2-p53 axis (see text for details). Stabilized p53 protein can upregulate the *NOXA, Puma and Bax genes* driving pro-apoptotic signals though the mitochondria to release Cytochrome c, APAF-1 and caspase-9 and subsequently activate death effector caspases 3, 7 and 8. This pathway is antagonized by the *Bcl-2* gene family. The ratio of BAX:BCL-2 protein expression or altered p53 function may affect the relative capacity for certain cell types to undergo apoptosis following radiotherapy. In normal or tumor-based vascular endothelium, radiation can initiate a sphingomyelin-dependent signaling cascade within the plasma cell membrane to generate ceramide by the action of acid sphingomyelinase (ASM). DNA damage can also activate *de novo* synthesis of ceramide through an ATM-dependent catalysis using the enzyme, ceramide synthase. Elevated levels of ceramide drive apoptosis through the intrinsic apoptosis pathway. The ceramide pathway can be antagonized by basic-fibroblast growth factor (b-FGF); this may be clinically useful in the treatment of radiation-induced endothelial damage in the gut following pelvic or whole body radiotherapy. The extrinsic pathway can also be activated by ionizing radiation in that the TRAIL receptor, a member of the TNF superfamily, induces apoptosis through the clustering of its receptors, DR4 and DR5 resulting in the formation of the death-inducing signaling complex (DISC). Subsequent recruitment of the adaptor molecule FADD, and activation of Caspases 8 and 10, activates Caspase 3 cleavage and the final death-effector proteins. The use of recombinant TRAIL and other gene therapies can potentiate radiation cell kill in a number of tumor models *in vitro* and *in vivo*.

In hematopoietic cells, radiation up-regulates the expression of pro-apoptotic genes such as *Fas, Bax* and *Caspase 3* and anti-apoptotic genes such as *Bcl-2* [30-33] and apoptosis-related genes may be useful targets during radiotherapy. In these, and other apoptosis-susceptible tumor cells, elevated BCL-2 levels predict for radioresistance *in vitro*. Furthermore, elevated BAX to BCL-2 protein ratios are associated with improved clinical radioresponse in rectal, bladder and prostate cancers [34-37]. Pre-clinical radiosensitization experiments following gene targeting (e.g. altering *Bcl-2, Bcl-xl, Bax, Bims* expression) show clinical promise given the observed increases in radiation-induced apoptosis within prostate, glioma and head and neck cancer xenografts [37-41]. Alternate radiosensitization approaches include the use of synthetic phospholipid derivatives that decrease anti-apoptotic *MAPK* and *PI3-K/AKT* signaling or increase pro-apoptotic *SAPK or JNK* signaling. These include cyclooxygenase-2 (COX-2) inhibitors, betulinic acid, and inhibitors of the proteasome. The latter drugs require validation within future Phase I/II radiotherapy trials [42, 43].

Major efforts in radiotherapy have also revolved around the clinical use of p53 and ceramide signaling pathways. Both pathways reflect a merging of basic science to clinical outcome and the development of novel cancer therapies and deserve special attention. During p53 signaling, the ATM kinase phosphorylates and stabilizes p53 [12]. The phosphorylated serine 15 and 20 residues inhibit p53 degradation by preventing the binding of MDM-2, a ubiquitin E3 ligase. Coupled to a number of post-translational events (e.g. phosporylation and de-phosphorylation, sumoylation and acetylation) p53 acquires the capability to modulate cell cycle arrest and cell death [18]. P53-dependent radiation-induced apoptosis is mediated in part by the transcription of pro-apoptotic genes such as *Bax, PUMA, Noxa* and *Killer/DR5* as well as a series of "p53-inducible genes" *(PIGs)* [18, 44]. The requirement for transactivation of p53 for the apoptotic phenotype is controversial however as other studies have shown that apoptosis mediated by p53 may can occur in the absence of *de novo* gene expression [45].

As stated previously, the level of radiation-induced apoptosis varies between tumors and normal tissues and may be superceded by p53-mediated tumor cell growth arrest responses [46]. However, in tumor cells that are prone to radiation-induced apoptosis (i.e. lymphomas) a wild type 53 genotype does predict for increased apoptosis and decreased clonogenic cell survival given apoptosis is the dominant mode of cell kill. Mutant p53-expressing cells will therefore acquire relative radioresistance, whereas in other tissues or tumors, radiation-induced apoptosis is p53-independent [2, 18]. P53-related clinical radiosensitization and prognostication studies using apoptosis as a target or endpoint is therefore reasonable if such strategies focus also on clonogenic cell kill [47].

Unfortunately, controversy continues to exist in the literature linking p53 status with clinical radioresponse due to differences in cell background and cell propensity for apoptosis or variability in co-expressed oncogenes (e.g. *ras*, *c-Myc,* etc ) in the model systems tested (reviewed in [18]). It is always difficult to translate experimental observations based on cell lines to scenarios relevant to human cancers irradiated *in situ* given the additional physiology and biology associated with the tumor microenvironment. In many retrospective cohort studies of radiotherapeutic outcome, a p53 mutation has been implied based on increased immunohistochemical (IHC) staining within human tissues (as mutant p53 proteins have increased protein half-lives (e.g. 1 to 2 hours) in the absence of DNA damage). However, there is no universal standard for scoring p53 expression using IHC and a variety of protocols, antibodies and endpoints have been used throughout the world. It is not surprising therefore large reviews and meta-analyses have failed to place p53 as a strong determinant of local control or disease-free survival when the IHC endpoint is used in isolation [18]. Nonetheless, in clinical studies where the tumor *p53* genotype was determined by direct DNA sequencing, tumors expressing a mutant p53 protein generally had decreased rates of local tumor control following radiotherapy. Whether this was strictly related to inhibition of radiation-induced apoptosis is unclear, given no tumor biopsies were taken for study during radiotherapy. The biologic reasons by which mutant p53 proteins cause clinical resistance remains unknown, but may relate to co-expression of p73 (a member of the p53 family) which can control relative chemo- and radiosensitivity [18, 48].

Within the tumor microenvironment, p53 may also control radiation-induced cell death as a function of intratumoral hypoxia [49]. Cellular anoxia leads to a 3-fold increase in tumor cell radioresistance [2] and activates an ATR-mediated S-phase arrest via p53 stabilization and CHK1 kinase activation. In contrast, tumor cell re-oxygenation, as a consequence of vascular instability, can activate ATM-p53 signaling and γH2AX due to the induction of DNA breaks in response to increased oxidative free radicals [50]. The process of re-oxygenation between daily radiation doses is an important factor in how fractionated radiotherapy accomplishes exponential cell killing over the duration of treatment [2]. One model suggests that sever hypoxia selects for mutant p53-expressing tumor cells resistant to apoptosis and which acquire increased genetic instability and aggressiveness [51], although strong clinical evidence for this is still lacking [18].

The studies above support the clinical activation of ATM-p53 *in vivo* to improve radiotherapy outcome. Reinstitution of wild type p53 function in tumor cells was used in a series of pre-clinical gene therapy studies to prove that this strategy can augment radiation-induced apoptosis and/or clonogenic cell kill [52, 53]. Promising results were observed in Phase II clinical trials of head and neck and lung cancers using p53 adenoviral gene therapy [54, 55]. P53-based radioprotection of normal tissues would allow for increased tumor dose-escalation using conformal of IMRT radiotherapy and also improve the therapeutic ratio. This may be achieved using chemical inhibitors of p53 (e.g. pifithrin-$\alpha$) as these agents can inhibit radiation-induced epithelial cell apoptosis and subsequent malabsorption within the villi of the small and large intestine post-irradiation [56, 57]. Such an agent, when clinically available, could be used to protect against the gastrointestinal tissue damage during whole-body or abdomino-pelvic irradiation.

Ceramide signaling is another pathway in which molecular pathways of radiation–induced apoptosis could soon alter clinical practice. Within the plasma membrane, ceramide is generated from sphingomyelin (SM) by the action of acid sphingomyelinase (ASM), or by *de novo* synthesis coordinated by ceramide synthase [4]. In certain cells such as endothelial, lymphoid and haematopoietic cells, ceramide is an important medaiator of radiation-induced apoptosis, while in other cells ceramide may serve only as a co-signal or play no role in cell death. The ceramide response to radiation is absent within tissues derived from mice deficient in ASM and can be inhibited *in vitro* by exogenous basic fibroblast growth factor (bFGF) [58, 59]. The latter may have clinical importance in that radiotherapy induces endothelial cell apoptosis within the vessels surrounding the clonogenic crypts within intestinal villi and this effect can be blocked using systemic bFGF [60]. In pre-clinical models, bFGF treatment protected animals against the gastrointestinal syndrome following whole-body radiotherapy (reviewed in [4]). More controversial is a recent study suggesting that tumor radioresponse is also a result of endothelial cell damage (i.e., tumor vasculature collapse) rather than direct killing of tumor cells; this provocative hypothesis will need further confirmation in human xenografts and primary tissues before being widely accepted by the radiation oncology community [61, 62]. Radiation-induced apoptosis also occurs within rodent salivary, lung and neural (brain and spinal cord) tissues, potentially relevant to predicting and preventing other types of acute and late radiation toxicity during therapy [63].

Although many clinical correlative studies have been published regarding radiation-induced apoptosis and radiotherapy response, apoptosis following irradiation appears to be insufficient to account for the therapeutic effect of anticancer agents against the common epithelial and mesenchymal tumors. In these tumors, the level of radiation-induced apoptosis rarely correlates with eventual clonogenic cell killing as measured by colony-forming assays [24-26, 64, 65]. For example, the radiosensitizing effects of androgen ablation in prostate cancer were initially thought to be secondary to increased apoptosis, but have now been linked to decreased proliferation and growth arrest [24, 66, 67]. It is therefore important to review induce two other anti-clonogen responses in tumor cells that can impact upon radiotherapy outcome: mitotic catastrophe and terminal growth arrest. These signaling pathways are discussed in the following two sections.

## 9.2 Mitotic catastrophe and the role of survivin

Faithful chromosome segregation is crucial for cells to maintain the integrity of their genome. Cells which enter mitosis with non-repaired DNA damage are prone to increased rates of genetic instability, carcinogenesis and cell death. Mitotic exit occurs after ubiquitination and proteolytic degradation of Cyclin B by the anaphase-promoting complex (APC) in concert with the AURORA, POLO, NIMA and spindle checkpoint kinases. The mitotic spindle checkpoint monitors the appropriate alignment of chromosomes by interrogating the interaction between chromosomes and microtubules at the chromosomal kinetochore [68]. This checkpoint normally delays chromosome segregation during anaphase to correct any defects in the mitotic spindle apparatus due to non-repaired DNA breaks or acentric DNA fragments. If defects persist, the cell could acquire abnormal spindle morphology and undergo cell death as "mitotic catastrophe.

For the majority of normal and tumor cells, death secondary to mitotic catastrophe accounts for most of the cell kill following irradiation [25, 63, 71, 72]. Mitotic catastrophe is the failure of mammalian cells to undergo mitosis after DNA damage resulting from defective DNA repair and cell cycle checkpoint control. This leads to chromosomal aberrations and abnormalities (tetraploidy or aneuploidy) due to erroneous chromosomal segregation. There is no consensus on the definition of mitotic catastrophe, although most investigators agree that it manifests as abnormal mitoses within enlarged cells containing spindle abnormalities (individual clusters of chromosomal mis-aggregation surrounded by nuclear envelopes), de-condensed chromatin and associated micronuclei (see Table 1).

The kinetochore-associated MAD2, BUBR1, BUB1 and BUB3 proteins are critical constituents of the spindle-checkpoint pathway: MAD2 and BUBR1 interact to regulate mitotic progression by temporal inhibition of the APC machinery and BUB1 and BUB3 mediate mitotic arrest in response to disruption of spindle microtubules [69]. Cells that lack either BUB1 or BUB3 do not undergo mitotic arrest and exhibit catastrophic mitoses when treated with spindle-disrupting agents such as the chemotherapy agents docetaxol or vinblastine [19]. There are few data pertaining to the effects of ionizing radiation on these mitotic spindle checkpoint kinases [69, 70]. However, cells with aberrant G2 checkpoint control due to a deficient activation of the ATM, ATR, p53, CHK1/2, PLK1 or 14-3-3σ signaling pathways are prone to radiation-induced mitotic catastrophe due to an uncoupling of DNA repair and G2 phase cell cycle progression [71].

Mitotic catastrophe has been used to describe cells with an apoptotic phenotype with prematurely-condensed chromatin, however the two cell death morphologies may be genetically distinct as inhibition of apoptosis by over-expressing BCL-2 leads to increased mitotic catastrophe [73]. Nevertheless, cells undergoing mitotic catastrophe may share some molecular characteristics of radiation-induced apoptosis such as caspase activation, phosphorylation of p53 and increased BAX expression which may lead to confusion in the literature [74]. It has been suggested that cells undergoing mitotic catastrophic represent a sub-population of tumor cells resistant to apoptosis signaling [74, 75]. Survivin is a protein member of the inhibitor of apoptosis (IAP) family, but is also a determinant of mitotic catastrophe [76, 77]. Survivin initially associates with centromeres at the prophase and metaphase, but later forms a complex with the Aurora B kinase during anaphase and telophase. Survivin becomes undetectable by the end of telophase [77, 78]. Small interfering RNA (siRNA) targeting survivin expression delays prometaphase and cells fail to properly align their chromosomes, forming a tetraploid nucleus [79] without the induction of apoptosis, similar to irradiated cells. These data are consistent with a greater role for survivin in the control of mitotic catastrophe than apoptosis.

Survivin is a potential radiotherapy target. Ionizing radiation activates transcriptional down-regulation of survivin in a p53-dependent and PI3K/AKT-dependent manner [80-83]. And yet, endogenous expression seems to be specific for human cancers given the observation that survivin is over-expressed in malignant, but not normal, human epithelium [76, 84]. The altered survivin expression in tumor cells may be a consequence of elevated promoter activity, gene amplification or transcriptional de-repression in the setting of mutant p53 [76].

A number of clinical studies have correlated the relative expression of survivin with decreased rates of survival [76] prompting investigators to target surviving in order to improve therapeutic sensitivity. Increased tumor cell radioresistance has been correlated to increased endogenous survivin levels *in vitro* and predicted for decreased disease-free survival following radio-chemotherapy in rectal cancer [85-87]. Genetic antagonists of survivin (e.g. antisense, siRNA ribozymes against survivin or expression of dominant-negative survivin mutants) leads to radiosensitization *in vitro* within pancreatic, melanoma, lung and colorectal carcinoma cell lines [80, 85, 87-89]. In other studies, survival was inhibited in H460 lung cancer cell lines growing either *in vitro* or as xenografts *in vivo,* when both survivin and XIAP were inhibited using antisense transfection [90]. Survivin is also a transcriptional target of vascular endothelial growth factor (VEGF) [91], suggesting that anti-survivin therapies could directly trigger mitotic cell death and intratumor vascular collapse [92, 93].

Increased mitotic catastrophe in tumor cells can also be achieved by inhibiting DNA repair during the G2 phase or decreasing its duration. Tumor cells often exhibit an aberrant G1, but not G2, checkpoint, which makes them vulnerable if a second checkpoint is abrogated. There are a number of candidate drugs which can inhibit the protein effectors of the G2 arrest in response to DNA damage including staurosporine, caffeine and UCN-01 which target CHK1 to elicit mitotic catastrophe. UCN-01 preferentially sensitizes p53-mutated, radioresistant tumor cells to ionizing radiation [94] and similar agents are being tested in clinical trials [1, 95].

## 9.3 Terminal growth arrest (tumor cell senescence)

Cellular senescence was first described by Hayflick and Moorhead in 1961 who observed a finite life-span for human diploid cells growing in culture and a resulting terminal growth arrest or *senescence* [96]. With the exception of certain cells (e.g. germline cells, hematopoietic stem cells, B or T cells), this irreversible proliferative arrest occurs in the vast majority of normal (non-immortalized) human cells. During senescence, cells acquire an enlarged, flattened shape with increased cytoplasmic and nuclear granularity and express senescence associated β-glactosidase (SA-β-gal)[17]. Although senescent cells have lost their ability to divide, they are still metabolically active for some time until undergoing necrosis. Indeed, senescing cells can continue to express degradative enzymes, inflammatory cytokines and growth factors which may be one factor in the acute reaction within normal tissues undergoing radiotherapy [97, 98]. The activation of senescence in culture within normal fibroblasts is directly linked to critical telomere shortening. Telomeres protect chromosomes from chromosomal fusion and the potential loss of genetic information and shorten 50-300 base pairs following each cell division. Telomeres are composed of repetitive sequences that can be maintained by the enzyme telomerase consisting of a reverse transcriptase (hTERT), a template RNA (hTERC) and accessory factors (EST1 and dyskerin in humans). Due to an endoreplication problem, a critical telomere length will finally be achieved after a given series of divisions and the cells will terminally arrest. The arrest is associated with the activation of DNA damage responses including the expression of several cyclin-dependent kinase inhibitors including $p21^{WAF}$ and $p16^{INK4a}$ [99].

In certain cells, this terminal arrest response overrides the apoptosis response that can be triggered by ionizing radiation, heat stress and antibodies to the Fas-ligand [100, 101]. For example, in fibroblasts and epithelial cells, p53 acts to induce cell cycle arrests in the G1 and G2 phases of the cell cycle to prevent cells from proliferating under conditions of DNA damage and repair. Depending on the cellular environment and context, this arrest can be either permanent (i.e. terminal) or temporary, dependent on whether DNA lesions are optimally repaired (see Figure 3). The capacity for p53-mediated cell cycle arrest versus apoptosis is not mutually exclusive as these two responses can be modified by certain growth factors or the initial level of DNA damage (i.e. number and repair of DNA-breaks) within a given cell line or type [18].

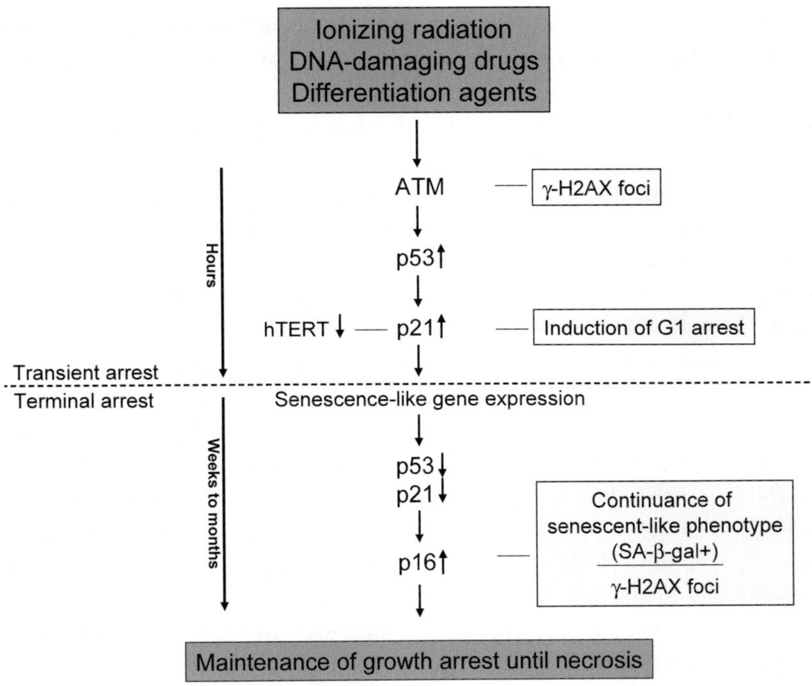


*Figure 3.* Pathways of reversible and irreversible cell cycle arrest following radiotherapy. Following genotoxic insult, DNA-breaks induced by radiotherapy can activate the ATM kinase which leads to the phosphorylation of the histone biomarker γH2AX at the sites of DNA damage. Coincident stabilization and activation of the p53 tumor suppressor protein initiates cell cycle checkpoint control through the elevated expression of the $p21^{WAF}$ cdk-inhibitor. Depending on the cellular environment and extent of DNA damage, the checkpoint arrest in fibroblasts or epithelial tissues can either be permanent (resulting in terminal growth arrest) or temporary, in which the damage is repaired and the cell proceeds to re-enter the cell cycle. Radiation-induced terminal growth arrest shares some hallmarks with replicative senescence, as $p21^{WAF}$ has also been show to downregulate hTERT, the catalytic subunit of the enzyme telomerase. Terminally arrested cells can also express elevated levels of the $p16^{INK4}a$ cdk-inhibitor. Both irradiated normal and tumor cells can undergo terminal cell arrest and acquire a senescent-like phenotype associated with increased expression of senescence-associated β-galactosidase (SA β-gal), increased nuclear granularity and continued metabolism until final necrosis. Modified from reference [19].

Telomerase is an RNA-dependent DNA polymerase that extends the ends of telomeres in germline and stem cells in order to avoid replicative senescence and telomere crisis. Telomerase is strongly up-regulated in most cancer cells [102] leading to unlimited proliferative potential. Indeed, over-expression of the human catalytic subunit of telomerase (hTERT) is sufficient to immortalize human cancer cells [103, 104]. Telomerase may therefore be a target for radiosensitization as telomerase activity and associated telomere length might be expected to relate to the cellular radiosensitivity of fibroblasts and/or epithelial cells. Indeed, senescent fibroblasts are more radiosensitive than their non-senescent counterparts with longer telomeres [105]. Furthermore, ionizing radiation increases the expression of hTERT [106] which relate to radiation-induced genetic instability given that chromosomal stability is reduced in hTERT-immortalized human fibroblasts after exposure to ionizing radiation [107]. Additionally, telomere length is correlated with differential cellular radiosensitivity in isogenic murine models, although this has not been confirmed in human tumor cell lines representing varying histopathologic types [108-111]. Nonetheless, clinico-pathologic studies of colorectal and head and neck cancers suggest that reduced hTERT activity is associated with reduced tumor cell survival and increased local tumor control following radiotherapy [103, 105, 112-115].

Most tumor cell lines have retained the capacity to undergo tumor cell arrest following therapy (i.e. tumor cell senescence) although it is not *p53*- and $p21^{WAF}$ -dependent [116-118]. In epithelial tumors, tumor cell senescence can be a rapid response to ionizing radiation or chemotherapeutic drugs. In 11 out of 14 tumor cell lines, induction of *SA-β-gal* and a senescent-like phenotype was observed following treatment with retinoids or ionizing radiation [116, 117]. A senescent-like morphology has also been documented within clinical breast cancer specimens treated with chemotherapy [116]. Normal cells can also undergo terminal arrest following chemotherapy or radiotherapy; exposure of human fibroblasts to mitomycin C and bleomycin results in accelerated senescence [119, 120]. Human fibroblasts exhibit a senescence-like morphology after exposure to doses as low as 0.1 Gy of ionizing radiation or in response to one non-repaired DNA-dsb [121]. The importance of tumor cell senescence over apoptosis in therapeutic response was confirmed in animal tumor models by Lowe and co-workers [122] in which the long-term chemocurability and anti-proliferative effects of cyclophosphamide in murine lymphomas was secondary to terminal growth arrest.

To translate the above observations to the clinic, further studies using specimens of treated tumors are required. Terminal arrest of tumor cells may explain the relatively slow resolution, yet ultimate cure, of some tumors following radiotherapy. This may be particularly true in the radioresponse of prostate cancer in which the final level of cell kill (using prostatic-specific antigen (PSA) levels as a surrogate for cell death) may not manifest itself until 12 to 16 months following radiotherapy [24]. If true, treatments which differentially increase terminal arrest in tumor cells may be useful adjuncts to conformal radiotherapy. For example, differentiation agents (e.g. retinoids), or HDAC (histone deacetylation) inhibitors (e.g. TSA, SAHA), have been used to induce a senescence-like phenotype and radiosensitize breast, prostate glioma and head and neck cancer cells, both *in vitro* and *in vivo* (for review see [1, 123-125]). HDAC inhibitors may provide a positive therapeutic ratio during radiotherapy as they were found to be potent radioprotectors against acute and late skin reactions following radiation treatment [126]. Future clinical studies should characterize and quantitate well-known and new biomarkers of terminal growth arrest (e.g. p53, $p21^{WAF}$, $p16^{INK4a}$, maspin, pTGF-β, BTG1/2; reviewed in [127]) within irradiated tissues. These could become new predictive biomarkers of radiocurability and reflect the extent of terminal growth arrest in tumor with following different drug-radiation combinations. Indeed, there is good evidence that radiotherapy outcome is related to the permanent inhibition of tumor cell proliferation [128-136]. These genes may be excellent new targets for novel cancer therapies utilizing radiotherapy.

## 10. CONCLUSIONS AND NEW CLINICAL DIRECTIONS

The debate regarding the relative role of apoptosis during radiotherapy cell kill has been dampened somewhat by the realization that cell death pathways will manifest themselves in a tissue- and time-specific manner. Currently, our ability to define radiotherapeutic outcome on the basis of specific cell death mechanism(s) is limited by the paucity of data pertaining to biopsy specimens derived from tumors undergoing radiotherapy. Pre-treatment specimens are usually available, yet these speak more to oncogenesis rather than radioresponse. This may partially explain the conflicting data regarding pre-treatment determination of intratumoral apoptosis and clinical radiotherapy response [137-150]. Non-invasive imaging may become increasingly powerful as a means to quantify apoptosis or terminal growth arrest. Initial studies *in vivo* have detected $p21^{WAF}$ gene expression following DNA damage[151, 152] or the level of therapy-induced apoptosis (99m-Tc Annexin-V) using scintigraphy or positron emission tomography (PET) [153, 154]. Similar approaches may soon be available to quantitate mitotic catastrophe in treated tumors.

Future therapies in radiation oncology will increasingly utilize molecular-based targeting of cell death pathways. Rapid introduction of these therapies into the clinical realm may be limited by the need to define appropriate scheduling and toxicity of the agents, inter-patient and intratumoral variability in the expression of the target (as a predictor of treatment response) and molecular cross-talk amongst redundant or parallel signaling pathways [1]. However, it is hoped that these limitations might be bypassed by simultaneous determination of multiple signaling pathways using genomic and proteomic analyses of human tissues. This will afford selection of the best molecular agents to be used in combination with radiotherapy in order to increase tumor cell kill, yet decrease normal cell kill during therapy.

## ACKNOWLEDGMENTS

The authors would like to thank members of the Bristow lab, Dr. R.P. Hill (PMH-Toronto) and Dr. N. Blin (University of Tuebingen) for helpful discussions. These studies were supported by a Canadian Prostate Cancer Research Initiative (CPCRI) IDEA grant, a US Army DOD Prostate Cancer Program operating grant and the PMH Foundation. RGB is a Canadian Cancer Society Research Scientist. OF is a Ph.D. Candidate at the University of Tuebingen. Cell photographs in Table 1 kindly provided by Bromfield and Bristow (Medical Biophysics, University of Toronto), Dr. A Evans (Pathology, Princess Margaret Hospital, Toronto) and with permission from Sato et al; Oncogene, 2000 Nov 2;19(46):5281-90.

## REFERENCES

1. Ma, B.B., et al., Combined-modality treatment of solid tumors using radiotherapy and molecular targeted agents. J Clin Oncol, 2003. **21**(14): p. 2760-76.
2. Bristow, R.G. and R. Hill, Molecular and Celullar Basis of Radiotherapy. Basic Science of Oncology, 1998: p. McGraw Hill Publishers, New York.
3. Ward, J.F., Radiation mutagenesis: the initial DNA lesions responsible. Radiat Res, 1995. **142**(3): p. 362-8.
4. Kolesnick, R. and Z. Fuks, Radiation and ceramide-induced apoptosis. Oncogene, 2003. **22**(37): p. 5897-906.
5. Morgan, W.F., et al., Genomic instability induced by ionizing radiation. Radiat Res, 1996. **146**(3): p. 247-58.
6. Snyder, A.R. and W.F. Morgan, Gene expression profiling after irradiation: clues to understanding acute and persistent responses? Cancer Metastasis Rev, 2004. **23**(3-4): p. 259-68.
7. Kim, I.A., et al., The influence of Ras pathway signaling on tumor radiosensitivity. Cancer Metastasis Rev, 2004. **23**(3-4): p. 227-36.

8. Gupta, A.K., et al., Local recurrence in head and neck cancer: relationship to radiation resistance and signal transduction. Clin Cancer Res, 2002. **8**(3): p. 885-92.
9. Harari, P.M. and S.M. Huang, Combining EGFR inhibitors with radiation or chemotherapy: will preclinical studies predict clinical results? Int J Radiat Oncol Biol Phys, 2004. **58**(3): p. 976-83.
10. Hahn, S.M., et al., A Phase I trial of the farnesyltransferase inhibitor L-778,123 and radiotherapy for locally advanced lung and head and neck cancer. Clin Cancer Res, 2002. **8**(5): p. 1065-72.
11. Pilch, D.R., et al., Characteristics of gamma-H2AX foci at DNA double-strand breaks sites. Biochem Cell Biol, 2003. **81**(3): p. 123-9.
12. Bakkenist, C.J. and M.B. Kastan, Initiating cellular stress responses. Cell, 2004. **118**(1): p. 9-17.
13. Willers, H., J. Dahm-Daphi, and S.N. Powell, Repair of radiation damage to DNA. Br J Cancer, 2004. **90**(7): p. 1297-301.
14. Rothkamm, K., et al., Pathways of DNA double-strand break repair during the mammalian cell cycle. Mol Cell Biol, 2003. **23**(16): p. 5706-15.
15. Olive, P.L. and J.P. Banath, Phosphorylation of histone H2AX as a measure of radiosensitivity. Int J Radiat Oncol Biol Phys, 2004. **58**(2): p. 331-5.
16. Lukas, J., C. Lukas, and J. Bartek, Mammalian cell cycle checkpoints: signalling pathways and their organization in space and time. DNA Repair (Amst), 2004. **3**(8-9): p. 997-1007.
17. Dimri, G.P., et al., A biomarker that identifies senescent human cells in culture and in aging skin in vivo. Proc Natl Acad Sci U S A, 1995. **92**(20): p. 9363-7.
18. Cuddihy, A.R. and R.G. Bristow, The p53 protein family and radiation sensitivity: Yes or no? Cancer Metastasis Rev, 2004. **23**(3-4): p. 237-57.
19. Roninson, I.B., Tumor cell senescence in cancer treatment. Cancer Res, 2003. **63**(11): p. 2705-15.
20. Ponten, F., et al., Induction of p53 expression in skin by radiotherapy and UV radiation: a randomized study. J Natl Cancer Inst, 2001. **93**(2): p. 128-33.
21. Pawlik, T.M. and K. Keyomarsi, Role of cell cycle in mediating sensitivity to radiotherapy. Int J Radiat Oncol Biol Phys, 2004. **59**(4): p. 928-42.
22. Kao, G.D., W.G. McKenna, and T.J. Yen, Detection of repair activity during the DNA damage-induced G2 delay in human cancer cells. Oncogene, 2001. **20**(27): p. 3486-96.
23. Wilson, G.D., Radiation and the cell cycle, revisited. Cancer Metastasis Rev, 2004. **23**(3-4): p. 209-25.
24. Bromfield, G.P., et al., Cell death in irradiated prostate epithelial cells: role of apoptotic and clonogenic cell kill. Prostate Cancer Prostatic Dis, 2003. **6**(1): p. 73-85.
25. Brown, J.M. and B.G. Wouters, Apoptosis: mediator or mode of cell killing by anticancer agents? Drug Resist Updat, 2001. **4**(2): p. 135-6.
26. Abend, M., Reasons to reconsider the significance of apoptosis for cancer therapy. Int J Radiat Biol, 2003. **79**(12): p. 927-41.
27. Shankar, S., T.R. Singh, and R.K. Srivastava, Ionizing radiation enhances the therapeutic potential of TRAIL in prostate cancer in vitro and in vivo: Intracellular mechanisms. Prostate, 2004. **61**(1): p. 35-49.
28. Abou El Hassan, M.A., et al., Overexpression of Bcl2 abrogates chemo- and radiotherapy-induced sensitisation of NCI-H460 non-small-cell lung cancer cells to adenovirus-mediated expression of full-length TRAIL. Br J Cancer, 2004. **91**(1): p. 171-7.
29. Marini, P. and C. Belka, Death receptor ligands: new strategies for combined treatment with ionizing radiation. Curr Med Chem Anti-Canc Agents, 2003. **3**(5): p. 334-42.

30. Chauhan, D., et al., Cytochrome c-dependent and -independent induction of apoptosis in multiple myeloma cells. J Biol Chem, 1997. **272**(48): p. 29995-7.
31. Yu, Y. and J.B. Little, p53 is involved in but not required for ionizing radiation-induced caspase-3 activation and apoptosis in human lymphoblast cell lines. Cancer Res, 1998. **58**(19): p. 4277-81.
32. Herr, I. and K.M. Debatin, Cellular stress response and apoptosis in cancer therapy. Blood, 2001. **98**(9): p. 2603-14.
33. Rudner, J., et al., Radiation sensitivity and apoptosis in human lymphoma cells. Int J Radiat Biol, 2001. **77**(1): p. 1-11.
34. Mackey, T.J., et al., bcl-2/bax ratio as a predictive marker for therapeutic response to radiotherapy in patients with prostate cancer. Urology, 1998. **52**(6): p. 1085-90.
35. Matsumoto, H., et al., Bax to Bcl-2 ratio and Ki-67 index are useful predictors of neoadjuvant chemoradiation therapy in bladder cancer. Jpn J Clin Oncol, 2004. **34**(3): p. 124-30.
36. Scopa, C.D., et al., bcl-2/bax ratio as a predictive marker for therapeutic response to radiotherapy in patients with rectal cancer. Appl Immunohistochem Mol Morphol, 2001. **9**(4): p. 329-34.
37. Wacheck, V., et al., Bcl-x(L) antisense oligonucleotides radiosensitise colon cancer cells. Br J Cancer, 2003. **89**(7): p. 1352-7.
38. Lowe, S.L., et al., Prostate-specific expression of Bax delivered by an adenoviral vector induces apoptosis in LNCaP prostate cancer cells. Gene Ther, 2001. **8**(18): p. 1363-71.
39. Belka, C. and W. Budach, Anti-apoptotic Bcl-2 proteins: structure, function and relevance for radiation biology. Int J Radiat Biol, 2002. **78**(8): p. 643-58.
40. Arafat, W.O., et al., An adenovirus encoding proapoptotic Bax synergistically radiosensitizes malignant glioma. Int J Radiat Oncol Biol Phys, 2003. **55**(4): p. 1037-50.
41. Yip, K.W., et al., Potential utility of bim(s) as a novel apoptotic therapeutic molecule. Mol Ther, 2004. **10**(3): p. 533-44.
42. Belka, C., et al., Apoptosis-modulating agents in combination with radiotherapy-current status and outlook. Int J Radiat Oncol Biol Phys, 2004. **58**(2): p. 542-54.
43. Wachsberger, P., R. Burd, and A.P. Dicker, Tumor response to ionizing radiation combined with antiangiogenesis or vascular targeting agents: exploring mechanisms of interaction. Clin Cancer Res, 2003. **9**(6): p. 1957-71.
44. Fridman, J.S. and S.W. Lowe, Control of apoptosis by p53. Oncogene, 2003. **22**(56): p. 9030-40.
45. Slee, E.A., D.J. O'Connor, and X. Lu, To die or not to die: how does p53 decide? Oncogene, 2004. **23**(16): p. 2809-18.
46. Fei, P., E.J. Bernhard, and W.S. El-Deiry, Tissue-specific induction of p53 targets in vivo. Cancer Res, 2002. **62**(24): p. 7316-27.
47. Lowe, S.W., et al., p53 status and the efficacy of cancer therapy in vivo. Science, 1994. **266**(5186): p. 807-10.
48. Gasco, M. and T. Crook, p53 family members and chemoresistance in cancer: what we know and what we need to know. Drug Resist Updat, 2003. **6**(6): p. 323-8.
49. Leo, C., A.J. Giaccia, and N.C. Denko, The hypoxic tumor microenvironment and gene expression. Semin Radiat Oncol, 2004. **14**(3): p. 207-14.
50. Hammond, E.M. and A.J. Giaccia, The role of ATM and ATR in the cellular response to hypoxia and re-oxygenation. DNA Repair (Amst), 2004. **3**(8-9): p. 1117-22.
51. Graeber, T.G., et al., Hypoxia-mediated selection of cells with diminished apoptotic potential in solid tumours. Nature, 1996. **379**(6560): p. 88-91.

52. Geoerger, B., et al., Potentiation of radiation therapy by the oncolytic adenovirus dl1520 (ONYX-015) in human malignant glioma xenografts. Br J Cancer, 2003. **89**(3): p. 577-84.
53. Nishizaki, M., et al., Synergistic inhibition of human lung cancer cell growth by adenovirus-mediated wild-type p53 gene transfer in combination with docetaxel and radiation therapeutics in vitro and in vivo. Clin Cancer Res, 2001. **7**(9): p. 2887-97.
54. Swisher, S.G., et al., Induction of p53-regulated genes and tumor regression in lung cancer patients after intratumoral delivery of adenoviral p53 (INGN 201) and radiation therapy. Clin Cancer Res, 2003. **9**(1): p. 93-101.
55. Fang, B. and J.A. Roth, The role of gene therapy in combined modality treatment strategies for cancer. Curr Opin Mol Ther, 2003. **5**(5): p. 475-82.
56. Komarova, E.A., et al., Dual effect of p53 on radiation sensitivity in vivo: p53 promotes hematopoietic injury, but protects from gastro-intestinal syndrome in mice. Oncogene, 2004. **23**(19): p. 3265-71.
57. Komarov, P.G., et al., A chemical inhibitor of p53 that protects mice from the side effects of cancer therapy. Science, 1999. **285**(5434): p. 1733-7.
58. Santana, P., et al., Acid sphingomyelinase-deficient human lymphoblasts and mice are defective in radiation-induced apoptosis. Cell, 1996. **86**(2): p. 189-99.
59. Pena, L.A., Z. Fuks, and R.N. Kolesnick, Radiation-induced apoptosis of endothelial cells in the murine central nervous system: protection by fibroblast growth factor and sphingomyelinase deficiency. Cancer Res, 2000. **60**(2): p. 321-7.
60. Paris, F., et al., Endothelial apoptosis as the primary lesion initiating intestinal radiation damage in mice. Science, 2001. **293**(5528): p. 293-7.
61. Brown, M., et al., Comment on "Tumor response to radiotherapy regulated by endothelial cell apoptosis" (II). Science, 2003. **302**(5652): p. 1894; author reply 1894.
62. Garcia-Barros, M., et al., Tumor response to radiotherapy regulated by endothelial cell apoptosis. Science, 2003. **300**(5622): p. 1155-9.
63. Hill, R.P., et al., Normal tissue radiobiology: from the laboratory to the clinic. Int J Radiat Oncol Biol Phys, 2001. **49**(2): p. 353-65.
64. Aldridge, D.R., M.J. Arends, and I.R. Radford, Increasing the susceptibility of the rat 208F fibroblast cell line to radiation-induced apoptosis does not alter its clonogenic survival dose-response. Br J Cancer, 1995. **71**(3): p. 571-7.
65. Sheridan, M.T. and C.M. West, Ability to undergo apoptosis does not correlate with the intrinsic radiosensitivity (SF2) of human cervix tumor cell lines. Int J Radiat Oncol Biol Phys, 2001. **50**(2): p. 503-9.
66. Pollack, A., et al., The early supra-additive apoptotic response of R3327-G prostate tumors to androgen ablation and radiation is not sustained with multiple fractions. Int J Radiat Oncol Biol Phys, 2000. **46**(1): p. 153-8.
67. Hintz, B.L., C. Koo, and J.F. Murphy, Pattern of proliferative index (Ki-67) after anti-androgen manipulation reflects the ability of irradiation to control prostate cancer. Am J Clin Oncol, 2004. **27**(1): p. 85-8.
68. Musacchio, A. and K.G. Hardwick, The spindle checkpoint: structural insights into dynamic signalling. Nat Rev Mol Cell Biol, 2002. **3**(10): p. 731-41.
69. Okada, H. and T.W. Mak, Pathways of apoptotic and non-apoptotic death in tumour cells. Nat Rev Cancer, 2004. **4**(8): p. 592-603.
70. Sudo, T., et al., Activation of Cdh1-dependent APC is required for G1 cell cycle arrest and DNA damage-induced G2 checkpoint in vertebrate cells. Embo J, 2001. **20**(22): p. 6499-508.
71. Roninson, I.B., Tumor senescence as a determinant of drug response in vivo. Drug Resist Updat, 2002. **5**(5): p. 204-8.

72. Jonathan, E.C., E.J. Bernhard, and W.G. McKenna, How does radiation kill cells? Curr Opin Chem Biol, 1999. **3**(1): p. 77-83.
73. Elliott, M.J., et al., Bcl-2 inhibits early apoptotic events and reveals post-mitotic multinucleation without affecting cell cycle arrest in human epithelial tumor cells exposed to etoposide. Cancer Chemother Pharmacol, 1999. **44**(1): p. 1-11.
74. Castedo, M., et al., Cell death by mitotic catastrophe: a molecular definition. Oncogene, 2004. **23**(16): p. 2825-37.
75. Nitta, M., et al., Spindle checkpoint function is required for mitotic catastrophe induced by DNA-damaging agents. Oncogene, 2004. **23**(39): p. 6548-58.
76. Altieri, D.C., Survivin, versatile modulation of cell division and apoptosis in cancer. Oncogene, 2003. **22**(53): p. 8581-9.
77. Li, F., et al., Control of apoptosis and mitotic spindle checkpoint by survivin. Nature, 1998. **396**(6711): p. 580-4.
78. Uren, A.G., et al., Survivin and the inner centromere protein INCENP show similar cell-cycle localization and gene knockout phenotype. Curr Biol, 2000. **10**(21): p. 1319-28.
79. Carvalho, A., et al., Survivin is required for stable checkpoint activation in taxol-treated HeLa cells. J Cell Sci, 2003. **116**(Pt 14): p. 2987-98.
80. Lu, B., et al., Survivin as a therapeutic target for radiation sensitization in lung cancer. Cancer Res, 2004. **64**(8): p. 2840-5.
81. Mirza, A., et al., Human survivin is negatively regulated by wild-type p53 and participates in p53-dependent apoptotic pathway. Oncogene, 2002. **21**(17): p. 2613-22.
82. Hoffman, W.H., et al., Transcriptional repression of the anti-apoptotic survivin gene by wild type p53. J Biol Chem, 2002. **277**(5): p. 3247-57.
83. Zhou, M., et al., DNA damage induces a novel p53-survivin signaling pathway regulating cell cycle and apoptosis in acute lymphoblastic leukemia cells. J Pharmacol Exp Ther, 2002. **303**(1): p. 124-31.
84. Adida, C., et al., Developmentally regulated expression of the novel cancer anti-apoptosis gene survivin in human and mouse differentiation. Am J Pathol, 1998. **152**(1): p. 43-9.
85. Asanuma, K., et al., A role for survivin in radioresistance of pancreatic cancer cells. Jpn J Cancer Res, 2002. **93**(9): p. 1057-62.
86. Rodel, F., et al., High survivin expression is associated with reduced apoptosis in rectal cancer and may predict disease-free survival after preoperative radiochemotherapy and surgical resection. Strahlenther Onkol, 2002. **178**(8): p. 426-35.
87. Rodel, C., et al., Spontaneous and radiation-induced apoptosis in colorectal carcinoma cells with different intrinsic radiosensitivities: survivin as a radioresistance factor. Int J Radiat Oncol Biol Phys, 2003. **55**(5): p. 1341-7.
88. Grossman, D., et al., Expression and targeting of the apoptosis inhibitor, survivin, in human melanoma. J Invest Dermatol, 1999. **113**(6): p. 1076-81.
89. Pennati, M., et al., Radiosensitization of human melanoma cells by ribozyme-mediated inhibition of survivin expression. J Invest Dermatol, 2003. **120**(4): p. 648-54.
90. Cao, C., et al., XIAP and survivin as therapeutic targets for radiation sensitization in preclinical models of lung cancer. Oncogene, 2004.
91. O'Connor, D.S., et al., Control of apoptosis during angiogenesis by survivin expression in endothelial cells. Am J Pathol, 2000. **156**(2): p. 393-8.
92. Mesri, M., et al., Suppression of vascular endothelial growth factor-mediated endothelial cell protection by survivin targeting. Am J Pathol, 2001. **158**(5): p. 1757-65.

93. Blanc-Brude, O.P., et al., Therapeutic targeting of the survivin pathway in cancer: initiation of mitochondrial apoptosis and suppression of tumor-associated angiogenesis. Clin Cancer Res, 2003. **9**(7): p. 2683-92.
94. Dixon, H. and C.J. Norbury, Therapeutic exploitation of checkpoint defects in cancer cells lacking p53 function. Cell Cycle, 2002. **1**(6): p. 362-8.
95. Koniaras, K., et al., Inhibition of Chk1-dependent G2 DNA damage checkpoint radiosensitizes p53 mutant human cells. Oncogene, 2001. **20**(51): p. 7453-63.
96. Hayflick, L. and P.S. Moorhead, The serial cultivation of human diploid cell strains. Exp Cell Res, 1961. **25**: p. 585-621.
97. Campisi, J., From cells to organisms: can we learn about aging from cells in culture? Exp Gerontol, 2001. **36**(4-6): p. 607-18.
98. Elenbaas, B. and R.A. Weinberg, Heterotypic signaling between epithelial tumor cells and fibroblasts in carcinoma formation. Exp Cell Res, 2001. **264**(1): p. 169-84.
99. Sedelnikova, O.A., et al., Senescing human cells and ageing mice accumulate DNA lesions with unrepairable double-strand breaks. Nat Cell Biol, 2004. **6**(2): p. 168-70.
100. Wang, E., M.J. Lee, and S. Pandey, Control of fibroblast senescence and activation of programmed cell death. J Cell Biochem, 1994. **54**(4): p. 432-9.
101. Spaulding, C., W. Guo, and R.B. Effros, Resistance to apoptosis in human CD8+ T cells that reach replicative senescence after multiple rounds of antigen-specific proliferation. Exp Gerontol, 1999. **34**(5): p. 633-44.
102. Harrington, L. and M.O. Robinson, Telomere dysfunction: multiple paths to the same end. Oncogene, 2002. **21**(4): p. 592-7.
103. Guo, C., et al., Inhibition of telomerase is related to the life span and tumorigenicity of human prostate cancer cells. J Urol, 2001. **166**(2): p. 694-8.
104. Orlando, C., et al., Telomerase in urological malignancy. J Urol, 2001. **166**(2): p. 666-73.
105. Rubio, M.A., S.H. Kim, and J. Campisi, Reversible manipulation of telomerase expression and telomere length. Implications for the ionizing radiation response and replicative senescence of human cells. J Biol Chem, 2002. **277**(32): p. 28609-17.
106. Pandita, T.K. and J.L. Roti Roti, Role of telomerase in radiocurability (review). Oncol Rep, 2003. **10**(2): p. 263-70.
107. Pirzio, L.M., et al., Human fibroblasts expressing hTERT show remarkable karyotype stability even after exposure to ionizing radiation. Cytogenet Genome Res, 2004. **104**(1-4): p. 87-94.
108. Bouffler, S.D., et al., Telomeric sequences, radiation sensitivity and genomic instability. Int J Radiat Biol, 2001. **77**(10): p. 995-1005.
109. Goytisolo, F.A., et al., Short telomeres result in organismal hypersensitivity to ionizing radiation in mammals. J Exp Med, 2000. **192**(11): p. 1625-36.
110. Hauguel, T. and F. Bunz, Haploinsufficiency of hTERT leads to telomere dysfunction and radiosensitivity in human cancer cells. Cancer Biol Ther, 2003. **2**(6): p. 679-84.
111. Wong, K.K., et al., Telomere dysfunction impairs DNA repair and enhances sensitivity to ionizing radiation. Nat Genet, 2000. **26**(1): p. 85-8.
112. Kim, H.R., et al., Change of telomerase activity in rectal cancer with chemoradiation therapy. J Korean Med Sci, 2000. **15**(2): p. 167-72.
113. Ogawa, Y., et al., Changes in telomerase activity of advanced cancers of oral cavity and oropharynx during radiation therapy: correlation with clinical outcome. Int J Mol Med, 1998. **2**(3): p. 301-307.
114. Sawant, S.G., et al., Telomerase activity as a measure for monitoring radiocurability of tumor cells. Faseb J, 1999. **13**(9): p. 1047-54.

115. Turriziani, M., et al., Residual telomerase activity: a marker of cell survival after exposure to gamma radiation in vitro. Anticancer Res, 2003. **23**(6C): p. 4561-9.
116. Chang, B.D., et al., A senescence-like phenotype distinguishes tumor cells that undergo terminal proliferation arrest after exposure to anticancer agents. Cancer Res, 1999. **59**(15): p. 3761-7.
117. te Poele, R.H., et al., DNA damage is able to induce senescence in tumor cells in vitro and in vivo. Cancer Res, 2002. **62**(6): p. 1876-83.
118. Wouters, B.G., et al., A p53 and apoptotic independent role for p21waf1 in tumour response to radiation therapy. Oncogene, 1999. **18**(47): p. 6540-5.
119. Rodemann, H.P., et al., Selective enrichment and biochemical characterization of seven human skin fibroblasts cell types in vitro. Exp Cell Res, 1989. **180**(1): p. 84-93.
120. Niggli, H.J., et al., Mitomycin C-induced postmitotic fibroblasts retain the capacity to repair pyrimidine photodimers formed after UV-irradiation. Mutat Res, 1989. **219**(4): p. 231-40.
121. Di Leonardo, A., et al., DNA damage triggers a prolonged p53-dependent G1 arrest and long-term induction of Cip1 in normal human fibroblasts. Genes Dev, 1994. **8**(21): p. 2540-51.
122. Schmitt, C.A., et al., Dissecting p53 tumor suppressor functions in vivo. Cancer Cell, 2002. **1**(3): p. 289-98.
123. Kim, J.H., J.H. Shin, and I.H. Kim, Susceptibility and radiosensitization of human glioblastoma cells to trichostatin A, a histone deacetylase inhibitor. Int J Radiat Oncol Biol Phys, 2004. **59**(4): p. 1174-80.
124. Biade, S., et al., Chemical agents that promote chromatin compaction radiosensitize tumour cells. Int J Radiat Biol, 2001. **77**(10): p. 1033-42.
125. Camphausen, K., et al., Enhanced radiation-induced cell killing and prolongation of gammaH2AX foci expression by the histone deacetylase inhibitor MS-275. Cancer Res, 2004. **64**(1): p. 316-21.
126. Chung, Y.L., A.J. Wang, and L.F. Yao, Antitumor histone deacetylase inhibitors suppress cutaneous radiation syndrome: Implications for increasing therapeutic gain in cancer radiotherapy. Mol Cancer Ther, 2004. **3**(3): p. 317-25.
127. Chang, B.D., et al., Effects of p21Waf1/Cip1/Sdi1 on cellular gene expression: implications for carcinogenesis, senescence, and age-related diseases. Proc Natl Acad Sci U S A, 2000. **97**(8): p. 4291-6.
128. Buffa, F.M., et al., Molecular marker profiles predict locoregional control of head and neck squamous cell carcinoma in a randomized trial of continuous hyperfractionated accelerated radiotherapy. Clin Cancer Res, 2004. **10**(11): p. 3745-54.
129. Gewirtz, D.A., Growth arrest and cell death in the breast tumor cell in response to ionizing radiation and chemotherapeutic agents which induce DNA damage. Breast Cancer Res Treat, 2000. **62**(3): p. 223-35.
130. Lera, J., et al., Tumor proliferation, p53 expression, and apoptosis in laryngeal carcinoma: relation to the results of radiotherapy. Cancer, 1998. **83**(12): p. 2493-501.
131. Li, R., et al., Ki-67 staining index predicts distant metastasis and survival in locally advanced prostate cancer treated with radiotherapy: an analysis of patients in radiation therapy oncology group protocol 86-10. Clin Cancer Res, 2004. **10**(12 Pt 1): p. 4118-24.
132. Pillai, M.R., P.G. Jayaprakash, and M.K. Nair, Tumour-proliferative fraction and growth factor expression as markers of tumour response to radiotherapy in cancer of the uterine cervix. J Cancer Res Clin Oncol, 1998. **124**(8): p. 456-61.

133. Qureshi, K.N., et al., Combined p21WAF1/CIP1 and p53 overexpression predict improved survival in muscle-invasive bladder cancer treated by radical radiotherapy. Int J Radiat Oncol Biol Phys, 2001. **51**(5): p. 1234-40.
134. Rigaud, J., et al., Expression of p21 cell cycle protein is an independent predictor of response to salvage radiotherapy after radical prostatectomy. Prostate, 2004. **58**(3): p. 269-76.
135. Santucci, M.A., et al., Radiation-induced gadd45 expression correlates with clinical response to radiotherapy of cervical carcinoma. Int J Radiat Oncol Biol Phys, 2000. **46**(2): p. 411-6.
136. Sohda, M., et al., Pretreatment evaluation of combined HIF-1alpha, p53 and p21 expression is a useful and sensitive indicator of response to radiation and chemotherapy in esophageal cancer. Int J Cancer, 2004. **110**(6): p. 838-44.
137. Adell, G.C., et al., Apoptosis in rectal carcinoma: prognosis and recurrence after preoperative radiotherapy. Cancer, 2001. **91**(10): p. 1870-5.
138. Chang, J., et al., Apoptosis and proliferation as predictors of chemotherapy response in patients with breast carcinoma. Cancer, 2000. **89**(11): p. 2145-52.
139. Chung, E.J., et al., Spontaneous apoptosis as a predictor of radiotherapy in patients with stage IIB squamous cell carcinoma of the uterine cervix. Acta Oncol, 1999. **38**(4): p. 449-54.
140. Gasinska, A., et al., A ratio of apoptosis to mitosis, proliferation pattern and prediction of radiotherapy response in cervical carcinoma. Neoplasma, 2002. **49**(6): p. 379-86.
141. Kim, J.Y., et al., Tumor apoptosis in cervical cancer: its role as a prognostic factor in 42 radiotherapy patients. Int J Cancer, 2001. **96**(5): p. 305-12.
142. Langendijk, H., et al., Cell proliferation and apoptosis in stage III inoperable non-small cell lung carcinoma treated by radiotherapy. Radiother Oncol, 2000. **56**(2): p. 197-207.
143. Lara, P.C., et al., Apoptosis in carcinoma of the bladder: relation with radiation treatment results. Int J Radiat Oncol Biol Phys, 1999. **43**(5): p. 1015-9.
144. Levine, E.L., et al., Apoptosis, intrinsic radiosensitivity and prediction of radiotherapy response in cervical carcinoma. Radiother Oncol, 1995. **37**(1): p. 1-9.
145. Marijnen, C.A., et al., High intrinsic apoptosis, but not radiation-induced apoptosis, predicts better survival in rectal carcinoma patients. Int J Radiat Oncol Biol Phys, 2003. **57**(2): p. 434-43.
146. Paxton, J.R., et al., Apoptosis in cervical squamous carcinoma: predictive value for survival following radiotherapy. J Clin Pathol, 2000. **53**(3): p. 197-200.
147. Ravi, D., et al., Apoptosis, angiogenesis and proliferation: trifunctional measure of tumour response to radiotherapy for oral cancer. Oral Oncol, 2001. **37**(2): p. 164-71.
148. Rodel, C., et al., Apoptosis, p53, bcl-2, and Ki-67 in invasive bladder carcinoma: possible predictors for response to radiochemotherapy and successful bladder preservation. Int J Radiat Oncol Biol Phys, 2000. **46**(5): p. 1213-21.
149. Shears, L.L., et al., Apoptosis in esophageal cancer following induction chemoradiotherapy. J Surg Res, 1998. **79**(1): p. 20-4.
150. Wheeler, J.A., et al., ASTRO Research Fellowship: apoptosis as a predictor of tumor response to radiation in stage IB cervical carcinoma. American Society for Therapeutic Radiology and Oncology. Int J Radiat Oncol Biol Phys, 1995. **32**(5): p. 1487-93.
151. Shah, R. and W.S. El-Deiry, p53-Dependent Activation of a Molecular Beacon in Tumor Cells Following Exposure to Doxorubicin Chemotherapy. Cancer Biol Ther, 2004. **3**(9).
152. Blasberg, R.G. and J.G. Tjuvajev, Molecular-genetic imaging: current and future perspectives. J Clin Invest, 2003. **111**(11): p. 1620-9.

153. Haas, R.L., et al., In vivo imaging of radiation-induced apoptosis in follicular lymphoma patients. Int J Radiat Oncol Biol Phys, 2004. **59**(3): p. 782-7.
154. Lahorte, C.M., et al., Apoptosis-detecting radioligands: current state of the art and future perspectives. Eur J Nucl Med Mol Imaging, 2004. **31**(6): p. 887-919.

Chapter 14

# IMAGING OF APOPTOTIC CELLS IN VIVO

Juhana Hakumäki
*Dept. of Biomedical NMR, A.I.Virtanen Institute for Molecular Sciences, University of Kuopio, P.O.Box 1627, and Dept. of Clinical Radiology, Kuopio University Hospital, P.O.Box 1777, FI-70211 Kuopio, Finland*

**Abstract**: The induction of apoptosis plays a key role in non-surgical cancer treatment – whether this occurs by irradiation, chemotherapy, or hormones. Recent advancements in imaging science and metabolic studies by nuclear magnetic resonance have painted an intriguing picture of the metabolic and biophysical processes involved with the progression of apoptosis *in situ*. It is now possible for us to detect and visualize previously inaccessible and even unrecognized biological phenomena in living cells and tissues undergoing therapeutically induced apoptosis. These new imaging techniques will have an increasing impact in the preclinical design of new anticancer agents and novel treatment protocols to enhance apoptotic cascades and to combat drug resistance. With the advent of molecular medicine and patient-tailored therapeutic options and drug molecules, rapid and accurate visualization of apoptotic response in the clinical settings can be of significant diagnostic and prognostic worth.

**Key words**: apoptosis, tumor, imaging, NMR, MRI, MRS, PET, SPECT, ultrasound

## 1. INTRODUCTION

Cell death in tissue takes place along a continuum spanning between necrotic cell death and apoptosis. In contrast to rather uncontrolled necrotic cell death, apoptosis is genetically controlled or "programmed" and cells become committed to a suicide process through numerous cascades of metabolic events. Although at tissue level, apoptosis may generally appear as a subtle process, the individual cells experience profound changes in morphology, structure and biochemistry brought forth by the activation of distinct proteases and lipases. Among the often-quoted hallmark phenomena



*M. Sluyser (ed.), Application of Apoptosis to Cancer Treatment*, 321-339.

are DNA fragmentation, nuclear condensation, cell shrinkage, and the formation of membrane-encapsulated apoptotic bodies. These concerted actions serve to confine all cellular material towards a recycling process by macrophages and/or neighboring cells, and to protect the surrounding tissue from harmful, uncontrolled release of intracellular messengers, metabolites, and electrolyte fluxes that could elicit inflammation and necrosis[1]. It is exactly this careful orchestration of cell events that provides such an array of markers and targets for visualizing apoptosis by modern imaging technologies (Figure 1).

As demonstrated elsewhere in this book, apoptosis acts as a counterbalance to cell proliferation. It is therefore not only required for normal development and tissue function, but also disturbances in its regulation can promote many disease states[2,3] such as cancer. The primary goal of modern chemotherapy in cancer treatment is tumor cell apoptosis, and due to the inherent cleanliness of the death process as described above, this mode of cell death is sought after by most other anti-cancer treatments, including hormonal agents, toxins, and radiotherapy.

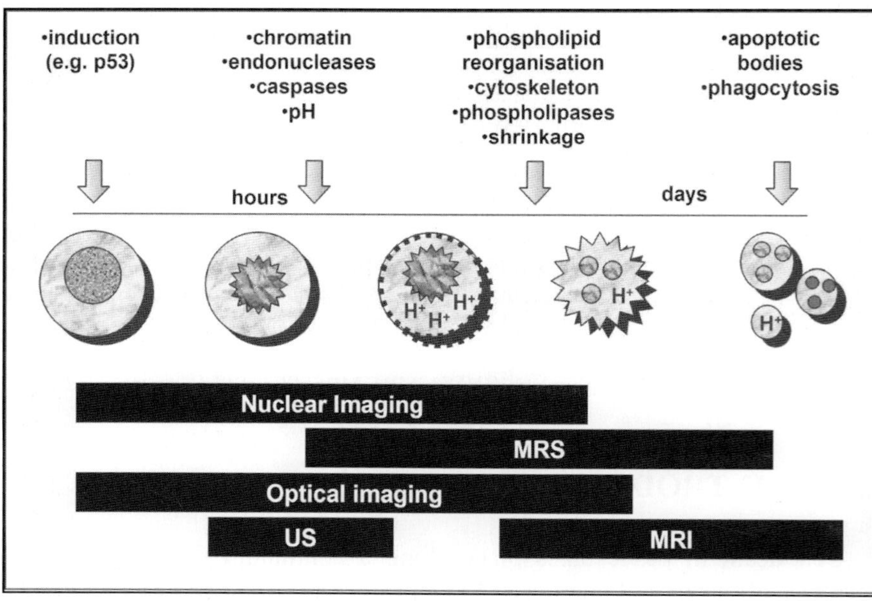


*Figure 1.* A cartoon and timeline of some key biochemical and biophysical changes occurring in cells that can be exploited for imaging purposes. Generally, molecular events precede those affecting the physical properties of cells, thereby rendering biochemical and molecular imaging markers (such as nuclear imaging and MRS) more sensitive in the early phases than for instance MRI.

Targeted facilitation of apoptosis *in vivo* has been shown to effectively increase the numbers of apoptotic cells in tumors[4-6], and the detected early apoptotic response correlates well with subsequent outcome[6-8]. As described in more detail elsewhere in this book, targeted therapies for modulating apoptosis in malignancies are finding many forms and approaches. Fortunately, non-invasive imaging techniques sensitized to apoptosis are already available, and could become important in future clinical assessment of apoptosis-inducing therapies *in situ*. In the optimal situation, the physician would be able to select the non-responding patients from the responders at much earlier time points than currently possible with anatomical imaging only. Likewise, these techniques could prove immensely useful in basic drug development and testing.

During the last ten years or so, many research groups have studied the possibility of using noninvasive imaging techniques for the purpose of imaging the effects of the apoptotic cascades *in vivo*. The research began with metabolic characterization of cells by nuclear magnetic resonance (NMR) spectroscopy (MRS)[9]. Nuclear imaging techniques together with magnetic resonance imaging (MRI) approaches[9] then followed together with optical imaging[10] and even ultrasound (US)[11,12]. The many papers have now shown beyond doubt that by imaging, phenomena as subtle as apoptosis can non-destructively be visualized by exploiting the myriad of changes involving membrane composition, protein synthesis, glycolysis, phosphatidylcholine, phosphatidylserine and cell fatty acid turnover, energy levels and even intracellular pH throughout the execution of the "apoptotic program" *in vivo*. In addition to the use of exogenous targeted markers for nuclear imaging, and more recently MRI, intrinsic image contrast in MRI and US can also be sensitized to actual biophysical changes in the cellular milieu. All of the mainstream *in vivo* imaging techniques (Table 1) with the exception of cell imaging by microscopy will be outlined here together with some examples of promising applications.

*Table 1.* A comparison of the imaging modalities for detecting apoptosis *in vivo*.

| | penetration | resolution | ionisation | sensitivity | receptors | metabolism |
|---|---|---|---|---|---|---|
| MRI | +++ | >0.02 mm | - | + | + | + |
| MRS | +++ | >2 mm | - | + | - | +++[a] |
| SPECT | +++ | >4 mm[b] | + | ++ | ++ | ++ |
| PET | +++ | >5 mm[b] | ++ | +++ | +++ | ++ |
| Ultrasound | ++ | >0.2 mm | | | | |
| Optical | + | >10 mm[c] | - | ++ | ++ | ++ |

[a] MRS is the only modality capable of detecting intrinsic metabolites.

[b] SPECT and PET values are typical for dedicated small-animal research microimaging systems. In clinical systems, spatial resolution is several millimeters poorer.

[c] Optical coherence tomography (OCT) can reach resolutions above 10 μm while limited to extreme surfaces (< 2 mm depth).

# 2. NUCLEAR AND OPTICAL IMAGING

At quick glance, it is apparent that the most numerous applications for *in vivo* imaging of apoptosis are based on marker molecules. This is not surprising given the fact that after all, marker molecules, i.e. biologically active dyes, are what also has made the histological detection of apoptosis possible[13]. Of particular interest here are targeted labels binding to phosphatidylserine, PS.

PS is an abundant phospholipid mostly residing on the cytoplasmic, inner leaflet of the cell membrane[14]. It is however translocated to the outer leaflet of the membrane during apoptosis, when aminophospholipid translocase becomes inactivated and another enzyme, scramblase, is activated to further enhance translocation. Interestingly, this PS externalization occurs very early in the apoptotic chain of events[14], preceding such hallmark events as nuclear condensation and DNA laddering, and perhaps serves as an important signal to neigboring and phagocytizing cells. Of the several proteins known to specifically bind to PS, annexin V is perhaps best known. Optimally up to eight annexin V moieties can bind to one exposed PS, which contributes to its efficiency as a cell label. Applications based on this approach will be discussed below with an introduction to the imaging technologies involved.

## 2.1 Nuclear imaging

As in X-ray based computed tomography (CT), modern nuclear imaging techniques rely on the rotation of detector arrays around the subject of investigation. With this approach, the position and concentration of a radionuclide marker introduced into the experimental animal or clinical patient can be calculated. Unlike in CT, however, the emission source is an unknown source within the body, and rigorous reconstruction algorithms are required to increase resolution[15].

Nuclear imaging techniques, especially positron emission tomography (PET), have roughly a million-fold sensitivity gain compared to other available imaging techniques. Since receptor proteins are normally present in the nanomolar range, PET has become the method of choice for mapping and studying the function of many receptors, particularly within the central nervous system[16]. However, the use of PET is somewhat limited for imaging apoptosis, since most positron emitting radionuclei are physically short-lived, and many single photon emission tomography (SPECT) radioligands, such as technetium ($^{99m}$Tc) are easier and less expensive to produce.

SPECT utilizes single photons (at energies of about 140 keV), instead of positron emissions (which result in two 511 keV photons traversing in opposite directions). Because only single photons are emitted from the radionuclides used for SPECT, collimators are required for source localization. This results in tremendous loss of acquisition efficiency. Also, the useful resolution of SPECT is slightly inferior to PET. However, it is much less expensive to perform SPECT scans, and the typically used SPECT radionuclei have long physical half lives (e.g. 6 h for $^{99}$Tc). Together these factors have made this mode of emission tomography an exciting modality for studying apoptosis *in vivo*.

When reviewing literature, it also becomes quite evident that $^{99m}$Tc-labeled Annexin V molecules are the most extensively investigated and used *in vivo* marker of apoptosis to date. Since the first *in vivo* imaging demonstration by Blankenberg and co-workers[17] (Figure 2), numerous $^{99m}$Tc-Annexin V –based radioligands have been developed and clinical trials have already been undertaken[12]. In part, their popularity is due to the manifold advantages of $^{99m}$Tc compared with many other radionuclides, such as low cost, relatively long physical half-life, and easy availability. For further reading on nuclear imaging of apoptosis, the reader is kindly directed to an exceedingly thorough and clearly written overview by Lahorte and co-workers[12]. This paper encompasses most currently existing nuclear labels and labeling methods for Annexin V and derivatives.

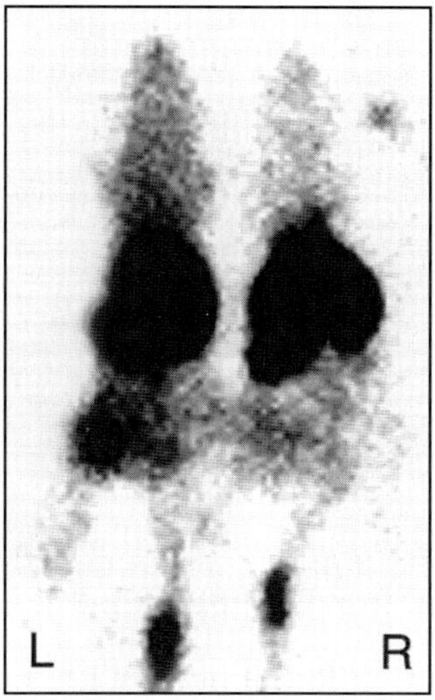


*Figure 2.* Imaging a treated murine lymphoma with radiolabeled annexin V. Mice bearing subcutaneous lef flank murine B-cell lymphomas were treated with 100 mg/kg of cyclophosphamide (i.p.) to induce apoptosis. Twenty hours after treatment mice were injected with 150 µCi of $^{99m}$Tc HYNIC annexin V (50 µg/kg of protein), and imaged one hour later. The treated tumor (mass on left flank) demonstrated compound uptake 363% above control levels. L, left; R, right. © The National Academy of Sciences of the USA 1998, reproduced with permission from reference 17.

## 2.2 Optical imaging

Optical imaging of apoptosis *in vivo* is an emerging field of study[10]. Since visible light such as produced by fluorescein, is not able to penetrate more than 1–2 mm into biological tissue, its use in biological imaging is severely compromised[18]. However, light in the near infrared (NIR) range (wavelengths 700-1200 nm) allows better penetration in tissue[19], but at the cost of considerable absorption and scattering. Optical tomograpy can however be used to reconstruct the probed structure and location by back-projecting the transmitted light through the object along multiple paths similarly to CT. Despite its shortcomings in terms of penetration and resolution, optical imaging offers the opportunity for real-time monitoring of biological phenomena such as apoptosis. The First *in vivo* data in irradiated tumor-bearing nude mice showed increased Cy5.5-Annexin V uptake in tumors over a period of ten days due to excellent fluorochrome stability[19]. It is worth noting, that this is also a considerable temporal advantage in comparison to almost all radionuclides.

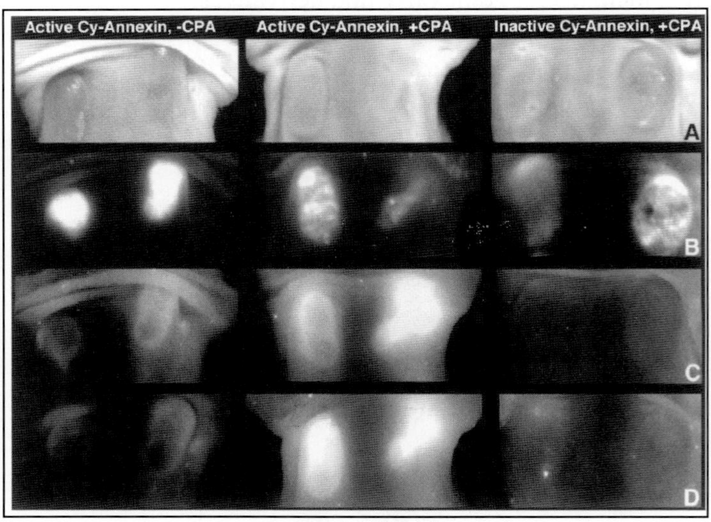


*Figure 3.* Imaging of apoptosis in CR-red fluorescent tumor model implanted in mammary pad. Animals were divided into three groups: (a) injected with active Cy-annexin and not treated with CPA; (b) injected with active Cy-annexin and treated with CPA; and (c) injected with inactive Cy-annexin and treated with CPA. A, visible light image of implanted tumors; B, expression of DsRed2 in CR tumor (red fluorescence channel); C, near-infrared signal measured in tumors at 75 min after the injection of Cy-annexin; D, near-infrared signal measured in tumors at 20 h after the injection of Cy-annexin. Animals received an i.p. injection of 170 mg/kg CPA. © American Association for Cancer Research 2003, reproduced with permission from reference 20.

Similar results (Figure 3) have been obtained by Petrovsky et al. by using Cy 5.5-annexin as a NIR fluorescence probe for apoptosis in different tumor types and therapies and imaged from outside an intact living animal[20].

Another optical imaging approach that deserves mention is bioluminescence imaging (BLI), which is based on the use of the luciferase enzyme. This technique has recently been used to investigate the spatial relationship between chemotherapy-induced tumour apoptosis and total tumour burden over time, as assessed by BLI together with microSPECT[21].

## 3. ULTRASOUND

Studies have surprisingly shown that ultrasound (US) imaging may also be able to detect apoptosis via the biophysical effects exerted by subcellular nuclear changes such as chromatin condensation and DNA fragmentation *in vitro*. Using apoptotic and control AML-3 leukemia cells, Czarnota and co-workers discovered that twenty-four hours from the onset of cisplatin-induced apoptosis (95% apoptotic cells), a 2 to 5-fold increase in US backscatter signal could be detected[22]. Interestingly, this increase appeared to correlate nicely with the actual progression of nuclear condensation[23]. In the latter study, the authors used high frequency US imaging to demonstrate epidermal apoptosis also *in vivo*. It remains to be seen however, whether US will become a truly useful tool for detecting changes in often deep-seated tumors *in vivo*, where also much lower percentages of apoptotic cells are manifest (typically < 10% at any time point) despite anticancer therapy.

## 4. NUCLEAR MAGNETIC RESONANCE

Certain atomic nuclei, such as the proton ($^{1}H$), possess what is called non-zero angular momentum, i.e. nuclear spin. This creates a magnetic dipole moment along the axis of rotation. When such nuclei are placed in an external magnetic field, net magnetization is created by the equilibrium difference between magnetic dipole moments existing in two different energy states. This net magnetization, which forms the basis of all nuclear magnetic resonance (NMR) experiments, is proportional to the strength of the external magnetic field, and can be perturbed by radio frequency (RF) radiation. When perturbed magnetization and its recovery (i.e. relaxation) to unperturbed equilibrium state is detected through receivers, information on the nuclear spin environment of the sample or tissue can be obtained. When this signal is used to detect and quantify the presence of chemically distinct compounds, it is called NMR spectroscopy (or clinically, MRS). When this information is used to create spatially resolved maps of quantities and/or relaxation properties of the observed nucleus, the procedure is called magnetic resonance imaging (MRI) and when the above techniques are combined, magnetic resonance spectroscopic imaging (MRSI). Because signal manipulation and detection in NMR occur in the radio frequency range of the electromagnetic spectrum, no ionizing radiation is involved. This makes NMR particularly attractive for noninvasive *in vivo* studies of living systems.

$^1$H is the most sensitive non-radioactive NMR nucleus with a 99.9% natural abundance. It is particularly well suited for NMR studies, because virtually all biological material and compounds possess hydrogen. MRI benefits greatly from the highly concentrated water protons (~80 mol/l in tissue). Typical $^1$H spectra can be seen in Figures 4A and Figure 5. Other naturally occurring nuclei can however be used for NMR spectroscopy: the most important are phosphorus ($^{31}$P) with 100% natural abundance, and a stable isotope of carbon ($^{13}$C) with 1,1% natural abundance. Although both of these nuclei are much less sensitive than proton, the particular power of $^{31}$P NMR spectroscopy lies in detecting cell pH and energy metabolites, such as shown in Figure 4B. To enhance sensitivity, and especially to decipher metabolic pathways, $^{13}$C labeling of substrates is often used.

NMR is sadly not a very sensitive technique. However, this is at least partly compensated by the fact that in metabolic studies dynamic changes over the whole range of metabolites can be observed and accurately quantified, as long as they exist above a practical detection threshold (millimolar range *in vivo*, and one tenth of that *in vitro*). For a more detailed account on the basics of biological NMR and MRI, the reader is kindly directed to David Gadian's excellent textbook[24].

## 4.1 Magnetic Resonance Spectroscopy (MRS)

Magnetic resonance research into apoptosis started in the mid-nineties when several groups began looking at relevant metabolic indicators of apoptosis. Adebodun and Post were the first to identify, by using $^{31}$P MRS, an aberrant metabolic profile in leukemia cells treated with dexamethasone to undergo apoptosis *in vitro*[25]. The researchers reported a clear reduction in phosphomonoesters (i.e. phosphoethanolamine and phosphocholine) and ATP. Following up on the study, Nunn and co-workers used $^1$H MRS to study metabolic changes in neutrophils undergoing apoptosis[26], and instead observed an increase in phosphocholine concentration, which the authors attributed to their model of Fas-receptor mediated apoptosis and phospholipase-C activation. Two years later, Williams et al.[27] set out to better characterize metabolic changes in Chinese hamster ovary cells (CHO-

K1) and HL-60 leukemia cells using $^{31}$P MRS, which were treated with several apoptosis-inducing drugs with functionally very different modes of action, namely farnesol, chelerythrine, etoposide, campothecin, and ceramide. Only two metabolites were observed to consistently increase during apoptosis: fructose-1,6-bisphosphate (FBP, a glycolytic pathway intermediate) and cytidylyldiphosphocholine (CDP-choline, an intermediate step metabolite in phosphatidylcholine biosynthesis). In this case, the increase in FBP concentration in HL-60 cells was be explained by either depletion of cellular NAD(H), or by the activation of 6-phosphofructo-1-kinase through AMP accumulation, an observation corroborated soon by the work of Ronen and co-workers[28]. The NMR-observed accumulation of CDP-choline in apoptotic cells (Figure 4B) was later linked by the same group to the inhibition of choline phosphotransferase (CPT) by isotope labeling studies[29]. CPT is likely to become inhibited in apoptotic cells, since it has an alkaline pH optimum (8 - 8.5), yet cellular acidosis (pH <6.5) appears to be a frequent event in apoptosis[27].

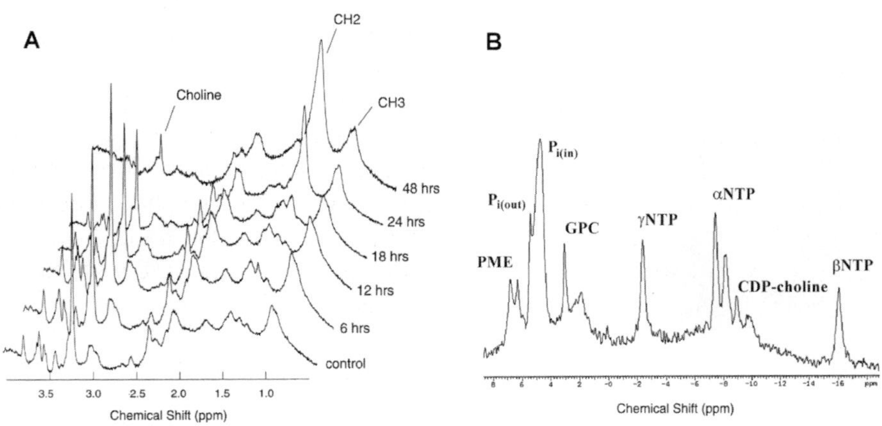


*Figure 4.* A) Time course of $^{1}$H NMR spectroscopy resonances in Jurkat leukemia cells. Note the resulting increase in lipid methylene (-$CH_2$) resonances over time and with increasing percentage of apoptotic cells (not shown). © American Society of Hematology 1997, reproduced with permission from reference 28. B) $^{31}$P NMR spectrum of apoptotic Chinese hamster ovary CHO-K1 cells is shown (CDP-choline, cytidylyldiphosphocholine; GPC, glycerophosphocholine; NTP, nucleotide triphosphate, $P_i$, inorganic phosphate; PME, phosphomonoesters. © Wiley & Sons 1998, reproduced with permission from reference 27.

Blankenberg et al. were the first to use $^{1}$H MRS of living cells *in vitro* to detect metabolic changes caused by apoptosis. In their seminal paper, large signals from mobile intracellular lipids were shown to accumulate over time

(Figure 4A) and a correlation between the intensity ratio of the 1.3 ppm (-$CH_2$-) to 0.9 ppm (-$CH_3$) -resonances and the fraction of apoptotic cells following induction of apoptosis was established[30]. At the time, it was thought that the signals originated from plasma membrane microdomains[31], and that these were perhaps connected to modified membrane microfluidity brought forth by apoptosis. However, the more recent studies have given enough grounds to discard this hypothesis: it is currently quite accepted that these resonances originate from a storage form of cell lipids within the cell, i.e. from triglycerides within cytoplasmic lipid droplets[32].

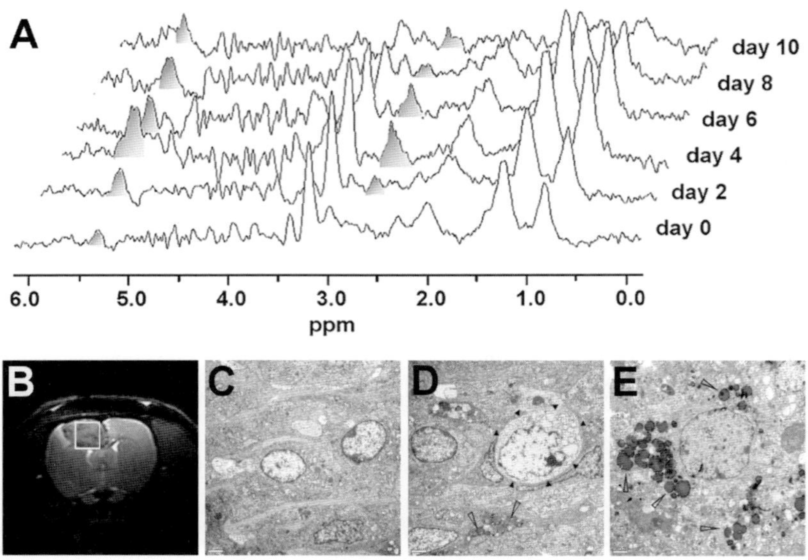


*Figure 5.* A typical time course of $^1$H MRS data from an HSV-tk transfected glioma treated with ganciclovir *in vivo* for a period of 10 days (A). Signals from polyunsaturated lipid proton moieties are shown shaded (at 5.3 and 2.8 ppm). An untreated glioma is shown on conventional MRI, with a white box to denote the localization from which spectroscopy signals were obtained (B). Transmission electron microscopy (magnification x 5000) shows untreated tumor cells (C), lipid droplets (open arrows) and apoptotic bodies (black triangles) (D), and densely osmiophilic, polyunsaturated fatty acid-rich droplets (E) in treated tumours after four days of treatment. © 1999 Nature publishing, reproduced with permission from reference 33.

Hakumäki et al. extended the previous approach by showing that $^1$H MRS could be used to noninvasively detect apoptosis *in vivo*[33]. In this study, herpes-simplex virus thymidine kinase (HSV-tk)- transfected BT4C gliomas were treated by systemic administration of ganciclovir to induce widespread apoptosis in tumor tissue. As a result, significant increases in polyunsaturated lipid signals could be observed following tumor treatment (Figure 5, previous page). The lipid signals also correlated with the number of apoptotic cells, concentration of cholesteryl esters and triglycerides, and the number of osmiophilic cytoplasmic lipid droplets in tumor cells.

Surpisingly, it is not yet quite clear why lipids accumulate in apoptotic cells[32]. There is however increasing evidence to support a role for phospholipase-A2 activity[33,34]. Recently, two groups have shown using Jurkat T-cells *in vitro*, that apoptosis is indeed accompanied by an quantitative relationship between cytoplasmic lipid droplets and lipid signal in $^{1}$H MRS[35,36], with more independent biochemical evidence suggesting a role for intracellular phospholipase-A2 activation after induction of apoptosis[37]. In fact, more recent work by Liimatainen et al. demonstrates a plausible stoichiometric relationship between membrane fatty acid release by phospholipase-A2 as determined by gas liquid chromatography and quantitative $^{1}$H MRS[38].

Due to the poor relative sensitivity and spatial resolution of MRS, it is an absolute necessity to use markers with sufficient signal intensity. If this requirement is fulfilled, molecular markers can easily be mapped with techniques such as MRSI (Figure 6). So far, the $^{1}$H MRS signals from cytoplasmic lipids appear the most robust NMR candidates for *in vivo* detection of apoptosis. However, they are not perfect since the metabolic factors governing their dynamics are poorly understood and theoretically, $^{1}$H such signals could also arise from other cellular processes, such as necrosis, and from other endogenous cell types[32]. It has also been documented that the spectral lipid baseline features may differ, especially with multidrug resistant cell lines[39]. However, many available *in vitro* assays for apoptosis can also be ambiguous. For instance, even DNA fragmentation, often considered the gold standard for apoptosis detection, cannot always be observed in cells undergoing apoptosis[40]. $^{31}$P markers such as CDP-choline and FBP could also act as apoptosis markers. However, $^{31}$P MRS is hampered by its poor sensitivity and spatial resolution. Also, it might be difficult to distinguish resonances such as FBP from the partly overlapping phosphomonoester signals (Figure 4B).

Scott and Adebodun have elegantly used $^{13}$C MRS *in vitro* to assess protein synthesis during dexamethasone-induce apoptosis in CEM-C7-14 human leukemia cells[41]. By measuring the incorporation of $^{13}$C-labeled amino acids into cell proteins, they could observe not cessation, but a reduction of protein synthesis by as much as 87%. Unfortunately $^{13}$C NMR is too insensitive to design any practical imaging applications.

Interestingly, $^{1}$H MRS can also be sensitized to molecular diffusion, in similar fashion as diffusion-weighted magnetic resonance imaging, or DWI[42]. This technology assesses the Brownian molecular motion *in vivo* and with the use of intracellular probes such as naturally occurring tumor metabolites, can be used to study the biophysical status of tissue. Hakumäki and co-workers showed that during apoptosis-inducing therapy, there is a striking ~50% reduction in the apparent diffusion coefficient (ADC) of intracellular choline *in vivo*, accompanied by a significant increase in the rapidly diffusing water component reflecting extracellular water[43].

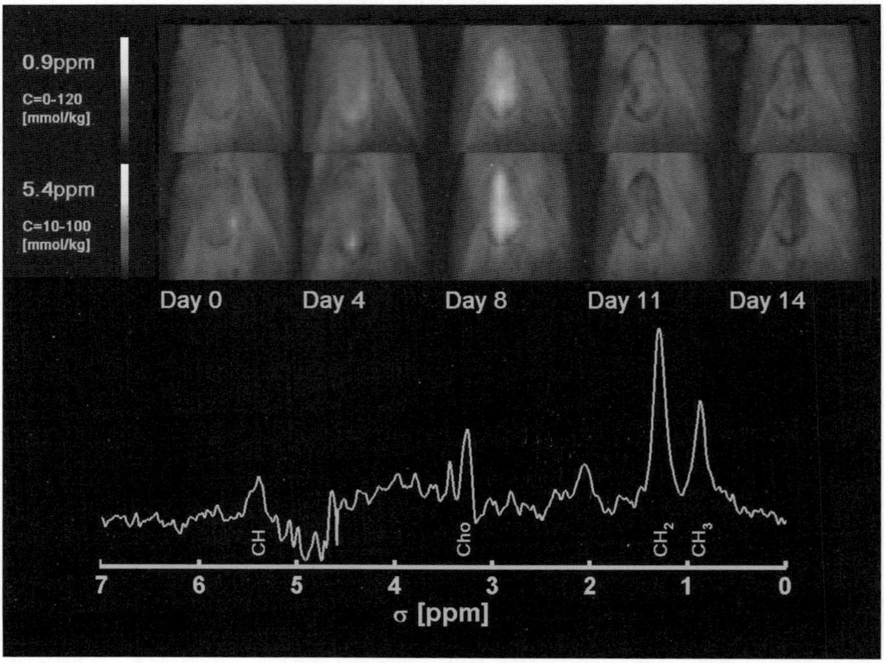


*Figure 6.* Magnetic resonance spectroscopic imaging (MRSI) of lipid accumulation in an apoptotic experimental rat glioma over 14 days time. In this hybrid technique, spectroscopic resonances unique to different chemical species (as exemplified by the spectrum below) can be mapped to produce magnetic resonance images such as for the lipid acyl chain terminal methyl (-$CH_3$, 0.9ppm) and unsaturated carbon vinyl (=CH-, 5.4 ppm) protons. Although the spatial resolution is considerably lower than in conventional MRI, advanced techniques actually allow volume resolutions similar to micro-PET (4.7 μl, such as here) without the use of any exogenous agents or marker molecules. The lipid concentrations are coded by the color bars on left, superimposed onto a anatomical MRI images. Figure courtesy of Mr. Timo Liimatainen, Dept. of Biomedical NMR, University of Kuopio, Finland.

These results directly imply decrease in cell size and number, together with increased intracellular viscosity as experienced by the intracellular choline metabolites. Recently, Hortelano et al. have shown similar MRS results in cells[44], supporting the idea that cells apoptotic shrinkage and loss of cells can be held accountable for the observed molecular diffusion phenomena.

## 4.2 Magnetic Resonance Imaging (MRI)

The information provided by MRI has long been mostly anatomical, and the technology clearly excels at this. MRI has in fact, become an indispensable imaging modality that can probe tissue and its biophysical properties with exceptional in-plane resolution *in vivo*. One of the key advantages is that this can be done noninvasively, without physically altering tissue and without the use of radioactive markers[24]. This is important clinically, where MRI is more widely available, accurate localization is required, and the levels of therapeutically induced apoptosis may still be low. Hakumäki et al. have recently shown in a BT4C rat glioma model, that in (HSV-tk)- transfected tumors, T1- and T2- relaxation times and water diffusion increase during ganciclovir therapy. However, T1- relaxation times in the rotating frame ($T1_{rho}$) are elevated even when tumor volumes still increase, but significant apoptosis is starting to take place[45]. Recently, Gröhn and co-workers used short interpulse intervals for adiabatic Carr-Purcell (CP) –pulse trains to produce a novel contrast mode which was sensitive to early therapeutic response, similarly to $T1_{rho}$[46].

Unfortunately, the biophysical phenomena governing intrinsic image contrast generation are rather poorly understood, and are likely to involve a number of processes related to tissue microarchitecture, protein/lipid content and composition, pH, flow, oxygenation and any combination of these. At best, this still leaves room for ambiguity reflecting differences in cancer tissue type and the chosen therapeutic approach. Also, one might argue that changes in MR image contrast or water diffusion due to cell shrinkage and membrane blebs are unlikely to observe any of the early phases in the apoptotic process. This has in fact been shown by Valonen and co-workers, who recently showed that the DWI increase in apoptotic BT4C gliomas directly correlates with cell loss[47]. T1 and T2 appear to behave similarly. However, $T1_{rho}$ and short interpulse interval-CP contrast appear to be a curious exceptions to this, and currently lack proper explanation[45,46].

In light of the above, it would be extremely beneficial to combine chemical specificity (such as that obtained by nuclear imaging or MRS) with the much better resolution (>50 μm) of MRI. Here in fact, a deep conceptual revolution has occurred in the recent years. MRI contrast has been shown to be adjustable by chemically malleable marker ligands linked to magnetically active compounds, such as superparamagnetic iron oxide nanoparticles, i.e. SPIO[48,49] or gadolinium (Gd) chelates that have been used to detect the overexpression of certain receptors in tumors *in vivo* [49,50]. Along this line of approach, Zhao and co-workers demonstrated success by labeling the C2-domain of a phosphatidylserine binding protein, synaptotagmin I, with an SPIO particle[51]. The SPIO-labeled protein was shown to bind to apoptotic cells *in vitro*, and when administered intravenously into mice bearing EL4 lymphomas, decreases in MR image intensities in tumor regions containing large numbers of apoptotic cells could be observed (Figure 7). In analogous fashion to Blankenberg et al.[17] and above, Schellenberger and co-workers have cross-linked SPIO nanoparticles with Annexin V, and demonstrated dose-dependent signal intensity reduction in T2-weighted images of camptothecin-treated apoptotic (65%) Jurkat cells versus untreated controls *in vitro*[52].

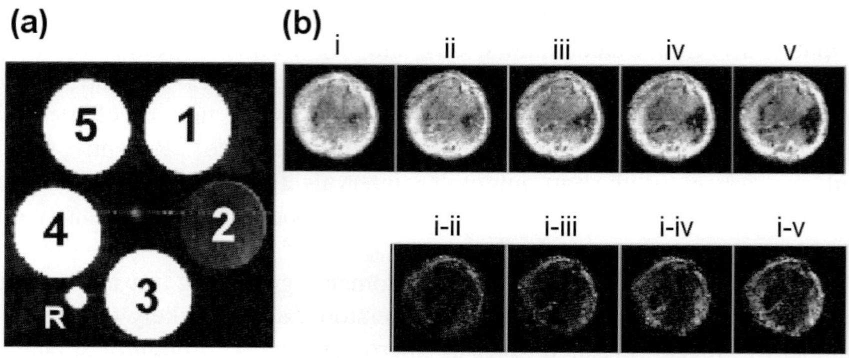


*Figure 7.* MR images of CHO-K1 cells in test tubes (a). The cells were treated with etoposide to induce apoptosis, incubated with C2-SPIO or plain SPIO (200 μM Fe), washed and then immobilized in 5 mm diameter NMR tubes and imaged. Apoptotic cells incubated with BSA conjugated SPIO (1), apoptotic cells incubated with C2-SPIO (2), control cells incubated with C2-SPIO (3), apoptotic cells with no label (4), control cells incubated with BSA-SPIO (5), reference capillary with water (R). The signal loss due to retention of targeted contrast agent is clearly visible in sample number 2. In (b), MR images of a tumor in a drug-treated mouse following injection of C2–SPIO (20 mg Fe/kg tissue), are shown. MR images i-v are from before and after injection of C2–SPIO. The first control (i) was acquired before injection of contrast agent and afterwards at time points of 11 (ii), 47 (iii), 77 (iv), and 107 (v) minutes postinjection. Contrast agent uptake (and apoptotic cells) become visible as areas of image intensity loss, further highlighted in the respective subtraction images (i-ii, i-iii, i-iv, i-v), as bright areas. © Nature publishing 2001, reproduced with permission from reference 51.

# 5. FUTURE DIRECTIONS

Although in may seem that the recent developments in imaging of apoptosis *in vivo* are indeed breathtaking, we are essentially very much at the beginning. Clinical imaging applications for cancer need to be tested and developed effectively. This requires much more experimental research with solid, consistent models, and relevant oncological problems. Flow cytometry, immunohistological stains, and the gold standard i.e. TUNEL assay need to be kept abreast to critically assess the localization, extent and kinetics of therapeutically induced apoptosis, and to confirm the imaging findings. Naturally, the optimal time frames for imaging therapeutically induced apoptosis will have to be addressed as well.

It is evident that many more markers for apoptosis, with better sensitivity and specificity must be explored. In this sense, highly specific, targeted apoptosis-detecting ligands would be preferable. Perhaps one should realize that there is a caveat even to the widely used Annexin-V approach: in some cases, necrotic cells undergoing lysis may also bind Annexin-V when the phosphatidylserine still residing on the inner membrane leaflets becomes exposed[53]. This could lead to overestimation or incorrect assignment of apoptosis in necrotic insults.

Among the most exciting novel targets are probably those directed at the very initial steps of the apoptotic program, such as DNA-damage reporting p53 expression[54] and at the apoptotic cascades themselves via radiolabeled caspase inhibitors[12]. Since caspase activation is known to occur early in the apoptotic program, such radiolabeled inhibitors would offer a highly specific tool detecting early events in apoptosis. It is also believed that these marker molecules might be more specific to apoptosis for reasons mentioned above. Overall in terms of sensitivity, nuclear imaging probes will fare far better than the other modalities.

Due to certain advantages such as the extremely high spatial resolution and true non-invasiveness, a small niche will remain for MRI agents however. Optimal MRI contrast generation and agent delivery will be the subject of future research. For instance, the above-described SPIO-labeling, although being sensitive, provides "negative" contrast i.e. signal intensity reduction (Figure 7). This is unfortunately more prone to artifacts and also more difficult to interpret due to the often seen intrinsic heterogeneity of tumor masses on MRI. The conjugation of marker proteins with paramagnetic gadolinium-based agents, might become a means to avert this problem by providing "positive" contrast, i.e. brightening on MRI, although this approach is likely to be plagued by severe sensitivity problems.

The biodistribution and metabolism of NMR spectroscopic markers can be efficiently probed and more information gained by using advanced MRSI techniques. Among the potential applications, lipid $^1$H MRSI is the most robust and practical (Figure 6). The use of $^{31}$P clinically appears to be unlikely due to its poor sensitivity, long acquisition times, and inferior spatial resolution. It is noteworthy however, that intracellular pH can be observed noninvasively only by $^{31}$P MRS. Intracellular pH on the other hand appears to be one of the critical factors at work in the apoptotic program, affecting even caspase activation during apoptosis.

Regardless of the technique employed, the generation of protein-based targeted imaging agents is expensive, and all compounds heading for clinical applications will need to be assessed for patient biodistribution and safety. While waiting for new developments at this front, we are likely to witness some of the other techniques, such as SPECT, MRS and diffusion-weighted MRI gain foothold within the clinic. However, for those involved with experimental studies and drug design, *in vivo* imaging of apoptosis has already become reality, with an ever-increasing and powerful array of technologies to choose from.

## ACKNOWLEDGEMENTS

The author would like to thank the Finnish Academy, the Finnish Cancer Institute and Instrumentarium Science Foundation for support.

## REFERENCES

1. J. Savill and V. Fadok, Corpse clearance defines the meaning of cell death, *Nature* **407**, 784-788 (2000).
2. R.A. Kinloch et al, The pharmacology of apoptosis, *Trends Pharmacol. Sci.* **20**, 35-42 (1999).
3. G.I. Evan and K.H. Vousden, Proliferation, cell cycle and apoptosis in cancer, *Nature* **411**, 342-348 (2001).
4. B. Dubray et al., In vitro radiation-induced apoptosis and early response to low-dose radiotherapy in non-Hodgkin's lymphomas, *Radiother. Oncol.* **46**, 185-191 (1998).
5. B. Jansen et al., Chemosensitisation of malignant melanoma by BCL2 antisense therapy, *Lancet* **356**, 1728-1733 (2000).
6. W.F. Symmans et al., Paclitaxel-induced apoptosis and mitotic arrest assessed by serial fine-needle aspiration: implications for early prediction of breast cancer response to neoadjuvant treatment, *Clin. Cancer Res.* **6**, 4610-4617 (2000).
7. R.E. Meyn et al., Apoptosis in murine tumors treated with chemotherapy agents, *Anti-Cancer Drugs* **6**, 443-450 (1995).
8. P.A. Ellis et al., Preoperative chemotherapy induces apoptosis in early breast cancer, *Lancet* **349**, 849 (1997).
9. J.M. Hakumäki J.M. and K.M. Brindle, Techniques: visualizing apoptosis by nuclear magnetic resonance, *Trends Pharmacol. Sci.* **24**, 146-149 (2003).

10. C. Bremer et al., Optical-based molecular imaging: contrast agents and potential medical applications, *Eur Radiol.* **13**, 231-243 (2003).
11. F.G. Blankenberg, K. Ohtsuki, H.W. Strauss, Dying a thousand deaths. Radionuclide imaging of apoptosis, *Q. J. Nucl. Med.* **43**, 170-176 (1999).
12. C.M. Lahorte et al., Apoptosis-detecting radioligands: current state of the art and future perspectives, *Eur. J. Nucl. Med. Mol. Imaging* **31**, 887-919 (2004).
13. C. Stadelmann and H. Lassmann, Detection of apoptosis in tissue sections, *Cell Tissue Res.* **301**, 19-31 (2000).
14. S.J. Martin et al., Early redistribution of plasma-membrane phosphatidylserine is a general feature of apoptosis regardless of the initiating stimulus - Inhibition by overexpression of Bcl-2 and Abl, *J. Exp. Med.* **182**, 1545-1556 (1995).
15. J.T. Kuikka et al., Future developments in nuclear medicine instrumentation: a review., *Nucl. Med. Commun.* **19**, 3-12 (1998).
16. A.F.Chatziioannou, Molecular imaging of small animals with dedicated PET tomographs, *Eur. J. Nucl. Med. Mol. Imaging* **29**, 98-114 (2002).
17. Blankenberg, F. G. et al., In vivo detection and imaging of phosphatidylserine expression during programmed cell death, *Proc. Natl. Acad. Sci. USA* **95**, 6349-6354 (1998).
18. T.J. Sweeney et al., Visualizing the kinetics of tumor-cell clearance in living animals, *Proc.Natl. Acad. Sci. USA* **96**, 12044-12049 (1999).
19. J.S. Reynolds et al., Imaging of spontaneous canine mammary tumors using fluorescent contrast agents, *Photochem Photobiol.* **70**, 87-94 (1999).
20. A. Petrovsky et al., Near-infrared fluorescent imaging of tumor apoptosis, *Cancer Res.* **63**, 1936-42 (2003).
21. S.J. Mandl et al., Multi-modality imaging identifies key times for annexin V imaging as an early predictor of therapeutic outcome, *Mol Imaging* **3**, 1-8 (2004).
22. G.J. Czarnota et al., Ultrasonic biomicroscopy of viable, dead and apoptotic cells, *Ultrasound Med. Biol.* **23**, 961-965 (1997).
23. G.J. Czarnota et al., Ultrasound imaging of apoptosis: high-resolution non-invasive monitoring of programmed cell death in vitro, in situ and in vivo, *Br. J. Cancer* **81**, 520-527 (1999).
24. D.G. Gadian, *NMR and its applications to living systems* (Oxford Science, Oxford, 1995).
25. F. Adebodun and J.F.M. Post, $^{31}$P NMR characterization of cellular metabolism during dexamethasone-induced apoptosis in human leukemic cell lines, *J. Cell. Physiol.* **158**, 180-186 (1994).
26. A.V.W. Nunn et al., Characterisation of secondary metabolites associated with neutrophil apoptosis, *FEBS Lett.* **392**, 295-298 (1996).
27. S.N.O. Williams et al., Induction of apoptosis in two mammalian cell lines results in increased levels of fructose-1,6-bisphosphate and CDP-choline as determined by $^{31}$P MRS, *Magn. Reson. Med.* **40**, 411-420 (1998).
28. S. Ronen et al., Magnetic resonance detects metabolic changes associated with chemotherapy-induced apoptosis, *Br. J. Cancer* **80**, 1035-1041 (1999).
29. M.L. Anthony et al., Inhibition of phosphatidylcholine biosynthesis following induction of apoptosis in HL-60 cells, *J. Biol. Chem.* **274**, 19686-19692 (1999).
30. F.G. Blankenberg et al., Quantitative analysis of apoptotic cell death using proton nuclear magnetic resonance spectroscopy, *Blood* **89**, 3778-3785 (1997).
31. C.E. Mountford CE and L.C. Wright, Organization of lipids in the plasma membranes of malignant and stimulated cells: a new model, *Trends Biochem Sci.* **13**, 172-7 (1988).

32. J.M. Hakumäki and R.A. Kauppinen, $^{1}$H NMR visible lipids in the life and death of cells, *Trends Biochem. Sci.* **25**, 357-362 (2000).
33. J.M. Hakumäki et al., $^{1}$H MRS detects polyunsaturated fatty acid accumulation during gene therapy of glioma: implications for the in vivo detection of apoptosis, *Nature Med.* **5**, 1323-1327 (1999).
34. J.L. Griffin et al., Assignment of $^{1}$H nuclear magnetic resonance visible polyunsaturated fatty acids in BT4C gliomas undergoing ganciclovir-thymidine kinase gene therapy-induced programmed cell death, *Cancer Res.* **63**, 3195-3201 (2003).
35. M. Di Vito et al., $^{1}$H NMR-visible mobile lipid domains correlate with cytoplasmic lipid bodies in apoptotic T-lymphoblastoid cells, *Biochim. Biophys. Acta.* **1530**, 47-66 (2001).
36. N.M. Al-Saffar et al, Apoptosis is associated with triacylglycerol accumulation in Jurkat T-cells, *Br. J. Cancer* **86**, 963-970 (2002).
37. D. De Valck et al., Differential activation of phospholipases during necrosis or apoptosis: a comparative study using tumor necrosis factor and anti-Fas antibodies, *J. Cell Biochem.* **71**, 392-399 (1998).
38. T. Liimatainen et al., $^{1}$H magnetic resonance spectroscopic imaging of phospholipase activity in rat gliomas in vivo, *Proc. 12$^{th}$ Ann. Sci. Meeting ISMRM, Kyoto, Japan* (2004).
39. L. Le Moyec et al., Proton nuclear magnetic resonance spectroscopy reveals cellular lipids involved in resistance to adriamycin and taxol by the K562 leukemia cell line, *Cancer Res.* **56**, 3461-3467 (1996).
40. D.S. Ucker et al., Genome digestion is a dispensable consequence of physiological cell-death mediated by cytotoxic T-lymphocytes, *Mol. Cell. Biol.* **12**, 3060-3069 (1992).
41. C.E. Scott and F. Adebodun, $^{13}$C-NMR investigation of protein synthesis during apoptosis in human leukemic cell lines, *J. Cell Physiol.* **181**, 147-152 (1999).
42. M.E. Moseley et al., Diffusion-weighted MR imaging of anisotropic water diffusion in cat central nervous system, *Radiology* **176**, 439-445 (1990).
43. J.M. Hakumäki et al., Quantitative $^{1}$H nuclear magnetic resonance diffusion spectroscopy of BT4C rat glioma during thymidine kinase-mediated gene therapy in vivo: identification of apoptotic response. *Cancer Res.* **58**, 3791-3799 (1998).
44. S. Hortelano et al., Intracellular water motion decreases in apoptotic macrophages after caspase activation. *Cell Death Differ.* **8**, 1022-1028 (2001).
45. J.M. Hakumäki et al., Early gene therapy-induced apoptotic response in BT4C gliomas by magnetic resonance relaxation contrast T1 in the rotating frame, *Cancer Gene Ther.* **9**, 338-345 (2002).
46. O.H.J. Gröhn et al., Novel magnetic resonance imaging contrasts for monitoring response to gene therapy in rat glioma, *Cancer Res.* **63**, 7571-7574 (2003).
47. P.K. Valonen et al., Water diffusion in a rat glioma during ganciclovir-thymidine kinase gene therapy-induced programmed cell death in vivo: correlation with cell density, *J. Magn. Reson. Imaging* **19**, 389-396 (2004).
48. R. Weissleder et al., Superparamagnetic iron oxide: pharmacokinetics and toxicity, *AJR Am J Roentgenol.* **152**, 167-173 (1989).
49. R. Weissleder et al., In vivo magnetic resonance imaging of transgene expression, *Nature Med.* **6**, 351-355 (2000).
50. D. Artemov et al., Magnetic resonance molecular imaging of the HER-2/neu receptor, *Cancer Res.* **63**, 2723-2727 (2003).
51. M. Zhao et al., Non-invasive detection of apoptosis using magnetic resonance imaging and a targeted contrast agent, *Nature Med.* **7**, 1241-1244 (2001).

52. E.A. Schellenberger et al., Annexin V-CLIO: a nanoparticle for detecting apoptosis by MRI, *Mol Imaging* **1**, 102-107 (2002).
53. A. Flotats and I. Carrio, Non-invasive in vivo imaging of myocardial apoptosis and necrosis. *Eur. J. Nucl. Med. Mol. Imaging* **30**, 615-30 (2003).
54. M. Doubrovin et al., Imaging transcriptional regulation of p53-dependent genes with positron emission tomography in vivo, *Proc. Natl. Acad. Sci. USA.* **98**, 9300-9305 (2001).

# Chapter 15

# APOPTOSIS-INDUCING ANTICANCER DRUGS IN CLINICAL TRIALS

Wei Hu and John J. Kavanagh
*The University of Texas M. D. Anderson Cancer Center, Department of Gynecologic Medical Oncology, 1515 Holcombe, Houston, TX 77030*

Abstract: Apoptosis, or programmed cell death, plays an essential role in controlling cell number in many developmental and physiologic settings as well as in chemotherapy-induced tumor cell killing. It is a genetically regulated biologic process that is regulated by the ratio of proapoptotic proteins over antiapoptotic proteins. Researchers have recently begun using apoptosis inducers in cancer therapy. In a number of studies, attempts have been made to induce apoptosis by triggering the tumor necrosis factor-related apoptosis-inducing ligand receptor and the Bcl-2 family of proteins, caspases, and inhibitors of apoptotic proteins. Most of these agents are still in the preclinical phase of development because of their low efficacy and potential for resistance. However, the results obtained with other agents have been promising. This chapter reviews the development of apoptosis inducing anticancer drugs in clinical trials.

Key words: apoptosis; anticancer drugs; clinical trials

## 1. INTRODUCTION

Apoptosis, or programmed cell death, plays an essential role in controlling cell number in many developmental and physiologic settings. Apoptosis is frequently impaired in many human tumors, suggesting that disruption of the apoptotic function contributes significantly to the transformation of a normal cell into a tumor cell. Apoptosis is also an important phenomenon in chemotherapy-induced tumor-cell killing. The switching on and off of apoptosis is determined by the ratio of proapoptotic



*M. Sluyser (ed.), Application of Apoptosis to Cancer Treatment,* 341-364.

proteins over antiapoptotic proteins. Apoptosis inducers are currently being used in cancer therapy. In a number of studies, attempts have been made to induce apoptosis directly by triggering core components of the cell-death machinery, such as Bcl-2 family proteins, intracellular anti-apoptotic proteins (IAPs), tumor necrosis factor (TNF)-related apoptosis-inducing ligand (TRAIL) and caspases, and modulating apoptosis indirectly by targeting protein kinases, phosphotases, transcription factors, cell-surface receptors, and proteasomes[1,2]. Most of these agents are still in the preclinical phase of development because of their low efficacy and because of their potential for resistance. However, some of these agents have been shown to have promise as apoptosis inducers. In this chapter, we review the development of apoptosis-inducing anticancer drugs in clinical trials.

## 2. MOLECULAR BASIS OF APOPTOSIS PATHWAY

Apoptosis is a genetically regulated biologic process. The apoptotic cell is characterized by a loss of cell volume, plasma-membrane blebbing, nuclear condensation, chromatin aggregation, and endonucleocytic degradation of DNA into nucleosomal fragments. These cell changes occur following a cascade of cell signaling and caspase-mediated events that regulate pro- and anti-apoptotic proteins. These changes may be triggered by 2 major pathways: the death receptor (DR)-induced pathway or extrinsic pathway and the mitochondria-apoptosome-mediated apoptotic pathway or intrinsic pathway. Both of these pathways lead to caspase activation and cleavage of specific cellular substrates. The receptor-triggered apoptosis pathway includes ligands and their receptors, such as Fas, TNF, TRAIL, and downstream molecules (i.e., caspases and Bcl-2 family members). The mitochondria-apoptosome–mediated pathway includes the apoptotic stimuli induced by radiation therapy and chemotherapy, mitochondria, apoptosome, and key effecter caspases. Caspases are activated in a cascade-like fashion. Initiator or upstream caspases (caspases 8, 9 and 10) can activate effectors or downstream caspases, including caspases 3, 6, and 7, which leads to induction of apoptosis. Cross talk also exists between the 2 apoptotic pathways. For example, Fas cross talks with the mitochondria-apoptosome–mediated pathway that is mediated through the activation of caspase 8 to cleave the Bid protein resulting in the release of cytochrome C from mitochondria[3,4]. The inhibitors of apoptosis proteins, including XIAP, cIAPs, survivin, the PI3K/AKT/NFκB pathway, and heat shock proteins (HSPs), can interact with the caspases leading to the inhibition of apoptosis[5,6] (Figure 1).

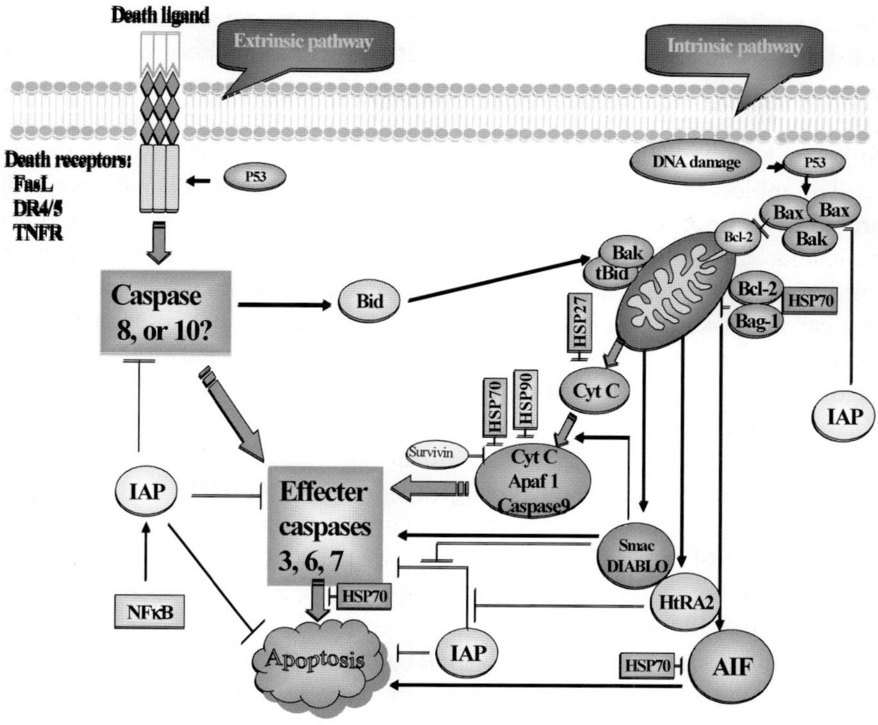


Fig.1

*Figure 1.* Death-receptor-mediated (extrinsic) and mitochondrial-mediated (intrinsic) apoptotic pathways. Both intrinsic and extrinsic pathways lead to caspase activation and cleavage of specific cellular substrates. A number of inhibitor of apoptosis proteins, including XIAP, cIAPs, survivin, the PI3K/AKT/NF□B pathway, and HSPs, can block both apoptotic pathways.

## 2.1 Death receptor-induced apoptotic pathway

The TNF gene superfamily includes TNF, Fas ligand (FasL), and Apo2 or Apo2L/TRAIL. CD95 (Apo-1/Fas) is a prototype DR characterized by the presence of an 80 amino acid death domain in its cytoplasmic tail. This domain when activated by either agonistic anti-CD95 antibodies or cognate CD95 ligand is essential in recruiting a number of signaling components that initiate apoptosis. The complex of proteins that form when CD95 is triggered is called the death-inducing signaling complex (DISC). The DISC consists of an adaptor protein (Fas-associated death domain-containing protein) and initiator caspases, and it is essential in inducing apoptosis. A number of proteins have been reported to regulate formation or activity of the DISC, including Raf-1[7] and protein kinase C (reviewed by Peter and Krammer in ref. 8). FasL initiates apoptosis by binding to its surface receptor Fas. As a consequence, there is sequential activation of caspases and the release of cytochrome C from the mitochondria, with additional caspase activation followed by cellular degradation and death.

The molecular mechanisms controlling the expression of death receptor are not well understood. Recent studies have shed important insight into the molecular mechanisms controlling FasL gene expression at the level of transcription. Nuclear factors, such as nuclear factor-inactivated T cells, nuclear factor-kappa B, specificity protein-1, early growth response factor, interferon regulatory factor, c-Myc, and the fork head transcriptional regulator, activate FasL expression alone or cooperatively (reviewed by Kavurma and Khachigian in ref. 9).

Apo2L/TRAIL is one of several members of the TNF gene superfamily that induces apoptosis through engagement of DRs. Apo2L/TRAIL interacts with an unusually complex receptor system that is comprised of 2 receptors, TRAIL-R1/DR4 and TRAIL-R2/DR5, which contain cytoplasmic death domains and signal apoptosis. Two other decoy receptors, TRAIL-R3/DcR1 and TRAIL-R4/DcR2, lack a functional death domain and do not inhibit TRAIL-induced apoptosis. Apo2L/TRAIL has garnered intense interest as a potential candidate for cancer therapy because as a trimmer it selectively induces apoptosis in many transformed cells but not in normal cells[10-12].

## 2.2 The mitochondria-apoptosome-mediated apoptotic pathway

Bcl-2 family proteins play a central role in controlling the mitochondria pathway. To date, more than 20 members of this family have been identified in humans, including proteins that suppress apoptosis (Bcl-2, Bcl-Xl, Mcl-1, A1, Bcl-W, and Bcl-G) and proteins that promote apoptosis (Bax, Bak, Bok, Bad, Bid, Bik, and Biml) (reviewed by Adams and Cory in ref. 13). The release of cytochrome c is the central gate in turning apoptosis on and off and is regulated by the proapoptotic proteins and antiapoptotic proteins of the Bcl-2 family. The release of cytochrome c from the mitochondria has been shown to promote the oligomerization of a cytochrome c/Apaf-1/procase-9 complex (apoptosome), which leads to the induction of apoptosis[14] (Figure 1). How do Bcl-2 family members regulate cytochrome c released from mitochondria? Several models have suggested that Bcl-2 members form channels that facilitate protein transport and interact with other mitochondrial proteins, such as the voltage-dependent anion channel, and also induce the rupture of the outer mitochondrial membrane (reviewed by Hengartner in ref. 15; 16–18). Overproduction of the Bcl-2 protein also prevents the induction of cell death by nearly all cytotoxic anticancer drugs and radiation therapy, thus contributing to drug resistance in patients with some types of cancer[19]. It has also been suggested that paclitaxel cytotoxicity may be mediated in part through phosphorylation and functional inhibition of Bcl-2, and upregulation of Bcl-2 expression may be a mechanism of resistance to paclitaxel[16].

Alteration of p53 is the most common mutation in human cancer. The p53 protein functions as a transcription factor regulating downstream genes involved in cell-cycle arrest, DNA repair, and programmed cell death. Activated p53 can positively regulate apoptotic-related genes, including BAX, which induces cytochrome c released from the mitochondria and leads to induction of apoptosis[17-19]. p53 can also trigger expression of death receptors, such as Fas, which results in activation of apoptosis.[20] Loss of p53 function confers genomic instability, diminished cell-cycle arrest, and impaired apoptosis (Figure 1)[21].

Apoptosis-inducing factor (AIF) is a novel caspase-independent death effecter that is released from the mitochondria. AIF can induce caspase-independent peripheral chromatin condensation and large-scale DNA fragmentation when added to the purified nuclei. In addition to its apoptogenic activity on nuclei, AIF can also participate in the regulation of apoptotic mitochondrial membrane permeabilization and can exhibit NADH oxidase activity (Figure 1)[22].

## 2.3 Inhibitors of apoptosis proteins

### 2.3.1 XIAP, c-IAP1, and c-IAP2

In recent years, a number of inhibitors of IAPs, including XIAP, c-IAP1, cIAP2, and survivin, have attracted the attention pharmaceutical companies because they play a key role in cell survival by modulating death signaling pathways[5,23-26]. XIAP, c-IAP1, and cIAP2 have been shown to directly bind and inhibit caspases 3, 7, and 9, but not caspases 1, 6, 8, or 10[5]. Two negative regulators of XIAP XAF1 or XIAP-associated factor and Smac/DIABLO or direct IAP binding protein with low pI) have been identified (reviewed by Korneluk)[27]. Activated Smac/DIABLO was found to bind to XIAP and to inhibit its caspase-binding activity. The more recently identified OMI/HtrA2 is a membrane-associated protein found in healthy cells with a significant proportion observed within the mitochondria. In vitro, HtrA2 shifts into cytosol in response to ultraviolet (UV) irradiation where it can interact with IAPs and prevent XIAP inhibition of active caspase, thus protecting cells from UV-induced death. In vivo, the proapoptotic activity of HtrA2 involves both IAP-binding and serine-protease activity (Figure 1)[28].

### 2.3.2 Survivin

Survivin is a bifunctional protein that suppresses apoptosis and regulates cell division. This protein has garnered great interest as a potential drug target because its expression is among the most tumor-specific of all human gene products. The majority of tumors express survivin protein at high levels, but not in the normal adult tissue[29]. Survivin was found to inhibit the 2 apoptotic enzymes, caspase 3 and caspase 7, thus protecting programmed cells from death. Survivin may act simultaneously with the bcl-2 family proteins, but has a different apoptosis inhibitory mechanism[30]. Numerous reports have demonstrated the expression of survivin in various tumors, such as neuroblastoma, melanoma, bladder carcinoma, breast tumors, lung non-small cell tumors, esophageal carcinomas, colorectal carcinomas, and leukemic cells[31-33]. The putative point of survivin intervention at caspase 9 as suggested by O'Connor and associates is that survivin phosphorylation on threonine 34 is necessary for association of survivin with caspase 9 (Figure 1)[34,35]. But the question of whether survivin directly binds and inhibits caspase 9 remains unclear. Several models are proposed that might explain how survivin directly or indirectly inhibits caspases. Survivin might bind and inhibit caspase directly, similarly to IAPs such as XIAP. Survivin might sequester SMAC, thus protecting IAPs from this inhibitory protein. As well, survivin might somehow enhance the function of IAPs, having a function opposite to SMAC[29].

A recent study showed that survivin also binds to the catalytic domain of Aurora-B, which plays a role in chromosome segregation. Whether survivin exerts its anti-apoptotic function through aurora kinase or other kinases is still unknown[36].

### 2.3.3 NFκB

NFκB is an important transcription factor, involving the regulation of transcription of genes mediating inflammation, carcinogenesis, and pro- or anti-apoptotic reactions, including cIAPs, Bcl-2, and Bcl-x. NFκB is sequestered in the cytoplasm in an inactive form through its association with one of several inhibitory molecules, including IκB-α, IκB-ß, IκB-γ, p105, and p100Activation of the NF-κB-signaling cascade results in a complete degradation of IκB or partial degradation of the carboxyl termini of p105 and p100 precursors, allowing the translocation of NFκB to the nucleus, where it induces transcription (Figure 2) (reviewed by Chen et al. in ref. 37).

NFκB can be inhibited by PS-431 through degradation of IkB[38] and by farnesyltransferase inhibitors (FTIs) and PI3K inhibitor through the inhibition of IkB phosphorylation.

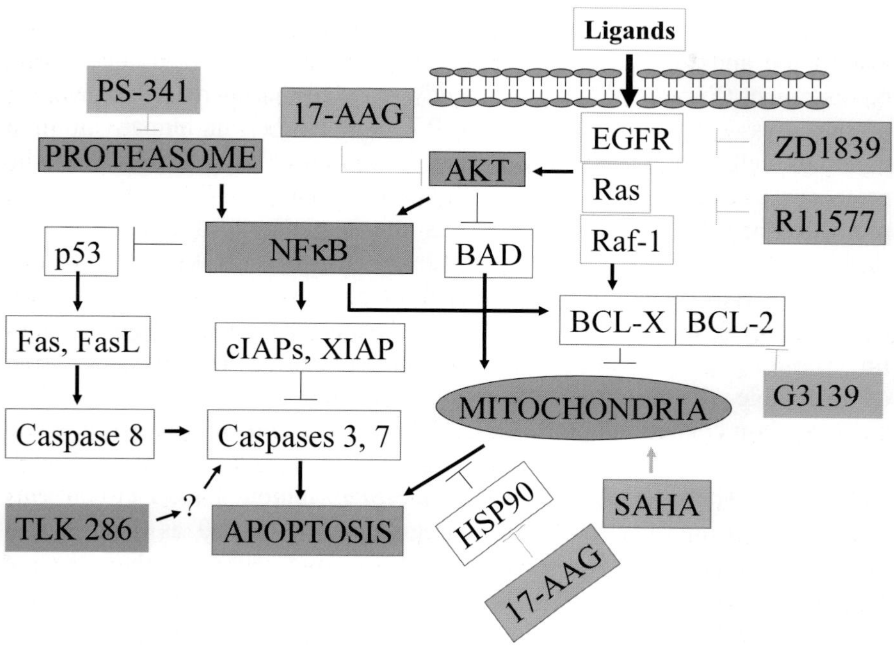


*Figure 2.* Induction of apoptosis machinery by targeting NFκB, AKT and Bcl-2 family.

### 2.3.4 Heat shock proteins (HSPs)

HSPs belong to the superfamily of molecular chaperones because of their ability to stabilize proteins and polypeptide and minimize protein misfolding and aggregation within the cells[39]. HSP70, as a major anti-apoptotic HSP, can protect cells from a variety of stressful conditions[40]. Although the mechanisms involved in HSP70 protection of cells in response to different stresses are not fully understood, a number of studies have shown that HSP70 can block the release of cytochrome C[41], prevent Apaf-1 oligomerization[42], and inhibit procaspase 9 recruitment[43]. Both p53 and c-myc have been linked to HSP70-mediated cell protection. The binding of HSP70 to the mutant p53 protein may stabilize p53 in its mutant form, preventing the pro-apoptotic function of wild-type p53, thus accelerating tumorigenesis[44]. HSP70 also co-localizes with c-Myc in phase-dense nuclear structures and modulates the functional levels of the c-Myc protein[45]. HSP70 may also mediate anti-apoptotic effects by chaperoning BAG-1, an intermediate protein in the association of HSP70 and Bcl-2, causing inhibition of apoptosis[40,46]. Increased expression of Bcl-2 has been reported to accompany the induction of HSPs by a number of heat shock–inducing agents[46]. HSP70 specifically interacted with AIF as shown by ligand blots and co-immunoprecipitation. Cells overexpressing HSP70 were protected against the apoptogenic effects of AIF targeted to the extra-mitochondrial compartment (Figure 1)[47]. A recent study reported that increased expression of the stress-response protein of HSP70 and HSP27 in human multiple myeloma cells was pronounced as early as 8 hours after exposure to the proteosome inhibitor PS-341[48], suggesting that HSP70-mediated cell protection occurs rapidly in cancer cells after exposure to the drugs. In our study, we found that the differentially expressed proteins were pronounced in 2774 ovarian cancer cells treated with manumycin (a farnesyl transferase inhibitor) for 3, 6 and 16 hours and were further identified by peptide mass fingerprinting as HSP70. Further functional studies revealed that inhibition of inducible HSP70 by quercetin correlates with enhancement of manumycin-mediated apoptosis in 2,774 cells and enhancement of specific cleavage of PARP into apoptotic fragment in 2,774 cells treated with manumycin. The interaction between the HSP70 inhibitor and FTI confirms the functional significance of the upregulation of HSP70 as a protective mechanism against FTI-induced apoptosis and support the use of combination treatment[49].

Taken together, a better understanding of cellular mechanisms regulating apoptosis induced by intrinsic and extrinsic apoptotic stimuli and by anticancer drugs might offer a strong rationale for the combination of chemotherapy and other biologic treatments.

# 3. APOPTOSIS-INDUCING ANTICANCER DRUGS IN CLINICAL TRIAL

Apoptosis inducers have recently been used in cancer therapy. Intrinsic and extrinsic pathways can be activated separately, but activation of caspases appears central to most apoptotic pathways. A number of studies have attempted to induce apoptosis using a portion of an apoptotic pathway that triggers the TRAIL receptor, Bcl-2 family of proteins, caspases, and IAPs. Most of them are still in preclinical development because of their low efficacy and potential for resistance. Following is a summary of reports from studies evaluating some of the more promising agents being tested in ongoing clinical trials targeting the core components of the cell-death machinery and modulating apoptosis by targeting protein kinases, phosphotases, transcription factors, and cell-surface receptors.

## 3.1 Direct effect on apoptosis by targeting the core components of the cell-death machinery

### 3.1.1 TRAIL-receptor 1 monoclonal antibody

TNF-a and FasL are effective apoptosis-inducers; however, severe toxicity, such as hypotension, abnormalities in liver function, leukopenia, chills, and thrombus formation, make TNF-a and FasL difficult to use systemically as antitumor drugs[50]. Unlike TNF and FasL, TRAIL preferentially induces apoptosis in tumor cells compared with normal cells suggesting that TRAIL could be a potentially powerful anticancer agent.[12] The TRAIL effecter pathway appears to be a vital component of immunosurveillance of spontaneous or resident tumor cells by both T cells and natural killer cells stimulating more hope that manipulating TRAIL activity is a natural path to improved cancer immunotherapy.

Recombinant forms of these ligands are capable of potentiating the effect of chemotherapeutic drugs in vitro and in vivo in the animal model[51]. Animal studies have demonstrated the antitumor activity of TRAIL and potentiation of the chemotherapeutic effect of TRAIL. Therefore, phase I studies with TRAIL are needed[52-54].

The TRAIL-receptor 1 (TRAIL-R1) monoclonal antibody (mAb) is an agonistic antibody that specifically recognizes the R1 protein, which is found on the surface of a number of solid and hematopoietic cancer cells. TRAIL-R1 mAb was developed by Takeda Chemical Industries (Nihonbashi 2-chome, Chuo-ku, Tokyo). Preclinical experiments have shown that TRAIL-R1 mAb has the ability to kill a variety of human tumor cell lines and that it exhibits effectiveness in treating human breast, colon, and uterine cancers. Human Genome Sciences (Rockville, MD; www.hgsi.com) is currently enrolling patients in a phase I clinical trial in the United States to evaluate the safety and pharmacology of TRAIL-R1 mAb in patients with advanced tumors. Because it has been suggested that some chemotherapy agents, such as STI-571, are able to potentiate TRIAL activity in vitro, it is possible that a combination treatment might activate both apoptotic pathways[55,56].

However, a recent study has suggested that TRAIL also induces apoptosis and cell death in human hepatocytes[57]. Despite TRAIL's preferential proapoptotic activity on cancer cells compared with its activity on normal cells, the agents that target DR4 or DR5 in cancer therapy must be used with utmost caution due to hepatocyte apoptosis.

### 3.1.2 G3139

Bcl-2 is known to prevent programmed cell death, enhance metastatic potential, and promote resistance to anticancer therapy. High expression of the Bcl-2 proto-oncogene is found in various solid tumors. G3139 is an antisense phosphorothioate oligodeoxynucleotide that suppresses Bcl-2 expression[58]. It was developed by Genta Inc, (Lexington, MA) as a potential anticancer agent. Preclinical studies of G3139 have shown that it enhances the antitumor effect of a broad range of cytotoxic agents, including docetaxel and taxanes[59,60].

Bcl-2 antisense therapy is well tolerated[61,62]. The first results of a phase I dose-escalation clinical trial evaluating subcutaneous administration of the bcl-2 antisense oligonucleotide G3139 were published as a preliminary report in 1997 and as a complete study in 21 patients with non-Hodgkin's lymphoma in 2000. Local inflammation (at the infusion site) was the most common side effect observed. The maximum-tolerated dosage was 147.2 $mg/m^2$/day and the dose-limiting toxicity was thrombocytopenia. A complete response (CR)/partial response (PR) rate of 14% and a stable disease (SD) rate of 43% were reported. Nine patients showed disease progression. Recently, phase I studies of G3139 in combination with conventional chemotherapeutic agents have been reported. A phase I study of G3139 combined with carboplatin and etoposide in patients with small cell lung cancer showed PR in 12 patients (86%) and stable disease in two. Median time to progression was 5.9 months. No evidence of bcl-2 suppression in mononuclear cells was observed. It has been suggested that the combination of G3139, carboplatin, and etoposide is well tolerated and results in an encouraging response rate and time to progression in patients with extensive-stage small cell lung cancer[63]. Van de Donk et al. (64) reported that G3139 induced a 40% partial response and a 30% minor response in vincristind, Adriamycin (doxorubicin), dexamethasone (VAD)-refractory myeloma[64], suggesting that G3139 may overcome classical resistance and restore sensitivity of myeloma tumor cells to VAD chemotherapy.

Phase III trials are currently ongoing in chronic lymphocytic eukemia, non-small cell lung cancer, advanced malignant melanoma, and multiple myeloma (http://www.genta.com/ctrials.aspx. However, in a recent report from the U. S. Food and Drug Administration (FDA), Genasense (Genta Incorporated, Berkley Heights, NJ) was rejected as treatment for malignant melanoma. Genta's submission data were based largely on an open-label 771-patient trial (GM-301) of oblimerson combined with dacarazine. No statistically significant effect on overall survival was noted. Patients on oblimersen/dacarbazine had a median survival duration of 274 days compared with 238 days for dacabazine alone (p=0.18). Secondary end points did show a significant effect on progression-free survival. The lessons learned from G3139 failure are that if primary end point data (overall survival rate) for cancer treatment are not met, secondary end point data, although positive, will not guarantee FDA approval[65].

Antisense approaches for decreasing the expression of a wide variety of antiapoptotic genes, including Bcl-2 and XIAP, are in various stages of preclinical development. However, the common problems for antisense therapy are the efficient delivery of the DNA to every tumor cell in vivo. It is also unclear whether these inhibitors will have specificity for tumor cells or will be useful in single-drug therapy.

## 3.2 Induction of apoptosis machinery by targeting protein kinases, transcription factors, and proteasome

### 3.2.1 TLK286

TLK286, (formerly known as TER286), is a modified glutathione analogue. The cytotoxic activity of TLK286 is mediated by metabolic activation of the drug by the enzyme glutathione S-transferase (GST), primarily by the isozyme GST P1-1. A cytotoxic fragment, tetrakis (chloroethyl) phosphorodiamidate, is released intracellularly following metabolism by GST P1-1[66]. The role of the GST P1-1 activation-dependent mechanism of TLK286 offers rationale for investigating this drug in the treatment of advanced ovarian cancer, particularly in drug-resistant disease[67]. Preclinical studies with TLK286 have shown a positive correlation, with the antitumor activity of TLK286 increasing with the levels of GST P1-1 through induction of apoptosis[68]. However, the exact mechanisms of the action of TLK286 in platinum-resistant cancer cells are not well understood. Possible mechanisms have been suggested to be associated with the stress-induced apoptotic pathway[69,70].

We are currently conducting a phase II trial with TLK286 in patients with advanced epithelial ovarian malignancies that are refractory to taxane- and platinum-based chemotherapies. Objective tumor responses have been observed in 5 (15%) of 34 evaluable patients (1 CR and 4 PR). All 5 responding patients had platinum- and paclitaxel-refractory tumors and bulky disease. The disease stabilization rate (CR+PR+SD/number of evaluable patients) was 50% (17/34). The Kaplan-Meier estimate of survival at 16 months was 63%. TLK286 is well tolerated and has significant single-agent activity in women with ovarian cancer whose tumors are refractory to platinum and paclitaxel therapies[71]. The results of this phase II study have shown that TLK286 is a promising candidate in patients with platinum-resistant ovarian cancer and patients who have failed one prior second-line therapy. TLK286 has demonstrated significant single-agent activity in 2 phase II trials of ovarian cancer.

Phase II trials of TLK286 in combination with chemotherapy in ovarian cancer as well as a phase III randomized study of TLK286 versus Doxil as third-line therapy in platinum- refractory or -resistant ovarian cancer are ongoing. A recent report presented at a meeting of the American Society of Clinical Oncology showed that TLK286 at 50% of the single agent dose (500 mg/m2) in combination with carboplatin (AUC 5) shows enhanced efficacy with an objective tumor response rate of 56% and disease stabilization rate of 88% in patients with platinum-refractory or -resistant (≥ third line) ovarian cancer. Patient follow-up and dose escalation continues. The TLK286/carboplatin combination is an active regimen for future studies in ovarian cancer, non-small cell lung cancer, and breast cancer[72].

In March 2003, it was announced that a phase III registration trial of TLK 286 as monotherapy had been initiated in patients with platinum-refractory ovarian cancer patients.

### 3.2.2 PS-341

The proteasome is the primary component of the protein degradation pathway of the cell and is present in all eukaryotic cells. By degrading regulatory proteins, the proteasome serves as a central element for many cellular regulatory signals and, thus, is a novel target for therapeutic drugs. PS-341, which was developed by Millenium Pharmaceuticals (Cambridge, MA), is a small molecule and a potent, selective inhibitor of the proteasome. In vitro and mouse xenograft studies of PS-341 have shown it to have antitumor activity in a variety of tumor types, including myeloma, chronic lymphocytic leukemia, prostate cancer, pancreatic cancer, and colon cancer[73].A number of cell-cycle–related and apoptotic-related proteins are degraded by the proteasome, including the inhibitory protein IkB, an inhibitor of NFκB that inhibits the apoptosis process through the interaction with IAP- and TNF-mediated apoptosis[74,75].

Bortezomib (formerly known as MLN341, LDP-341, and PS-341) is the first proteasome inhibitor to be investigated in clinical trial as a potential anticancer drug. It is currently being studied in a number of clinical trials[74,76]. Phase I trials in a variety of tumor types have shown PS-341 to be well tolerated. A recent report from a phase I trial of PS-341 in patients with advanced solid tumors with observations in androgen-independent prostate cancer (AIPC) demonstrated that 2 patients with AIPC had prostate-specific antigen response, and 2 patients had PR in lymph nodes. Biologic activity (inhibition of NFκB-related markers) and antitumor activity was seen in AIPC at tolerated doses of PS-341[77].

Phase II trials in several hematologic malignancies and solid tumor types have been initiated (reviewed by Adams in ref. 76; 78-85). The multi-center phase II study by Richardson et al. presents compelling evidence that PS-341 has a role in the treatment of multiple myeloma. In patients with highly refractory diseases, there was a 35% overall response rate, including a 10% CR as defined by a normal serum protein concentration on electrophoresis using a regimen of 1.3 $mg/m^2$ given twice weekly for 2 weeks (days 1, 4, 8, 11; rest days 12-21). Toxic effects included thrombocytopenia, peripheral neuropathy, and neotropenia, but severe toxic effects were uncommon. On the basis of these studies, the FDA has very recently approved PS-341 as treatment for patients with refractory multiple myeloma. Until now, the most advanced study of PS-341 has been a phase III trial in patients with relapsed and refractory multiple myeloma (www.mlnm.com/rd/oncology) [86, 87].

Currently, phase I trials in advanced solid tumors in combination with chemotherapy and radiation are also undergoing[77]. PS-341 has been shown to have promise in the treatment refractory multiple myeloma; however, clinically significant activity of PS-341 was not evident in metastatic renal cell cancer. Further evaluation in this disease setting is not recommended[88].

### 3.2.3 Suberoylanilide hydroxamic acid

It has been suggested that suberoylanilide hydroxamic acid (SAHA) induces apoptosis that is characterized by mitochondrial stress. So far, however, the critical elements of this apoptotic program remain poorly defined[89]. Numerous antiproliferative effects have been reported for SAHA, including induction of $G_0/G_2$ cell-cycle arrest, differentiation, and selective apoptosis of transformed cells[83,90,]91. SAHA results in translocation of Bcl-2 family member Bid into the mitochondria and causes the release of cytochrome C. In particular, SAHA-induced cell death is regulated by p53[92]. SAHA shows strong antiproliferative effects but low toxicity in normal tissues and is currently being investigated in clinical trials as therapy for solid and hematologic tumors[92,93]. In a phase I study of SAHA as treatment for hematologic malignancies and solid tumor, Daily i.v. SAHA is well tolerated and has shown inhibition of the biological target in vivo, and has antitumor activity in solid and hematological tumors[94]. SAHA is currently in phase I and II trials for treatment of advanced multiple myeloma, cutaneous T cell lymphoma, peripheral T-cell lymphoma and recurrent or metastatic squamous cell cancer of the head and neck (www.atonpharma.com).

### 3.2.4 P53 gene therapy

**INGN201 (Ad5CMV-p53)**

One of the advanced gene therapy agents used in the treatment of human cancer is the replication-incompetent adenovirus that delivers a p53 expression cassette, RPR/INGN 201. Preclinical studies, both in human cell lines and animal models of head and neck cancers, have demonstrated that the p53 gene contained in RPR/INGN 201 is efficiently transcribed and translated into p53 protein. Treatment with RPR/INGN 201 inhibited cell growth in human squamous cell carcinoma of the head and neck (SCCHN) cell lines of diverse p53 status and suppressed tumor growth in animal xenografts of human SCCHN through the apoptotic pathway and other effects, such as inhibition of angiogenesis[95, 96].

The results of phase II trials in SCCHN have indicated that multiple courses of intralesional injections are well tolerated, with injection-site pain being the most frequent adverse event (57% of patients). Transient fever and headache were also noted. Evaluation of clinical activity in 17 patients (1 patient withdrew before treatment) in the nonresectable arm showed 2 patients with tumor regression greater than 50%, and 6 patients with SD for up to 3.5 months duration. Of the 15 patients in the resectable arm, 2 patients remained disease-free at 24 months. Although p53 status was not an entry criterion, p53 gene integrity was measured in all patients, and objective tumor response was observed in lesions with wild-type p53 and lesions with defective p53. This observation suggests that RPR/INGN 201 exerts antitumor effects on SCCHN lesions regardless of their p53 status[97].

However, the first interim analysis of the large international p53 gene-therapy study in patients with primary stage III ovarian cancer showed that addition of p53 gene therapy to standard treatment did not improve therapeutic effectiveness in patients with optimally debulked morbidity. This unexpected result led to the study being closed[98]. The reasons for the failure of p53 gene therapy are not clear. It may be due to multiple genetic changes in cancer and epigenetic dysregulations leading to aberrant silencing of genes.

**ONYX-015 (CI-1042)**

The tumor suppressor gene p53 can induce the tidy destruction of cancer cells via apoptosis. Preclinical data has shown p53 gene therapy to be promising for clinical application. However, clinical trials of p53 gene replacement have had limited success[98-101]. The reasons are not yet clear. It may include the multiple genetic changes in cancer and epigenetic dysregulations leading to aberrant silencing of gene[98]. There have been many other attempts to design a treatment that targets p53 such as ONYX-015.

Mutations in the p53 tumor suppressor gene are the most common type of genetic abnormality in cancer. ONYX-015 (CI-1042), is an adenovirus that is modified selectively to replicate in and kill cells that harbor p53 mutations. It is a recombinant adenovirus that carries a loss-of-function mutation at the E1B locus, the product of which is a 55-kDa protein that binds to and inactivates the p53 tumor suppressor protein. Wild-type adenoviruses must disable this gene before viral replication can occur. The ONYX-015 adenovirus will leave normal cells unaffected. ONYN-015 is under development by Onyx Pharmaceuticals (Richmond, CA) for the potential treatment of various solid tumors, including head and neck, gastrointestinal, and pancreatic tumors. ONYN-015 is being investigated in phase II trials for colorectal, ovarian, pancreatic, and oral tumors; in phase I trials for digestive tract, esophageal, and liver tumors; and in phase III trials for the recurrent head and neck cancer[102].

### 3.2.5 FTI-R115777 (tipifarnib)

FTIs were initially developed to inhibit cancer cell growth by blocking farnesylation of Ras and preventing its required localization to the plasma membrane. However, it is increasingly apparent that inhibition of farnesylation of other proteins may contribute to the growth-inhibitory and apoptotic effects of FTIs through the inhibition of the activity of AKT. FTIs show in vitro activity against a range of tumor cell lines. The in vivo antitumor activity of FTIs (including regression of some tumors) has been observed against a number of tumor types, including Bcr-Abl–expressing leukemias, glioma, pancreatic, colorectal cancers, and melanoma. Several FTIs are being testing in the clinical trials. R115777 is under clinical development at Janssen Pharmaceutica, Inc. In a phase I trial of R115777 twice daily for 21 days in adults with refractory and relapsed leukemia, dose-limiting toxicity occurred at 1,200 mg bid, with central neurotoxicity evidenced by ataxia, confusion, and dysarthria. Clinical responses occurred in 10 of 34 (29%) evaluable patients, including 2 CRs. R115777 also induced responses in adult patients with chronic myelogenous leukemia and myelodysplastic syndrome. The phase II and III trials are ongoing.[103] The results from a phase III study of the combination of gemcitabine and tipifarnib showed that the combined treatment does not prolong overall survival in advanced pancreatic cancer compared with single-agent gemcitabine[104].

### 3.2.6 17-AAG

HSP90 is a molecular chaperone protein whose association is required for stability and function of multiple mutated, chimeric, and overexpressed signaling proteins that promote cancer cell growth and survival (Figure 1). HSP90 client proteins include mutated p53, Bcr-Abl, Raf-1, Akt, HER-2/neu, and HIF-1alpha. HSP90 inhibitors, by interacting specifically with a single molecular target, cause the destabilization and eventual degradation of HSP90 client proteins. They have also shown promising antitumor activity in preclinical model systems. Geldanamycin is the first HSP90 inhibitor to be identified. Unfortunately, it has limited clinical potential because it causes liver toxicity[105]. 17AAG is a derivative of geldanamycin with reduced liver toxicity that retains the potent antitumor activity of the parent compound. This agent has entered phase I clinical trial as the first-in-class HSP90 inhibitor at the Cancer Research Campaign Center for Cancer Therapeutics in the United States[106]. HSP90 inhibitors are unique in that, although they are directed toward a specific molecular target, they simultaneously inhibit multiple signaling pathways on which cancer cells depend for growth and survival. Small molecule inhibitors of HSP90 have been very useful in understanding HSP90 biology and in validating this protein as a molecular target for anticancer drug development[106-109].

### 3.2.7 ZD1839

The overexpression of epidermal growth factor receptor (EGFR) occurs frequently in human cancers. Activation of the EGFR-tyrosine kinase (TK) provides signals that drive dysregulated proliferation, invasion and metastasis, angiogenesis, and enhanced cell survival. EGFR-TK is a promising drug target for many types of solid tumors. ZD1839 (Iressa; AstraZeneca Pharmaceuticals LP, Wilmington, DE) is the EGFR-TK inhibitor that blocks signaling pathways responsible for driving proliferation, invasion, and survival of cancer cells, and it is currently being testing in phase II and III trials in a variety of solid tumors. The promising results are observed from non-small cell lung cancer[110]. On May 5, 2003, gefitinib (Iressa; ZD1839) 250-mg tablets (AstraZeneca Inc.) received accelerated approval by the FDA as monotherapy for patients with locally advanced or metastatic non-small cell lung cancer after failure of both platinum-based and docetaxel chemotherapies[111]. The EGFR has emerged in recent years as a viable therapeutic target in human cancer. However, clinical trials have revealed significant variability in the response to gefitinib, with higher response seen in Japanese patients than in a predominantly European-derived population. In the United States, partial clinical responses to gefitinib have been observed most frequently in women, in nonsmokers, and in patients with adenocarcinomas. Recent study published in Science demonstrated that EGFR mutations in lung cancer correlated with clinical response to Gefitinib therapy [112]. It has been suggested that EGFR mutations may predict sensitivity to gefitinib and will be useful marker for selecting the right patients for targeted therapy.

### 3.2.8 LY2181308

LY2181308 targets survivin, a molecule that promotes cell survival. In preclinical tests, the antisense drug LY2181308 has been shown to successfully inhibit tumor growth in animal models. Preclinical results with LY2181308 are among the first to demonstrate antitumor activity in vivo with a drug that specifically inhibits survivin expression. Isis licensed LY2181308 to Lilly in April 2003. Lilly plans to initiate clinical trials of LY2181308 in 2004 (www.bio.com/industryanalysis).

## 4. FUTURE DIRECTIONS

Because apoptosis is frequently impaired in many human tumors and because it is an important phenomenon in chemotherapy-induced tumor cell killing, modulation of apoptosis by targeting proapoptotic- and anti-apoptotic proteins will be important for a better treatment of cancer. Though the use of apoptosis inducers has recently appeared in cancer therapy, nontumor cytotoxic effects are still present in normal cells due largely to the similarity of genes that evoke apoptosis in both normal and tumor cells. Therefore, the development of more selective apoptosis inducers is needed to minimize side effects and maximize efficacy. Recently, a number of inhibitors of intracellular antiapoptotic proteins has attracted the attention of pharmaceutical companies because they play a key role in cell survival by modulating signaling pathways. Much work needs to be done to better understand the network of IAPs involved in the modulation of apoptosis. Studies targeting IAPs with small molecules, antisense technology, and small interfering RNAs for gene silencing hold promise for developing future molecular targeted therapy. In the postgenomic era, the future of cancer therapy will be characterized by personalized treatment and careful selection of therapeutic targets. Proteomics might help to rapidly identify novel apoptotic-related proteins in preferred pathways by using good-quality patient samples. A better understanding of cellular mechanisms, regulating apoptosis induced by intrinsic and extrinsic apoptotic stimuli, and anticancer drugs might offer a strong rationale for the personalized combination of chemotherapy drugs.

## ACKNOWLEDGEMENTS

We thank Vickie J. Williams in the Department of Scientific Publication and Kim Williamson in the Department of Gynecologic Medical Oncology for editing our manuscript.

## REFERENCES

1. M. Jaattela. Programmed cell death: many ways for cells to die decently. *Ann Med*, 34: 480-488 (2002).
2. J. C. Reed. Apoptosis-based therapies. *Nat Rev Drug Discov*, 1: 111-121 (2002).
3. S. J. Korsmeyer, Wei M. C., Saito M., Weiler S., Oh K. J., and Schlesinger P. H. Pro-apoptotic cascade activates BID, which oligomerizes BAK or BAX into pores that result in the release of cytochrome c. *Cell Death Differ*, 7: 1166-1173 (2000).
4. X. M. Sun, MacFarlane M., Zhuang J., Wolf B. B., Green D. R., and Cohen G. M. Distinct caspase cascades are initiated in receptor-mediated and chemical-induced apoptosis. *J Biol Chem*, 274: 5053-5060 (1999).
5. Y. L. Yang and Li X. M. The IAP family: endogenous caspase inhibitors with multiple biological activities. *Cell Res*, 10: 169-177 (2000).
6. R. Kim, Tanabe K., Uchida Y., Emi M., Inoue H., and Toge T. Current status of the molecular mechanisms of anticancer drug-induced apoptosis. The contribution of molecular-level analysis to cancer chemotherapy. *Cancer Chemother Pharmacol*, 50: 343-352 (2002).
7. S. J. Park, Kim Y. Y., Ju J. W., Han B. G., Park S. I., and Park B. J. Alternative splicing variants of c-FLIP transduce the differential signal through the Raf or TRAF2 in TNF-induced cell proliferation. *Biochem Biophys Res Commun*, 289: 1205-1210 (2001).
8. M. E. Peter and Krammer P. H. The CD95(APO-1/Fas) DISC and beyond. *Cell Death Differ*, 10: 26-35 (2003).
9. M. M. Kavurma and Khachigian L. M. Signaling and transcriptional control of Fas ligand gene expression. *Cell Death Differ*, 10: 36-44 (2003).
10. H. N. LeBlanc and Ashkenazi A. Apo2L/TRAIL and its death and decoy receptors. *Cell Death Differ*, 10: 66-75 (2003).
11. A. Trauzold, Wermann H., Arlt A., Schutze S., Schafer H., Oestern S., Roder C., Ungefroren H., Lampe E., Heinrich M., Walczak H., and Kalthoff H. CD95 and TRAIL receptor-mediated activation of protein kinase C and NF-kappaB contributes to apoptosis resistance in ductal pancreatic adenocarcinoma cells. *Oncogene*, 20: 4258-4269 (2001).
12. T. M. Baetu and Hiscott J. On the TRAIL to apoptosis. *Cytokine Growth Factor Rev*, 13: 199-207 (2002).
13. J. M. Adams and Cory S. The Bcl-2 protein family: arbiters of cell survival. *Science*, 281: 1322-1326 (1998).

14. P. Li, Nijhawan D., Budihardjo I., Srinivasula S. M., Ahmad M., Alnemri E. S., and Wang X. Cytochrome c and dATP-dependent formation of Apaf-1/caspase-9 complex initiates an apoptotic protease cascade. *Cell*, 91: 479-489 (1997).
15. M. O. Hengartner. The biochemistry of apoptosis. *Nature*, 407: 770-776 (2000).
16. R. K. Srivastava, Mi Q. S., Hardwick J. M., and Longo D. L. Deletion of the loop region of Bcl-2 completely blocks paclitaxel-induced apoptosis. *Proc Natl Acad Sci U S A*, 96: 3775-3780 (1999).
17. A. Gross, McDonnell J. M., and Korsmeyer S. J. BCL-2 family members and the mitochondria in apoptosis. *Genes Dev*, 13: 1899-1911 (1999).
18. H. Henry, Thomas A., Shen Y., and White E. Regulation of the mitochondrial checkpoint in p53-mediated apoptosis confers resistance to cell death. *Oncogene*, 21: 748-760 (2002).
19. T. Miyashita and Reed J. C. Tumor suppressor p53 is a direct transcriptional activator of the human bax gene. *Cell*, 80: 293-299 (1995).
20. M. R. Hussein, Haemel A. K., and Wood G. S. Apoptosis and melanoma: molecular mechanisms. *J Pathol*, 199: 275-288 (2003).
21. N. D. Smith, Rubenstein J. N., Eggener S. E., and Kozlowski J. M. The p53 tumor suppressor gene and nuclear protein: basic science review and relevance in the management of bladder cancer. *J Urol*, 169: 1219-1228 (2003).
22. S. A. Lipton and Bossy-Wetzel E. Dueling activities of AIF in cell death versus survival: DNA binding and redox activity. *Cell*, 111: 147-150 (2002).
23. Q. L. Deveraux, Leo E., Stennicke H. R., Welsh K., Salvesen G. S., and Reed J. C. Cleavage of human inhibitor of apoptosis protein XIAP results in fragments with distinct specificities for caspases. *Embo J*, 18: 5242-5251 (1999).
24. Q. L. Deveraux, Roy N., Stennicke H. R., Van Arsdale T., Zhou Q., Srinivasula S. M., Alnemri E. S., Salvesen G. S., and Reed J. C. IAPs block apoptotic events induced by caspase-8 and cytochrome c by direct inhibition of distinct caspases. *Embo J*, 17: 2215-2223 (1998).
25. N. Roy, Deveraux Q. L., Takahashi R., Salvesen G. S., and Reed J. C. The c-IAP-1 and c-IAP-2 proteins are direct inhibitors of specific caspases. *Embo J*, 16: 6914-6925 (1997).
26. J. Q. Cheng, Jiang X., Fraser M., Li M., Dan H. C., Sun M., and Tsang B. K. Role of X-linked inhibitor of apoptosis protein in chemoresistance in ovarian cancer: possible involvement of the phosphoinositide-3 kinase/Akt pathway. *Drug Resist Updat*, 5: 131-146 (2002).
27. M. Holcik, Gibson H., and Korneluk R. G. XIAP: apoptotic brake and promising therapeutic target. *Apoptosis*, 6: 253-261 (2001).
28. A. M. Verhagen, Silke J., Ekert P. G., Pakusch M., Kaufmann H., Connolly L. M., Day C. L., Tikoo A., Burke R., Wrobel C., Moritz R. L., Simpson R. J., and Vaux D. L. HtrA2 promotes cell death through its serine protease activity and its ability to antagonize inhibitor of apoptosis proteins. *J Biol Chem*, 277: 445-454 (2002).
29. J. C. Reed. The Survivin saga goes in vivo. *J Clin Invest*, 108: 965-969 (2001).
30. F. Li, Ambrosini G., Chu E. Y., Plescia J., Tognin S., Marchisio P. C., and Altieri D. C. Control of apoptosis and mitotic spindle checkpoint by survivin. *Nature*, 396: 580-584 (1998).
31. G. Ambrosini, Adida C., and Altieri D. C. A novel anti-apoptosis gene, survivin, expressed in cancer and lymphoma. *Nat Med*, 3: 917-921 (1997).
32. T. Yamamoto and Tanigawa N. The role of survivin as a new target of diagnosis and treatment in human cancer. *Med Electron Microsc*, 34: 207-212 (2001).

33. B. Sela. Survivin: anti-apoptosis protein and a prognostic marker for tumor progression and recurrence. *Harefuah*, 141: 103-107, 123 (2002).
34. D. S. O'Connor, Grossman D., Plescia J., Li F., Zhang H., Villa A., Tognin S., Marchisio P. C., and Altieri D. C. Regulation of apoptosis at cell division by p34cdc2 phosphorylation of survivin. *Proc Natl Acad Sci U S A*, 97: 13103-13107 (2000).
35. M. Mesri, Wall N. R., Li J., Kim R. W., and Altieri D. C. Cancer gene therapy using a survivin mutant adenovirus. *J Clin Invest*, 108: 981-990 (2001).
36. J. Chen, Jin S., Tahir S. K., Zhang H., Liu X., Sarthy A. V., McGonigal T. P., Liu Z., Rosenberg S. H., and Ng S. C. Survivin enhances Aurora-B kinase activity and localizes Aurora-B in human cells. *J Biol Chem*, 278: 486-490 (2003).
37. F. Chen, Castranova V., and Shi X. New insights into the role of nuclear factor-kappaB in cell growth regulation. *Am J Pathol*, 159: 387-397 (2001).
38. N. Mitsiades, Mitsiades C. S., Richardson P. G., Poulaki V., Tai Y. T., Chauhan D., Fanourakis G., Gu X., Bailey C., Joseph M., Libermann T. A., Schlossman R., Munshi N. C., Hideshima T., and Anderson K. C. The proteasome inhibitor PS-341 potentiates sensitivity of multiple myeloma cells to conventional chemotherapeutic agents: therapeutic applications. Blood, 101: 2377-2380 (2003).
39. C. Sarto, Binz P. A., and Mocarelli P. Heat shock proteins in human cancer. *Electrophoresis*, 21: 1218-1226. (2000).
40. E. M. Creagh, Sheehan D., and Cotter T. G. Heat shock proteins--modulators of apoptosis in tumour cells. *Leukemia*, 14: 1161-1173 (2000).
41. D. D. Mosser, Caron A. W., Bourget L., Meriin A. B., Sherman M. Y., Morimoto R. I., and Massie B. The chaperone function of hsp70 is required for protection against stress-induced apoptosis. *Mol Cell Biol*, 20: 7146-7159 (2000).
42. A. Saleh, Srinivasula S. M., Balkir L., Robbins P. D., and Alnemri E. S. Negative regulation of the Apaf-1 apoptosome by Hsp70. *Nat Cell Biol*, 2: 476-483 (2000).
43. H. M. Beere, Wolf B. B., Cain K., Mosser D. D., Mahboubi A., Kuwana T., Tailor P., Morimoto R. I., Cohen G. M., and Green D. R. Heat-shock protein 70 inhibits apoptosis by preventing recruitment of procaspase-9 to the Apaf-1 apoptosome. *Nat Cell Biol*, 2: 469-475 (2000).
44. P. Hainaut and Milner J. Interaction of heat-shock protein 70 with p53 translated in vitro: evidence for interaction with dimeric p53 and for a role in the regulation of p53 conformation. *Embo J*, 11: 3513-3520. (1992).
45. P. J. Koskinen, Sistonen L., Evan G., Morimoto R., and Alitalo K. Nuclear colocalization of cellular and viral myc proteins with HSP70 in myc-overexpressing cells. *J Virol*, 65: 842-851. (1991).
46. S. Takayama, Bimston D. N., Matsuzawa S., Freeman B. C., Aime-Sempe C., Xie Z., Morimoto R. I., and Reed J. C. BAG-1 modulates the chaperone activity of Hsp70/Hsc70. *Embo J*, 16: 4887-4896. (1997).
47. L. Ravagnan, Gurbuxani S., Susin S. A., Maisse C., Daugas E., Zamzami N., Mak T., Jaattela M., Penninger J. M., Garrido C., and Kroemer G. Heat-shock protein 70 antagonizes apoptosis-inducing factor. *Nat Cell Biol*, 3: 839-843 (2001).
48. N. Mitsiades, Mitsiades C. S., Poulaki V., Chauhan D., Fanourakis G., Gu X., Bailey C., Joseph M., Libermann T. A., Treon S. P., Munshi N. C., Richardson P. G., Hideshima T., and Anderson K. C. Molecular sequelae of proteasome inhibition in human multiple myeloma cells. *Proc Natl Acad Sci U S A*, 99: 14374-14379 (2002).
49. W. Hu, Wu W., Verschraegen C. F., Chen L., Mao L., Yeung S. C., Kudelka A. P., Freedman R. S., and Kavanagh J. J. Proteomic identification of heat shock protein 70 as a candidate target for enhancing apoptosis induced by farnesyl transferase inhibitor. Proteomics, 3: 1904-1911 (2003).

50. T. Nakamoto, Inagawa H., Takagi K., and Soma G. A new method of antitumor therapy with a high dose of TNF perfusion for unresectable liver tumors. *Anticancer Res*, 20: 4087-4096 (2000).
51. H. Walczak, Miller R. E., Ariail K., Gliniak B., Griffith T. S., Kubin M., Chin W., Jones J., Woodward A., Le T., Smith C., Smolak P., Goodwin R. G., Rauch C. T., Schuh J. C., and Lynch D. H. Tumoricidal activity of tumor necrosis factor-related apoptosis-inducing ligand in vivo. *Nat Med*, 5: 157-163 (1999).
52. S. M. Kornblau, Konopleva M., and Andreeff M. Apoptosis regulating proteins as targets of therapy for haematological malignancies. *Expert Opin Investig Drugs*, 8: 2027-2057 (1999).
53. R. K. Srivastava. TRAIL/Apo-2L: mechanisms and clinical applications in cancer. *Neoplasia*, 3: 535-546 (2001).
54. S. de Jong, Timmer T., Heijenbrok F. J., and de Vries E. G. Death receptor ligands, in particular TRAIL, to overcome drug resistance. *Cancer Metastasis Rev*, 20: 51-56 (2001).
55. S. Lacour, Hammann A., Wotawa A., Corcos L., Solary E., and Dimanche-Boitrel M. T. Anticancer agents sensitize tumor cells to tumor necrosis factor-related apoptosis-inducing ligand-mediated caspase-8 activation and apoptosis. *Cancer Res*, 61: 1645-1651 (2001).
56. H. Xiang, Fox J. A., Totpal K., Aikawa M., Dupree K., Sinicropi D., Lowe J., and Escandon E. Enhanced tumor killing by Apo2L/TRAIL and CPT-11 co-treatment is associated with p21 cleavage and differential regulation of Apo2L/TRAIL ligand and its receptors. *Oncogene*, 21: 3611-3619 (2002).
57. M. Jo, Kim T. H., Seol D. W., Esplen J. E., Dorko K., Billiar T. R., and Strom S. C. Apoptosis induced in normal human hepatocytes by tumor necrosis factor-related apoptosis-inducing ligand. *Nat Med*, 6: 564-567 (2000).
58. F. Ciardiello and Tortora G. Inhibition of bcl-2 as cancer therapy. *Ann Oncol*, 13: 501-502 (2002).
59. A. W. Tolcher. Preliminary phase I results of G3139 (bcl-2 antisense oligonucleotide) therapy in combination with docetaxel in hormone-refractory prostate cancer. *Semin Oncol*, 28: 67-70 (2001).
60. S. Leung, Miyake H., Zellweger T., Tolcher A., and Gleave M. E. Synergistic chemosensitization and inhibition of progression to androgen independence by antisense Bcl-2 oligodeoxynucleotide and paclitaxel in the LNCaP prostate tumor model. *Int J Cancer*, 91: 846-850 (2001).
61. M. J. Morris, Tong W. P., Cordon-Cardo C., Drobnjak M., Kelly W. K., Slovin S. F., Terry K. L., Siedlecki K., Swanson P., Rafi M., DiPaola R. S., Rosen N., and Scher H. I. Phase I trial of BCL-2 antisense oligonucleotide (G3139) administered by continuous intravenous infusion in patients with advanced cancer. *Clin Cancer Res*, 8: 679-683 (2002).
62. C. M. Rudin, Otterson G. A., Mauer A. M., Villalona-Calero M. A., Tomek R., Prange B., George C. M., Szeto L., and Vokes E. E. A pilot trial of G3139, a bcl-2 antisense oligonucleotide, and paclitaxel in patients with chemorefractory small-cell lung cancer. *Ann Oncol*, 13: 539-545 (2002).
63. C. M. Rudin, Kozloff M., Hoffman P. C., Edelman M. J., Karnauskas R., Tomek R., Szeto L., and Vokes E. E. Phase I study of G3139, a bcl-2 antisense oligonucleotide, combined with carboplatin and etoposide in patients with small-cell lung cancer. *J Clin Oncol*, 22: 1110-1117 (2004).
64. N. W. van de Donk, de Weerdt O., Veth G., Eurelings M., van Stralen E., Frankel S. R., Hagenbeek A., Bloem A. C., and Lokhorst H. M. G3139, a Bcl-2 antisense

oligodeoxynucleotide, induces clinical responses in VAD refractory myeloma. *Leukemia*, 18: 1078-1084 (2004).
65. S. Frantz. Lessons learnt from Genasense's failure. Nat Rev Drug Discov, 3: 542-543 (2004).
66. A. Satyam, Hocker M. D., Kane-Maguire K. A., Morgan A. S., Villar H. O., and Lyttle M. H. Design, synthesis, and evaluation of latent alkylating agents activated by glutathione S-transferase. *J Med Chem*, 39: 1736-1747 (1996).
67. L. A. Rosario, O'Brien M. L., Henderson C. J., Wolf C. R., and Tew K. D. Cellular response to a glutathione S-transferase P1-1 activated prodrug. *Mol Pharmacol*, 58: 167-174 (2000).
68. E. Izbicka, Lawrence R., Cerna C., Von Hoff D. D., and Sanderson P. E. Activity of TER286 against human tumor colony-forming units. *Anticancer Drugs*, 8: 345-348 (1997).
69. R. J. Davis. Signal transduction by the JNK group of MAP kinases. *Cell*, 103: 239-252 (2000).
70. C. Hagemann and Blank J. L. The ups and downs of MEK kinase interactions. *Cell Signal*, 13: 863-875 (2001).
71. J. Kavanagh, Kudelka, AP, Spriggs, DR, Michael A Bookman, MA, Lewis, L, Henner, WD. Phase 2 study of TLK 286 (GST p1-1 activated glutathione analog) in patients with platinum and paclitaxel refractory/resistant advanced epithelial ovarian cancer. *ASCO*, Abstract # 831: (2002).
72. J. Kavanagh, L. Lewis, H. Choi, R. Iyer, C. Moore, A. Jameson, J. Keck, R. Parra, K. Patel, G. L. Brown. Phase 1-2a study of TLK286 (a novel glutathione analog) in combination with carboplatin in platinum refractory or resistant (= 3rd line) ovarian cancer. *ASCO* abstract No: 5060 (2004).
73. R. LeBlanc, Catley L. P., Hideshima T., Lentzsch S., Mitsiades C. S., Mitsiades N., Neuberg D., Goloubeva O., Pien C. S., Adams J., Gupta D., Richardson P. G., Munshi N. C., and Anderson K. C. Proteasome inhibitor PS-341 inhibits human myeloma cell growth in vivo and prolongs survival in a murine model. *Cancer Res*, 62: 4996-5000 (2002).
74. J. Adams, Palombella V. J., Sausville E. A., Johnson J., Destree A., Lazarus D. D., Maas J., Pien C. S., Prakash S., and Elliott P. J. Proteasome inhibitors: a novel class of potent and effective antitumor agents. *Cancer Res*, 59: 2615-2622 (1999).
75. J. Adams. The development of proteasome inhibitors as anticancer drugs. *Cancer Cell*, 5: 417-421 (2004).
76. J. Adams. Development of the proteasome inhibitor PS-341. *Oncologist*, 7: 9-16 (2002).
77. C. N. Papandreou, Daliani D. D., Nix D., Yang H., Madden T., Wang X., Pien C. S., Millikan R. E., Tu S. M., Pagliaro L., Kim J., Adams J., Elliott P., Esseltine D., Petrusich A., Dieringer P., Perez C., and Logothetis C. J. Phase I trial of the proteasome inhibitor bortezomib in patients with advanced solid tumors with observations in androgen-independent prostate cancer. *J Clin Oncol*, 22: 2108-2121 (2004).
78. B. Barlogie, Shaughnessy J., Zangari M., and Tricot G. High-dose therapy and immunomodulatory drugs in multiple myeloma. *Semin Oncol*, 29: 26-33 (2002).
79. J. Adams. Proteasome inhibition: a novel approach to cancer therapy. *Trends Mol Med*, 8: S49-54 (2002).
80. G. L'Allemain. [Update on . . . the proteasome inhibitor PS341]. *Bull Cancer*, 89: 29-30 (2002).
81. J. Adams. Proteasome inhibition in cancer: development of PS-341. *Semin Oncol*, 28: 613-619 (2001).

82. P. J. Elliott and Ross J. S. The proteasome: a new target for novel drug therapies. *Am J Clin Pathol*, 116: 637-646 (2001).
83. T. Owa, Yoshino H., Yoshimatsu K., and Nagasu T. Cell cycle regulation in the G1 phase: a promising target for the development of new chemotherapeutic anticancer agents. *Curr Med Chem*, 8: 1487-1503 (2001).
84. J. Wright, Hillsamer V. L., Gore-Langton R. E., and Cheson B. D. Clinical trials referral resource. Current clinical trials for the proteasome inhibitor PS-341. *Oncology (Huntingt)*, 14: 1589-1590, 1593-1584, 1597 (2000).
85. J. Adams, Palombella V. J., and Elliott P. J. Proteasome inhibition: a new strategy in cancer treatment. *Invest New Drugs*, 18: 109-121 (2000).
86. P. Richardson. Clinical update: proteasome inhibitors in hematologic malignancies. *Cancer Treat Rev*, 29 Suppl 1: 33-39 (2003).
87. P. G. Richardson, Barlogie B., Berenson J., Singhal S., Jagannath S., Irwin D., Rajkumar S. V., Srkalovic G., Alsina M., Alexanian R., Siegel D., Orlowski R. Z., Kuter D., Limentani S. A., Lee S., Hideshima T., Esseltine D. L., Kauffman M., Adams J., Schenkein D. P., and Anderson K. C. A phase 2 study of bortezomib in relapsed, refractory myeloma. *N Engl J Med*, 348: 2609-2617 (2003).
88. N. B. Davis, Taber D. A., Ansari R. H., Ryan C. W., George C., Vokes E. E., Vogelzang N. J., and Stadler W. M. Phase II trial of PS-341 in patients with renal cell cancer: a University of Chicago phase II consortium study. *J Clin Oncol*, 22: 115-119 (2004).
89. C. Henderson, Mizzau M., Paroni G., Maestro R., Schneider C., and Brancolini C. Role of caspases, bid and p53 in the apoptotic response triggered by histone deacetylase inhibitors TSA and SAHA. *J Biol Chem* (2003).
90. V. Medina, Edmonds B., Young G. P., James R., Appleton S., and Zalewski P. D. Induction of caspase-3 protease activity and apoptosis by butyrate and trichostatin A (inhibitors of histone deacetylase): dependence on protein synthesis and synergy with a mitochondrial/cytochrome c-dependent pathway. *Cancer Res*, 57: 3697-3707 (1997).
91. V. M. Richon, Emiliani S., Verdin E., Webb Y., Breslow R., Rifkind R. A., and Marks P. A. A class of hybrid polar inducers of transformed cell differentiation inhibits histone deacetylases. *Proc Natl Acad Sci U S A*, 95: 3003-3007 (1998).
92. A. Melnick and Licht J. D. Histone deacetylases as therapeutic targets in hematologic malignancies. *Curr Opin Hematol*, 9: 322-332 (2002).
93. P. Marks, Rifkind R. A., Richon V. M., Breslow R., Miller T., and Kelly W. K. Histone deacetylases and cancer: causes and therapies. *Nat Rev Cancer*, 1: 194-202 (2001).
94. W. K. Kelly, Richon V. M., O'Connor O., Curley T., MacGregor-Curtelli B., Tong W., Klang M., Schwartz L., Richardson S., Rosa E., Drobnjak M., Cordon-Cordo C., Chiao J. H., Rifkind R., Marks P. A., and Scher H. Phase I clinical trial of histone deacetylase inhibitor: suberoylanilide hydroxamic acid administered intravenously. Clin Cancer Res, 9: 3578-3588 (2003).
95. G. L. Clayman and Dreiling L. Local delivery for gene therapy. *Curr Oncol Rep*, 1: 138-143 (1999).
96. I. Brukner. INGN-201. Introgen Therapeutics. *Curr Opin Investig Drugs*, 2: 1776-1785 (2001).
97. G. L. Clayman, el-Naggar A. K., Lippman S. M., Henderson Y. C., Frederick M., Merritt J. A., Zumstein L. A., Timmons T. M., Liu T. J., Ginsberg L., Roth J. A., Hong W. K., Bruso P., and Goepfert H. Adenovirus-mediated p53 gene transfer in patients with advanced recurrent head and neck squamous cell carcinoma. *J Clin Oncol*, 16: 2221-2232 (1998).

98. A. G. Zeimet and Marth C. Why did p53 gene therapy fail in ovarian cancer? *Lancet Oncol*, 4: 415-422 (2003).
99. F. F. Lang, Bruner J. M., Fuller G. N., Aldape K., Prados M. D., Chang S., Berger M. S., McDermott M. W., Kunwar S. M., Junck L. R., Chandler W., Zwiebel J. A., Kaplan R. S., and Yung W. K. Phase I trial of adenovirus-mediated p53 gene therapy for recurrent glioma: biological and clinical results. *J Clin Oncol*, 21: 2508-2518 (2003).
100. INGN 201: Ad-p53, Ad5CMV-p53, Adenoviral p53, INGN 101, p53 gene therapy--Introgen, RPR/INGN 201. *BioDrugs*, 17: 216-222 (2003).
101. M. Bucci. Unique strategy restores p53 function in mice. *J Natl Cancer Inst*, 96: 581 (2004).
102. E. E. Cohen and Rudin C. M. ONYX-015. Onyx Pharmaceuticals. *Curr Opin Investig Drugs*, 2: 1770-1775 (2001).
103. L. R. Kelland. Farnesyl transferase inhibitors in the treatment of breast cancer. *Expert Opin Investig Drugs*, 12: 413-421 (2003).
104. E. Van Cutsem, van de Velde H., Karasek P., Oettle H., Vervenne W. L., Szawlowski A., Schoffski P., Post S., Verslype C., Neumann H., Safran H., Humblet Y., Perez Ruixo J., Ma Y., and Von Hoff D. Phase III trial of gemcitabine plus tipifarnib compared with gemcitabine plus placebo in advanced pancreatic cancer. *J Clin Oncol*, 22: 1430-1438 (2004).
105. J. G. Supko, Hickman R. L., Grever M. R., and Malspeis L. Preclinical pharmacologic evaluation of geldanamycin as an antitumor agent. *Cancer Chemother Pharmacol*, 36: 305-315 (1995).
106. I. Hostein, Robertson D., DiStefano F., Workman P., and Clarke P. A. Inhibition of signal transduction by the Hsp90 inhibitor 17-allylamino-17-demethoxygeldanamycin results in cytostasis and apoptosis. *Cancer Res*, 61: 4003-4009 (2001).
107. L. Neckers. Development of small molecule hsp90 inhibitors: utilizing both forward and reverse chemical genomics for drug identification. *Curr Med Chem*, 10: 733-739 (2003).
108. L. Neckers, Schulte T. W., and Mimnaugh E. Geldanamycin as a potential anti-cancer agent: its molecular target and biochemical activity. *Invest New Drugs*, 17: 361-373 (1999).
109. E. B. Agnew, Wilson R. H., Grem J. L., Neckers L., Bi D., and Takimoto C. H. Measurement of the novel antitumor agent 17-(allylamino)-17-demethoxygeldanamycin in human plasma by high-performance liquid chromatography. *J Chromatogr B Biomed Sci Appl*, 755: 237-243 (2001).
110. M. Los, Burek C. J., Stroh C., Benedyk K., Hug H., and Mackiewicz A. Anticancer drugs of tomorrow: apoptotic pathways as targets for drug design. *Drug Discov Today*, 8: 67-77 (2003).
111. M. H. Cohen, Williams G. A., Sridhara R., Chen G., McGuinn W. D., Jr., Morse D., Abraham S., Rahman A., Liang C., Lostritto R., Baird A., and Pazdur R. United States Food and Drug Administration Drug Approval summary: Gefitinib (ZD1839; Iressa) tablets. *Clin Cancer Res*, 10: 1212-1218 (2004).
112. J. G. Paez, Janne P. A., Lee J. C., Tracy S., Greulich H., Gabriel S., Herman P., Kaye F. J., Lindeman N., Boggon T. J., Naoki K., Sasaki H., Fujii Y., Eck M. J., Sellers W. R., Johnson B. E., and Meyerson M. EGFR mutations in lung cancer: correlation with clinical response to gefitinib therapy. Science, 304: 1497-1500 (2004).

# INDEX

## *1*



## *A*



## *B*

## *C*



## *D*

## E



## F



## G



## H



## I



## J

## K



## L



## M



## N



## O



## P

## *R*



## *S*



## *T*

## *U*



## *V*



## *X*



## *Z*